Student Resource Manual
to Accompany

Differential Equations

A Modeling Perspective

Second Edition

Robert L. Borrelli

Courtney S. Coleman

·

Harvey Mudd College

WILEY

To order books or for customer service call 1-800-CALL-WILEY (225-5945).

ISBN 0-471-43333-0

Printed in the United States of America

10 9 8 7 6 5 4 3 2 1

Printed and bound by Hamilton Printing Company

Preface

The problems in the text are arranged roughly by topic, and roughly by increasing challenge in each topical group. Many of the problems are solvable by following an example in the section. This manual has solutions and graphs for every other problem. Your instructor will expect you to write your own solutions, perhaps based on what you see here.

Getting the Most out of the Problems

- Look over the assigned problems and identify the related examples in the text.

- Try to do the problems on your own, but if you aren't successful then you may want to look at solutions of similar problems.

- Explain your answers, how you got them, and why they provide solutions of the problems.

- The figures in the manual give you an idea of what your own figures should look like.

- Put enough detail in your figures to be informative, but not so much that it is confusing. Use the graphs in the text and this manual as your guides. To improve the appearance of your figures you may want to change the scales on the axes, the solve time, or the initial data.

- Explain how your figures answer the questions posed in the problem.

Getting the Most Out of Your ODE Solver

- Check out the ODE Architect numerical solver that comes with the text or any other numerical solvers available to you. Practice using your solver on simple ODEs. Explore the capabilities and limitations of your solver by applying it to some of the figures in the text.

- Some features in the text and manual figures may not be available on your numerical solver. If that happens, check with your instructor about ways to work around the difficulties.

- In some cases a CAS (Computer Algebra System) can be used to get a solution formula, which can then be graphed.

Getting the Most Out of a Group Project

●

- A group project requires the efforts of a team so that it can be completed in a reasonable length of time.

- The way you will work on a group project is different from the way you work on an individual problem. Communication within the group is critical in order to stay focused on the goal. Be open to new ideas and approaches, but at the same time be honest in expressing your opinions and evaluations of your colleagues' suggestions.

- The write-up of your group project is expected to be longer and more thoroughly documented than the usual write-up of an individual problem, so be prepared to spend more time.

Acknowledgements

●

We particularly want to thank Marie Vanisko (California State University-Stanislaus) and Timothy Comer (Benedictine University) for their painstaking efforts in checking the accuracy of the solutions to the problems. We would like to express our gratitude to the many people who have helped to bring this second edition to fruition. We are especially indebted to Jenny Switkes whose help has been invaluable. Her mathematical skills and talented programming ability contributed significantly at every level of our project. Kudos also to our LaTeX gurus, David Richards (Synopsys) and Lisa Wice (Harvey Mudd College) whose phenomenal facility with this typesetting package created the clean layout and formatting of this text. Thanks also to the following students for typesetting help and in suggesting improvements: Steven Avery, Ben Brooks, Katherine Bryant, Melissa Federowicz, Avani Gadani, Lena Kaloostian, Meredith McDonald, Stuart Mershon, Imad Muhi El-Ddin, and Azusa Yabe. Our thanks also to Paul Damikolas and Bijan Dehbozorgi for testing ODE Architect on examples and Spotlights. Our special thanks to the Wiley editoral staff, especially to Laurie Rosatone, Anne Scanlan-Rohrer, and Kelly Boyle, for help in guiding this project to completion, no mean feat. Any mistakes in the manual, though, are ours, and we would appreciate it if you send in your corrections and suggestions.

A Final Word

●

Keep an open mind while doing your assignment, and have fun along the way using your computer. If you have comments or suggestions about anything in this manual or in the text, send e-mail to John Wiley & Sons at math@wiley.com.

R. L. Borrelli
C. S. Coleman
Claremont, January 2004

Contents

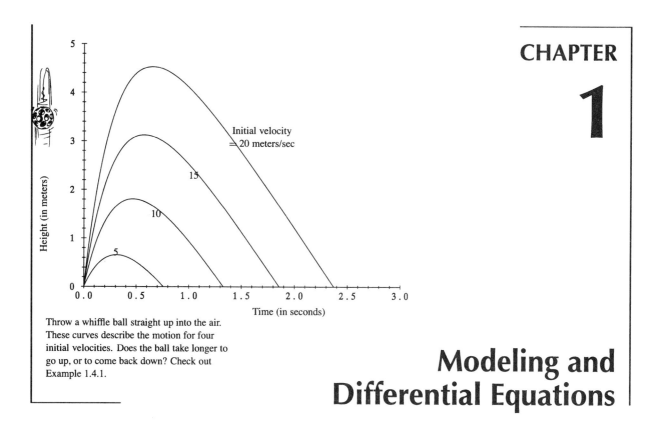

Throw a whiffle ball straight up into the air.
These curves describe the motion for four
initial velocities. Does the ball take longer to
go up, or to come back down? Check out
Example 1.4.1.

CHAPTER

1

Modeling and Differential Equations

The relationship between ODEs and dynamic processes are so basic that we explore the connections right from the start. Throughout the chapter we use simple processes to create models with ODEs. Differential systems come up in Section 1.4.

1.1 The Modeling Approach

Comments:

In this introductory section we touch upon several of the main themes of our approach to modeling and differential equations. We construct an initial value problem that models a vertically moving body. The reasoning process leading to that IVP is called *modeling* and the resulting IVP is called a *model*. We use the Antiderivative Theorem to solve the IVP. In the figures we visualize the behavior of the solution curves of the data. We examine the sensitivity of the structure of these solution curves as the initial value y_0 changes, although we don't yet explicitly use the term "sensitivity." This scenario is one that will be repeated many times in this book. So if you didn't grasp everything this first time around, just wait for the next opportunity. As in any field, there are a lot of terms to be defined. We have opted to include some of the definitions in this opening section, but postpone others till later.

PROBLEMS

Verify a Solution Formula. Verify by direct substitution that the function $y(t)$ is a solution of the given ODE. [*Hint*: see Example 1.1.1.]

1. $y(t) = 4e^{2t}$; $\quad y' = 2y$ **2.** $y(t) = 4e^{3t} + 2/3$; $\quad y' = 3y - 2$

3. $y(t) = 2e^{5t}$; $\quad y' = 3y + 4e^{5t}$ **4.** $y(t) = e^{-t} + 5e^{t}$; $\quad y' = y - 2e^{-t}$

5. $y(t) = 5\sin t - \cos t$; $\quad y'' = -y$ **6.** $y(t) = e^{t} - e^{-t}$; $\quad y'' = y$

Answer 1: Since $y' = \dfrac{d(4e^{2t})}{dt} = 8e^{2t}$, and $2y = 8e^{2t}$, it follows that $y'(t) = 2y(t)$, for all t.

Answer 3: Since $y' = \dfrac{d(2e^{5t})}{dt} = 10e^{5t}$, and $3y + 4e^{5t} = 3(2e^{5t}) + 4e^{5t} = 10e^{5t}$, it follows that $y' = 3y + 4e^{5t}$, for all t.

Answer 5: Since $y'' = \dfrac{d^2}{dt^2}(5\sin t - \cos t) = \dfrac{d(5\cos t + \sin t)}{dt} = -5\sin t + \cos t$, and $-y = -5\sin t + \cos t$, it follows that $y'' = -y$, for all t.

Find Solutions. Simple exponential functions such as $y = e^{rt}$, where r is a constant, are often solutions of ODEs. Find all values of r so that $y = e^{rt}$ is a solution of the given ODE. [*Hint*: insert $y = e^{rt}$ into the ODE and find values of r that yield a solution. For example, $y = e^{rt}$ solves $y' - y = 0$ if $re^{rt} - e^{rt} = (r - 1)e^{rt} = 0$. Since $e^{rt} \neq 0$, we must have $r = 1$. So $y = e^{t}$ solves the given ODE. The symbol $y^{(n)}$ denote the nth derivative of $y(t)$.]

☞ Underscoring indicates an answer at the back of the book.

7. $y' + 3y = 0$ **8.** $y' - 5y = 0$ **9.** $y'' + 5y' + 6y = 0$

10. $y^{(5)} - 3y^{(3)} + 2y' = 0$ **11.** $y'' + 2y' + 2y = e^{-t}$ **12.** $y'' - y = 3e^{2t}$

Answer 7: $y' + 3y$ becomes $re^{rt} + 3e^{rt} = (r + 3)e^{rt}$. So, $y' + 3y = 0$ for all t if and only if $r = -3$ since e^{rt} is never zero. Hence, $y = e^{-3t}$ is a solution.

Answer 9: $y'' + 5y' + 6y$ becomes $r^2 e^{rt} + 5re^{rt} + 6e^{rt} = (r^2 + 5r + 6)e^{rt} = (r + 2)(r + 3)e^{rt}$, which is clearly zero for all t if and only if $r = -2$ or $r = -3$. Hence, $y = e^{-2t}$ and $y = e^{-3t}$ are both solutions.

Answer 11: $y'' + 2y' + 2y$ becomes $(r^2 + 2r + 2)e^{rt}$, which is equal to e^{-t} if and only if $r = -1$. Hence, $y = e^{-t}$ is a solution.

Find Solutions. Sometimes multiples of a power of t solve an ODE. Find all values of the constant r so that $y = rt^3$ is a solution of the given ODE.

13. $t^2 y'' + 6ty' + 5y = 0$ **14.** $t^2 y'' + 6ty' + 5y = 2t^3$ **15.** $t^4 y' = y^2$

Answer 13: The ODE $t^2 y'' + 6ty' + 5y = 0$ becomes $t^2(6rt) + 6t(3rt^2) + 5(rt^3) = 29rt^3 = 0$, which holds for all t if and only if $r = 0$.

Answer 15: The ODE $t^4 y' = y^2$ becomes $3rt^6 = r^2 t^6$, and so $3r = r^2$. Thus, $r = 0$, $r = 3$.

Find Solutions. Find all values of the constant r so that $y = t^r$ is a solution of the given ODE.

16. $t^2 y'' + 4ty' + 2y = 0$ **17.** $t^4 y^{(4)} + 7t^3 y''' + 3t^2 y'' - 6ty' + 6y = 0$

Answer 17: $t^4 y^{(4)} + 7t^3 y''' + 3t^2 y'' - 6ty' + 6y = 0$ becomes

$$[r(r-1)(r-2)(r-3) + 7r(r-1)(r-2) + 3r(r-1) - 6r + 6]t^r = 0$$

The terms in the bracketed expression may be multiplied out to obtain $r^4 + r^3 - 7r^2 - r + 6$ which factors to $(r - 1)(r + 1)(r - 2)(r + 3)$. So, $r = 1, -1, 2$, or -3 are the values of r for which $y = t^r$ is a solution of the given ODE.

Families of Solutions. Find an ODE that has the given family of functions as solutions. Verify your result by substituting the family of solutions into your ODE. [*Hint*: start by differentiating the family of functions at least once with respect to t. Then look for a way to eliminate the constant C. The resulting ODE is not unique.]

18. $y(t) = Ct$, all C, all t **19.** $y(t) = 4Ct^2$, all C, all t

20. $y(t) = \left(t^{3/2} + C\right)^2$, all C, all $t \geq 0$ **21.** $y(t) = Ce^{-t}\sin 2t$, all C, all t

Answer 19: Since $y = 4Ct^2$, we have $y' = 8Ct$. Eliminating C we obtain the ODE $ty' - 2y = 0$ which has $y = 4Ct^2$ as a family of solutions. We may verify our results by direct substitution: $ty' = 8Ct^2 = 2y$.

Answer 21: Differentiating $y = Ce^{-t} \sin 2t$, we have $y' = 2Ce^{-t} \cos 2t - Ce^{-t} \sin 2t = 2Ce^{-t} \cos 2t - y$. Differentiating again, $y'' = -2Ce^{-t} \cos 2t - 4Ce^{-t} \sin 2t - y' = -2Ce^{-t} \cos 2t - 4y - y'$. Since $2Ce^{-t} \cos 2t = y + y'$, it follows that $y'' = -(y + y') - 4y - y'$. Hence $y = Ce^{-t} \sin 2t$ satisfies the ODE $y'' + 2y' + 5y = 0$, for all values of the constant C.

Use the Antiderivative Theorem. Use the Antiderivative Theorem to find all solutions of each ODE. [*Hint:* for Problems 25–30 write y'' as $(y')'$, y''' as $(y'')'$, and so on.]

☞ For Problem 24 check the Integral table on the inside back cover.

22. $y' = 5 + \cos t$ **23.** $y' = t^2 + t + e^{-t}$ **24.** $y' = e^{-t} \cos 2t$

25. $y'' = 0$ **26.** $y''' = 2$ **27.** $y'' = 6t^2 + 2$

28. $y''' = t - 2$ **29.** $y^{(4)} = 1$ **30.** $y^{(4)} = \sin t$

Answer 23: Antidifferentiating $y' = t^2 + t + e^{-t}$ once, $y = t^3/3 + t^2/2 - e^{-t} + C$, where C is an arbitrary constant.

Answer 25: Antidifferentiating $y'' = 0$ twice, we obtain $y = C_1 t + C_2$, where C_1 and C_2 are arbitrary constants.

Answer 27: Antidifferentiating $y'' = 6t^2 + 2$ twice, $y = t^4/2 + t^2 + C_1 t + C_2$, where C_1 and C_2 are arbitrary constants.

Answer 29: Antidifferentiating $y^{(4)} = 1$ four times, $y = t^4/24 + C_1 t^3 + C_2 t^2 + C_3 t + C_4$, where C_1, C_2, C_3 and C_4 are arbitrary constants.

Solve IVPs. Find the solution formula for each IVP. Select appropriate axis ranges and sketch the solution curve. [*Hint:* see Examples 1.1.2 and 1.1.3.]

31. $y' = (2t + 1)^2$, $y(0) = 1$ **32.** $y'' = \sin t$, $y(0) = 0$, $y'(0) = 1$

33. $y' = 2t/(t^2 + 1)$, $y(0) = 2$ **34.** $y' = 1/(t^2 + 1)$, $y(1) = \pi/2$

35. $y' + t \sin t = 1$, $y(0) = -1$ **36.** $y' = \sec^2 t + 3$, $y(\pi/4) = -1$

Answer 31: Antidifferentiating $y' = (2t + 1)^2$, $y(t) = (2t + 1)^3/6 + C$, where C is a constant. Applying the initial condition, we obtain $C = 5/6$, so $y(t) = (2t + 1)^3/6 + 5/6$. See figure.

Answer 33: Antidifferentiating $y' = 2t/(t^2 + 1)$, $y = \ln(t^2 + 1) + C$, where C is a constant. Applying the initial condition, we obtain $C = 2$, so $y(t) = \ln(t^2 + 1) + 2$. See figure.

Answer 35: Antidifferentiating $y' = 1 - t \sin t$, $y = t - \sin t + t \cos t + C$, where C is a constant. Applying the initial condition, we obtain $C = -1$, so $y(t) = t - \sin t + t \cos t - 1$. See figure.

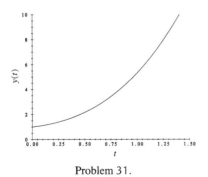

Problem 31.

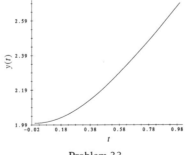

Problem 33.

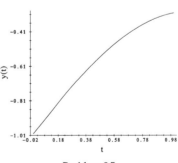

Problem 35.

Modeling Problems: Vertical Motion. Use $g = 9.8$ **meters/sec^2 and ignore air resistance.**

37. A body is thrown vertically upward from the ground with an initial velocity of 20 meters/sec.

(a) Find the highest point and the time to reach that point. [*Hint*: see Example 1.1.3.]

(b) Find the height of the body and the velocity after 3 sec.

(c) When does the body hit the ground?

Answer 37:   The modeling IVP is

$$y''(t) = -g, \qquad y(0) = 0, \quad y'(0) = 20 \text{ meters/sec}$$

where $y(t)$ is the height of the body above the ground at time t (note that the mass of the body is irrelevant). The model equations are valid for $0 \leq t \leq T$, where T is the positive time of impact at which $y(T) = 0$. The value of g is 9.80 meters/sec^2.

(a) Integrating each side of $y'' = -g$ from $t = 0$ to a general value of t, $0 \leq t \leq T$, we have that $y'(t) - y'(0) = -gt$. We know that $v_0 = y'(0) = 20$ meters/sec and that the velocity $y'(t)$ is 0 at the highest point, giving $0 - 20 = -gt$ so that $t = 20/g \approx 2.04$ sec is the time required for the body to reach its highest point. Integrating $y'' = -g$ twice, we have $y(t) = y_0 + v_0t - (1/2)gt^2$. With the highest point reached at time $t \approx 2.04$ sec, an initial velocity v_0 of 20 meters/sec, and a zero initial height y_0, we find that the highest point reached is about 20.41 meters.

(b) Using the equation $y(t) = y_0 + v_0t - (1/2)gt^2$ from the solution of **(a)** with $y_0 = 0$, $t_0 = 0$, $v_0 = 20$ meters/sec, $g = 9.80$ meters/sec^2, and $t = 3$ sec, we have $y(3) \approx 15.90$ meters. Using the equation $y'(t) - y'(0) = -gt$ in the solution of **(a)**, we have $y'(3) = -9.80 \cdot 3 + 20 = -9.40$ meters/sec (the body is falling).

(c) Again from **(a)**, $y(t) = y_0 + v_0t - (1/2)gt^2$ with $y_0 = 0$, so at the time T of impact we have $y(T) = 0 = 20T - 4.90T^2$, or $T = 20/4.90 \approx 4.08$ sec.

38.   *Longer to Rise or to Fall?*   A ball is thrown vertically upward with initial velocity v_0 from ground level (no air resistance). Answer the following questions:

(a) Does the ball spend as much time going up as it does coming down?

(b) What is the time required for the ball to reach ground level again?

(c) What is the velocity of the ball on impact with the ground?

39.   A person drops a stone from the top of a building, waits 1.5 sec, then hurls a baseball downward with an initial speed of 20 meters/sec.

(a) If the ball and stone hit the ground at the same time, how high is the building?

(b) Show that if too much time passes before the ball is thrown downward, then the ball can't catch up with the stone. Show that the maximum waiting time for a catch-up is independent of the building's height.

Answer 39:   The model IVPs are $s'' = -g$, $s(0) = H$, $s'(0) = 0$ for the position $s(t)$ of the stone until impact, and $b'' = -g$, $b(1.5) = H$, $b'(1.5) = -20$ for the position $b(t)$ of the ball (valid until impact), where $s(t)$ and $b(t)$ are the heights of the stone and the ball above the ground at time t, $g = 9.8$ meters/sec^2, H is the height of the building. The solution of the IVP $y'' = -g$, $y(t_0) = H$, $y'(t_0) = 0$, is $y(t) = H - gt^2/2$. The solution of the IVP $b'' = -g$, $b(1.5) = H$, $b'(1.5) = -20$ is given by the formula $b(t) = H - 20(t - 1.5) + g(t - 1.5)^2/2$.

(a) Let T be the time that the stone hits the ground. Then at impact, $s(T) = H - 4.9T^2$. On the other hand, $b(T) = H - 20(T - 1.5) - 4.9(T - 1.5)^2$. Since both the ball and the stone hit the ground at time T, we have $s(T) = b(T)$, so $H - 4.9T^2 = H - 20(T - 1.5) - 4.9(T - 1.5)^2$. Canceling H and $-4.9T^2$ and solving for T, we have that $T \approx 3.58$ sec. Now since $0 = s(T) = H - gT^2/2$ we have $H = 4.9T^2 \approx 4.9(3.58)^2 \approx 62.8$, which is the height of the building in meters.

(b) The model IVP for the ball is $b'' = -g$, $b(\tau) = H$, $b'(\tau) = v_0$, where τ is the waiting time and v_0 is the initial velocity of the ball (v_0 is negative since the ball is hurled downward). We require that $s(T) = b(T)$ at the time $T > 0$ of impact, and that $0 < \tau < T$. From **(a)**, $s(T) = H - 4.9T^2$ and $b(T) = H - v_0(T - \tau) - 4.9(T - \tau)^2$. Set $H - 4.9T^2 = H + v_0(T - \tau) - 4.9(T - \tau)^2$. This equation may be solved for T to obtain $T = \tau(4.9\tau + v_0)/(9.8\tau + v_0)$ or $T/\tau = 1 - 4.9\tau/(9.8\tau + v_0)$. Now since $T/\tau > 1$ we must have that $9.8\tau + v_0 < 0$, or $\tau < -v_0/9.8$. Hence, for any $v_0 < 0$ the waiting time τ cannot be larger than $-v_0/9.8$. Of course, given any pair $\tau > 0$ and $v_0 < 0$ such that $\tau < -v_0/9.8$ there is precisely one building height H which makes $s(T) = b(T)$ for some $T > \tau$. To see this proceed as follows: for the given pair $\tau > 0$, and $v_0 < 0$ such that $\tau < -v_0/9.8$ determine T via the formula $T = \tau(1 - 4.9\tau/(9.8\tau + v_0))$ and H via the formula $H = 4.9T^2$. Then $s(T) = b(T)$ which proves our claim.

1.2 A Modeling Adventure

Comments:

Modeling is a complex process, and you become adept at it mostly by extensive practice rather than by memorizing modeling rules or schematics. For this reason, it is probably best to work through a specific modeling process such as the population models of this section rather than to think about modeling principles in the abstract. In this section, we also show how to find the general solution of the ODE, $y' = ay - H$ where a and H are constants.

PROBLEMS

Find General Solution Formulas.   Use the Antiderivative Theorem to find the general solution of each ODE. Give the interval on which each solution lives.

1. $y' + y = 1$ [*Hint*: multiply each side of the ODE by e^t.]

2. $y' + y = t$ **3.** $y' + y = t + 1$ **4.** $2yy' = 1$ [*Hint*: $(y^2)' = 2yy'$]

5. $2yy' = t$ **6.** $y'' + y' = e^t$ [*Hint*: note that $(e^t y')' = e^t y'' + e^t y'$.]

Answer 1:   Following the hint, multiply by e^t and observe that the ODE, $y' + y = 1$, becomes $e^t(y' + y) = (ye^t)' = e^t$. So, $ye^t = e^t + C$, and $y = 1 + Ce^{-t}$, C any constant, for all t.

Answer 3:   Multiplying by e^t, $y' + y = t + 1$ becomes $(ye^t)' = (t+1)e^t$. So, $ye^t = \int^t (se^s + e^s)ds = (t-1)e^t + e^t + C = te^t + C$, and $y = t + Ce^{-t}$, C any constant, for all t.

Answer 5:   The ODE, $2yy' = t$, may be rewritten as $(y^2)' = t$, so $y^2 = t^2/2 + C$ and $y = \pm\sqrt{t^2/2 + C}$, C any constant and defined on any t-interval where $t^2/2 > -C$.

7. *Not the General Solution*   The example outlined below shows that not every family of solutions of an ODE involving an arbitrary constant qualifies as a general solution of that ODE.

(a) Verify that for any constant C, $y = Ct^2$ is a solution of the ODE $ty' - 2y = 0$, for all t.

(b) Show that $y = Ct^2$ is *not* the general solution of this ODE by verifying that the function
$$y = \begin{cases} t^2, & t \geq 0 \\ -t^2, & t < 0 \end{cases}$$
is also a solution of the ODE for all t.

Answer 7:

(a) $y = Ct^2$ is a solution of the ODE $ty' - 2y = 0$ for all t since $ty' - 2y = t(2Ct) - 2Ct^2 = 2Ct^2 - 2Ct^2 = 0$ for all t and all constants C.

(b) To show that $y = Ct^2$ is *not* the general solution we verify that these are not the *only* solutions to the ODE. Notice that, for $t \geq 0$, $y = t^2$ is a solution because $ty' - 2y = t(2t) - 2t^2 = 0$, and that for $t < 0$, $y = -t^2$ is a solution because $ty' - 2y = t(-2t) - 2(-t^2) = 0$. Hence, $y = \begin{cases} t^2, & t \geq 0 \\ -t^2, & t < 0 \end{cases}$ is also a solution of the ODE $ty' - 2y = 0$, for all t, because $y(0) = y'(0) = 0$. Note that this solution does not have the form Ct^2 for some constant C.

Qualitative Behavior of Solutions.

8. *Solutions That Escape to Infinity at Finite Times*   Carry out the steps below to show that almost all solutions of the ODE $y' = y^2$ escape to infinity in finite time.

(a) First show that $y(t) = 0$ is a solution that *is* defined for all t.

(b) Suppose that a function $y(t)$ is nonzero throughout some t-interval I. Show that $y(t)$ solves the ODE $y' = y^2$ on I if and only if $y(t)$ solves the ODE $\left[y^{-1}\right]' = -1$ on I.

(c) Find a family of solutions of the ODE $y' = y^2$ by solving the ODE $\left[y^{-1}\right]' = -1$. What is the maximal t-interval on which each of these solutions is defined?

(d) Using **(a)** and **(c)**, plot representative solution curves for the ODE $y' = y^2$.

(e) Show that $y(t) = 0$ is the only solution of $y' = y^2$ that is defined for all t and that any other solution escapes to infinity in finite time.

Modeling Problems.

9. *Exponential Growth*   Say that the model IVP for a fish population is given by $y'(t) = ay(t)$, $y(0) = y_0$, where a and y_0 are positive constants (no harvesting).

(a) Find a solution formula for $y(t)$.

(b) What happens to the population as time advances? Is this a realistic model? Explain.

Answer 9:

(a) We can solve the IVP, $y' = ay$, $y(0) = y_0$, by replacing H by 0 in formula (9), the formula that gives the solution of initial value problem (4). So $y = y_0 e^{at}$ for $t \geq 0$. We make the restriction $t \geq 0$, not because there is any difficulty in defining the exponential for negative values of t, but because we only formulated the fishing model for future projections.

   We could also repeat the steps given in this section for getting from IVP (4) to formula (9), after replacing H by 0. This step-by-step derivation is an aid to understanding just why the formula $y = y_0 e^{at}$ gives the unique solution for each value of y_0. This solution formula makes sense if you remember from calculus that the exponential function e^{at} is the only nonconstant function whose derivative is a multiple of itself.

(b) If the initial population y_0 is positive, then the solution formula $y = y_0 e^{at}$ tells us that the tonnage of fish grows exponentially in time since a is positive. Such rapid growth cannot be sustained for any length of time, because food and other supporting resources would soon be used up. The model is unrealistic, except possibly for short periods of time.

10. *Extinction Time*   Suppose that $y' = ay - H$, $y(0) = y_0$, models a fish population, where a and H are positive constants and $0 < y_0 < H/a$. Find the time t^* when the population dies out. [*Hint*: set $y = 0$ in solution formula (9) and solve for t.]

11. *Exponential Decay with Restocking*   This problem deals with a first-order rate law for a population $y(t)$ where the death rate exceeds the birth rate, but this aspect is counteracted in part by a constant restocking rate. R and k are positive constants.

(a) Explain the terms in the modeling ODE, $y' = -ky + R$.

(b) If the population at time $t_0 = 0$ is y_0, find a formula for $y(t)$ for $t \geq 0$. [*Hint*: adapt the solution process in the text for solving IVP (4) for exponential growth with harvesting.]

(c) Interpret the solution formula in terms of the long-term behavior of the population. Treat the cases $0 < y_0 < R/k$, $y_0 = R/k$, $y_0 > R/k$ separately.

(d) If $k = 0.5$, $R = 2$, graph several solutions with $0 < y_0 < R/k$, $y_0 = R/k$, $y_0 > R/k$.

Answer 11:

(a) The positive constant k is the difference between the death and birth rates per population unit. The actual rate at which organisms are lost, then, is $-ky$, the minus sign because the death rate exceeds the birth rate. The rate at which the organisms are restocked is constant, and independent of the population size, so it is a positive constant, R.

(b) Add ky to each side of the ODE in (a), and then multiply each side of the equation by e^{kt} to get $e^{kt}(y' + ky) = Re^{kt}$, which becomes $(e^{kt}y)' = Re^{kt}$. Antidifferentiate each side, and multiply by e^{-kt}, to get $y(t) = R/k + Ce^{-kt}$. Knowing that $y(0) = y_0$, this equation can be used to find that $C = y_0 - R/k$. Substituting that in for C, we get the final answer, $y(t) = R/k + (y_0 - R/k)e^{-kt}$.

(c) If $0 < y_0 < R/k$, then $(y_0 - R/k)e^{-kt}$ is a negative term increasing towards 0. As t increases, and the term gets nearer to 0, the population $y(t) = R/k + (y_0 - R/k)e^{-kt}$ will increase and get nearer to R/k. If $y_0 = R/k$, then the term $(y_0 - R/k)e^{-kt}$ is always 0, and the population will remain at R/k forever. If $y_0 > R/k$, then $(y_0 - R/k)e^{-kt}$ is positive, and will decrease towards zero as t increases. So the population decreases toward R/k with time t.

(d) The figure shows three solution curves of the ODE $y' = -.5y + 2$. The initial conditions at $t_0 = 0$ are $y_0 = 2, 4, 6$. Since $R/k = 2/0.5 = 4$ in this case, the initial conditions correspond to $y_0 < R/k$, $y_0 = R/k$ and $y_0 > R/k$, respectively. Note how the curves tend to $R/k = 4$ as t increases.

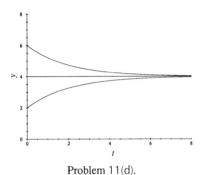

Problem 11(d).

12. *Population Growth*   A population grows exponentially for T months with growth constant 0.03 per month. Then the growth constant suddenly increases to 0.05 per month. After a total of 20 months, the population has doubled. At what time T did the growth constant change? [*Hint*: solve $y' = 0.03y$, $y(0) = y_0$, over the interval $0 \le t \le T$. Then use $y(T)$ as the initial value for $y' = 0.05y$, solving over the interval $T \le t \le 20$.]

13. *Periodic Harvesting*   In the fish population model defined by IVP (4) assume that $a = 1$ and that the fish are harvested periodically according to the harvesting term $H(t) = 1 + \cos t$. Find the solution formula for this IVP and describe the effects of harvesting on the fish population for various values of the initial population y_0. [*Hint*: follow the outline used to produce the solution formula (9). Use the fact that $\left(e^{-t}(\sin t - \cos t)/2\right)' = e^{-t}\cos t$.]

Answer 13: Using $a = 1$, $H(t) = 1 + \cos(t)$ in the ODE $y' = ay - H$, we have the ODE $y' = y - [1 + \cos t]$. Moving all terms to the left-hand side and multiplying through by e^{-t}, we have the equivalent ODE $e^{-t}[y' - y + 1 + \cos t] = 0$. Since, by the Product Rule, $[e^{-t}f(t)]' = e^{-t}f'(t) - e^{-t}f(t)$ and $\left[e^{-t}(\sin t - \cos t)/2\right]' = e^{-t}\cos t$, it follows that

$$e^{-t}[y' - y + 1 + \cos t] = e^{-t}y' - e^{-t}y + e^{-t} + e^{-t}\cos t$$
$$= \left[e^{-t}y - e^{-t} + e^{-t}\left[\frac{\sin t - \cos t}{2}\right]\right]'$$
$$= \left[e^{-t}\left[y - 1 + \frac{\sin t - \cos t}{2}\right]\right]'$$

So we now have the equivalent ODE

$$\left[e^{-t}\left[y - 1 + \frac{\sin t - \cos t}{2}\right]\right]' = 0$$

Since the only function whose derivative is zero everywhere is a constant function, there is a constant C such that

$$e^{-t}\left[y - 1 + \frac{\sin t - \cos t}{2}\right] = C$$

Solving this ODE for $y(t)$ we see that any solution must look like

$$y(t) = Ce^t + 1 + \frac{\cos t - \sin t}{2}$$

To solve the IVP is now just a matter of finding a constant C for which the formula for $y(t)$ satisfies the initial condition. Set $t = 0$, $y(0) = y_0$, and solve for C. We get $C + 1 + 1/2 = y_0$ so that $C = y_0 - 3/2$. So the IVP has the unique solution

$$y(t) = \left(y_0 - \frac{3}{2}\right)e^t + 1 + \frac{\cos t - \sin t}{2}, \quad \text{for } t \geq 0$$

If $y_0 = 3/2$ then the population curve traces out the periodic curve

$$1 + \frac{\cos t - \sin t}{2}$$

However, if the initial population is smaller than $3/2$, then the population dies off because the term $(y_0 - 3/2)e^t \to -\infty$ as $t \to +\infty$. If the initial population is larger than $3/2$, then the population grows without bound because $(y_0 - 3/2)e^t \to +\infty$ as $t \to +\infty$.

1.3 Models and Initial Value Problems

Comments:

The Radioactive Decay Law and half-lives are introduced and modeled. The harvested logistic population model is introduced. ODEs are classified as linear or nonlinear. Qualitative analysis is used to describe solution behavior. At the end of this section there is background and review material on population processes.

The SPOTLIGHT ON MODELING: RADIOCARBON DATING treats radioactive decay and its application to dating. You might get some radioactive substances and test them with a Geiger counter from the physics lab. The notion of a half-life is important here.

PROBLEMS

Order. Give the order of each ODE, and state your reason.

1. $(t^2 - 1)y'' + (\cos t)y' = y\ln\left(1 - t^2\right)$ **2.** $(y')^2 - ty = 1$

3. $ty'' - 2yy' + (\sin t)y^3 = e^t$ **4.** $y''' + y\cot t = e^{-t/2}$

5. $y'y'' = y^3\cos t$ **6.** $y^2t^3 = 1/(1 + y')$

Answer 1: The order is two since y'' is the highest order derivative in the equation.

Answer 3: The order is two because y'' is the highest order derivative that appears in the equation.

Answer 5: The order is two, because y'' is the highest order derivative that appears in the equation. It does not matter if it's being multiplied by y'; that does not affect the order of the derivative.

Linear or Nonlinear ODEs. State whether each of the following differential equations is linear in y or nonlinear. If linear, write the ODE in normal linear form. If nonlinear, write it in normal form.

7. $y' = 3y(1 - y)$

8. $3\sin t = (1 - t^2)y' - ty$

9. $(t + \cos t)y - \sin(1 + y') = e^t$

10. $(2y' - 1)/t^2 - 5ty = t^4$

11. $yy' = 1$

12. $t/y = \cos t/y'$

13. $y' = e^t$

14. $y' = e^y$

Answer 7: The ODE $y' = 3y(1 - y)$ can be rewritten in normal form as $y' = 3y - 3y^2$. This ODE is nonlinear because the y^2 term prevents it from being written in normal linear form. [Recall that a first order ODE is linear if it can be written in the normal linear form $y' + p(t)y = q(t)$ where $p(t)$ and $q(t)$ are functions that do not depend on y, but may depend on t.]

Answer 9: This ODE is nonlinear because the $\sin(1 + y')$ term prevents it from being written in normal linear form. The ODE's normal form is $y' = \sin^{-1}((t + \cos t)y - e^t) - 1$.

Answer 11: This ODE is nonlinear because the yy' term prevents it from being written in normal linear form. The ODE's normal form is $y' = 1/y$.

Answer 13: This ODE is in linear form, with $p(t) = 0$ and $q(t) = e^t$.

Solutions. Find all solutions of each nonlinear ODE, and describe the largest intervals on which they live. Sketch representative solution curves. [*Hint*: write the ODE as $(F(y))' = G(t)$ for some functions $F(y)$ and $G(t)$, then use the Antiderivative Theorem. See Example 1.3.3.]

15. $yy' = t$

16. $y' = y^3/2$

17. $ty' + y = 2t$

18. $y' = |t|$

19. $y' = -(1/t)y + |t|/t$

20. $y' = t|t|$

21. $y' = 4t^2y^2$

Answer 15: The chain rule implies that $(y^2/2)' = yy'$ and so the ODE assumes the form $(y^2/2)' = t$. Applying the Antiderivative Theorem, we obtain the formula $y^2/2 = t^2/2 + C$, so $y = \sqrt{t^2 + 2C}$, or $y = -\sqrt{t^2 + 2C}$ for any constant C with t restricted to an interval such that $t^2 > -2C$. See figure for some solution curves, where C takes on various positive values. The solution curves are branches of hyperbolas. See figure.

Answer 17: The ODE is linear, but we solve it anyway. Following Example 1.3.3, notice that $(ty)' = ty' + y$, so we have that $(ty)' = 2t$. Applying the Antiderivative Theorem we obtain the formula $ty = t^2 + C$, where C is a constant. Solving for y, we have $y = t + C/t$. See figure.

Answer 19: The ODE is linear, but we solve it anyway. Following Example 1.3.3, for $t > 0$ note that the ODE becomes $ty' + y = t$, and that $(ty)' = ty' + y$. Applying the Antiderivative Theorem for $t > 0$ we have $y = t/2 + C/t$, where C is any constant. For $t < 0$, note that the ODE becomes $ty' + y = -t$. Applying the Antiderivative Theorem for $t < 0$ we have $y = -t/2 + C/t$, where C is any constant. So the largest interval on which solutions live is either $t > 0$ or $t < 0$. See figure.

Answer 21: Following Example 1.3.3, $y(t) = 0$ is a solution to the problem. To find solutions that are not zero everywhere, divide the ODE by y^2 to obtain the equivalent ODE $(1/y)' = -y^{-2}y'$. Applying the Antiderivative Theorem we obtain the solution $y = -3/(4t^3 + C)$ which is defined on any interval for which $4t^3 + C$ has the same sign. See figure.

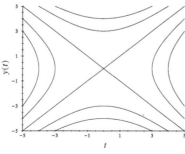

Problem 15.

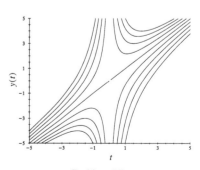

Problem 17.

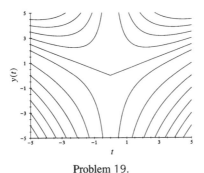

Problem 19.

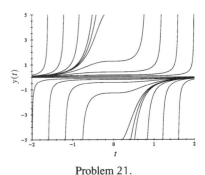

Problem 21.

22. *Smoothness Properties of Solutions*   Show that all solutions of the ODE $y' = \sin y$ are infinitely differentiable. [*Hint*: note that $\sin y$ is an infinitely differentiable function of y.]

Families of Solutions.   In each case, find an ODE that does *not* involve arbitrary constants and has the given family of functions as solutions over the interval $-\infty < t < \infty$. Verify your result by substituting the family of solutions into your ODE. [*Hint*: start by differentiating the family of functions at least once. Then look for a way to eliminate the two arbitrary constants C_1, C_2. The resulting ODE is not unique.]

23. $y(t) = C_1 e^{-t} + C_2 e^t$ **24.** $y(t) = (C_1 + C_2 t)\, e^{2t}$

     Answer 23:   Differentiating twice, we have $y'' = C_1 e^{-t} + C_2 e^t = y$, i.e., the ODE is $y'' = y$.

Modeling Problems: Radioactive Decay.

25.   A radioactive substance has a half-life of 1000 years; what percent remains after 100 years?

     Answer 25:   Since $k = (\ln 2)/\tau$ [text formula (4)], we have $k = 0.693/1000 = 0.000693$ (years)$^{-1}$. Since $N(t) = N(0)e^{-kt}$, we have that $N(100)/N(0) = e^{-100k} = e^{-0.0693} \approx 0.933$; approximately 93% of the original amount $N(0)$ remains after 100 years.

26.   A substance $y(t)$ decays at a rate proportional to the amount present, and in 25 years 1.1% of the initial amount y_0 has decomposed. What is the half-life of the substance?

27.   Radioactive phosphorus with a half-life of 14.2 days is used as a tracer in biochemical studies. After an experiment with 8 grams of phosphorus, researchers must safely store the material until only 10^{-5} grams remain. How long must the contents be stored?

     Answer 27:   If $N(t)$ is the amount of the phosphorus at time t (in days), then $N(0) = 8$, and $N(t) = 8e^{-kt}$ [from text formula (3)]. Now $k = (\ln 2)/14.2 = 0.0488$/day from text formula (4). To find the time t for which $N(t) = 1.0 \times 10^{-5}$, we have $1 \times 10^{-5} = 8e^{-0.0488t}$. Taking logarithms and solving for t, we have $t \approx 278.5$ days.

Harvested Logistic Population Models.   For each harvested logistic fish population model, sketch the equilibrium lines in the first quadrant of the ty-plane: $t \geq 0$, $y \geq 0$. Then use the technique described in Example 1.3.2 to sketch solution curves above and below the equilibrium lines. Estimate the minimal value y_0 of the fish population at time $t = 0$ that does not lead to extinction. What seems to be the ultimate fish population as t gets large if y_0 exceeds that minimal value?

28. $y' = y - y^2/6 - 5/6$ **29.** $y' = y - y^2/6 - 9/8$

30. $y' = y - y^2/6 - 3/2$ **31.** $y' = y - y^2/6 - 2$

32. $y' = y - y^2/6$ **33.** $y' = y - y^2/6 + 9/2$

     Answer 29:   Factoring the right-hand side of the equation, we get that $y' = -(y - 4.5)(y - 1.5)/6$, so our equilibrium lines occur when $y = 4.5$ and when $y = 1.5$. If $0 < y < 1.5$, y' is negative, so the solution curves fall towards $-\infty$. If $1.5 < y < 4.5$, y' is positive, so the solution curves rise towards $y = 4.5$. If $y > 4.5$, y' is negative, so the solution curves fall towards $y = 4.5$. At this rate of harvesting (9/8 tons/year), there needs to be an initial amount of at least 1.5 tons of fish so that the population does not die out. With any initial amount greater than this the population over time will approach 4.5 tons. See figure.

Answer 31: The roots of $y - y^2/6 - 2$ are complex, so y' is never zero, but always has fixed sign (negative in this case). This means that there are no equilibrium points, and since y' is negative, no matter what the initial fish population is, the population will over time be extinct. See figure.

Answer 33: Note that this is a case of restocking at the rate of 9/2 tons/year. Factoring the right-hand side we have $y' = -(y+3)(y-9)/6$ which means that we have only one equilibrium point that actually means anything to us, $y = 9$. If $0 < y < 9$, y' is positive, and so the solution curves rise towards $y = 9$. If $y > 9$, y' is negative, so all solution curves fall towards $y = 9$. At this rate of restocking (9/2 tons/year), the fish population always reaches an equilibrium of nine tons over time. See figure.

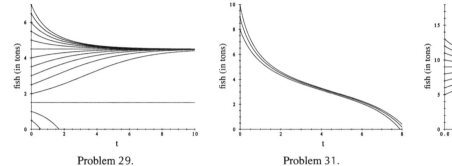

Problem 29.   Problem 31.   Problem 33.

Qualitative Analysis. In producing the graph of solution curves for the IVP $y' = y - 5/3$, $y(0) = y_0$, which appears in Figure 1.2.2, we used the solution formula (9) from Section 1.2 and graphing software. We could have produced a rough sketch of that graph by observing that the rate function $f(y) = y - 5/3$ has the property that $f(5/3) = 0$, and $f(y) > 0$ if $y > 5/3$, $f(y) < 0$ if $y < 5/3$. So solution curves that originate at $y_0 > 5/3$ must rise and never fall. The opposite is true if $y_0 < 5/3$. The straight line $y = 5/3$ is an equilibrium solution since the rate function always vanishes on that line. One other fact is needed: solution curves of an ODE $y' = f(y)$ never intersect if the rate function $f(y)$ and its derivative $\partial f/\partial y$ are continuous. Apply these ideas to sketch by hand representative solution curves in the region $t \geq 0$, $y \geq 0$ for each ODE. Refer to Example 1.3.2 for some hints on how to extract qualitative solution information from each ODE.

The handshake icon denotes a problem suitable for a team project.

34. $y' = y(1 - y)$ **35.** $y' = y - y^2/12 - 5/3$

36. $y' = 6y - y^2 - 5$ 37. $y' = (y-2)(y-4)(y-6)$

38. $y' = \sin y$ 39. $y' = \cos y$

40. $y' = y \cos y$ 41. $y' = (y^2 - 2y + 1)(y - 3)$

Answer 35: This is the same as Example 1.3.2. So the equilibrium lines are $y = 10$ and $y = 2$, and the solution curves in the region $y < 2$ will fall and cut the extinction line $y = 0$, while the solution curves in the region $y > 2$ will gravitate towards $y = 10$, as will the solution curves in the region $y > 10$ as t increases. See figure.

Answer 37: $y' = 0$ when $y = 2$, $y = 4$, and $y = 6$. When $y < 2$, y' is negative, so solution curves in this region will fall and cut the extinction line $y = 0$. When $2 < y < 4$, y' is positive, so solution curves will rise towards $y = 4$. When $4 < y < 6$, solution curves will fall towards $y = 4$, and when $y > 6$, solution curves will escape to infinity. See figure.

Answer 39: There will be equilibrium lines where $\cos y = 0$, which are at $y = \pi/2$, $y = 3\pi/2$, and all other lines where $y = \pi/2 + n\pi$. When $0 < y < \pi/2$, y' is positive, so solution curves in this region will rise towards $y = \pi/2$ as t increases. When $3\pi/2 > y > \pi/2$, y' is negative, so solution curves will fall towards $y = \pi/2$ as t increases. The other regions between the positive zeros are treated similarly, solution curves rising in one region and falling in adjacent regions. See figure.

Answer 41:   First notice that there is going to be an equilibrium line at $y = 3$. Second, $(y^2 - 2y + 1)$ can be factored into $(y - 1)^2$, so that $y = 1$ is another equilibrium point. When $y < 1$, y' is negative, so solution curves will fall off and cut $y = 0$. When $1 < y < 3$, solution curves will curve down towards $y = 1$, and when $y > 3$, y' is positive, so solution curves will curve up towards infinity. See figure.

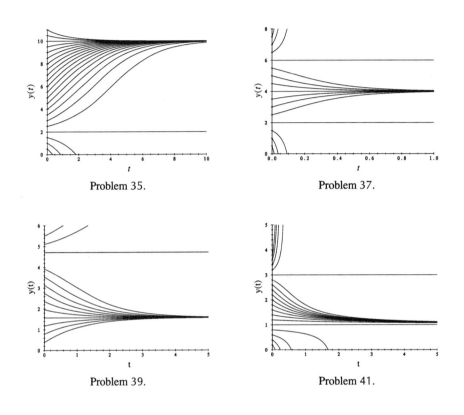

Problem 35.   Problem 37.

Problem 39.   Problem 41.

Background material/Growth Processes:

The U.S. Bureau of the Census frequently predicts population trends. To do this, demographers use known population data to formulate "laws" of population change. Each law is converted into a mathematical formula that can then be used to make the predictions. Since there is a good deal of uncertainty about which law best describes actual population shifts, several are derived, each proceeding from different assumptions. Similarly, ecologists make predictions about changes in the numbers of fish and animal populations, and biologists formulate laws for changing densities of bacteria growing in cultures. Here we introduce some general laws of population change, solve the corresponding mathematical models, and interpret the results in terms of the growth, stabilization, or decline of a population.

Background material/Growth Models:

Let $y(t)$ denote the population at time t of a species. The values of $y(t)$ are integers and change by integer amounts as time goes on. However, for a large population an increase by one or two over a short time span is infinitesimal relative to the total, and we may think of the population as changing continuously instead of by discrete jumps. Once we assume that $y(t)$ is continuous, we might as well smooth off any corners on the graph of $y(t)$ and assume that the function is differentiable. If we had let

$y(t)$ denote the population *density* (i.e., the number per unit area or volume of habitat), the continuity and differentiability of $y(t)$ would have seemed more natural. However, we shall continue to interpret $y(t)$ as the size of the population, rather than density.

The underlying principle

$$\text{Net rate of change} = \text{Rate in} - \text{Rate out} \tag{i}$$

applies to the changing population $y(t)$. For a population, the "Rate in" term is the sum of the birth and the immigration rates, while the "Rate out" term is the sum of the death and the emigration rates. Let's regroup the rates into an internal rate (birth minus death) and external rate (immigration minus emigration). Averaged over all classes of age, sex, and fertility, a typical individual makes a net contribution R to the internal rate of change. The internal rate of change at time t is, then, $Ry(t)$, where

$$Ry = (\text{Individual's contribution}) \times (\text{Number of individuals})$$

The *intrinsic rate coefficient* R will differ from species to species, but it always denotes the average individual's contribution to the rate. Given the rate coefficient R and the migration rate M, the Balance Law becomes

$$y'(t) = Ry(t) + M \tag{ii}$$

and the study of population changes becomes the problem of solving ODE (ii).

In many cases R and M can be approximated by constants or else by simple functions of the population levels and time. If simple models suffice for observed growth processes, there is little need for more complex assumptions. In modeling, it's considered better to have fewer assumptions.

Background material/Exponential Growth:

It may safely be pronounced, therefore, that population, when unchecked, goes on doubling itself every twenty-five years.

Malthus

Thomas Robert Malthus (1766–1834) was a professor of history and political economy in England. The quotation is from "An Essay on the Principle of Population As It Affects the Future Improvement of Society." Malthus's views have had a profound effect on Western thought. Both Darwin and Wallace have said that it was reading Malthus that led them to the theory of evolution.

The Malthusian principle of explosive growth of human populations has become one of the classic laws of population change. The principle follows directly from (ii) if we set $M = 0$ and let $R = r$, a positive constant whose units are $(\text{time})^{-1}$. ODE (ii) becomes the linear ODE $y' = ry$ which has the exponentially growing solution

$$y(t) = y_0 e^{rt}, \quad \text{for all } t \geq 0 \tag{iii}$$

where y_0 is the population at the time $t = 0$. We see from (iii) that the *doubling time* of a species is given by $T = (\ln 2)/r$ since if $y(t + T) = 2y(t)$, then

$$y_0 e^{r(t+T)} = 2 y_0 e^{rt}$$

$$e^{rT} = 2$$

$$T = \frac{1}{r} \ln 2$$

Note the close connection with the half-life of a radioactive element.

Malthus claimed a doubling time of 25 years for the human population, which implies that the corresponding rate coefficient $r = (1/T) \ln 2 = (1/25) \ln 2 \cong 0.02777$. Let's use this information to calculate the annual percentage increase in a population with a 25 year doubling time. The solution formula (iii) implies that

$$y(t+1)/y(t) = \frac{y_0 e^{r(t+1)}}{y_0 e^{rt}} = e^r$$

and so Malthus's value of $r = 0.02777$ gives us that $y(t+1)/y(t) \approx 1.0282$. This corresponds to a 2.8% annual increase in population. Malthus's figure for r is too high for our late-twentieth-century world (use an Almanac to calculate the intrinsic rate coefficient for the U.S.A). However, individual countries, for example, Mexico and Sri Lanka, have intrinsic rate coefficients which exceed 0.02777 and may be as high as 0.033 with a corresponding doubling time of 21 years.

Background material/logistic growth:

The positive checks to population are extremely various and include … all unwholesome occupations, severe labor and exposure to the seasons, extreme poverty, bad nursing of children, great towns, excesses of all kinds, the whole train of common diseases and epidemics, wars, plague, and famine.

Malthus

The unbridled growth of a population as predicted by the simple Malthusian law of exponential increase cannot continue forever. Malthus claimed that resources grow at most arithmetically, i.e., the net increase in resources each year does not exceed a fixed constant. An exponential increase in the size of a population must soon outstrip available resources. The resulting hardships would surely put a damper on growth.

The simplest way to model restricted growth with no net migration is to account for overcrowding by requiring the rate coefficient R to have the form $r_0 - r_1 y$, where r_0 and r_1 are positive constants. Since $r_0 - r_1 y$ is negative when y is large, $y' = (r_0 - r_1 y)y$ is negative and the population declines. It is customary to write the rate coefficient as $r(1 - y/K)$, where r and K are positive constants, rather than as $r_0 - r_1 y$. We then have the *logistic equation* with initial condition,

$$y' = r\left(1 - \frac{y}{K}\right)y, \qquad y(0) = y_0 \tag{iv}$$

Observe that $y(t) = 0$ and $y(t) = K$ are constant solutions of the logistic equation, the so-called *equilibrium* solutions. The coefficient r is the *logistic rate coefficient* and K is the *carrying capacity* of the species.

The logistic equation and its solutions were introduced in the 1840s by the Belgian statistician Pierre-François Verhulst (1804–1849). He predicted that the population of Belgium would eventually level off at 9,500,000. The estimated 2000 population of a little more than 10,200,000 is remarkably close to the predicted value.

Logistic laws are well suited to laboratory or other isolated populations for which there is reason to believe that K and r are constants. Beginning with the experiments of the Soviet biologist G.F. Gauze in the early 1930s, there have been numerous experiments with colonies of protozoa growing under controlled laboratory conditions. The results of these experiments generally confirm the logistic model. See J.H. Vandermeer, "The Competitive Structure of Communities: An Experimental Approach with Protozoa," *Ecology* **50** (1969), pp. 362–371.

Background Material: Radioactive Decay Law

A primitive form of the decay law would go something like this:

> Radioactive Decay Law. In a sample containing a large number of radioactive nuclei the decrease in the number over time is directly proportional to the elapsed time and to the number of nuclei at the start.

Denoting the number of radioactive nuclei in the sample at time t by $N(t)$, and a time interval by Δt, the law translates to the mathematical equation

$$N(t + \Delta t) - N(t) = -kN(t)\Delta t \qquad \text{(v)}$$

where k is a positive coefficient of proportionality. Observation suggests that in most decay processes k is independent of t and N. A decay process of this type is said to be of *first order* with *rate constant k*.

The mathematical model (v) of the decay law helps us to spot flaws in the law itself. $N(t)$ and $N(t + \Delta t)$ must be integers, but $k\Delta t$ need not be an integer, or even a fraction. If we want to keep the form of the law, we must transform the real phenomenon into an idealization in which a continuous rather than a discrete amount $N(t)$ undergoes decay. The number of nuclei can be calculated from the mass of the sample at any given time. So there is no need to be specific about the units for N or t at this point. Even if $N(t)$ is continuous, (v) could not hold for arbitrarily large Δt since $N(t + \Delta t) \to 0$ as $\Delta t \to \infty$. Nor does (v) make sense if Δt is so small that no nucleus decays in the time span Δt. The failure of the law for large Δt can be ignored since we are interested only in local behavior in time. The difficulty with small Δt is troublesome. We can only hope that the mathematical procedures we now introduce will lead to a mathematical model from which accurate predictions can be made.

If we divide both sides of formula (v) by Δt and let $\Delta t \to 0$ (ignoring the difficulty with small Δt mentioned above), we have the linear ODE

$$N'(t) = \lim_{\Delta t \to 0} \frac{N(t + \Delta T) - N(t)}{\Delta t} = -kN(t) \qquad \text{(vi)}$$

Note that k is measured in units of reciprocal time.

The accuracy of the dating process depends on a knowledge of the exact ratio of radioactive ^{14}C to the carbon in the atmosphere. The ratio is now known to have changed over the years. Volcanic eruptions and industrial smoke dump radioactively dead ^{14}C into the atmosphere and lower the ratio. But the most drastic change in recent times has occurred through the testing of nuclear weapons, which releases radioactive ^{14}C, resulting in an increase of 100% in the ratio in some parts of the Northern Hemisphere. These events change the ratio in the atmosphere and so in living tissue. The variations are now factored into the dating process.

Since the experimental error in determining the rate constants for decay processes with long half-lives is large, we would not expect to use this process to date events in the very recent past. Recent events can be dated if a radioactive substance with a short half-life is involved, e.g., white lead contains a radioactive isotope with a half-life of only 22 years. At the other extreme, radioactive substances such as uranium, with half-lives of billions of years, can be used to date the formation of the earth itself.

1.4 The Modeling Process: Differential Systems

Comments:

The notions of state variables and dynamical systems are introduced so that the reader can see that a finite collection of variables that change in time often is adequate to describe the evolution of a natural process. In this book dynamical systems are mostly modeled with differential equations. Discrete dynamical systems are considered in the WEB SPOTLIGHT ON CHAOS IN NUMERICS. Any one-step numerical ODE solver algorithm (e.g., the Euler or Runge-Kutta methods of the SPOTLIGHT ON APPROXIMATE NUMERICAL SOLUTIONS) can be considered to be a discrete dynamical simulation of the ODE. One may even treat the discrete solver algorithm as a discrete model of the original natural phenomenon.

A secondary aim of this section is to introduce the notion of a differential system in a mild way and to establish the connection between differential systems, modeling, and dynamical systems. The vertical motion model and simple compartment models do that in familiar environments.

An experiment

Throw a whiffle ball up into the air; does it take longer to go up or to come down? Ask a group this question, and we guarantee that the responses will divide four ways: longer going up, longer coming down, equal times, or it all depends. If you are in a room with a high ceiling, throw a real whiffle ball up and ask everyone the same question. You will probably get the same four responses, because it is hard to tell the times by just watching the ball. You may want to devise some way to do alternate experiments with a whiffle ball using equipment from a physics lab. Even in this simple setting, the proof that the fall time is longer is hard [Problem 9(**c**)], but the computer simulation (Figures 1.4.2, 1.4.3) shows the answer quite dramatically. See the article, "Projectile Motion with Arbitrary Resistance," Tilak de Alwis, *The College Mathematics Journal*, **26** (1995), 361–367, for a general discussion (with additional references) of vertical motion against resistance.

PROBLEMS

Solutions of a System. Check whether the given pair of functions $x(t)$, $y(t)$ forms a solution to the differential system $dx/dt = 2x + 3y$, $dy/dt = -4y$.

1. $x(t) = e^{-4t}$, $y(t) = -2e^{-4t}$

2. $x(t) = 2e^{2t}$, $y(t) = -e^{2t}$

3. $x(t) = e^{2t} + 3e^{-4t}$, $y(t) = -6e^{-4t}$

4. $x(t) = 2e^{2t} - 4e^{-4t}$, $y(t) = 8e^{-4t}$

Answer 1: This pair of functions forms a solution to the system since

$$\frac{dx(t)}{dt} = -4e^{-4t} = 2e^{-4t} - 6e^{-4t} = 2x(t) + 3y(t) \quad \text{and} \quad \frac{dy(t)}{dt} = 8e^{-4t} = -4y(t)$$

Answer 3: This pair of functions forms a solution to the system since

$$\frac{dx(t)}{dt} = 2e^{2t} - 12e^{-4t} = 2e^{2t} + 6e^{-4t} - 18e^{-4t} = 2x(t) + 3y(t)$$

$$\frac{dy(t)}{dt} = 24e^{-4t} = -4y(t)$$

Convert a Second-Order ODE to a System. Write each ODE as a normal first-order system.

5. $y'' + 5y' + 6y = 0$

6. $y'' + 8y' + 7y = 2\cos(t)$

7. $y'' = -k/(y + R)^2$

8. $y'' + \sin y' - (1 + y^2) = \cos t$

9. $e^t y'' + ty' + \sin y = \sin t$

10. $e^y y'' + y' \cos y = yt$

Answer 5: $y' = x$, $x' = -5x - 6y$

Answer 7: $y' = x$, $x' = -k/(y + R)^2$

Answer 9: $y' = x$, $x' = -tx/e^t - (\sin y)/e^t + (\sin t)/e^t$

Linear Cascades.

11. Find the general solution of the linear cascade in Example 1.4.3 when $a = c$.

Answer 11: To find the general solution to the linear cascade $x' = cx$, $y' = bx + cy$, find the general solution of $x' = cx$ by first writing the ODE as $x' - cx = 0$, and then multiply through by the integrating factor e^{-ct} to get $e^{-ct}x' - ce^{-ct}x = 0$, or $(e^{-ct}x)' = 0$. Use the Antiderivative Theorem to get $e^{-ct}x = C_1$, so $x = C_1 e^{ct}$, where C_1 is any constant. We can now find $y(t)$ by using $x(t)$ and the same method: replace x by $C_1 e^{ct}$ in the ODE $y' = bx + cy$, and rewrite it as $y' - cy = bC_1 e^{ct}$. Then multiply through by the integrating factor e^{-ct} to get $e^{-ct}y' - ce^{-ct}y = bC_1$, or $(e^{-ct}y)' = bC_1$. Now use the Antiderivative Theorem again to get $e^{-ct}y = bC_1 t + C_2$, or $y = bC_1 t e^{ct} + C_2 e^{ct}$, where C_2 is any constant. So, the general solution to the linear cascade is $x(t) = C_1 e^{ct}$ and $y(t) = bC_1 t e^{ct} + C_2 e^{ct}$, where C_1 and C_2 are any constants.

12. Find the general solution of the linear cascade $x' = x + 1$, $y' = x - y$.

13. *Second-Order ODEs and Systems* Find the general solution of the ODE $y'' + ay' = 1$, where a is a nonzero constant by carrying out the following steps:

(a) Rewrite the second-order ODE as a linear cascade. [*Hint*: set $y' = x$.]

(b) Find the general solution of the resulting cascade. [*Hint*: see Example 1.4.3.]

Answer 13:

(a) Using the substitution $y' = x$ we have the first order system of ODEs in normal form

$$y' = x$$
$$x' = 1 - ax$$

where a is a nonzero constant.

(b) In order to solve this first order system, multiply both sides of the second equation by e^{at} to get $e^{at}(x' + ax) = e^{at}$, or $(e^{at}x)' = e^{at}$. Use the Antiderivative Theorem to get $e^{at}x = (1/a)e^{at} + C_1$, or $x = 1/a + C_1 e^{-at}$, where C_1 is any constant. Now, since $x = y'$, we have

$$y' = 1/a + C_1 e^{-at}$$

Use the Antiderivative Theorem again to obtain $y = t/a - (C_1/a)e^{-at} + C_2$. Replace $-C_1/a$ by C_1 (okay since C_1 is arbitrary) to get the general solution

$$y(t) = \frac{t}{a} + C_1 e^{-at} + C_2, \qquad C_1 \text{ and } C_2 \text{ any constants.}$$

Modeling Problems.

14. *Approaching the Limiting Velocity* The velocity of the whiffle ball in Example 1.4.1 is described by the ODE $v' + (c/m)v = -g$. Recall that $g = 9.8$ meters/sec^2, $c/m = 10$ sec^{-1}, and $v_0 = 20$ meters/sec.

(a) How much time must elapse for the whiffle ball to reach 98% of its limiting velocity?

(b) Comment on your result in **(a)**. Is the time more or less than you expected?

15. *Viscous Damping: Longer to Rise or to Fall?* Suppose that a whiffle ball of mass m is thrown straight up with velocity v_0 from height h and is subject to viscous air resistance.[1]

---

[1]See the following articles for interesting alternative proofs that a ball takes longer to fall than to rise if a damping force is present:

1. F. Brauer, "What goes up must come down eventually," *American Mathematical Monthly*, **108** (2001), 437–440.

2. Peter Macgregor and F. Brauer, Commentary on "What goes up must come down eventually," *American Mathematical Monthly*, **109** (2002), 774–775.

(a) Show that the IVP, $y' = v$, $y(0) = h$, $v' = -g - (c/m)v$, $v(0) = v_0$, governs the whiffle ball's height y and velocity v. Find the solution formulas for $y(t)$ and $v(t)$ for this IVP. How much time does it take for the ball to reach maximum height?

(b) Suppose that $h = 2$ meters, $g = 9.8$ meters/sec^2, and $c/m = 5$ sec^{-1} for the system in (a). Use a grapher to estimate the rise time and the fall time (back to the initial height) for $v_0 = 10, 30, 50, 70$ meters/sec. Repeat with $c/m = 1, 10$.

☞ A proof that it takes longer to fall than to rise.

(c) Suppose that $c/m = 1$ in the IVP in (a). Find the time T_{max} for the whiffle ball to reach its highest point. Let the time $\tau > 0$ be such that $y(\tau) = h$. Show that $\tau > 2T_{max}$. [*Hint*: let $C = y(2T_{max}) - h$. Express C explicitly in terms of g and v_0 and show that $C = 0$ if $v_0 = 0$, $dC/dv_0 > 0$ for $v_0 > 0$.]

Answer 15:

(a) With the origin at ground level and up as the positive direction, the whiffle ball is initially at $y(0) = h$ with initial velocity $y'(0) = v_0$. The whiffle ball undergoes viscous damping due to air resistance, and so we see from text formula (1) that $my'' = -mg - cy'$. Recalling that $y' = v$, we have $y'' = v'$, so $mv' = -mg - cv$. Solving for v', we get $v' = -g - (c/m)v$. The initial conditions are $y(0) = h$ and $v(0) = v_0$. To find the time required for the ball to reach maximum height, use text formula (3) for the solution $v(t)$ and then set $v(t) = 0$:

$$v(t) = (v_0 + \frac{mg}{c})e^{-ct/m} - \frac{mg}{c} = 0$$

$$e^{ct/m} = \frac{v_0 + mg/c}{mg/c}$$

Taking logarithms, we have

$$ct/m = \ln\left(\frac{v_0 + mg/c}{mg/c}\right)$$

So $\dfrac{m}{c} \ln\left(\dfrac{v_0 + mg/c}{mg/c}\right)$ seconds is the time it takes the whiffle ball to reach its maximal height.

(b) The curves of Graphs 1–3 for $y_0 = 2$, $v_0 = 10, 30, 50, 70$ clearly suggest that it takes longer for the whiffle ball to fall back to the initial height of 2 meters than to rise from that height. In the case $c/m = 5$ (Graph 2), the drag on the whiffle ball is five times what it is in the first case, $c/m = 1$ (Graph 1), so it is reasonable to expect that the ball will not rise as high. Similarly if $c/m = 10$ (Graph 3), the maximum height is even smaller. Visual inspection of these graphs suggests that the rise and fall times are as given below (in seconds). The values of v_0 are in bold, the respective rise times and fall times follow:

Graph 1 ($c/m = 1$): **10**, 0.6, 1.15; **30**, 1.4, 2.6; **50**,1 1.9, 4.35; **70**, 2.25, 5.95
Graph 2 ($c/m = 5$): **10**, 0.2, 0.5; **30**, 0.5, 2.5; **50**, 0.6, 4.65; **70**, 0.7, 6.8.
Graph 3 ($c/m = 10$): **10**, 0.1, 0.4; **30**, 0.2, 3.3; **50**, 0.25, 5.25; **70**, 0.3, 7.7

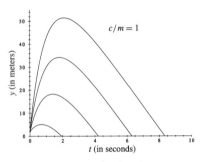

Problem 15(b), Graph 1.

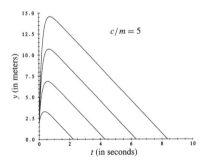

Problem 15(b), Graph 2.

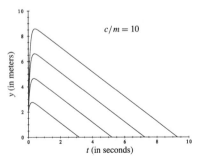

Problem 15(b), Graph 3.

(c) If T is the time it takes to reach maximum height, the strategy is to show that $y(2T) > h$, which implies that the time to fall must be longer than the time to rise.

Solving the IVP, $v' + v = -g$, $v(0) = v_0$, we obtain $v(t) = -g + (v_0 + g)e^{-t}$. Setting $v(T) = 0$ and solving for T,

$$T = \ln(1 + v_0/g)$$

and so this value of T is the time it takes the whiffle ball to reach maximum height (where $v = 0$).

Now $y'(t) = v(t) = -g + (v_0 + g)e^{-t}$, and $y(0) = h$. Integrating, we see that

$$y(t) = -gt - (v_0 + g)e^{-t} + v_0 + g + h$$

So evaluating $y(t)$ at $2T$ and substituting in our expression for T, we have

$$y(2T) = -2gT - (v_0 + g)e^{-2T} + v_0 + g + h$$
$$= -2g\ln(1 + v_0/g) - g^2/(v_0 + g) + v_0 + g + h$$

Define the function $C(v_0) = y(2T) - h$, and so

$$C(v_0) = -2g\ln(1 + v_0/g) - g^2/(v_0 + g) + v_0 + g$$

If $C(v_0) > 0$ for $v_0 > 0$, then $y(2T) > h$. Note that $C(0) = -2g\ln 1 - g^2/g + 0 + g = 0$. To show that $C(v_0) > 0$ for $v_0 > 0$ all we need to do is to show that $dC/dv_0 > 0$ for $v_0 > 0$:

$$\frac{dC}{dv_0} = \frac{-2g}{v_0 + g} + \frac{g^2}{(v_0 + g)^2} + 1$$
$$= \frac{-2g(v_0 + g) + g^2 + (v_0 + g)^2}{(v_0 + g)^2} = \frac{v_0^2}{(v_0 + g)^2} > 0$$

and so $C(v_0)$ is an increasing function of v_0 and is positive if $v_0 > 0$. This shows that $y(2T) > h$ for $v_0 > 0$ and it takes longer for the whiffle ball to fall through a given distance than it does to rise the same distance. This is reasonable because the gravitational and resistance forces oppose each other during the fall (and so the magnitude of the acceleration is lower), but these two forces act in the same direction during the rise. By contrast, in the case of no damping the rise and fall times are equal.

16. *Uranium-Thorium Compartment Model: Linear Cascade*   Use the model equations in Example 1.4.2 and the solution process in Example 1.4.3 to answer the following questions:

(a) How long does it take for the amount of uranium-234 to be reduced by 90%?

(b) If initially there is no thorium-230, what is the maximal amount of thorium-230 and when does it reach that level?

(c) How long does it take the maximum amount of thorium-230 to drop by 90%?

Spotlight on Modeling: Radiocarbon Dating

Comments:

In the problems we give informal definitions of a general compartment system and a linear cascade compartment system, and note how they can be represented by labeled boxes and arrows. By examination of the assemblage, a system of rate equations can be written for the flow of the substance being tracked through the compartments. See the SPOTLIGHT ON MODELING: COLD MEDICATION I. There is a different interpretation of a compartmental system where there is just one physical compartment containing a substance which decays over time into other substances (e.g., radioactive decay, see Problem 5).

PROBLEMS

1. *Dating Stonehenge*   In 1977, the rate of ^{14}C radioactivity of a piece of charcoal found at Stonehenge in southern England was 8.2 disintegrations per minute per gram of carbon. Given that in 1977 the rate of ^{14}C radioactivity of a living tree was 13.5 disintegrations per minute per gram, estimate the date of construction by the two methods suggested below.

   (a) Date the Stonehenge charcoal by using the formula technique outlined in this Spotlight.

   (b) Estimate the date by scaling the measure of ^{14}C and using a grapher. [*Hint*: follow the procedure that led to Figure 1.]

   (c) Estimate the date by scaling time as was done in the text for the Lascaux cave paintings, and then using a grapher. [*Hint*: follow the procedure that led to Figure 2.]

   Answer 1:

   (a) Following the modeling techniques in the text for dating the cave painting, we want $q(t)$ to be the amount of ^{14}C per gram of carbon in the Stonehenge charcoal at time t, where time is measured from 1977. This gives the set of modeling equations in IVP (1), i.e. $q'(t) = 0$ for $t \le T$, $q'(t) = -kq(t)$ for $T \le t \le 0$, and $q(0) = q_0$. Now T is the time when the charcoal was living wood, and k is to be determined via the method outlined in the text. $T \le t \le 0$ represents the span of years up to 1977 when the residual radioactivity of the charcoal was measured, and $q(0)$ is the amount of ^{14}C per gram of charcoal in 1977. Solving the rate equation, we have that $q(t) = q(T)e^{-k(t-T)}$, $T \le t \le 0$; so that $q(0) = q(T)e^{kT}$. Solving for T, $T = (1/k)\ln(q(0)/q(T))$. According to the text, the half-life of ^{14}C is roughly 5568 years, so $k = \ln(2)/5568$. We know that in 1977 the rate of disintegrations in living wood was 13.5 per minute per gram. We then assume that this rate is the same at $t = 0$ as it was at $t = T$, namely $13.5 = q'(T) = -kq(T)$. For the Stonehenge charcoal fragment we are given that $8.2 = q'(0) = -kq(0)$ and since $T = (1/k)\ln[q(0)/q(T)]$, we have $T = (5568/\ln 2)\ln(8.2/13.5) \approx -4005$ years. The tree from which the charcoal was made was living in approximately the year 2028 B.C.E. The date is only approximate because of the various uncertainties mentioned in the text.

   (b) Following the graphical age-dating technique described in this Spotlight and pictured in Figure 1, we set $Q(t) = q(t)/q(0)$, where $q(t)$ is the amount of carbon-14 in the charcoal sample. Using the IVP $q' = -kq, q(0) = q_0$, we see that $Q'(t) = -kQ$, $Q(0) = 1$. As noted in the text, $k \approx 1.24 \times 10^{-4}$/year. According to the text, the time T when the wood in the sample was living solves the equation $Q(T) = q(T)/q(0) = q'(T)/q'(0) = \frac{13.5}{8.2} \approx 1.646$ (using the given data). So we follow the graph of $Q(t) = e^{-kt}$ backward in time until the graph crosses the line $Q = 1.646$; then we read the time T at the crossing point. The figure below shows the horizontal line $Q = 1.646$. The vertical line through the crossing point suggests that $T \approx -4000$, that is, the wood was living about 2000 B.C.E.

   (c) We use the "rescaling time" technique discussed in the text; $t = \tau s$, where $\tau \approx 5568$ is the half-life of carbon-14 and s is a new dimensionless measure of time. By text formula (10), we see that $dQ/ds = -(\ln 2)Q$, $Q(0) = 1$, and $S \le s \le 0$ where $Q(t) = q(t)/q(0)$ from (b) and S is the value of s for which $Q = 13.5/8.2 \approx 1.646$. The figure below shows the graph of $Q(t)$ crossing the horizontal line $Q = 1.646$ when $s \approx 0.72$, so $S \approx 0.72$ (see the vertical line). So $T + \tau S \approx 5568 \times 0.72 \approx 4009$ years ago, that is about 2009 B.C.E. for the date the wood in the charcoal sample was alive.

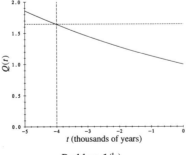

Problem 1(b).

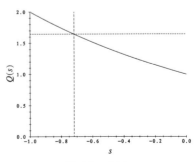

Problem 1(c).

2. *Dating a Sea Shell* An archaeologist finds a sea shell that contains 60% of the ^{14}C of a living shell. Approximately how old is the shell?

☞ A
three-compartment
model.

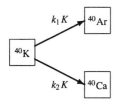

3. *The Bones of Olduvai: Potassium-Argon Dating*[2] Olduvai Gorge, in Kenya, cuts through volcanic flows, volcanic ash, and sedimentary deposits. It is the site of bones and artifacts of early hominids, considered by some to be precursors of man. In 1959, Mary and Louis Leakey uncovered a fossil hominid skull and primitive stone tools of obviously great age. ^{14}C dating methods for the skull being inappropriate for a specimen of that age and nature (because the half-life of ^{14}C is too short), dating had to be based instead on the ages of the underlying and overlying volcanic strata. The method used was that of potassium-argon decay. The potassium-argon clock is an accumulation clock, in contrast to the ^{14}C dating method. The approach described below models this clock.

The potassium-argon method depends on measuring the accumulation of "daughter" argon atoms, which are decay products of radioactive potassium atoms. Specifically, potassium-40 (^{40}K) decays to argon-40 (^{40}Ar) and to calcium-40 (^{40}Ca) at rates proportional to the amount of potassium present with respective constants of proportionality k_1 and k_2, where the values of k_1 and k_2 in units of yr^{-1} are very small (see the figure in the margin). This means the corresponding half-lives are very long.

The model for this decay process may be written in terms of the amounts $K(t)$, $A(t)$, and $C(t)$ of potassium, argon, and calcium in a sample of rock. Using the Balance Law we get

$$K' = -(k_1 + k_2)K, \qquad A' = k_1 K, \qquad C' = k_2 K$$

where t is measured *forward* from the time $t = 0$ at which the volcanic ash was deposited around the skull.

(a) Solve the system to find $K(t)$, $A(t)$, and $C(t)$ in terms of k_1, k_2, and $k = k_1 + k_2$. Set $K(0) = K_0$, $A(0) = C(0) = 0$. Why is $K(t) + A(t) + C(t) = K_0$ for all $t \geq 0$? Show that $K(t) \to 0$, $A(t) \to k_1 K_0 / k$, and $C(t) \to k_2 K_0 / k$ as $t \to \infty$.

(b) The age T of the volcanic strata is the current value of the time variable t because the potassium-argon clock started when the volcanic material was laid down. This age is estimated by measuring the ratio of argon to potassium in a sample. Show that this ratio is $A/K = (k_1/k)(e^{kT} - 1)$. Show that the age of the sample (in years) is $(1/k) \ln[(k/k_1)(A/K) + 1]$.

(c) When the actual measurements were made at the University of California at Berkeley, the age T of the bones was estimated to be 1.75 million years. The values of the constants of proportionality k_1 and k_2 are known to be (approximately) 5.3×10^{-10}/yr. Using the formula $k\tau = \ln 2$, where τ is the half-life, we see that $\tau_1 \approx \tau_2 \approx 1.3 \times 10^9$ years, a value that is appropriate for dating items from one million years of age on up. What was the value of the measured ratio A/K?

Answer 3:

(a) Setting $k = k_1 + k_2$, $K(0) = K_0$, $A(0) = 0$, $C(0) = 0$ and solving, we have that $K(t) = K_0 e^{-kt}$. Substituting this into the second ODE, we have $A' = k_1 K_0 e^{-kt}$. Integrating and setting $A(0) = 0$, we obtain $A(t) = (K_0 k_1/k)(1 - e^{-kt})$. Similarly, $C'(t) = k_2 K_0 e^{-kt}$ and with $C(0) = 0$ we obtain $C(t) = (K_0 k_2/k)(1 - e^{-kt})$. (see Section 1.2), total amounts are conserved, so

$$K(t) + A(t) + C(t) = K_0 e^{-kt} + (K_0 k_1/k)(1 - e^{-kt}) + (K_0 k_2/k)(1 - e^{-kt})$$

$$= K_0 e^{-kt} + K_0 - K_0 e^{-kt} = K_0$$

for all t. Another way to see this is that since $[K(t) + A(t) + C(t)]' = 0$, we have that $K(t) + A(t) + C(t) = K(0) + A(0) + C(0) = K(0) = K_0$. Equilibrium solutions may be obtained by analyzing the solution as $t \to +\infty$. As $t \to +\infty$, the exponential e^{-kt} approaches zero due to the negative exponent,

[2]For further reading consult the book: B. Dalrymple and M. A. Lanphere, *Potassium-Argon Dating: Principles, Techniques, and Applications to Geochronology* (W.H. Freeman, San Francisco, 1969). It should be noted that the half-life of the potassium-argon process is not yet known with any precision. The value of 1.3 billion years given in Problem 3 seems to be the most widely used estimate.

and $K(t) \to 0$. Similarly, the exponential terms in $A(t)$ and $C(t)$ approach zero, and we have that $A(t) \to (K_0k_1/k)$ and $C(t) \to (K_0k_2/k)$ as $t \to +\infty$.

(b) We see that $A/K = (K_0k_1/k)(1 - e^{-kT})/K_0e^{-kT} = (k_1/k)(e^{kT} - 1)$ after factoring e^{-kT} from the numerator and from the denominator. We have $(Ak/Kk_1) = e^{kT} - 1$, so $T = (1/k)\ln[(Ak/Kk_1) + 1]$.

(c) Substitute $k_1 = 5.76 \times 10^{-11}\text{yr}^{-1}, k_2 = 4.85 \times 10^{-10}\text{yr}^{-1}$ into the equation $A/K = (k_1/k)(e^{kT} - 1)$ in **(b)** and recall that $k = k_1 + k_2$ to get

$$\frac{A}{K} = \frac{5.76 \times 10^{-11}}{5.76 \times 10^{-11} + 4.85 \times 10^{-10}} \left(e^{[(5.76\times10^{-11} + 4.85\times10^{-10})\cdot 1.75\times10^6]} - 1 \right)$$

$$\approx 1.01 \times 10^{-4}$$

 4. *Parameter Identification* Describe an experiment that would lead to an approximate determination of the decay constant k for ^{14}C, or equivalently, to the half-life of ^{14}C. [*Hint*: say you have just come into possession of a wood fragment from the sarcophagus of the Egyptian king Tutankhamen who is known from historical sources to have died in 1325 B.C. The same wood still grows in Egypt, and a Geiger counter held to a living fragment reads 13.5 disintegrations per minute. A Geiger counter held to the ancient wood fragment reads 8.9 disintegrations/min.] Check out the world-wide Web to see how others have solved this problem.

☞ A
four-compartment
model.

 5. *Uranium Series: A Linear Cascade* Suppose that element A undergoes a series of radioactive decays: $A \to B \to C \to D$. The rates of decay of elements A, B, and C are proportional to the respective amounts present (first-order rate laws). Suppose that the distinct rate constants k_A, k_B, and k_C correspond to the decays $A \to B$, $B \to C$, and $C \to D$, respectively.

(a) Write out the system of four rate equations for the amounts $A(t)$, $B(t)$, $C(t)$, and $D(t)$ of each element. If the amounts A_0, B_0, C_0, and D_0 are initially present at $t = 0$, find the amount of element B for any later time t. [*Hint*: the system of rate equations is a linear cascade.]

☞ See also
Example 1.4.2.

(b) Now let A, B, C, and D represent Uranium-234, Thorium-230, Radium-226, and Lead-206, respectively. Uranium-234 has a half-life of approximately 2×10^5 years, Thorium-230 has a half-life of approximately 8×10^4 years, and Radium-226 a half-life of approximately 1600 years. How much thorium and radium will be present after 1×10^6 years if initially 238 grams of uranium are present but no thorium, radium, and lead? [*Hint*: scale the system to units of 10^5 years and then calculate $B(10)$ and $C(10)$.]

Answer 5:

(a) There are four ODEs with four initial conditions in this problem.

$$\begin{cases} (1) & A'(t) = -k_A A(t), & A(0) = A_0 \\ (2) & B'(t) = k_A A(t) - k_B B(t), & B(0) = B_0 \\ (3) & C'(t) = k_B B(t) - k_C C(t), & C(0) = C_0 \\ (4) & D'(t) = k_C C(t), & D(0) = D_0 \end{cases}$$

To find $B(10)$ and $C(10)$ we need not find $D(t)$, but we do so anyway.

Solving ODE (1), we see that $A(t) = A_0e^{-k_At}$. Substituting this term into ODE (2), we get

$$B'(t) = k_A A_0 e^{-k_At} - k_B B(t)$$

$$(e^{k_Bt}B(t))' = k_A A_0 e^{-k_At}e^{k_Bt}$$

$$e^{k_Bt}B(t) = \frac{k_A A_0}{k_B - k_A}e^{-k_At}e^{k_Bt} + K_1$$

$$B(t) = \frac{k_A A_0}{k_B - k_A}e^{-k_At} + K_1e^{-k_Bt}$$

for some constant K_1. Since $B(0) = B_0 = (k_A A_0)/(k_B - k_A) + K_1$, we find K_1 in terms of B_0:

$$B(t) = \left(\frac{k_A A_0}{k_B - k_A}\right) e^{-k_A t} + \left(B_0 - \frac{k_A A_0}{k_B - k_A}\right) e^{-k_B t}$$

Now we can plug this into ODE (3) to get

$$C'(t) + k_C C(t) = \left(\frac{k_A k_B A_0}{k_B - k_A}\right) e^{-k_A t} + \left(k_B B_0 - \frac{k_A k_B A_0}{k_B - k_A}\right) e^{-k_B t}$$

$$(e^{k_C t} C(t))' = \left(\frac{k_A k_B A_0}{k_B - k_A}\right) e^{-k_A t} e^{k_C t} + \left(k_B B_0 - \frac{k_A k_B A_0}{k_B - k_A}\right) e^{-k_B t} e^{k_C t}$$

$$e^{k_C t} C(t) = \left(\frac{k_A k_B A_0}{(k_B - k_A)(k_C - k_A)}\right) e^{-k_A t} e^{k_C t}$$
$$+ \left(\frac{k_B B_0}{k_C - k_B} - \frac{k_A k_B A_0}{(k_B - k_A)(k_C - k_B)}\right) e^{-k_B t} e^{k_C t} + K_2$$

$$C(t) = \left(\frac{k_A k_B A_0}{(k_B - k_A)(k_C - k_A)}\right) e^{-k_A t}$$
$$+ \left(\frac{k_B B_0}{k_C - k_B} - \frac{k_A k_B A_0}{(k_B - k_A)(k_C - k_B)}\right) e^{-k_B t} + K_2 e^{-k_C t}$$

for some constant K_2. The initial value $C(0) = C_0$ determines the constant K_2 to be

$$K_2 = C_0 - \frac{k_A k_B A_0}{(k_B - k_A)(k_C - k_A)} - \frac{k_B B_0}{k_C - k_B} + \frac{k_A k_B A_0}{(k_B - k_A)(k_C - k_B)}$$

Hence, $\quad C(t) = \left(\frac{k_A k_B A_0}{(k_B - k_A)(k_C - k_A)}\right) e^{-k_A t}$

$$+ \left(\frac{k_B B_0}{k_C - k_B} - \frac{k_A k_B A_0}{(k_B - k_A)(k_C - k_B)}\right) e^{-k_B t}$$

$$+ \left(C_0 - \frac{k_A k_B A_0}{(k_B - k_A)(k_C - k_A)} - \frac{k_B B_0}{k_C - k_B} + \frac{k_A k_B A_0}{(k_B - k_A)(k_C - k_B)}\right) e^{-k_C t}$$

Finally, we can substitute this into ODE (4) to get

$$D'(t) = \left(\frac{k_A k_B k_C A_0}{(k_B - k_A)(k_C - k_A)}\right) e^{-k_A t} + \left(\frac{k_B k_C B_0}{k_C - k_B} - \frac{k_A k_B k_C A_0}{(k_B - k_A)(k_C - k_B)}\right) e^{-k_B t}$$

$$+ \left(k_C C_0 - \frac{k_A k_B k_C A_0}{(k_B - k_A)(k_C - k_A)} - \frac{k_B k_C B_0}{k_C - k_B} + \frac{k_A k_B k_C A_0}{(k_B - k_A)(k_C - k_B)}\right) e^{-k_C t}$$

Antidifferentiating, we get

$$D(t) = - \left(\frac{k_B k_C A_0}{(k_B - k_A)(k_C - k_A)}\right) e^{-k_A t} + \left(-\frac{k_C B_0}{k_C - k_B} + \frac{k_A k_C A_0}{(k_B - k_A)(k_C - k_B)}\right) e^{-k_B t}$$

$$+ \left(-C_0 + \frac{k_A k_B A_0}{(k_B - k_A)(k_C - k_A)} + \frac{k_B B_0}{k_C - k_B} - \frac{k_A k_B A_0}{(k_B - k_A)(k_C - k_B)}\right) e^{-k_C t} + K_3$$

for some constant K_3. Use the initial condition $D(0) = D_0$ to solve for K_3 and rearrange terms to get the final answer

$$D(t) = \left(\frac{k_B k_C A_0}{(k_B - k_A)(k_C - k_A)}\right)(1 - e^{-k_A t}) + \left(\frac{k_C B_0}{k_C - k_B} - \frac{k_A k_C A_0}{(k_B - k_A)(k_C - k_B)}\right)(1 - e^{-k_B t})$$

$$+ \left(C_0 - \frac{k_A k_B A_0}{(k_B - k_A)(k_C - k_A)} - \frac{k_B B_0}{k_C - k_B} + \frac{k_A k_B A_0}{(k_B - k_A)(k_C - k_B)}\right)(1 - e^{-k_C t}) + D_0$$

(b) Since we know the half-lives of the elements, we can determine their rate constants by use of the formula $k = (\ln 2)/\tau$. Hence, $k_A \approx 3.47 \times 10^{-6}$, $k_B \approx 8.66 \times 10^{-6}$, $k_C \approx 4.33 \times 10^{-4}$. Note that $A_0 = 238$, $B_0 = C_0 = D_0 = 0$. Substituting these values back into our equation for $B(t)$, we get:

$$B(10^6) = \frac{3.47 \times 10^{-6} \times 238}{(8.66 - 3.47) \times 10^{-6}}\left(e^{-3.47} - e^{-8.66}\right)$$

$$\approx 5.09 \text{ grams of thorium}$$

Doing the same for $C(10^6)$:

$$C(10^6) = \left(\frac{3.47 \times 10^{-6} \times 8.66 \times 10^{-6} \times 238}{(8.66 - 3.47) \times 10^{-6} \times (433 - 3.47) \times 10^{-6}}\right)e^{-3.47}$$

$$+ \left(0 - \frac{3.47 \times 10^{-6} \times 8.66 \times 10^{-6} \times 238}{(8.66 - 3.47) \times 10^{-6} \times (433 - 8.66) \times 10^{-6}}\right)\left(e^{-8.66}\right)$$

$$+ \left(0 - \frac{7152}{2229} - 0 + \frac{7152}{2202}\right)e^{-433}$$

$$\approx .102 \text{ grams of radium}$$

Doing the same for $D(10^6)$:

$$D(10^6) = \left(\frac{8.66 \times 10^{-6} \times 4.33 \times 10^{-4} \times 238}{(8.66 - 3.47) \times 10^{-6} \times (433 - 3.47) \times 10^{-6}}\right)(1 - e^{-3.47})$$

$$+ \left(0 - \frac{3.47 \times 10^{-6} \times 4.33 \times 10^{-4} \times 238}{(8.66 - 3.47) \times 10^{-6} \times (433 - 8.66) \times 10^{-6}}\right)(1 - e^{-8.66})$$

$$+ \left(0 - \frac{7152}{2229} - 0 + \frac{7152}{2202}\right)e^{-433} + 0$$

$$\approx 388 - 162 + 0 = 226 \text{ grams of lead}$$

Background material/Radiocarbon Dating:

The accuracy of the dating process depends on a knowledge of the exact ratio of radioactive ^{14}C to the carbon in the atmosphere. The ratio is now known to have changed over the years. Volcanic eruptions and industrial smoke dump radioactively dead ^{14}C into the atmosphere and lower the ratio. But the most drastic change in recent times has occurred through the testing of nuclear weapons, which releases radioactive ^{14}C, resulting in an increase of 100% in the ratio in some parts of the Northern Hemisphere. These events change the ratio in the atmosphere and so in living tissue. The variations are now factored into the dating process.

Since the experimental error in determining the rate constants for decay processes with long half-lives is large, we would not expect to use this process to date events in the very recent past. Recent events can be dated if a radioactive substance with a short half-life is involved, e.g., white lead contains a radioactive isotope with a half-life of only 22 years. At the other extreme, radioactive substances such as uranium, with half-lives of billions of years, can be used to date the formation of the earth itself.

Spotlight on Modeling: Cold Medication I

Comments:

On a modeling level, the first-order systems of ODEs that represent the flow of medications through the compartments of the body seem quite natural to many people, who find these systems easier to understand than the mechanical, gravitational, or electrical systems traditionally used in ODE courses. On a technique level, the method used here to solve a linear cascade is that used in Example 1.4.3. The rate constants for the flow of medication may depend on the age and the state of health of the individual. This sensitivity to the parameter values in the model equations is explored.

PROBLEMS

Medication Levels. Problems 1–4 have to do with antihistamine levels in the GI tract and the bloodstream. See Examples 1 and 2.

1. Solve IVP (3) but with $x(0) = 1$, $y(0) = 1$, $k_1 = 0.6931$, $k_2 = 0.0231$, $0 \le t \le 6$. Graph the levels of antihistamine in the GI tract and in the blood. Estimate the highest level of the antihistamine in the blood and the time it takes to reach that level.

2. *Time of Maximum Medication Level* Show that after a single dose the antihistamine level in the bloodstream reaches its maximum when $t = (\ln k_1 - \ln k_2)/(k_1 - k_2)$, $k_1 \ne k_2$. Use the model developed in Examples 1 and 2. [*Hint*: when is $y'(t) = 0$?]

3. Suppose that A units of antihistamine are present in the GI tract and B units in the blood at time 0. Solve IVP (3) with the condition $y(0) = 0$ replaced by $y(0) = B$.

4. *Sensitivity to Changes in* k_1 Let $A = 1$, keep $k_2 = 0.0231$ as in Example 3, but let k_1 vary.

   (a) Display the effects on the antihistamine levels in the bloodstream if $k_1 = 0.06931, 0.11, 0.3, 0.6931$, 1.0, and 1.5. Plot the graphs over a 24-hour period. Why do the graphs for larger values of k_1 cross the graphs for smaller values?

   (b) You need to keep medication levels within a fixed range so that the medication is both therapeutic and safe. Suppose that the desired range for antihistamine levels in the blood is from 0.2 to 0.8 for a unit dose taken once. With $k_2 = 0.0231$, find upper and lower bounds on k_1 so that the antihistamine levels in the blood reach 0.2 within 2 hours and stay below 0.8 for 24 hours.

   Answer 1: Using the method of Example 2, first solve $dx/dt = -k_1 x$ and then solve $dy/dt = k_1 x - k_2 y$. Since $x(0) = 1$, we have that $x = e^{-k_1 t}$. Substituting this into $dy/dt = k_1 x - k_2 y$, we obtain the ODE $y' + k_2 y = k_1 e^{-k_1 t}$. Multiply through by the factor $e^{k_2 t}$ to obtain $e^{k_2 t} y' + k_2 e^{k_2 t} y = \left(y e^{k_2 t} \right)' = k_1 e^{(k_2 - k_1)t}$. Now antidifferentiate. Since $y(0) = 1$ we have $y = k_1 e^{-k_1 t}/(k_2 - k_1) + C e^{-k_2 t}$, where $C = 1 - k_1/(k_2 - k_1)$ and $k_1 = 0.6931$, $k_2 = 0.0231$, $k_2 - k_1 = -0.6700$. See the figure for antihistamine levels. For the antihistamine in the blood, the maximum of about 1.75 units is reached in about four hours.

   Answer 3: Using the method of Example 2, solve the first ODE of system (3), and then the other. The IVP, $dx/dt = -k_1 x$, $x(0) = A$, gives $x = A e^{-k_1 t}$. Substituting this into the ODE, $dy/dt = k_1 x - k_2 y$, and rearranging, $y' + k_2 y = A k_1 e^{-k_1 t}$. Multiply through by the factor $e^{k_2 t}$ to obtain $e^{k_2 t} y' + k_2 e^{k_2 t} y = (y e^{k_2 t})' = A k_1 e^{(k_2 - k_1)t}$, giving $y = A k_1 e^{-k_1 t}/(k_2 - k_1) + C e^{-k_2 t}$, where C is constant. Using $y(0) = B$, we get $B = A k_1/(k_2 - k_1) + C$, so $C = B - A k_1/(k_2 - k_1)$. So, $x = A e^{-k_1 t}$ and $y = A k_1 (e^{-k_1 t} - e^{-k_2 t})/(k_2 - k_1) + B e^{-k_2 t}$.

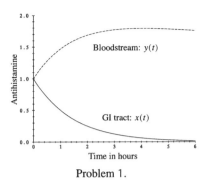

Problem 1.

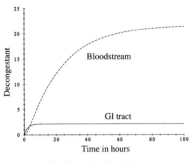

Continuous Dosage at a Constant Rate.   Some cold pills dissolve continuously in the GI tract at a constant rate I (see the figure in the margin). Problems 5–10 have to do with this model.

5.   Use the compartment diagram and the Balance Law to construct IVPs for $x(t)$ and $y(t)$.

6.   Solve the system found in Problem 5 (assume that $k_1 \neq k_2$).

7.   *Approach to Equilibrium*   What happens to the levels of medication in the GI tract and in the blood as $t \to \infty$?

Answer 5:   The model IVP for the amounts $x(t)$ and $y(t)$ of continuous-acting medication in the GI tract and bloodstream, respectively, is $x'(t) = I - k_1 x$, $x(0) = 0$, and $y'(t) = k_1 x - k_2 y$, $y(0) = 0$.

Answer 7:   For the system found in Problem 5, as $t \to +\infty$, $x(t) \to I/k_1$ and $y(t) \to I/k_2$.

Continuous Dosage at a Constant Rate: Decongestant and Antihistamine.   Many cold medications contain a decongestant as well as an antihistamine. Problems 8–10 deal with this situation.

8.   *Antihistamines and Decongestants*   Use a numerical solver to plot $x(t)$ and $y(t)$ for 200 hours. Use $I = 1$ unit/hour, $k_1 = 0.6931$ (hour)$^{-1}$, and $k_2 = 0.0231$ (hour)$^{-1}$ for antihistamine. Are the antihistamine levels in the blood close to equilibrium as $t \to 200$? Repeat with the decongestant: $k_1 = 1.386$ (hour)$^{-1}$, $k_2 = 0.1386$ (hour)$^{-1}$, $0 \leq t \leq 100$.

9.   *The Old and the Sick*   The coefficients k_1 and k_2 for the old and sick may be much less than those for the young and healthy. Plot decongestant and antihistamine levels in the GI tract and the blood if the values of k_1 and k_2 are one-third of those in Problem 8.

10.   *Safe and Effective Zone*   Assume that the clearance coefficients k_1 and k_2 have one-third of the values given in Problem 8. Ideally, the levels of antihistamine in the blood should reach and remain between 25 and 50 units for a dose taken continuously at the rate of 1 unit per hour. Does this happen?

Answer 9:   See Graph 1 for decongestant levels and Graph 2 for antihistamine levels. As expected, the levels of both medications in the blood continue to rise, particularly the antihistamine.

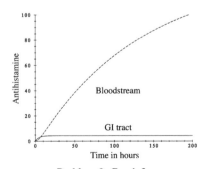

Problem 9, Graph 1.   Problem 9, Graph 2.

Multicompartment Medication Models. The body is made up of many compartments, not just the GI tract and the bloodstream. Here is a three-compartment model for the flow of an antibiotic through the body.

 Here is a three-compartment model that leads to a linear cascade.

11. *Tetracycline in the Body* The antibiotic tetracycline is prescribed for ailments ranging from acne to acute infections. The drug is taken orally and absorbed through the intestinal wall into the bloodstream, and eventually is removed from the blood by the kidneys and excreted. Assume all movement of the drug follows first-order rate laws.

(a) If the compartments of the system are the intestinal tract, the bloodstream, and the urinary system, derive the three-compartment model IVP, $x' = -ax$, $x(0) = x_0$; $y' = ax - by$, $y(0) = 0$; $z' = by$, $z(0) = 0$. The variables $x(t)$, $y(t)$, and $z(t)$ represent the amounts of tetracycline in milligrams per cubic centimeter of solution at time t in the intestinal tract, bloodstream, and urinary system, respectively. Explain each term in the compartment model.

(b) It has been experimentally shown that $a = 0.72$ per hour and $b = 0.15$ per hour. Show that the amount of tetracycline in the bloodstream reaches a maximum about 2.75 hours after absorption from the intestine begins (assuming that $x(0) > 0$, but that ingestion has ceased by time 0). Sketch the graph of $y = y(t)$ if $x_0 = 0.0001$.

(c) Show that the amount of tetracycline in the bloodstream declines to about 20% of the maximum after about 15 hours.

Answer 11:

(a) In the ODE $x' = -ax$ that predicts the amount of tetracycline in the intestinal tract at time t, the ax term accounts for the rate at which the antibiotic leaves the intestinal tract. The rate is proportional to the amount of tetracycline present in the intestinal tract and is negative since this compartment is *losing* the substance at this rate. In the ODE $y' = ax - by$, the term ax represents the entrance rate of tetracycline into the bloodstream from the intestinal tract. The term $-by$ denotes the exit rate out of the bloodstream and into the urinary system. The term by in the ODE $z' = by$ is the entrance rate into the urinary system from the bloodstream. The initial data $x(0) = x_0$ indicates that there are x_0 milligrams of tetracycline in the intestinal tract at time 0, $y(0) = z(0) = 0$ implies no antibiotic in bloodstream and urinary tract at $t = 0$.

(b) Since $a = 0.72$ per hour and $b = 0.15$ per hour, then the system becomes:

$$x' = -0.72x$$
$$y' = 0.72x - 0.15y$$
$$z' = 0.15y$$

Now $y(t)$ represents the amount of tetracycline in the bloodstream at time t, so that the maximum amount of the antibiotic is present when $y'(t) = 0$. To show that $y'(t) = 0$ at time $t = 2.75$ we must first find the solutions $x(t)$ and $y(t)$. We find $x(t)$ first:

$$x' = -0.72x, \quad \text{so} \quad x(t) = C_1 e^{-0.72t}$$

where C_1 is any positive constant.

Substituting what we got for $x(t)$ into $y' = 0.72x - 0.15y$, we can solve for $y(t)$ in a similar way:

$$y' = 0.72C_1 e^{-0.72t} - 0.15y$$
$$y' + 0.15y = 0.72C_1 e^{-0.72t}$$
$$e^{0.15t}(y' + 0.15y) = 0.72C_1 e^{0.15t} e^{-0.72t}$$
$$(e^{0.15t} y)' = 0.72C_1 e^{0.57t}$$
$$e^{0.15t} y = (0.72/0.57)C_1 e^{-0.57t} + C_2$$
$$y(t) = (0.72/0.57)C_1 e^{-0.72t} + C_2 e^{-0.15t}, \quad C_1 \text{ and } C_2 \text{ any positive constants}$$

Now we have the general solutions $x(t)$ and $y(t)$!

$$x(t) = C_1 e^{-0.72t}$$

$$y(t) = -(0.72/0.57)C_1 e^{-0.72t} + C_2 e^{-0.15t}$$

Using the initial conditions $x(0) = x_0$ and $y(0) = 0$, we find unique values for the constants C_1 and C_2 and get unique solutions for the IVP:

$$x(0) = x_0 = C_1 e^0 \quad y(0) = 0 = -(0.72/0.57)C_1 e^0 + C_2 e^0$$

Hence,

$$C_1 = x_0, \quad C_2 = (0.72/0.57)C_1 \approx 1.3x_0$$

and so,

$$x(t) = x_0 e^{-0.72t} \quad \text{and} \quad y(t) = 1.3x_0 (e^{-0.15t} - e^{-0.72t})$$

Taking the derivative of the equation for $y(t)$ gives us

$$y'(t) = 1.3x_0(-0.15e^{-0.15t} + 0.72e^{-0.72t})$$

Setting $y'(t) = 0$ and solving for t to find the time when $y(t)$ is maximal, we get

$$0.72e^{-0.72t} = 0.15e^{-0.15t}$$

$$\ln(0.72) - 0.72t = \ln(0.15) - 0.15t$$

$$0.57t = \ln\left(\frac{0.72}{0.15}\right)$$

$$t \approx 2.75$$

Thus, we have shown that $y(t)$, the amount of tetracycline in the bloodstream, has a critical point at about $t = 2.75$; there are no other critical points.

Now all that is left to do is make sure that this extreme is in fact a maximum value. To do this, evaluate $y''(2.75) = 1.3x_0(0.0225e^{-0.4125} - 0.51842e^{-1.98}) \approx 1.3x_0 \cdot (-.06) < 0$ since $x_0 > 0$. Since $y''(2.75)$ is negative, the critical point is a maximum. The figure shows a graph of $y(t), 0 \le t \le 15$ hrs.

(c) The maximum value of y, which occurs at about $t = 2.75$, is

$$y(2.75) = -1.3x_0 e^{-0.72(2.75)} + 1.3x_0 e^{-0.15(2.75)} = 0.6811x_0$$

We want to check that at time $t = 15$ the amount of antibiotic left in the bloodstream is 20% of this maximum value. That is, we want to show that $y(15) = 0.2y(2.75) = 0.2(0.6811x_0) \approx 0.14x_0$.

$$y(15) = -1.3x_0 e^{-0.72(15)} + 1.3x_0 e^{-0.15(15)} \approx 0.14x_0$$

Thus, the amount of tetracycline in the bloodstream declines to about 20% of the maximum after about 15 hours.

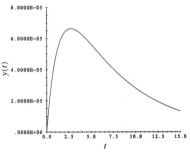

Problem 11(b).

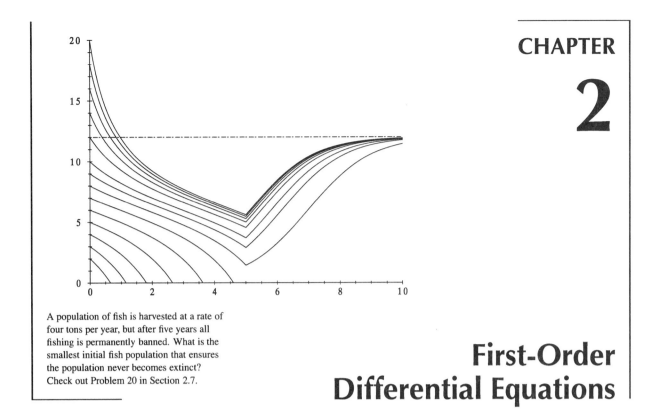

A population of fish is harvested at a rate of four tons per year, but after five years all fishing is permanently banned. What is the smallest initial fish population that ensures the population never becomes extinct? Check out Problem 20 in Section 2.7.

CHAPTER

2

First-Order
Differential Equations

The main thrust of this chapter is to find solution formulas for first-order ODEs, discuss some ODE models, introduce the geometry of solution curves, and show the role of numerical solvers in understanding the behavior of solutions. Solution methods are tabulated on pages 120 and 121 for reference purposes. Questions of the existence and uniqueness of solutions of initial value problems (IVPs) and the sensitivity of IVPs to changes in the data can be answered from these formulas. We hope users of this text will use numerical solvers to approximate and graph solutions of ODEs; a computer icon designates problems for which a numerical solver is appropriate. Most CAS packages have numerical solvers, but can also produce solution formulas for many first-order ODEs (even ODEs with parameters), not to mention graphs and direction fields.

Systems of ODEs are introduced early in this chapter for several reasons. Systems arise naturally in modeling. Some systems can be solved using the first-order techniques of the chapter. All numerical ODE solvers available on the market today are designed to solve systems of first-order ODEs in normal form.

We prepare a first-order linear ODE in normal form for antidifferentiation by multiplying by an integrating factor (Sections 2.1 and 2.2). The separable ODE, $N(y)y' + M(x) = 0$, is already in shape for antidifferentiation (Section 2.5), as are the exact ODEs of the WEB SPOTLIGHT ON EXACT ODEs. The changes of variables in the SPOTLIGHT ON CHANGE OF VARIABLES: PURSUIT MODELS are applied to convert an ODE to a form that can then be prepared for antidifferentiation. So, it isn't surprising that antidifferentiation lies at the heart of finding solution formulas for ODEs.

The goal of this chapter and its Spotlights is to deepen and extend some topics that were introduced in Chapter 1: existence and uniqueness of a solution for a first-order IVP, numerical approximation of a solution of an IVP, and a study of sensitivity via computer simulation. The early sections of the chapter have a bare-bones treatment of the topics, with many examples.

We are mindful that although numerical methods are important because they drive the engine that produces graphs of solutions, they are not the main object of interest in the course. Therefore we give numerical methods a featured spot as early as possible in the text, but we limit our coverage to concepts that would affect the ability to produce decent graphs. There are many numerical solvers available on the market today which can generate most of the graphs of solutions of ODEs which appear in this text. In particular, the ODE Architect (included with this text) will handle any of the IVPs in the text. Most CAS packages contain numerical ODE solvers as well, but because of their symbolic manipulation facilities they are also capable of producing solution formulas for some classes of ODEs. See Professor Switkes's Technology Resource Manuals which outline the use of Maple, Mathematica, and Matlab coordinated with this text. The web is also a good place to look for information on ODE solvers.

2.1 Linear Differential Equations

Comments:

The effective use of integrating factors to find the general solution of a first-order linear ODE is the central topic here. The two model ODEs that use the technique are a pollution model and the continuously compounded interest model. Solution formulas are used to tell us how solutions behave as time advances.

PROBLEMS

Integrating Factors.   Use the five steps of the Method of Integrating Factors to find the general solution of each linear ODE. [*Hint*: see Example 2.1.3 and write the ODE in normal linear form.]

1. $y' + y = 3$	**2.** $y' - 2y = e^t$	**3.** $y' + ty = t$
4. $y' - y = e^{2t} - 1$	**5.** $y' - 2ty = t$	**6.** $y' + y = te^{-t} + 1$
7. $y' = \sin t - y \sin t$	**8.** $2y' + 3y = e^{-t}$	**9.** $t(2y - 1) + 2y' = 0$

Answer 1:   The integrating factor for $y' + y = 3$ is e^t. We obtain $e^t(y' + y) = (e^t y)' = 3e^t$. Integrating, $e^t y = 3e^t + C$ and the general solution is given by $y = Ce^{-t} + 3$, $-\infty < t < \infty$, C any constant.

Answer 3:   The integrating factor for $y' + ty = t$ is $e^{t^2/2}$. We obtain $e^{t^2/2}y' + e^{t^2/2}ty = (e^{t^2/2}y)' = te^{t^2/2}$. Integrating, $e^{t^2/2}y = e^{t^2/2} + C$ and the general solution is given by $y = Ce^{-t^2/2} + 1$, $-\infty < t < \infty$, C any constant.

Answer 5:   For $y' - 2ty = t$, the integrating factor is $e^{\int -2t\,dt} = e^{-t^2}$. So, after multiplying by e^{-t^2}, we have that $e^{-t^2}(y' - 2ty) = (e^{-t^2}y)' = te^{-t^2}$. Integrating, we have that $e^{-t^2}y(t) = \int te^{-t^2}\,dt = -e^{-t^2}/2 + C$. The general solution is $y = Ce^{t^2} - 1/2$, where C is any constant and $-\infty < t < \infty$.

Answer 7:   The normal linear form is $y' + (\sin t)y = \sin t$, so the integrating factor is $e^{\int \sin t\,dt} = e^{-\cos t}$. We obtain $e^{-\cos t}(y' + (\sin t)y) = (e^{-\cos t}y)' = e^{-\cos t}\sin t$. Integrating, $e^{-\cos t}y = e^{-\cos t} + C$ and the general solution is $y = Ce^{\cos t} + 1$ since $\int e^{-\cos t}\sin t\,dt = e^{-\cos t}$, C any constant and $-\infty < t < \infty$.

Answer 9:   The normal linear form is $y' + ty = t/2$, so the integrating factor is $e^{\int t\,dt} = e^{t^2/2}$. We obtain $e^{t^2/2}(y' + ty) = (e^{t^2/2}y)' = (t/2)e^{t^2/2}$. Integrating, $e^{t^2/2}y = (1/2)e^{t^2/2} + C$ and the general solution is $y = Ce^{-t^2/2} + 1/2$, C any constant and $-\infty < t < \infty$.

Solving an IVP, Long-Term Behavior of the Solution.   First find the general solution of the linear ODE in each IVP by following the steps of the procedure. Then use the initial condition to find the solution of the IVP. Discuss that solution's qualitative behavior as $t \to +\infty$. Give the largest t-interval on which the solution is defined. [*Hint*: see Examples 2.1.3 and 2.1.4.]

10. $y' + 2y = 3$, $y(0) = -1$ **11.** $y' + 2ty = 2t$, $y(0) = 1$

12. $y' + (\cos t)y = \cos t$, $y(\pi) = 2$ **13.** $y' + y = e^{-t}/t^2$, $y(1) = 0$

Answer 11: The integrating factor for $y' + 2ty = 2t$ is e^{t^2}, so the ODE becomes $(e^{t^2}y)' = 2te^{t^2}$. Integrating, $e^{t^2}y = e^{t^2} + C$, so $y(t) = Ce^{-t^2} + 1$. Since $y(0) = 1$, $1 = C + 1$, so $C = 0$. The solution is $y(t) = 1$, $-\infty < t < \infty$. As $t \to +\infty$, $y(t) \to 1$.

Answer 13: The ODE, $y' + y = e^{-t}/t^2$, is already in normal linear form. The integrating factor is e^t, so the ODE becomes $(ye^t)' = 1/t^2$. Integrating, $ye^t = -1/t + C$, so $y = Ce^{-t} - e^{-t}/t$. Applying the initial condition, we have $0 = C - 1$. Thus $C = 1$, so the solution is $y(t) = e^{-t}(1 - 1/t)$, $t > 0$. As $t \to +\infty$, the factor e^{-t} forces $y(t) \to 0$. The largest interval on which the solution is defined is $t > 0$ since the initial point $t_0 = 1$ is located in that interval and the solution formula is not continuous at $t = 0$.

The General Solution: Long-Term Behavior. Find the general solution of each ODE over the indicated t-interval by using an integrating factor (after writing the ODE in normal linear form). Discuss the qualitative behavior of the solutions as $t \to 0^+$. Discuss the qualitative behavior of the solutions as $t \to +\infty$. [*Hint:* write the ODE in normal linear form.]

14. $ty' + 2y = t^2$, $\;\; t > 0$ **15.** $(3t - y) + 2ty' = 0$, $\;\; t > 0$

16. $ty' + y = \cos t$, $\;\; t > 0$ **17.** $y' = y/t + t^n$, $\;\; t > 0$,   integer n

Answer 15: The normal linear form for $(3t - y) + 2ty' = 0$ is $y' - (2t)^{-1}y = -3/2$, $t > 0$. The integrating factor is $e^{-(1/2)\ln t} = t^{-1/2}$, so the ODE becomes $(t^{-1/2}y)' = -3t^{-1/2}/2$. Integrating, $t^{-1/2}y = -3t^{1/2} + C$, so $y = Ct^{1/2} - 3t$, $t > 0$. Note that although $y(t)$ is defined at $t = 0$, it has no derivative there, so $y(t)$ does not satisfy the ODE at $t = 0$. As $t \to 0^+$, $y(t) \to 0$. As $t \to +\infty$, $y(t) \to -\infty$, since $\lim_{t\to\infty}(-3t + Ct^{1/2}) = \lim_{t\to\infty} t(-3 + Ct^{-1/2}) = -\infty$, C is an arbitrary constant and $t > 0$.

Answer 17: In normal linear form, the ODE $y' - (y/t) = t^n$, $t > 0$, has integrating factor $e^{-\ln t} = 1/t$. So, $(y/t)' = t^{n-1}$. The two cases, $n = 0$ and $n \neq 0$, must be considered. If $n = 0$, $(y/t)' = 1/t$ and $y = Ct + t\ln t$. If $n \neq 0$, $(y/t)' = t^{n-1}$, so $y/t = t^n/n + C$ and the general solution is $y = Ct + (t^{n+1}/n)$, where $t > 0$. In either case, C is any constant and $t > 0$.

$$\text{As } t \to 0^+, \; y = \begin{cases} \lim_{t\to 0^+}\left[Ct + t^{n+1}/n\right] = -1/1 + 0 = -1 & n = -1 \\ \lim_{t\to 0^+}\left[Ct + t^{n+1}/n\right] = -\infty \text{ since } n + 1 < 0 & n < -1 \\ \lim_{t\to 0^+}\left[0 + t^{n+1}/n\right] = 0/n + 0 = 0, & n > -1 \end{cases}$$

$$\text{As } t \to +\infty, \; y = \begin{cases} \lim_{t\to+\infty}\left[Ct + t^{n+1}/n\right] = +\infty, & n > 0, \quad \text{all C} \\ \lim_{t\to+\infty}\left[Ct + t\ln t\right] = \lim_{t\to+\infty} t(\ln t + C) = +\infty, & n = 0, \quad \text{all C} \\ \lim_{t\to+\infty}\left[Ct + 1/nt^{-n}\right] = \begin{cases} 0, & C = 0 \\ -\infty, & C < 0 \\ +\infty, & C > 0 \end{cases} & n < 0, n \neq -1 \\ \lim_{t\to+\infty}\left[Ct + 1/-1\right] = \begin{cases} -1, & C = 0 \\ -\infty, & C < 0 \\ +\infty, & C > 0 \end{cases} & n = -1. \end{cases}$$

Solving an IVP: Qualitative Behavior. Find the general solution of the ODE over the indicated t-interval. Then use the initial condition to find the solution of the IVP. Finally, discuss the qualitative behavior of the solution as t tends to the given value. [*Hint:* see Example 2.1.4.]

18. $ty' + 2y = \sin t$, $t > 0$; $y(\pi) = 1/\pi$; $t \to +\infty$

19. $(\sin t)y' + (\cos t)y = 0$, $0 < t < \pi$; $y(3\pi/4) = 2$; $t \to 0^+$

20. $y' + (\cot t)y = 2\cos t$, $0 < t < \pi$; $y(\pi/2) = 3$; $t \to 0^+$

21. $y' + (2/t)y = (\cos t)/t^2$, $t > 0$; $y(\pi) = 0$; $t \to 0^+$, $t \to +\infty$

Answer 19: The normal linear form for $(\sin t)y' + (\cos t)y = 0$ is $y' + (\cot t)y = 0$, and the integrating factor is $e^{\int \cot t\, dt} = e^{\ln(\sin t)} = \sin t$. We obtain $((\sin t)y)' = 0$. Integrating, $(\sin t)y = C$, so $y = C \csc t$. Since $y(3\pi/4) = 2$, $2 = C \csc 3\pi/4 = C\sqrt{2}$, so $C = \sqrt{2}$. The solution is $y = \sqrt{2} \csc t$, $0 < t < \pi$, where the t-interval is the largest interval including $t_0 = 3\pi/4$ for which $\csc t$ is defined. As $t \to 0^+$, $y(t) \to +\infty$.

Answer 21: The integrating factor for $y' + 2y/t = (\cos t)/t^2$, $t > 0$, is $e^{2\ln t} = t^2$. We obtain $(t^2 y)' = \cos t$. Integrating, $t^2 y = \sin t + C$, so $y = t^{-2}\sin t + Ct^{-2}$. Since $y(\pi) = 0$, $0 = C/\pi^2$ so $C = 0$. The solution is $y = t^{-2}\sin t$, $t > 0$. As $t \to 0^+$, $y \to +\infty$ since $(\sin t)/t \to 1$ and $1/t \to +\infty$. As $t \to +\infty$, $y \to 0$.

Modeling Problems.

22. *Pollution* A contaminant is pumped at the rate of 1 gal/min into a tank containing 1000 gal of clean water, and the well-stirred mixture leaves the tank at the same rate.

   (a) Find the amount $y(t)$ of waste in the tank at time t. What happens to y(t) as $t \to +\infty$?

   (b) How long does it take the concentration of waste to reach 20% of its maximum level?

23. *Pollution* Water with pollutant concentration c_0 lb/gal starts to run at a rate of 1 gal/min into a vat holding 10 gal of water mixed with 10 lb of pollutant. The mixture runs out at a rate of 2 gal/min. How much pollutant is in the vat at time t? What happens as $t \to 10$ min?

Answer 23: There are two amounts to consider in this model: the total amount Y of liquid in the vat, and the amount y of pollutant in the vat. The rate at which the total amount of liquid changes, Y', can be found with the Balance Law: $Y' = 1 - 2 = -1$, $Y(0) = 10$ gallons. Antidifferentiating, we see that $Y(t) = C - t$, where $C = 10$, so that $Y(t) = 10 - t$. Note that $0 \le t \le 10$; the tank runs dry after 10 minutes.

   Now, applying the Balance Law to the amount of pollutant in the vat, we get

$$y' = [c_0(\text{lbs/gal})(1\,\text{gal/min})] - [(y/Y)(\text{lbs/gal})(2\,\text{gal/min})]$$

$$= c_0 - \frac{2y}{10 - t}$$

$$y' + \left(\frac{2}{10 - t}\right)y = c_0, \quad y(0) = 10$$

With this ODE in normal linear form, we see that $p(t) = 2/(10 - t)$. The integrating factor is $e^{\int p(t)} = e^{-2\ln(10-t)} = (10 - t)^{-2}$. Multiply both sides by the integrating factor to get $(y(10 - t)^{-2})' = c_0(10 - t)^{-2}$. Antidifferentiating, we get $y(10 - t)^{-2} = c_0/(10 - t) + C$, or $y(t) = c_0(10 - t) + C(10 - t)^2$. The initial condition implies that $C = (1 - c_0)/10$, so $y(t) = c_0(10 - t) + (1 - c_0)(10 - t)^2/10$.

   As $t \to 10$ min, $y(t) \to 0$ lbs. This makes sense, because as $t \to 10.$, $Y(t)$ also goes to 0, meaning all the water has run out of the tank, and along with it all the pollutant.

24. *Validity of the Pollutant Model* Make a list of all the factors you can think of that would affect the validity of the pollutant model in Example 2.1.1. Justify the items on your list.

> Continuously Compounded Interest. If $A(t)$ denotes the money in an account at time t and the account accrues interest continuously at the rate of $r\%$, then $A(t)$ follows a first-order rate law where the constant of proportionality is $r/100$, so $A' = 0.01rA$.

Continuous Compounding. Some banks compound savings continuously instead of daily. Use the ODE for computing compound interest (see box) to answer the following questions.

25. What interest rate payable annually is equivalent to 9% interest compounded continuously?

Answer 25: The amount $A(1)$ at the end of one year with 9% continuous compounding is $A(1) = A_0 e^{0.09}$ since the model equation is $A' = 0.09A$, $A(0) = A_0$. Setting this equal to the amount after one year with interest payable annually at a rate $r\%$, we have $A_0e^{0.09} = A_0(1 + r/100)$, or $r/100 = e^{0.09} - 1 \approx 0.0942$. That is, the annual interest rate would have to be 9.42% to yield the same amount as continuous compounding at 9%.

26. What is the interest rate if continuously compounded funds double in eight years?

27. How long will it take A dollars invested in a continuously compounded savings account to double if the interest rate is 5%, 9%, 12%?

Answer 27: Find t such that $2A_0 = A_0e^{kt}$. Canceling A_0, taking logarithms, and solving, $t = \ln 2/k$ for $k = 0.05$, 0.09, and 0.12. The doubling times are approximately 13.86, 7.70, and 5.78 years.

28. *Rule of 72* Two common rules of thumb for bankers are the "Rule of 72" and the "Rule of 42." These rules say that the number of years for money invested at $r\%$ interest to increase by 100% or 50% is given by $72/r$ or $42/r$, respectively. Assuming continuous compounding, show that these rules overestimate the time required.

29. *Budgeting for a Car* At the end of every month Wilbert deposits a fixed amount in a savings account. He wants to buy a car that costs $12,400. Currently, he has $5800 saved. If he earns 7% interest compounded continuously on his savings and has an income of $2600 per month, what is the limit on Wilbert's monthly expenses if he wants to have enough saved to buy the car in one year?

Answer 29: Wilbert already has $5800 in a saving account earning 7% per annum interest. In one year's time, i.e. $t = 1$ he will earn $5800e^{0.07} = 6220.50$ from that source (rounding to the half-dollar), so he only has to raise $ 12,400 - $ 6220.50 = $ 6179.50 by depositing x dollars per month in his savings account for each of 12 consecutive months. Let S_n be the amount in Wilbert's account from this source at the end of n months. Then we see that $S_{n+1} = S_n e^{(0.07/12)} + x$, where $S_0 = 0$ and $n = 0, 1, 2, \ldots, 11$. Let $z = 0.07/12$. Then by iteration we find that

$$S_{12} = x + xe^z + xe^{2z} + \cdots + xe^{11z} = x(1 + e^z + e^{2z} + \cdots + e^{11z}).$$

Summing the truncated geometric series we have that

$$S_{12} = x\frac{e^{12z} - 1}{e^z - 1} \approx 12.394x.$$

Hence we must choose x such that

$$S_{12} = 12.394x = \$6179.50, \quad \text{or} \quad x = \$498.59$$

which represents the monthly deposit to his savings account in order to have enough money to buy the car after one year. Thus he has to limit his monthly expenses to $2600 - \$498.59 = \$2101.41$.

2.2 Linear Differential Equations: Qualitative Analysis

Comments:

Other than doing a "lucky guessing" example, we don't make a big deal about the structure of the solution set of a first order linear ODE because it will come up again in Chapter 3 when we treat second-order linear ODEs, and again in Chapter 6, where linear systems are handled. So why have we put the structure stuff here at all? Mostly for the readers who want to see and understand the theory of a linear ODE, and because it does serve as a neat way to look at the solution formula produced by the procedure in Section 2.1 for constructing a general solution. We also use the formulas to give us clues about long-term solution behavior.

PROBLEMS

Structure of Solutions. Find the general solution of each ODE in the form $y = y_u + y_d$, where y_u is the general solution of the undriven ODE and y_d is a particular solution of the driven ODE. [*Hint*: to find y_d use a lucky guess approach or the Method of Integrating Factors.]

1. $y' + 5y = t + 1$

2. $y' + y = \sin t$

3. $y' = ty + t$

4. $y' + y = (1 + e^t)^{-1}$

5. $y' - 2y = t^2 e^{2t}$

6. $y' - 3y/t = t^3$, $t > 0$

7. $y' + 2y/t = t^{-2} \cos t$, $t > 0$

8. $y' + y \cot t = 2t \csc t$, $0 < t < \pi$

Answer 1: For this ODE, $p(t) = 5$ and $q(t) = t + 1$. Thus, $P(t) = \int p(t)dt = 5t$. To find the general solution y_u of the undriven form of the equation, $y' + 5y = 0$, multiply both sides by $e^{P(t)} = e^{5t}$. This becomes $(e^{5t}y)' = 0$, which can be antidifferentiated to produce $y_u = Ce^{-5t}$.

Now find a solution y_d of the driven equation, $y' + 5y = t + 1$. Use the Method of Integrating Factors to set up and solve the ODE:

$$e^{5t}y' + 5e^{5t}y = e^{5t}(t+1), \quad (e^{5t}y)' = e^{5t}(t+1), \quad e^{5t}y = te^{5t}/5 + 4e^{5t}/25, \quad y_d = t/5 + 4/25$$

where the Integral Table on the inside back cover has been used. A constant of integration is not needed because you only need to find one particular solution. From this you can find the general solution $y(t) = y_u + y_d = Ce^{-5t} + t/5 + 4/25$, C any constant, $-\infty < t < \infty$.

Answer 3: First rearrange the equation into normal linear form, so that $y' - ty = t$. For this ODE $p(t) = -t$ and $q(t) = t$. Thus, $P(t) = \int p(t)dt = -t^2/2$. Now, to find the general solution y_u of the undriven equation, $y' - ty = 0$, multiply both sides by $e^{-t^2/2}$. This becomes $(e^{-t^2/2}y)' = 0$, which can be antidifferentiated to show that $y_u = Ce^{t^2/2}$.

Now find a solution y_d of the driven equation, $y' - ty = t$. Use the Method of Integrating Factors to set up the equation $(e^{-t^2/2}y)' = te^{-t^2/2}$. Antidifferentiate both sides to produce $e^{-t^2/2}y = -e^{t^2/2}$ (the constant of integration, C, is not needed because you only need one particular solution). From this you can find that $y_d = -1$. So the general solution is $y(t) = Ce^{t^2/2} - 1$, C any constant, $-\infty < t < \infty$.

Answer 5: In this ODE, $p(t) = -2$ and $q(t) = t^2 e^{2t}$. Thus $P(t) = \int p(t)dt = -2t$. To find the general solution y_u of the undriven equation $y' - 2y = 0$ multiply both sides by $e^{P(t)} = e^{-2t}$. This gives you $(e^{-2t}y)' = 0$, which can be antidifferentiated to show that $y_u = Ce^{2t}$.

Now find a solution y_d of the driven equation, $y' - 2y = t^2 e^{2t}$. Use the Method of Integrating Factors to set up the equation $(e^{-2t}y)' = e^{-2t}t^2 e^{2t} = t^2$. Antidifferentiate both sides to get $e^{-2t}y = t^3/3$. The integration constant C is not needed because you only need to find one particular solution. From this you can show that $y_d = t^3 e^{2t}/3$. So the general solution to the ODE is $y(t) = Ce^{2t} + t^3 e^{2t}/3$, C any constant, $-\infty < t < \infty$.

Answer 7: Here $p(t) = 2/t$ and $q(t) = t^{-2} \cos t$. Thus $P(t) = \int 2/t dt = 2\ln t$. To find the general solution y_u to the undriven equation $y' + 2y/t = 0$, multiply both sides by $e^{P(t)} = e^{2\ln t} = t^2$. This gives you $(t^2 y)' = 0$, which can be antidifferentiated to show that $y_u = Ct^{-2}$.

Now find a solution y_d of the driven equation $y' + 2y/t = t^{-2} \cos t$. Use the Method of Integrating Factors to set up the equation $(t^2 y)' = t^2(t^{-2} \cos t) = \cos t$. Antidifferentiate both sides to produce the equation $t^2 y = \sin t$. The integration constant C is not needed, because you only need to find one particular solution. From this you can obtain $y_d = t^{-2} \sin t$. So the general solution to the ODE is $y(t) = Ct^{-2} + t^{-2} \sin t$, C any constant, $t > 0$.

Response to Data and Input. Solve each IVP below and write the solution as the sum of the *response to the initial data* and the *response to the input function* (see Theorem 2.2.1). Give the largest t-interval on which the solution is defined.

9. $y' + y = e^{3t}$, $y(0) = y_0$

10. $3y' + 12y = 4$, $y(0) = y_0$

11. $y' + 3t^2 y = t^2$, $y(0) = 2$

12. $(\cos t)y' + (\sin t)y = 1$, $y(0) = 1$

13. $t^2 y' - ty = 1$, $y(1) = y_0$

14. $ty' + y = t \sin t$, $y(\pi) = -1$

15. $y' + ay = e^{at}$, $y(0) = y_0$, $a \neq 0$

16. $y' + ay = e^{bt}$, $y(0) = y_0$, $a \neq b$, neither is 0

Answer 9: Here $p(t) = 1$ and $q(t) = e^{3t}$. Thus $P(t) = \int p(t)dt = t$. The response to the initial data is the solution of the system if it were not driven, i.e. the solution to $y' + y = 0$. In this case, then, the response to the initial data $y(0) = y_0$ is $y_u = e^{P(t_0)}y_0e^{-P(t)} = e^0 y_0 e^{-t} = y_0 e^{-t}$. The response to the input is the solution of the driven system if $y_0 = 0$. By Theorem 2.2.1 this is $y_d = e^{-t}\int_0^t e^s e^{3s}ds = e^{3t}/4 - e^{-t}/4$. So the total response is $y(t) = y_u + y_d = [y_0 e^{-t}] + [e^{3t}/4 - e^{-t}/4]$. The solution is defined on the entire t-axis.

Answer 11: Here $p(t) = 3t^2$ and $q(t) = t^2$. Thus $P(t) = \int p(t)dt = t^3$. The response to the initial data is the solution of the system if it were not driven, i.e. the solution to $y' + 3t^2 y = 0$. In this case, then, the response to the initial data is $y_u = y_0 e^{-P(t)} = 2e^{-t^3}$.

The response to the input is the solution of the driven system if $y_0 = 0$. By Theorem 2.2.1 this is $y_d = e^{-t^3}\int_0^t e^{s^3}s^2 ds = 1/3 - e^{-t^3}/3$. So the total response is $y(t) = y_u + y_d = 2e^{-t^3} + 1/3 - e^{-t^3}/3$. The solution is defined on the entire t-axis.

Answer 13: After dividing through by t^2, $p(t) = -1/t$ and $q(t) = 1/t^2$. Thus $P(t) = \int p(t)dt = -\ln t$. The response to the initial data is the solution of the system if it were not driven, i.e. the solution to $y' - (1/t)y = 0$. In this case, then, the response to the initial data is $y_u = y_0 e^{\ln t} = y_0 t$.

The response to the input is the solution of the driven system if $y_0 = 0$. By Theorem 2.2.1 this is $y_d = e^{\ln t}\int_1^t e^{-\ln s}/s^2 ds = t\int_1^t (1/s^3)ds = t/2 - 1/(2t)$. So the total response is $y(t) = y_u + y_d = y_0 t + t/2 - 1/(2t)$. The solution is defined for all $t > 0$.

Answer 15: Here $p(t) = a$ and $q(t) = e^{at}$. Thus $P(t) = \int p(t)dt = at$. The response to the initial data is the solution of the system if it were not driven, i.e. the solution to $y' + ay = 0$. In this case, then, the response to the initial data is $y_u = y_0 e^{-at}$.

The response to the input is the solution of the driven system if $y_0 = 0$. By Theorem 2.2.1 this is $y_d = e^{-at}\int_0^t e^{as}e^{as}ds = e^{at}/2a - e^{-at}/2a$. So the total response is $y(t) = y_u + y_d = y_0 e^{-at} + e^{at}/2a - e^{-at}/2a$. The solution is defined on the entire t-axis.

Verification of Response Form of Solution. For each IVP below, write the solution in the form of the sum of the *response to the initial data* and the *response to the input*. Then verify directly that the response to the initial data satisfies IVP (6) and the response to the input satisfies IVP (8).

17. $y' + 2y = \sin t$, $y(0) = y_0$

18. $y' + 0.1y = 10c(t)$, $y(0) = y_0$. [*Hint*: follow Example 2.2.1. In your verification use the Fundamental Theorem of Calculus to differentiate the integral in formula (9).]

Answer 17: Here $p(t) = 2$ and $q(t) = \sin t$. Thus $P(t) = \int p(t)dt = 2t$. The response to the initial data is the solution of the system if it were not driven, i.e. the solution to $y' + 2y = 0$. In this case, then, the response to the initial data is $y_u = y_0 e^{-2t}$. Let's use formula (7) to find the response to the input, where in this case $P(t) = 2t$ and $g(t) = \sin t$. Using the table of integrals on the inside back cover we have the response to the input $= e^{-2t}\int_0^t e^{2s}\sin s\, ds$

$$= e^{-2t}\left[\frac{e^{2s}}{5}(2\sin s - \cos s)\right]_{s=0}^{s=t}$$

$$= \frac{2}{5}\sin t - \frac{1}{5}\cos t + \frac{1}{5}e^{-2t}$$

So the solution to the IVP is

$$y = y_0 e^{-2t} + \left(\frac{2}{5}\sin t - \frac{1}{5}\cos t + \frac{1}{5}e^{-2t}\right)$$

Find All Solutions: Long-Term Behavior. Use the lucky guess approach to find a particular solution y_d for each ODE. (Guess the form of the solution, and then determine the coefficients.) Then find the general solution of the driven ODE. Finally, describe the qualitative behavior of all solutions as $t \to +\infty$.

19. $y' + y = t^2$ [*Hint:* try $y_d = At^2 + Bt + C$.]   **20.** $y' + ty = t^2 - t + 1$

21. $y' + 2y = e^{-2t}$ [*Hint:* try $y_d = Ate^{-2t}$.]   **22.** $y' - 2y = 3e^{-t}$

23. $y' + y = 5\cos 2t$ **24.** $y' + y = e^{-t} + \cos t$

Answer 19: Substitute $y_d = At^2 + Bt + C$ into the ODE $y' + y = t^2$ to obtain $At^2 + [2A + B]t + C + B = t^2$. Equating the coefficients of like powers of t, we must have $A = 1$, $2A + B = 0$, $C + B = 0$, which satisfy these conditions and so $y_d = t^2 - 2t + 2$ is a particular solution of the ODE. Since $y_u = Ce^{-t}$ is the general solution of the undriven ODE, $y' + y = 0$, the general solution of the driven ODE is $y = y_u + y_d = Ce^{-t} + t^2 - 2t + 2$. As $t \to +\infty$, all solutions tend to the parabolic solution described by $y_d = t^2 - 2t + 2$.

Answer 21: Substitute $y_d = Ate^{-2t}$ into the ODE $y' + 2y = e^{-2t}$ to obtain $te^{-2t}(-2A + 2A) + e^{-2t}(A) = e^{-2t}$, which holds only if $A = 1$, and so $y_d = te^{-2t}$ is a particular solution of the ODE. Since $y_u = Ce^{-2t}$ is the general solution of the undriven ODE, $y' + 2y = 0$, the general solution of the driven ODE, $y' + 2y = e^{-2t}$, is $y = y_u + y_d = Ce^{-2t} + te^{-2t}$. As $t \to +\infty$, all solutions tend to 0, since e^{-2t} and te^{-2t} have this property.

Answer 23: Guess the solution $y_d = A\cos 2t + B\sin 2t$. Substitute y_d into the ODE $y' + y = 5\cos 2t$ to obtain $(-2A + B)\sin 2t + (2B + A)\cos 2t = 5\cos 2t$. So $-2A + B = 0$, and $2B + A = 5$. Solve to get $A = 1$ and $B = 2$, and so $y_d = \cos 2t + 2\sin 2t$ is a particular solution of the ODE. Since $y_u = Ce^{-t}$ is the general solution of the undriven ODE, $y' + y = 0$, the general solution of the driven ODE is $y = y_u + y_d = Ce^{-t} + \cos 2t + 2\sin 2t$. As $t \to +\infty$, all solutions tend to the periodic solution of period π, $y_d = \cos 2t + 2\sin 2t$.

More Guessing and Undetermined Coefficients. If a, b, c are nonzero constants, find a particular solution of each ODE that does not contain exponential functions. What is the long-term qualitative behavior (as $t \to +\infty$) of these solutions?

25. $y' + ay = b\cos t + c\sin t$ **26.** $y' + ay = bt + c$

Answer 25: Guess a particular solution of the form $y_d = A\cos t + B\sin t$. Substitute y_d into the ODE $y' + ay = b\cos t + c\sin t$ to obtain $(aA + B)\cos t + (-A + aB)\sin t = b\cos t + c\sin t$. So $aA + B = b$ and $-A + aB = c$. Solve for A, B to get $A = (ab - c)/(a^2 + 1)$ and $B = (ac + b)/(a^2 + 1)$.

So a particular solution is

$$y_d = \frac{1}{a^2 + 1}[(ab - c)\cos t + (ac + b)\sin t]$$

This solution is periodic; it continues to oscillate as t increases.

Behavior of Solutions.

27. *Adjusting the Input*   Consider the IVP $y' + p(t)y = q(t)$, $y(0) = y_0$. Determine an input $q(t)$ such that $y(t) = y_0$, for all $t \geq 0$, is the solution of the IVP.

Answer 27: Since $y(t) = y_0$, for $t \geq 0$, $y'(t) = 0$ for $t \geq 0$. Plug into the ODE $y' + p(t)y = q(t)$ to get $0 + p(t)y_0 = q(t)$. Choose the input $q(t)$ to be $y_0 p(t)$.

28. *Exponent Law*   Show that $e^{s+t} = e^s e^t$, for all s and t. [*Hint:* show that $y_1(t) = e^s e^t$ and $y_2(t) = e^{s+t}$ are both solutions of the IVP $y'(t) = y(t)$, $y(0) = e^s$, and apply Theorem 2.2.1.]

29. *Unbounded Solutions*   Suppose that $q(t)$ is continuous for all t, and that $y' + y = q(t)$ has a particular solution $y_d(t)$ with the property that $y_d(t) \to +\infty$ as $t \to +\infty$. Explain why *all* solutions have this property of unboundedness.

Answer 29: The general solution of the linear ODE, $y' + y = q(t)$ is $y = Ce^{-t} + y_d(t)$, where y_d is any particular solution. In this problem we assume that there is some solution $y_d(t)$ that "blows up" as $t \to +\infty$, [i.e., $y_d(t) \to +\infty$ as $t \to +\infty$]. Then, we see from the general solution formula that all solutions have this property since $Ce^{-t} \to 0$ as $t \to +\infty$, and so this term has no effect on the asymptotic behavior as $t \to +\infty$.

30. *Long-Term Behavior*   The examples in this section seem to imply that the general solution $y(t)$ of the driven ODE $y' + p(t)y = q(t)$ would tend to a particular solution $y_d(t)$ as $t \to +\infty$. Show that this is *not* always the case by creating your own linear ODE $y' + p(t)y = q(t)$ where there is some solution of the undriven ODE ($q = 0$) that tends to $+\infty$ as $t \to +\infty$, and there is some solution $y_d(t)$ of the driven ODE such that $y_d(t) \to 0$ as $t \to +\infty$.

Modeling Problems.

31. *Salt Solution*   At $t = 0$ a solution containing 2 lb of salt/gal starts to flow into a tank containing 50 gal of pure water; the inflow rate of salt water is 3 gal/min. After 3 min the mixture starts to flow out at a rate of 3 gal/min.

    (a) How much salt is in the tank at $t = 2$ min? At $t = 25$ min? [*Hint*: solve two IVPs: one IVP for $t_0 = 0$ and another for $t_0 = 3$.]

    (b) How much salt is in the tank as $t \to +\infty$? Can you guess without any calculation?

    Answer 31:

    (a) Two different ODEs describe the system, up to, then after $t = 3$ min. If $S(t)$ denotes the amount of salt in the tank at time t, then for $t \leq 3$ min, $S' = (2\text{lb/gal})(3\text{gal/min}) = 6$ lb/min. So $S = 6t + C$. Since no salt is present initially, $S(0) = 0 + C$ and $C = 0$. $S = 6t$ is the amount of salt in the tank for $t < 3$ min. At 3 min, $S = 6 \cdot 3 = 18$ lb, which becomes the initial condition for the next time interval ($t \geq 3$).

    For $t \geq 3$ min, $S' = (2\text{lb/gal})(3\text{gal/min}) - (S/V)(3\text{gal/min}) = 6 - 3S/V$, where V is the volume of water in the tank. V is constant for $t \geq 3$ since inflow and outflow rates are equal. Using $t_0 = 3$ and an integrating factor $\exp(3t/V)$, $[S\exp(3t/V)]' = 6\exp(3t/V)$, so $S = 2V + C\exp(-3t/V)$, $t \geq 3$. C is any constant. Since $V = 50 + (3\text{gal/min})(3\text{min}) = 59$ gal and $S(3) = 18$ lbs, we must have $S(3) = 18 = 2(59) + C\exp(-3 \cdot 3/59) = 118 + C\exp(-9/59)$. So, $C = -100\exp(9/59)$ and

    $$S = 118 - 100\exp(-3(t-3)/59), \quad t \geq 3$$

    When $t = 2$ min, $S(2) = 6 \cdot 2 = 12$ lbs salt. When $t = 25$ min, $S = 118 - 100\exp(-3 \cdot 22/59) \approx 85.3$ lbs salt.

    (b) As $t \to +\infty$, the exponential term in the formula for $S(t)$ approaches zero and the salt content approaches 118 lbs, which is two pounds of salt per gallon. This is expected, because in the long term the initial condition is irrelevant, and the concentration of salt in the tank is that of the inflow stream, (i.e., 2 lb of salt per gallon). Since there are 59 gallons of brine in the tank from $t = 3$ minutes on, the total amount of salt should indeed approach 118 lbs.

32. *Pollution*   Use the lucky guess approach of Example 2.2.3 to find the solution formula (11) for the pollution model in Example 2.2.2. [*Hint*: to find a particular solution of $y' + ay = f(t) + g(t)$, find a particular solution of $y' + ay = f(t)$ and a particular solution of $y' + ay = g(t)$ and add these two solutions.]

Newton's Law of Cooling. According to *Newton's Law of Cooling* (or *Warming*), the rate of change of the temperature of a body is proportional to the difference between the body's temperature and the surrounding medium's temperature.

Newton's Law of Cooling.   Problems 33–36 are based on this law.

33. Write a model ODE for a body's temperature, given the medium's temperature $m(t)$. Is the proportionality constant positive or negative? [*Hint*: if the surrounding temperature is warmer than the body's temperature then the temperature of the body rises.]

    Answer 33:   A direct translation into mathematical symbols of the word description of Newton's Law of Cooling (or Warming) is given by $y'(t) = k[y(t) - m(t)]$, where $y(t)$ is the body's temperature, $m(t)$ is the temperature of the medium, and k is the proportionality constant. The constant k must be negative because if the body is warmer than its surroundings (i.e., if $y > m$), then the body should cool down and so $y' < 0$. A similar argument applies if the body is cooler than the surrounding medium.

34. *A Sick Horse*   A veterinarian wants to find the temperature of a sick horse. The readings on the thermometer follow Newton's Law. At the time of insertion the thermometer reads 82°F. After 3 min the reading is 90°F, and 3 min later 94°F. A sudden convulsion destroys the thermometer before a final reading can be obtained. What is the horse's temperature?

35. *Cooling an Egg*   A hard-boiled egg is removed from a pot of hot water and set on the table to cool. Initially, the egg's temperature is 180°F. After an hour its temperature is 140°F. If the room's temperature is 65°F, when will the egg's temperature be 120°F, 90°F, 65°F?

    Answer 35:   The modeling IVP for the cooling egg is

    $$y' = k(y - 65), \qquad y(0) = 180$$

    where k is a negative constant of proportionality, the room temperature is 65° F, and the initial temperature of the egg is 180° F. In addition $y(1) = 140° F$, where "1" means "one hour," so time is measured in hours. Solve the IVP in terms of k, then use the extra information to find the value of k, and finally determine the time when the egg has cooled down to 120° F, 90° F, 65° F.

    The linear ODE $y' - ky = -65y$, has integrating factor e^{-kt} and the general solution is $y = Ce^{kt} + 65$. Since $y(0) = 180$, $C = 115$ and so $y = 115e^{kt} + 65$ Since $y(1) = 140$, $140 = 115e^k + 65$ and so $75 = 115e^k$, giving $k = \ln(75/115) \approx -0.427$. This gives the solution formula for the temperature of the egg

    $$y = 115e^{-0.427t} + 65$$

    To find the time T when $y = 120$, solve $120 = 115e^{-0.427T} + 65$ for T. Then $-0.427T = \ln(55/115)$, so $T \approx 1.727$ hours, or about 1 hour and 44 minutes. Repeat for $y = 90, 65$ to show that the egg cools down to 90° F after about 3 hours and 34 minutes and asymptotically approaches 65° F as $t \to \infty$.

36. *Cold Body, Hot Medium*   A cold egg is placed in an insulated pot of hot water. As the egg warms up, the water cools down. Create a model ODE for the temperature of the egg and another for the temperature of the water. What happens to the two temperatures as $t \to +\infty$?

2.3   Existence and Uniqueness of Solutions

Comments:

The five Basic Questions posed in this section are at the heart of the theory of differential equations: Does an IVP have a solution? Is it unique? How can a solution be described? How sensitive is a solution to changes in parameters and data? How far can a solution be extended; what happens to solutions as time advances? The focus so far has been on obtaining solution formulas. With the formulas in hand, it is often a simple matter to answer the questions. Sometimes solution formulas are too complicated to be of much help in understanding solution behavior. In any case, the questions have to be considered even in the absence of solution formulas. In this section we introduce examples and counterexamples needed to clarify the meaning of the questions, and we show approaches to answering them. When you read this section, you might tie the questions with a specific physical phenomenon that has already been discussed, for example, the "pollution in a tank" problem of Section 2.2. Existence: pollution enters the tank and the pollution level in the tank does build up (so the model IVP for tank pollution levels does have a solution). Uniqueness: if there are two identical tanks, identical amounts of pollutant flowing into the tanks, identical initial amounts of pollutant in the tanks, identical outflow rates, then pollution levels in the tanks are identical (uniqueness of the IVP's solution). Description of solution: words, formulas, pictures, graphs, computer output. Continuity: change the pollutant levels a little in the inflow stream and the levels in the tank change a little (the solution of the IVP varies continuously with the data). Extension and Long-Term Behavior: after a long time one expects the concentration of pollutant in the tank to be pretty much the same as the concentration of pollutant in the inflow stream (and the formula for the solution of the IVP will mirror that expectation). So the Basic Questions are just the mathematicians' way of encoding in a few words and symbols the everyday experience of people who work with dynamical systems. In addition, the answers to the five Basic Questions are crucial to the use of numerical solvers. Without the Existence, Uniqueness, Extension, and Continuity properties, we would never be sure whether the approximation our numerical solver generates has anything to do with the IVP being addressed, nor would we have any idea about its "life-span."

The question of uniqueness is also a central issue, especially these days when the meaning of chaotic dynamics is a popular topic for scientific debate. If an IVP has an unique solution, then any dynamical system accurately modeled by the IVP is deterministic (not chaotic) in the sense that the present state of the system completely determines its future (and its past) states, at least over the time span where the model is valid. Nonuniqueness introduces an element of uncertainty; which path does a solution follow, and how can the system decide which path to follow? This is a good reason for reading some of the nonuniqueness examples: you might well contrast determinism and uncertainty in this connection. Even when the conditions of uniqueness are satisfied, some systems of ODEs or their numerical approximations appear to have chaotic solutions, but we postpone that discussion to the WEB SPOTLIGHT ON CHAOS IN NUMERICS (numerical approximation) and Chapter 9 (chaos in ODEs).

Questions of continuity/sensitivity are left to the SPOTLIGHT ON CONTINUITY IN THE DATA and the WEB SPOTLIGHT ON SENSITIVITY OF SOLUTIONS TO THE DATA. The use of numerical computer approximations is taken up in the SPOTLIGHT ON APPROXIMATE NUMERICAL SOLUTIONS and the SPOTLIGHT ON COMPUTER IMPLEMENTATION.

PROBLEMS

The Existence and Uniqueness Theorem Applies. Verify that there exists exactly one solution to each of the following IVPs by checking the hypotheses of the Existence and Uniqueness Theorem. [Don't bother finding solution formulas for the IVPs in Problems 1–6.]

1. $y' = e^t y - y^3$, $y(0) = 0$

2. $y' = ty$, $y(0) = 1$

3. $y' = ty^2 - 1/(3y + t)$, $y(0) = 1$

4. $y' = -t^2/(1 - y^2)$, $y(-1) = 1/2$

5. $y' = 1 - y|y|$, $y(0) = 0$

6. $y' = y/\sin t$, $y(\pi/2) = 1$

Answer 1: The IVP is $y' = e^t y - y^3$, $y(0) = 0$. The function $f = e^t y - y^3$ is continuous for all t and y since e^t, y, and y^3 are continuous. Moreover, $\partial f/\partial y = e^t - 3y^2$ is also continuous for all t and y. So, the hypotheses of the Existence and Uniqueness Theorem are satisfied in every rectangle, and the given initial value problem has exactly one solution.

Answer 3: The IVP is $y' = ty^2 - 1/(3y + t)$, $y(0) = 1$. The function $f = ty^2 - 1/(3y + t)$ and its partial derivative $\partial f/\partial y = 2ty + 3/(3y + t)^2$ are continuous in t and y throughout any region not intersected by the straight line $3y + t = 0$. Since the initial point $(0, 1)$ does not lie on that line, there is a rectangle containing the initial point in which the hypotheses of the Existence and Uniqueness Theorem hold (for example, any rectangle in the region $3y + t > 0$). The initial value problem has exactly one solution in any of these rectangles that contains $(0, 1)$.

Answer 5: The function $f = 1 - y|y|$, $y(0) = 0$, is continuous on any rectangle containing the initial point $(0, 0)$. To verify that this function is differentiable at $(0, 0)$, take the limit of the function's slope as it approaches 0 from either side:

$$\lim_{h \to 0^+} \frac{f(0 + h) - f(0)}{h} = \lim_{h \to 0^+} \frac{1 - h^2 - 1}{h} = 0$$

$$\lim_{h \to 0^{-1}} \frac{f(0 - h) - f(0)}{h} = \lim_{h \to 0^-} \frac{1 + h^2 - 1}{h} = 0$$

Because the limit is identical from either side, the function is differentiable at $(0, 0)$. The partial derivatives $\partial f/\partial y = -2y$ for $y \geq 0$ and $\partial f/\partial y = 2y$ for $y \leq 0$, are continuous on any rectangle containing the point $y_0 = 0$, $t_0 = 0$. So, by the Existence and Uniqueness Theorem, the IVP has exactly one solution.

7. Show that the Existence and Uniqueness Theorem implies the IVP $y' = 2|t|y$, $y(0) = 1$, has a unique solution. Then show that the solution lives on the entire t-axis.

   Answer 7: The ODE is

$$y' = f(t, y) = 2|t|y = \begin{cases} 2ty, & t \geq 0 \\ -2ty, & t < 0 \end{cases}$$

and $\partial f/\partial y$ is given by

$$\partial f/\partial y = \begin{cases} 2t, & t \geq 0 \\ -2t, & t < 0 \end{cases}$$

So f and $\partial f/\partial y$ are continuous on any rectangle containing $t_0 = 0$ and $y_0 = 1$, and hence the IVP $y' = 2|t|y$, $y(0) = 1$ has a unique solution.

   To show that the solution is defined for all t, solve $y' = 2ty$, $y(0) = 1$ for $t \geq 0$ and $y' = -2ty$, $y(0) = 1$ for $t \leq 0$. Both ODEs are linear and can be solved by the use of integrating factors. For the first one, $y' - 2ty = 0$, $p(t) = -2t$ and the integrating factor is e^{-t^2}. So $(ye^{-t^2})' = 0$, and the solution is $y = Ce^{t^2}$ with $C = 1$. This solution is defined for all $t \geq 0$. Similarly, $y = e^{-t^2}$ solves the second ODE for $t \leq 0$. So the combined formulas give a continuously differentiable solution for all t.

Investigating Existence and Uniqueness When Theorem 2.3.1 Does Not Apply.

8. Why does Theorem 2.3.1 not apply to the IVP, $2tyy' = t^2 + y^2$, $y(0) = 1$?

9. Show that $y_1 = t^2$ and $y_2 = t^2 \operatorname{Step}(t, 0)$ are both solutions of the IVP, $ty' = 2y$, $y(0) = 0$, $-\infty < t < \infty$. Why doesn't this contradict Theorem 2.3.1?

   Answer 9: The function $y_1 = t^2$ is a solution of $ty' = 2y$, $y(0) = 0$, as we see by a direct substitution. The function $y_2(t) = t^2 \operatorname{Step}(t, 0)$ is also a solution of the differential equation for $t < 0$ and for $t > 0$, and satisfies the initial condition $y(0) = 0$. Note that $y_2(t) = 0$, $t \leq 0$. To verify that $y_2(t)$ is differentiable at $t = 0$, show that $y_2'(0)$ exists. Specifically, show that $\lim_{t \to 0^-} y_2'(t) = \lim_{t \to 0^+} y_2'(t)$: $\lim_{t \to 0^-} (y_2(t) - y_2(0))/t = \lim_{t \to 0^-} (0 - 0)/t = 0$, while $\lim_{t \to 0^+} (y_2(t) - y_2(0))/t = \lim_{t \to 0^+} (t^2 - 0)/t = \lim_{t \to 0^+} t = 0$. Thus, $y_2'(0)$ exists, and so $y_2(t)$ is a second solution of the initial value problem. This does not contradict the uniqueness part of Theorem 2.3.1 since the rate function $f = 2y/t$ is not continuous at the initial point (0,0), and so Theorem 2.3.1 doesn't apply.

10. Find all the solutions of the IVP, $ty' - y = t^{n+1}$, $y(0) = 0$, where n is any positive integer.

11. Show that the IVP $y' = |y|$, $y(0) = 0$, has a unique solution even though the hypotheses of Theorem 2.3.1 are not satisfied. [*Hint*: first find the unique solution for $y \leq 0$ and the unique solution for $y \geq 0$, then splice them together.]

   Answer 11:   For $y \geq 0$, any solution of $y' = |y|$, $y(0) = 0$ satisfies $y' = y$, $y(0) = 0$. Since every solution of $y' = y$ has the form $y = Ce^t$ for some nonnegative constant C, the initial condition implies that $C = 0$ and $y = 0$ for $t \geq 0$. Similarly, for $y \leq 0$, $y' = -y$, $y(0) = 0$, and, hence, $y = 0$. Thus, $y = 0$, all t, is the unique solution. Note, however, that $\partial|y|/\partial y = -1$ if $y < 0$, $+1$ if $y > 0$, and so $\partial f/\partial y$ is discontinuous on any region containing $y = 0$. In this case, a unique solution exists even though Theorem 2.3.1 cannot be used.

12. *Discontinuity in the Normalized ODE*   The linear ODE $ty' + y = 2t$ has the normal linear form $y' + y/t = 2$. Note that the coefficient $p(t) = 1/t$ is discontinuous at $t = 0$.

   (a) Find all the solutions. Graph several solutions over $|t| \leq 5$.

   (b) Show that the IVP $ty' + y = 2t$, $y(0) = 0$, has exactly one solution, but that if $y(0) = y_0 \neq 0$, there is no solution at all. Why doesn't this contradict Theorem 2.3.1?

Solvability of an IVP.   The IVPs below do not satisfy the conditions of Theorem 2.3.1. If the IVP has solutions, find them all. If there are none, why not?

13. $2yy' = -1$, $y(1) = 0$ [*Hint*: $(y^2)' = 2yy'$.]

14. $y' = 4\sqrt{y}$, $y(t_0) = 0$ [*Hint*: see Example 2.3.2.]

15. $y' = (1 - y^2)^{1/2}$, $y(0) = 1$ [*Hint*: the function arcsin y is an antiderivative of $(1 - y^2)^{-1/2}$.]

Answer 13: The IVP is $2yy' = -1$, $y(1) = 0$. Antidifferentiate each side of $2yy' = (y^2)' = -1$ to get $y^2 = -t + C$; $C = 1$ since $y = 0$ when $t = 1$. So there are, apparently, *two* solutions, $y_1 = \sqrt{1-t}$ and $y_2 = -\sqrt{1-t}$ valid for $t \leq 1$. The problem is that the derivative of each solution at $t = 1$ is infinite. So, the derivative of neither "solution" exists at the initial point, and there is no solution at all with a finite slope at $(0, 1)$.

Answer 15: The IVP is $y' = (1 - y^2)^{1/2}$, $y(0) = 1$. The rate function is $f(y) = (1 - y^2)^{1/2}$. Since the partial derivative f_y of f with respect to y is $-y/(1 - y^2)^{1/2}$, which is undefined at $y = 1$, Theorem 2.3.1 does not apply. The constant function $y_1(t) = 1$, all t, is a solution, but the problem may have additional solutions. In fact, separate variables to get $(1 - y^2)^{-1/2} y' = (\arcsin y)' = 1$. Integrating, $\arcsin y = t + C$, so $y = \sin(t + C)$. Using the initial condition $y(0) = 1$ get $C = \pi/2$, and so $y = \sin(t + \pi/2) = \cos t$. Since y' is nonnegative, restrict t to the interval $-\pi \leq t \leq 0$ [or to any interval of the form $(2n - 1)\pi \leq t \leq 2n\pi$] on which $y = \cos t$ has a nonnegative slope. Notice that the constant function $y(t) = -1$ is also a solution to the ODE (but not to the IVP). Construct another solution of the initial value problem by setting $y = -1$ for $t < -\pi$, then $y = \cos t$ for $-\pi \leq t \leq 0$, and finally $y = +1$ for $t \geq 0$. Such a function is continuous for all t, differentiable for all t (since $y' = -\sin t$ for $-\pi \leq t \leq 0$ and $y'(t) = 0$ for $t \leq -\pi$ and for $t \geq 0$), and satisfies the rate equation and initial condition. In fact, infinitely many solutions can be defined by linking the horizontal line $y(t) = -1$ for $t < (2n - 1)\pi$, to the horizontal line $y(t) = 1$ for $t > 2n\pi$, by means of the graph of $y(t) = \cos t$, $(2n - 1)\pi \leq t \leq 2n\pi$, n a nonpositive integer.

Jump Discontinuity in the ODE. Find the formula for the generalized solution of each IVP on the indicated interval. Use a grapher to plot the solution. Explain any odd features of the graph in terms of a feature of the IVP. [*Hint*: follow the approach used in Example 2.3.4.]

16. $y' = y \, \text{Step}(t, 0)$, $y(0) = 1$, all t

17. $y' + 2y = 1 - \text{Step}(t, 1)$, $y(0) = 0$, $0 \leq t \leq 2$

Answer 17: The IVP is $y' + 2y = \text{Step}(1 - t)$, $y(0) = 0$, where

$$1 - \text{Step}(t, 1) = \begin{cases} 1, & 0 \leq t \leq 1 \\ 0, & t \geq 1 \end{cases}$$

Use the integrating factor approach to find the general solution to the differential equation $y' + 2y = q(t)$, with $q(t) = 1 - \text{Step}(t, 1)$: $y(t) = Ce^{-2t} + e^{-2t} \int_0^t q(s)e^{2s} \, ds$. Apply the initial conditions $y(0) = 0$ to get $C = 0$. So,

$$y(t) = e^{-2t} \int_0^t q(s)e^{2s} \, ds = \begin{cases} e^{-2t} \int_0^t e^{2s} \, ds, & t \leq 1 \\ e^{-2t} \int_0^1 e^{2s} \, ds, & t \geq 1, \end{cases} \quad \text{since } q(s) = 0 \text{ for } s \geq 1$$

$$= \begin{cases} (1 - e^{-2t})/2, & t \leq 1 \\ e^{-2t}(e^2 - 1)/2, & t \geq 1 \end{cases}$$

The solution graph rises as t increases toward 1, but falls for $t \geq 1$. Since $q(t)$ has a discontinuity at $t = 1$, the solution graph has a sharp corner at $t = 1$, and this is visible in the graph. See the figure.

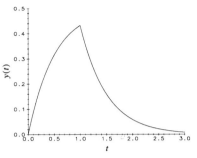

Problem 17.

18. $y' + p(t)y = 0$, $y(0) = 2$,   where $p(t) = 2 - \text{Step}(t, 1)$, $0 \le t \le 2$

19. $y' + y = \text{Step}(t, 1) - \text{Step}(t, 2)$, $y(0) = 1/2$, $0 \le t \le 4$ [*Hint*: obtain formulas for $y(t)$ on the intervals $0 \le t \le 1$, $1 \le t \le 2$, $2 \le t \le 4$, in that order.]

Answer 19: Define the IVP as

$$y' + y = \text{Step}(t, 1) - \text{Step}(t, 2) = \begin{cases} 0, & 0 \le t < 1 \\ 1, & 1 \le t < 2 , \\ 0, & 2 \le t \le 4 \end{cases} \quad y(0) = 1/2$$

To solve this IVP, use the integrating factor e^t to obtain

$$y(t) = e^{-t}/2 + e^{-t} \int_0^t e^s [\text{Step}(s, 1) - \text{Step}(s, 2)] \, ds$$

$$y(t) = \begin{cases} e^{-t}/2, & 0 \le t \le 1 \\ e^{-t}\left[1/2 + \int_1^t e^s \, ds\right] = e^{-t}/2 + 1 - e^{1-t} & 1 \le t \le 2 \\ e^{-t}/2 + e^{-t}\int_1^2 e^s ds = e^{-t}(1/2 + e^2 - e), & 2 \le t \le 4 \end{cases}$$

See the figure. Note the corners at $t = 1$ and $t = 2$ when the input turns on and then off.

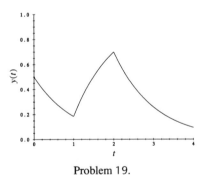

Problem 19.

Trouble with the Rate Function.   Problems 20 and 21 have IVPs with difficulties of their own. [*Hint*: see Example 2.3.2.]

20. Consider the IVP $y' = 3y^{2/3}$, $y(0) = 0$. Find formulas for infinitely many solutions on $-\infty < t < \infty$. Why doesn't the Existence and Uniqueness Theorem apply?

21. Show that the IVP, $y' = 3y^{2/3} \text{Step}(t, 0)$, $y(0) = 0$ has exactly one solution on the interval $t \le 0$, but infinitely many on the interval $-\infty < t < \infty$.

Answer 21: For $t \leq 0$, the IVP is $y' = 0$, $y(0) = 0$, and the unique solution is $y = 0$, $t \leq 0$. However, for $t \geq 0$, the IVP is $y' = 3y^{2/3}$, $y(0) = 0$. Separate variables to get $y^{-2/3}y'/3 = 1$. Integrate to obtain $y^{1/3} = t + C$, $y = (t + C)^3$, C any constant, $t \geq 0$. Then for any $s > 0$, let

$$y_s(t) = \text{Step}(t, s)(t - s)^3 = \begin{cases} 0, & t \leq s \\ (t - s)^3, & t \geq s \end{cases}$$

Since $s > 0$, $y(0) = 0$, so $y_s(t)$, $-\infty < t < \infty$, solves the IVP for every $s > 0$. So there are infinitely many solutions for $-\infty < t < \infty$.

Modeling Problems. Find a solution formula for IVP (13) with the given concentration $c(t)$. Use a grapher to visualize the solution curve. [*Hint*: see Example 2.3.6.]

22. $c(t) = 2(1 - \text{Step}(t, 10))$ 23. $c(t) = 0.2[1 - \text{Step}(t, 20) + \text{Step}(t, 40)]$

Answer 23: The IVP is

$$y' + .1y(t) = 10c(t) = 2\big[1 - \text{Step}(t, 20) + \text{Step}(t, 40)\big], \quad y(0) = y_0$$

Therefore

$$y(t) = y_0 e^{-.1t} + e^{-.1t} \int_0^t 2e^{.1s}\big[1 - \text{Step}(s, 20) + \text{Step}(s, 40)\big]\, ds$$

$$= y_0 e^{-.1t} + \begin{cases} 20\left(1 - e^{-0.1t}\right), & 0 \leq t \leq 20 \\ 20\left(1 - e^{-2}\right), & 20 \leq t \leq 40 \\ 20\left(1 - e^{-2}\right) + 20\left(1 - e^{4-.1t}\right), & t \geq 40 \end{cases}$$

As in Example 2.3.6 let $y_0 = 15$, and get the solution curve in the figure.

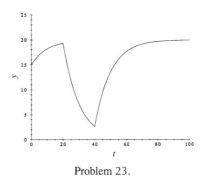

Problem 23.

2.4 Visualizing Solution Curves: Slope Fields

Comments:

The graphs your solver produces may look different from the graphs in the text. There are many reasons for this: sometimes the reasons are platform dependent, sometimes algorithm dependent, and sometimes maybe you (or we) just hit the wrong keys on the keyboard. The way your machine does arithmetic and function evaluations is an example of the former. Memory may also be a problem (the data sets for some of our graphs are huge). If you have an adaptive solver, then your default settings may cause you to miss some important features in the graph. In any case, there are some things you can do to improve the appearance of your graphs. For Runge-Kutta solvers (see the SPOTLIGHT ON APPROXIMATE NUMERICAL SOLUTIONS) try changing the step size and shortening the solve-time interval. For adaptive solvers try changing the default settings for the absolute and relative error bounds, the number of plotted points, the maximum number of function evaluations, the maximum and minimum interval step size, etc. Please keep in mind that the graphs that appear in this text are not the result of a first pass. Often, we had to adjust the time span, the intervals for the state variables, the initial data, or the solver parameters before we could get a picture that was both informative and good-looking. Another reason why your graphs may not look as good as those in the text: the printing technology that produced the graphs in the text is equivalent to a 1200 dpi laser printer. But do the best you can with the equipment you have—a good picture may be worth more than many words and formulas. If you are asked to "sketch a curve," we mean "draw without a computer." Otherwise we use the word "plot."

At the end of the solutions for Section 2.4 you will find some tips on using numerical solvers.

PROBLEMS

Nullclines, Slope Fields, Solution Curves. Here are two slope fields and the graphs (dashed) of nullclines. In each case, find the appropriate ODE from the list of six ODEs. Make an enlarged photocopy and sketch four distinct solution curves. Each solution curve should reach from edge to edge and cut the nullcline at least once. [*Hint*: to shorten the list of possible ODEs, consider the nature of the nullcline of each ODE. Then take a look at slope field elements above and below the nullclines.]

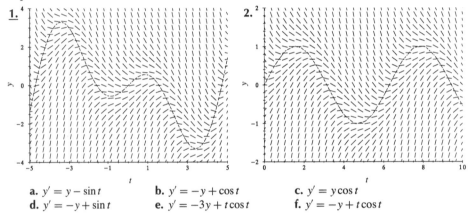

a. $y' = y - \sin t$ b. $y' = -y + \cos t$ c. $y' = y \cos t$
d. $y' = -y + \sin t$ e. $y' = -3y + t \cos t$ f. $y' = -y + t \cos t$

Answer 1:   **f.** The equation of the nullcline is either $3y = t \cos t$ (**e.**) or $y = t \cos t$ (**f.**), but the coordinates of the max and the min points of the nullcline (as read from the graph) suggest the latter.

Match the Slope Field with an ODE. Here are four slope fields and nine ODEs. For each slope field select the ODE that determines the field. Give reasons for your choices. [*Hint*: first look for any equilibrium solution lines (the rate function must be zero on these lines). Then locate the other nullclines and look at the sign of the rate function on horizontal and vertical lines in the ty-plane.]

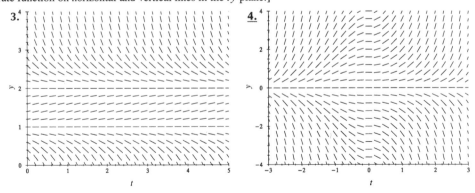

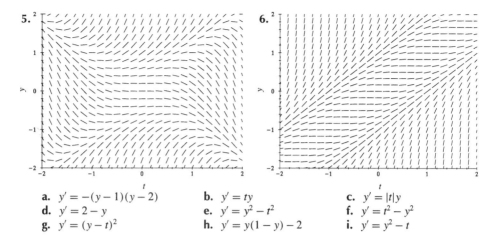

a. $y' = -(y-1)(y-2)$
b. $y' = ty$
c. $y' = |t|y$
d. $y' = 2 - y$
e. $y' = y^2 - t^2$
f. $y' = t^2 - y^2$
g. $y' = (y-t)^2$
h. $y' = y(1-y) - 2$
i. $y' = y^2 - t$

Answer 3: **a.** Notice the equilibrium lines at $y = 1$ and $y = 2$. The only ODE that has these equilibrium lines is $y' = -(y-1)(y-2)$.

Answer 5: **e.** There are nullclines at $y = t$ and $y = -t$. The two ODE's $y' = y^2 - t^2$ (**e.**) and $y' = t^2 - y^2$ (**f.**) have this property. Looking along the y-axis (where $t = 0$) notice that the slopes are positive. So $y' = y^2 - t^2$ is the corresponding ODE.

☞ This icon indicates that this problem is designed for the ODE Architect Tool.

Slope Fields. Enter the ODE into the **Editing Pane** of ODE Architect's solver tool and fix the scales to the indicated values as follows: depress the top button of the **Graph Mode Controls** to the right of a graphics screen and select **Scales**. In the **Graph Scales** panel click off **Auto Scale** and insert values for the **X-scale**. Repeat for the **Y-scale**. Create a slope field as follows: depress the top button of the **Graph Mode Controls** and select **Direction Fields**. Modify your slope field as follows: depress the top button of the **Graph Mode Controls** button and select **Edit**. On the **Edit 2D Graph** panel hit **Dir Field** and make your changes on the panel that comes up. Print your final slope field.

7. $y' = (1-t)y - t;$ $-2 \le t \le 4,$ $|y| \le 3$ **8.** $y' = y^2 - 4ty + t^2;$ $|t| \le 3,$ $|y| \le 5$

9. $y' = t^2/(y^2 - 1);$ $|t| \le 2,$ $|y| \le 2$ [*Hint*: the field segments are vertical at $y = \pm 1$.]

Answer 7: See the figure.

Answer 9: See the figure. Only the solution curves where $|y| < 1$ are plotted; you may want to plot curves outside that bound. No solution curve can touch $y = \pm 1$, where the slope is infinite.

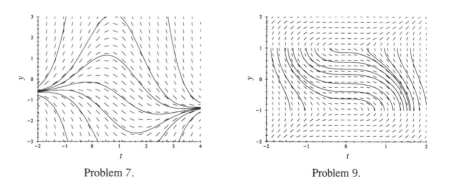

Problem 7.   Problem 9.

Sketching Solution Curves. Using nullclines, slope fields, and the sign of the rate function, sketch some solution curves for the ODEs. Verify your sketch by using a numerical solver.

10. $y' = (y+3)(y-2)$. Why can't the solution curve with initial point $t_0 = 0$, $y_0 = 0$, fall below the line $y = -3$ as t increases?

11. $y' = 2t - y$. Show that $y = 2t - 2$ is a solution. As t increases, what happens to all the solution curves that don't touch the straight line $y = 2t - 2$?

12. $y' = ty - 1$. As $t \to +\infty$, what happens to the solution curve passing through the point $t = 1$, $y = 1$? Through the point $t = -1$, $y = -1$? Through the point $t = 0$, $y = 0$?

13. $y' = (1-t)y$. What is the behavior of solutions as t increases? What happens as $t \to +\infty$?

14. $y' + (\sin t)y = t \cos t$. What is the long-term behavior of every solution that passes through the y-axis, both as $t \to +\infty$ and as $t \to -\infty$?

15. $y' = y - t^2$. For what values of the constants A, B, and C is the parabola defined by $y = At^2 + Bt + C$ a solution curve. What happens as $t \to -\infty$ to every solution curve that starts out at $t = 0$ above this parabola? As $t \to +\infty$?

Answer 11: The nullcline of the ODE $y' = 2t - y$ is $y = 2t$. If you plug in $y = 2t - 2$, then $y' = 2 = 2t - 2t + 2$, so $y = 2t - 2$ is a solution. Any solution curve that doesn't touch the line $y = 2t - 2$ will be attracted toward that line as $t \to +\infty$ since the general solution of the linear ODE $y' + y = 2t$ is $y = Ce^{-t} + 2t - 2$, which tends to $2t - 2$ as $t \to +\infty$. See the figure.

Answer 13: The ODE is $y' = (1-t)y$. The nullclines are $y = 0$, $t = 1$. Notice that $y = 0$ is also a solution. As $t \to +\infty$, solution curves approach $y = 0$. See the figure; note the symmetry about the lines $y = 0$ and $t = 1$. Each nonconstant solution reaches its extreme value as it crosses the y-axis and then falls (if $y > 0$) or rises (if $y < 0$) and is asymptotic to the t-axis as $t \to \pm\infty$.

Answer 15: The ODE is $y' = y - t^2$. The nullcline is $y = t^2$. If some solution curve has the equation $y = At^2 + Bt + C$, then $y' = 2At + B$, and the ODE becomes $2At + B = At^2 + Bt + C - t^2$. From this it can be found that $A = 1$, $B = 2$, and $C = 2$, making the parabolic solution curve formula $y = t^2 + 2t + 2$. See the figure, where the nullcline $y = t^2$ is the dashed parabola. The general solution of the linear ODE $y' - y = -t^2$ can be found by the integrating factor technique: $y = t^2 + 2t + 2 + Ke^t$, K an arbitrary constant. If $K \geq 0$, then $y \to +\infty$ as $t \to +\infty$ and to the parabolic solution curve as $t \to -\infty$. If $K = 0$, then $y \to +\infty$ as $t \to \pm\infty$. If $K < 0$, then $y \to -\infty$ as $t \to +\infty$ and to the parabolic solution curve as $t \to -\infty$.

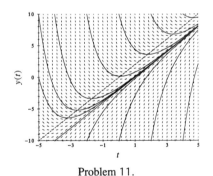

Problem 11.

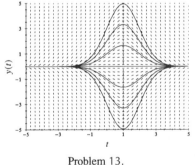

Problem 13.

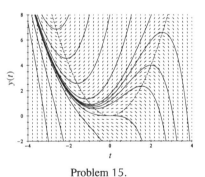

Problem 15.

Nullclines, Slope Fields, and Solution Curves.

16. *Rise and Fall of Solution Curves*   Imagine the rectangle R, $|t| \leq 6$, $|y| \leq 8$, to be filled with solution curves of the ODE $y' = -y \cos t$. Sketch the nullclines. Where are the solution curves rising? Falling? Make a rough sketch in R of the solution curves. Are the solution curves periodic; if so, what is the period? [*Hint*: find a formula for the general solution.]

17. *Slope Field*   For the nonlinear ODE $y' = 3y \sin y + t$ of Figure 2.4.3, there is no known explicit solution formula, so we must turn to a slope field or to a numerical ODE solver to see how solution curves behave.

☞ See the WEB
SPOTLIGHT ON
SLOPE FIELDS II

(a) Plot a slope field over the rectangle, $-6 \le t \le 3$, $-2 \le y \le 4$. Sketch some solution curves suggested by the slope field. Redraw your slope field with a finer grid, and see if your curves need any adjustment. Use a numerical solver to check the accuracy of your sketches. Record your observations about the use of slope fields to sketch solution curves.

(b) Take initial points $t_0 = -6$, $1.8 \le y_0 \le 2$, solve forward 10 units of time, and try to hit the target point $(0, 2)$. Is there a difficulty? Is there another way to "win"?

18. $y' = -2y(y - 3)$. Suppose that $y_0 = y(0)$ is chosen as follows: $y_0 > 3$, $0 < y_0 < 3$, $y_0 < 0$. What happens to every solution curve with $y(0) = y_0$ as t increases? As t decreases? [*Hint*: $y = 0$ and $y = 3$ are nullclines.]

19. $y' = y^2 - t$. What happens to every solution curve as t increases?

20. $y' = -(5t - 4y)/(t + y)$. Describe the behavior of the solution curves above the line $y = -t$, as t increases. What happens to solution curves below the line as t increases?

21. $y' = 1 - e^{ty}$. Explain why no solution curve with $y(0) > 0$ rises as t increases beyond $t = 0$. Why do these solution curves never fall below the t-axis?

Answer 17:

(a) The figure shows the slope field and the solution curves of $y' = 3y \sin y + t$ corresponding to the initial conditions $y(-6) = 0, 1, 1.5, 1.9, 1.95, 1.955, 1.956, 1.9565, 1.957, 1.96, 1.97, 2, 3, 4$. Wherever solutions curves spread apart, the solutions may be very sensitive to the initial conditions. When sketching solution curves, a small error in one region can therefore lead to large errors further along. Note the big blank region in the figure. It seems that the only way to fill that region with solution curves is to put the initial point inside the region and solve forward and backward in time. You can be pretty certain that for points in this region as $t \to -6^+$, $y(t) \to 1.95\ldots$, and as t increases, $y(t)$ gets "trapped" on the slanted curve at the upper right. Remember that solution curves cannot cross each other, so the solution curve is trapped and must behave as indicated. You might zoom on the upper right and pull apart the apparently merging curves.

(b) The "expansion" of solution curves of $y' = 3y \sin y + t$ starting at $t_0 = -6$, $1.8 \le y_0 \le 2$ makes it very hard to hit the point $(0, 2)$. Note in the figure how these curves later seem to squeeze together and exit at the upper left of the rectangle. The figure shows several of these solution curves which fail to hit $(0, 2)$. In the last figure we cheated in order to hit the point $(0, 2)$ from $(-6, y_0)$, where $1.8 \le y_0 \le 6$. We started at the target $(0, 2)$ and solved backward. The computer then gave us the number $y_0 = 1.956198501618254$, but this value can't be exactly right since data compression and expansion always cause uncertainties (not to mention the fact that numerical solvers only approximate true solution values).

Answer 19: The nullcline is the parabola $t = y^2$ which lies in the first and fourth quadrants of the ty-plane, passes through the origin and opens to the right. See the dashed curve in the figure. Solution curves touching the parabola remain inside it as t increases and appear to approach the lower branch. All curves tend to $-\infty$ as t decreases. Curves outside the parabola tend to $+\infty$ as t increases.

Answer 21: If $y_0 > 0$ and $t > t_0 = 0$, then $ty_0 > 0$ and $e^{ty_0} > 1$, so $y' = 1 - e^{ty_0}$ is negative. In fact, $1 - e^{ty}$ is negative for all $t > 0$, $y > 0$. So all the corresponding solution curves are falling in the first quadrant. On the other hand, $y = 0$, all t is a solution since $y' = 1 - e^{t \cdot 0} = 1 - 1 = 0$. So solution curves cannot touch the t axis because that would violate uniqueness.

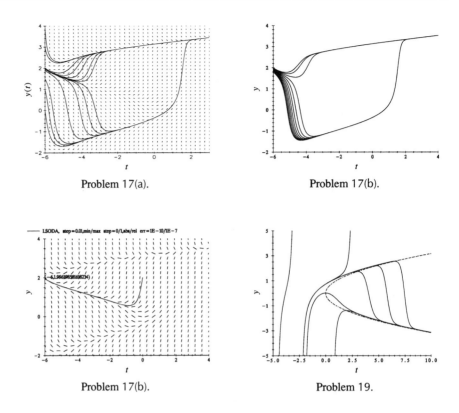

Problem 17(a).   Problem 17(b).

Problem 17(b).   Problem 19.

Qualitative Analysis.

22. Show that the solution of the IVP $y' = 1 - t \sin y$, $y(1) = 1$, remains in the half-strip $t \geq 1$, $0 < y < \pi/2$, for all $t \geq 1$. Plot a direction field for $1 \leq t \leq 6$, $-0.5 \leq y \leq 2$, and the solution curve of the IVP.

23. Show that the solution of the IVP $y' = -t^2 \sin y$, $y(0) = \pi/4$, stays in the half-strip $t \geq 0$, $0 < y < \pi/2$, for all $t \geq 0$. Plot a direction field and the solution curve for $0 \leq t \leq 3$, $-1 \leq y \leq 2$.

Answer 23: The solution curve stays in the half strip $t \geq 0$, $0 < y < \pi/2$ because $y = 0$ is a solution (so the solution curve can't touch the line $y = 0$) and $y' = -t^2 \sin(\pi/2) = -t^2$ is negative for $t > 0$ (so the solution curve can't rise to touch the line $y = \pi/2$ as t increases). See the figure.

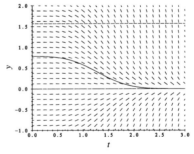

Problem 23.

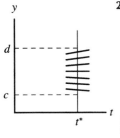

24. *Diverging and Converging Flows*   Think of the solution curves of the ODE $y' = f(t, y)$ as paths traced out by particles moving along *flow lines* of a fluid whose velocity at (t, y) is $f(t, y)$. Answer the questions below:

(a) If $\partial f/\partial y > 0$ at some $t = t^*$ and all y, $c < y < d$, explain why the flow slopes across the line $t = t^*$ diverge as indicated in the margin diagram. If $\partial f/\partial y > 0$ on some region in the ty-plane, what can be said about the solution curves in that region as t increases?

(b) Formulate an analogous property of flows in regions where $\partial f/\partial y < 0$.

 (c) Illustrate the properties in **(a)** and **(b)** with the ODE $y' = 0.1(y - 3)(y^2 - 1)$. Since the ODE is autonomous, any t^* will do, so take $t^* = 0$. Use a numerical solver to plot some of the diverging or converging solution curves for $0 \le t \le 5$, $-2 \le y \le 4$.

25. *Using Slope Fields to Bound Solutions*   Let $f(t, y)$ and $\partial f/\partial y$ be continuous on the whole ty-plane. Let $g(t)$ be a continuously differentiable function defined for all t.

(a) Suppose $g'(t) > f(t, g(t))$ for $t \ge T$, for some T. Also suppose that $y(t)$ is a solution of the ODE $y' = f(t, y)$ and that $y(T) < g(T)$. Show that the solution curve $y = y(t)$ always remains below the curve $y = g(t)$ for all $t \ge T$.

(b) Suppose $g'(t) < f(t, g(t))$ for $t \ge T$, for some T. Formulate and prove a property of solution curves of the ODE $y' = f(t, y)$ analogous to the property in **(a)**.

(c) Find constants k_1 and k_2 such that the solution curves of the ODE $y' = y - y^2 - 0.2\sin t$ with initial data $y(0)$ in the interval $k_1 < y(0) < k_2$ remain in that strip for all time. Find the smallest interval with that property. Is your result consistent with Figure 2.4.5? [*Hint*: take $g(t) = k_2$ in **(a)**, and $g(t) = k_1$ in **(b)**.]

Answer 25:   Since this is a group problem the solution is not given, but take a look at Problems 22–24.

Tips on using numerical solvers

Like any other tool, you will need a "break-in" period to use your numerical solver effectively. The documentation that comes with the solver is a good place to start, but there is no substitute for hands-on experience. Some solvers are command-line driven with a precise (and often complex) syntax for entering commands, but many solvers have a graphical interface which allows the user to select items from a menu and makes it easy to enter data and display graphs. Our advice is to get started right away in using your solver, even before you completely understand the basic concepts behind the differential equations you are studying. The reason is that the graphical displays created by the solver will give you a better feeling for solutions of ODEs and how they behave when the underlying model changes.

From time to time we will give practical tips that will help you avoid some of the frustrations in using your solver. Here is some advice you can use right now:

- Numerical solvers are designed to solve initial value problems where the ODE is in normal form. Before using your solver, write your ODE in normal form and identify the rate function. Next, examine the initial condition and identify the initial time t_0 and the initial value y_0. Finally, choose a solve-time interval for either forward or backward solving. Your solver may ask you for these items. Some solvers allow users the option to enter initial points graphically by clicking directly on the screen. Most solvers allow users to insert settings for the solver itself, but these are often set by default and not adjusted until something goes wrong.

- If your solver does not give you the option of setting your own scale on the axes at the outset, then not to worry: go ahead and let your solver set the scales automatically and then look at the result. It may be just what you wanted. If not, then at this stage your solver will let you select your own scales on the axes.

- Your solver may continue to crank away and never seem to come to an end. In that case you may want to abort and take a shorter solve-time interval. This difficulty usually happens when rate functions are very large, and so examining the rate function at the outset will give you a clue that you may run into a problem with the choice of a solve-time interval.

- Many solvers allow the user to choose the number of computed points (equally spaced in time) to display on the screen; these points are then connected with straight line segments. If the graph produced by your solver looks partly like a broken-line graph, then you should go back and increase the number of displayed points and solve again.

- Some solvers have trouble dealing with IVPs like

$$y' = ay - cy^2 - H(t), \qquad y(t_0) = y_0$$

if the harvesting function H is not continuous (for example, a step function). One way to cope with the problem is to choose a very large number of displayed points, but this doesn't work with all solvers. In any case, your solver will have to work hard (or be clever) in order to handle on-off harvesting functions (the Engineering Functions table on the inside front cover of the text describes these on-off functions). Here's a way that some solvers deal with this problem: The solver computes values at the equally spaced time points displayed on the computer screen by using internal time-steps which are selected automatically in order to stay within specified error bounds (selected by default), but even this process can be defeated by discontinuous harvesting functions. The problem appears to be that internal time-steps become so large that they miss the on-off points of the harvesting function. Some solvers have a setting that allows the user to select a maximum internal step size, and if it is set low enough then the solver will be better able to "see" the on-off points of the harvesting function. Some really sophisticated solvers will handle this fairly well when on-off functions are detected.

2.5   Separable Differential Equations: Planar Systems

Comments:

If you have trouble separating the ODE into the form $N(y)y' + M(x) = 0$, try to move x and y onto different sides of the equals sign to get $N(y)dy = -M(x)dx$ and then integrate each side. You may still have trouble after getting the ODE separated because you can't find the antiderivatives $G(y)$ of N and $F(x)$ of M. Because the solution is written implicitly as $G(y) + F(x) = C$, it may be hard to find $y(x)$ explicitly (sometimes impossible). Consequently, we encourage the use of integral tables, or a CAS (if available), and checking (and rechecking) your algebra.

In this section, we explain the difference between an integral curve [the graph of the level set $G(y) + F(x) = C_0$] and a solution curve [the graph of $y = y(x)$, x in some interval, where $G(y(x)) + F(x) = C_0$]. Because a solution curve is the graph of a differentiable function $y = y(x)$, the curve cannot double back on itself and cannot have vertical tangents, even though the integral curve on which the solution curve lies may have these properties.

PROBLEMS

Separation of Variables: Losing Solutions.   Sometimes when you rewrite an ODE to separate the variables, you may divide by zero and inadvertently lose solutions. Find all solutions $y = y(x)$ of the ODEs below, but watch out that you don't lose any.

1. $y' = 2xy^2$ **2.** $dy/dx = -4xy$ **3.** $2y\,dx + 3x\,dy = 0$

4. $y' = -xe^{-x+y}$ **5.** $(1-x)y' = y^2$ **6.** $y' = -y/(x^2 - 4)$

7. $y' = xe^{y-x^2}$ **8.** $yy' = (1 - y^2)\sin x$

Answer 1: The ODE $y' = 2xy^2$ has the constant solution $y = 0$, for all x. In addition, note that the ODE is separable and can be written in the separable form as $y^{-2}\,dy = 2x\,dx$. For $y \neq 0$, we can integrate the separated ODE, obtaining $-y^{-1} = x^2 + C$, or $y = -(x^2 + C)^{-1}$ where C is any constant. So the general solution of $y' = 2xy^2$ is given by $y = 0$ and the family of solutions $y = -1/(x^2 + C)$, C any constant. If $C \leq 0$, x must lie in an interval on which $x^2 + C \neq 0$. So for example if $C = -1$, we have three solutions on the three intervals, $x < -1$, $|x| < 1$, and $x > 1$.

Answer 3: The ODE $2y\,dx + 3x\,dy = 0$ separates to $2x^{-1}\,dx = -3y^{-1}\,dy$ which integrates to $2\ln|x| + K = -3\ln|y|$. Exponentiating both sides, $x^2 e^K = |y^{-3}|$. Solving for y and replacing $\pm e^{-K/3}$ with $C \neq 0$, $y = Cx^{-2/3}$, $x < 0$ or $x > 0$. Also, $y = 0$ for all x is a solution of $2y\,dx + 3x\,dy = 0$, as is $x = 0$ for all y. These two solutions were lost when the ODE was written in separated form.

Answer 5: The ODE $(1 - x)y' = y^2$ separates to $y^{-2}\,dy = (1 - x)^{-1}\,dx$, so $-y^{-1} = -\ln|1 - x| - C$. Solutions are given by $y = [C + \ln|1 - x|]^{-1}$, where x is restricted to an interval (not containing 1) for which $C + \ln|1 - x| \neq 0$. Also, $y = 0$ for all x is a solution that was lost when the ODE was written in separated form.

Answer 7: The ODE $y' = xe^{y - x^2}$ separates to $e^{-y}\,dy = xe^{-x^2}\,dx$ which integrates to $-e^{-y} = -e^{-x^2}/2 + K$. Solving for y gives $y = \ln[2/(C + e^{-x^2})]$ after replacing $-2K$ by C; x is in an interval for which $C + e^{-x^2} > 0$.

Solution Formulas. For each IVP find an implicit solution formula $H(x, y) = C$, where the constant C is determined by the initial data. Then find the explicit solution $y = y(x)$ and the largest x-interval on which the solution is defined. Graph the explicit solution. [*Hint*: see Example 2.5.4.]

9. $y' = (y + 1)/(x + 1)$, $y(1) = 1$

10. $y' = y^2/x$, $y(1) = 1$

11. $y' = ye^{-x}$, $y(0) = e$

12. $y' = 3x^2/(1 + x^3)$, $y(0) = 1$

13. $y' = -2x/y$, $y(1) = 2$

14. $2xyy' = 1 + y^2$, $y(2) = 3$

Answer 9: Separating variables in $y' = (y + 1)/(x + 1)$, we have $(y + 1)^{-1}y' - (x + 1)^{-1} = 0$, which integrates to $\ln|y + 1| - \ln|x + 1| = K$. So, $\ln(|y + 1|/|x + 1|) = K$ and $|y + 1| = |x + 1|e^K$. That is, $y + 1 = \pm e^K(x + 1)$, so $y = -1 + C(x + 1)$, where we have renamed $\pm e^K$ by C. Since $y(1) = 1$, $1 = -1 + 2C$, so $C = 1$. The solution becomes $y = x$, $-1 < x < \infty$, since y' is undefined at $x = -1$. See the figure for the solution line lying to the right of the dashed line at $x = -1$.

Answer 11: The ODE $y' = ye^{-x}$ separates to $y^{-1}\,dy = e^{-x}\,dx$, which integrates to $\ln|y| = -e^{-x} + C$, so $y = \pm\exp(C - e^{-x})$. The initial data $y(0) = e$ imply that $C = 2$, and we take the positive solution. So, $y = \exp(2 - e^{-x})$ for all x. See the figure.

Answer 13: The ODE $y' = -2x/y$ separates to $y\,dy + 2x\,dx = 0$, so $y^2 + 2x^2 = C$. Since $y(1) = 2$, $C = 6$ and (solving for y) the solution is $y = \sqrt{6 - 2x^2}$, $|x| < \sqrt{3}$. See the figure for the solution curve.

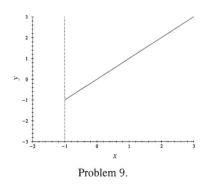

Problem 9.

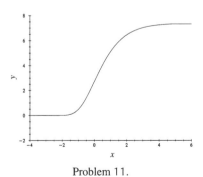

Problem 11.

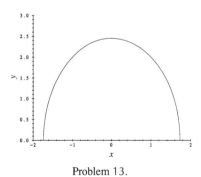

Problem 13.

Find the Implicit General Solution. For each ODE find the implicit general solution formula, but don't try to solve that formula for y as an explicit function of x.

15. $y' = (x^2 + 2)(y + 1)/xy$ **16.** $(1 + \sin x)\,dx + (1 + \cos y)\,dy = 0$

17. $(\sec^2 y)\,dy = (\sin x(1 - \cos^2 x))\,dx$ **18.** $(3y^2 + 2y + 1)y' = x\sin(x^2)$

19. $dx + (y - x)dy = 0$ [*Hint*: think of x as a function of y.]

> **Answer 15:** The ODE $y' = (x^2 + 2)(y + 1)/xy$ separates to $[y/(1 + y)]\,dy = [(x^2 + 2)/x]\,dx$, or $[1 - 1/(1 + y)]\,dy = (x + 2/x)\,dx$. Integrating, we see that the solutions are implicitly defined by $y - \ln|1 + y| = x^2/2 + 2\ln|x| + C$.

> **Answer 17:** The solutions of the separated ODE $(\sec^2 y)dy = (\sin x(1 - \cos^2 x))dx$ are given implicitly by $\tan y + \cos x(\sin^2 x + 2)/3 = C$, where y is restricted to an interval of the form $(2k - 1)\pi/2 < y < (2k + 1)\pi/2$ because $\tan((2k \pm 1)\pi/2) = \pm\infty$, respectively.

> **Answer 19:** Let's look at the ODE in differential form
>
> $$dx - (-y + x)dy = 0 \qquad\qquad\qquad (i)$$
>
> If we think of y as a function of x, then (i) takes the form $dy/dx = 1/(-y + x)$ which is neither linear (in y) nor separable.
>
> On the other hand, if we think of x as a function of y then (i) takes the form $dx/dy = x - y$, which is a linear ODE (in x). Use Theorem 2.2.2 to find the general solution of ODE and get
>
> $$x = Ce^y + y + 1, \qquad C \text{ an arbitrary constant}$$
>
> Finding y explicitly as a function of x from this formula is not a trivial matter, but it is not too hard to graph this function in the yx-plane, for any value of C.

20. *Finding Solution Curves* Graph (in the yx-plane) the solution curve $x = -e^y + y + 1$ for the linear ODE, $dx/dy = x - y$ (note the reversal of the roles of the variables x and y). Use a highlighter to trace out the solution curve of the IVP, $dy/dx = 1/(x - y)$, $y(2 - e) = 1$, where $2 - e \approx -.71828$.

Use a System to Plot Integral Curves. Find an integral for each ODE and plot representative integral curves in the indicated rectangle. [*Hint*: follow the system method used in Example 2.5.6 to plot curves. Choose initial points in the rectangle, and solve forward and backward in time.]

21. $(1 - y^2)y' + x^2 = 1$; $|x| \leq 3$, $|y| \leq 3$

22. $(1 - y^2)y' + x^2 = 0$; $|x| \leq 1.5$, $|y| \leq 2$

> **Answer 21:** Our basic strategy in Problems 21 and 22 is to find integral curves of $N(x, y)y' + M(x, y) = 0$ by finding solutions of the system, $dx/dt = N(x, y)$, $dy/dt = -M(x, y)$. A solver can be used to generate orbits of the system. $\int N(y)\,dy + \int M(x)\,dx$ is an integral of the ODE when $N(x, y) = N(y)$ and $M(x, y) = M(x)$. For this problem we see that $y - 1/3y^3 + 1/3x^3 - x$ is an integral of the separable ODE $(1 - y^2)y' + x^2 - 1 = 0$. Plots of some integral curves are shown in the figure. Here $N = 1 - y^2$, $M = x^2 - 1$ and the curves are computed solutions of the equivalent system $x' = 1 - y^2$, $y' = 1 - x^2$.

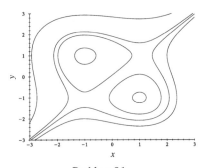

Problem 21.

🖱️ **Use a System to Plot Solution Curves.** Plot the solution curve $y = y(x)$ of each IVP on the largest possible x-interval. [*Hint*: see Example 2.5.6.]

23. $x^2 - 1 + (1 - y^2)y' = 0$; $y(-1) = -2$ **24.** $(1 - y^2)y' + x^2 = 0$; $y(-1) = 0.5$

25. $(2x^2 + 2xy)y' = 2xy + y^2$; $y(-0.5) = 1$

Answer 23: Integrating and using the initial condition, we see that the solution $y(x)$ satisfies the equation $x^3/3 - x + y - y^3/3 = 4/3$, or $x^3 - 3x + 3y - y^3 = 4$. In normal form the ODE is $y' = (1 - x^2)/(1 - y^2)$. The figure shows the solution curve through the initial point $x_0 = -1$, $y_0 = -2$. Since y' is undefined if $y = \pm 1$, and the initial value $y_0 = -2$ is below $y = -1$, the solution curve stays below $y = -1$ and is defined for $x < 2.36$, where 2.36 is the approximate value of x where $y(x) \approx -1$.

Answer 25: In normal form the ODE is $y' = (2xy + y^2)/(2x^2 + 2xy)$. The figure shows the solution curve through the initial point $x_0 = -0.5$, $y_0 = 1$. Since y' is not defined if $x = 0$ or if $x = -y$, the solution curve stops whenever it reaches either the line $x = 0$ or the line $x = -y$. The numerical solution suggests that one or the other of these events occurs at $x \approx -0.737$ and at $x = 0$. So the solution curve is defined for $-0.737 < x < 0$.

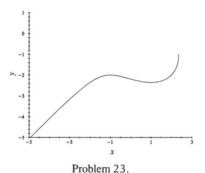

Problem 23.   Problem 25.

The System Way to Plot a Level Curve. Suppose that $f(x, y)$ is continuously differentiable in a region R where $\partial f/\partial x$ and $\partial f/\partial y$ have no common zeros. Here's a way to plot level curves of f.

26. For any point (x_0, y_0) in R show that the orbit of the autonomous planar IVP

☞ This approach beats most contour plotting software.

$$x' = \partial f/\partial y, \qquad x(0) = x_0$$
$$y' = -\partial f/\partial x, \qquad y(0) = y_0$$

is the branch of the level set, $f(x, y) = f(x_0, y_0)$, that contains the point (x_0, y_0). [*Hint*: show that $H(x, y) = f(x, y)$ is an integral of the above system.]

🖱️ **27.** Consider the function $f(x, y) = -x + 2xy + x^2 + y^2$. Use the method of Problem 26 to plot several level curves for f in the rectangle $|x| \le 6$, $|y| \le 6$.

Answer 27: Since $f = -x + 2xy + x^2 + y^2$, $\partial f/\partial x = -1 + 2y + 2x$, $\partial f/\partial y = 2x + 2y$, so the system to be solved is $x' = 2x + 2y$, $y' = 1 - 2y - 2x$. The figure shows solutions of this system with $t_0 = 0$ and $x_0 = -5, -3, -1$ and $y_0 = 5, 3, 1$. The corresponding values of f are 5, 3, 1, respectively. The level sets are tilted ellipses.

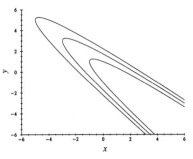

Problem 27.

28. *Orbits and Integral Curves* Show that any solution $x(t)$, $y(t)$ of the ODEs in IVP (20) generates an orbit that lies on a level set of an integral of the ODE $(1 - y^2)y' + x^2 = 0$. [*Hint:* see Examples 2.5.5 and 2.5.6.]

> Torricelli's Law. The speed at time t of water discharged through a sharp-edged hole at the bottom of a tank is the same as the speed that a body would acquire when water falls freely from the water's height (at time t) in the tank.

Modeling Problem. Torricelli's Law describes the discharge rate of water from a tank.

29. Use Torricelli's Law to determine how long it takes to drain a tank.

(a) If a body falls freely from a height h above the ground, what is its velocity when it hits the ground? Ignore air resistance.

(b) If a tank of water is box-shaped with cross-sectional area A_c, is filled to height h at time t, and has a hole of area A_0 in the bottom, what is the rate of change of the volume of water in the tank at time t?

(c) A box-shaped tank has cross-sectional area A_c and height H with a hole of area A_0 in the bottom. Let $h(t)$ denote the height of the water in the tank at time t. If the tank is initially filled to the top with water, find an IVP for the height $h(t)$ of water in the tank at time t.

(d) How long does it take the tank in (c) to empty?

Answer 29:

(a) The IVP for a body falling from height h to the ground is $y''(t) = -g$, $y(0) = h$, $y'(0) = 0$, where $0 \le t \le T$, T is the time of impact and $y(t)$ is the distance from the ground to the body at time t. Solving, we see that

$$y' = -gt, \quad y = -gt^2/2 + h$$

Setting $y = 0$ in order to find T, we see that $-gT^2/2 + h = 0$, so $T = \sqrt{2h/g}$. The velocity $y'(T)$ at impact is $y'(T) = -gT$, so $y'(T) = -\sqrt{2gh}$.

(b) According to (a), water exits the tank through the hole at the bottom of the tank with a velocity of $-\sqrt{2gh}$. So the volume of the water in the tank when $y(t) = h$ is $A_c h$ and it is decreasing at the rate $A_0\sqrt{2gh}$.

(c) The initial volume of water in the tank is $A_c H$. At time $t > 0$ as the water leaks out through the hole of cross-sectional area A_0, the volume is $V(t) = h(t)A_c$, where $h(t)$ is the height of the water in the tank. Since $V'(t) = h'(t)A_c = -A_0\sqrt{2gh(t)}$ [from (b)] we see that the IVP for $h(t)$ is

$$h' = -\left(\frac{A_0}{A_c}\sqrt{2g}\right)h^{1/2}, \quad h(0) = H$$

(d) Separating variables in the ODE in (c) for h and then solving, we see that

$$h^{-1/2}h' = -K, \quad 2h^{1/2} = -Kt + 2H^{1/2}, \quad [K = \frac{A_0}{A_c}\sqrt{2g}]$$

When the tank empties $h = 0$, so $T = \frac{2}{K}H^{1/2} = \frac{2A_c}{A_0\sqrt{2g}}H^{1/2}$ and

$$T = \frac{A_c}{A_0}\sqrt{\frac{2H}{g}}$$

where T is the time the tank runs dry.

2.6　A Predator-Prey Model: the Lotka–Volterra System

Comments:

Planar systems were introduced in Section 1.4. The simplest planar systems are linear cascades, which can be solved using the Method of Integrating Factors presented in Section 2.1. Nonlinear planar systems were presented in Section 2.5 after we encouraged the use of numerical solvers. Predator-prey models are nonlinear planar systems and we use our solver to visualize the solutions of these systems.

The Lotka-Volterra model is certainly an over simplification of reality, but it still has its uses; see also the background material at the end of the Problems.

PROBLEMS

Lotka–Volterra Systems. Which is the predator and which is the prey? Find the average predator and prey populations. Do the population cycles turn clockwise or counterclockwise around the equilibrium point in the first quadrant? [*Hint*: use Theorem 2.6.2. To determine cycle orientation, find the sign of x' if on the positive x-axis and the sign of y' on the positive y-axis.]

1. $x' = -x + xy, \quad y' = y - xy$　　　　　　　2. $x' = 0.2x - 0.02xy, \quad y' = -0.01y + 0.001xy$

3. $x' = (-1 + 0.09y)x, \quad y' = (5 - x)y$

> **Answer 1:** The system is $x' = -x + xy$, $y' = y - xy$. We see that $x(t)$ is the predator population and $y(t)$ is the prey population, as we can tell from the signs of the interaction terms xy in the rate equations. The average predator and prey populations are 1 since the coordinates of the internal equilibrium point are $x = 1$, $y = 1$ (see Theorem 2.6.2). Because $x' < 0$ if $x > 0$ and $y = 0$, while $y' > 0$ if $y > 0$ and $x = 0$, the orbits turn clockwise about $(1, 1)$ as time increases.

> **Answer 3:** The system is $x' = (-1 + 0.09y)x$, $y' = (5 - x)y$. We see that $x(t)$ is the predator population and $y(t)$ is the prey population at time t (look at the signs of the xy terms). The average predator and prey populations are 5 and 100/9, respectively, because $(5, 100/9)$ is the internal equilibrium point (see Theorem 2.6.2). Because $x' < 0$ if $x > 0$ and $y = 0$, while $y' > 0$ if $y > 0$ and $x = 0$, the orbits turn clockwise about $(5, 100/9)$ as time increases.

Estimating the Periods of Cycles. Plot slope fields, the equilibrium point, and the cycles that pass through the points $(5, 10)$ and $(10, 5)$ for the given systems. Plot component graphs for the orbits, and use these graphs to estimate periods.

4. Problem 1　　　　　　　5. Problem 2　　　　　　　6. Problem 3

> **Answer 5:** See the figure for the graphs of the cycles through $(5, 10)$ and $(10, 5)$. The periods of the cycles are about 145 and 190, as estimated from the component graphs. The graphs also show the constant component curves $x = 10$, $y = 10$.

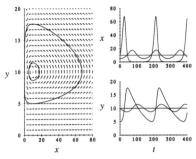

Problem 5.

Harvesting Strategies.

7. *Law of the Percentages* Verify Theorem 2.6.4 for the predator percentages. [*Hint*: the harvested system is $x' = -ax + bxy - H_1 x$, $y' = cy - kxy - H_2 y$, with $H_2 < c$. Explain why the predator fraction F of the total average catch is

$$F = \left[1 + \frac{k(a + H_1)}{b(c - H_2)}\right]^{-1}$$

Explain why F decreases as H_1 and H_2 increase.]

Answer 7: Rewriting the harvested system $x' = -ax + bxy - H_1 x$, $y' = cy - kxy - H_2 y$, we have $x' = -(a + H_1)x + bxy$, $y' = (c - H_2)y - kxy$. By the Law of the Averages, we know the average predator population is $\bar{x} = (c - H_2)/k$ and the average prey population is $\bar{y} = (a + H_1)/b$. The predator fraction F of the total average catch is

$$F = \frac{\bar{x}}{\bar{x} + \bar{y}} = \frac{(c - H_2)/k}{(c - H_2)/k + (a + H_1)/b} = \frac{1}{1 + k(a + H_1)/b(c - H_2)}$$

The quantity $k(a + H_1)/b(c - H_2)$ increases as H_1 increases (because the numerator increases), and also as H_2 increases but stays smaller than c (because the denominator decreases). Since this ratio is in the denominator of F, F itself must decrease as H_1 and H_2 increase.

8. *Harvesting a Predator-Prey Community to Extinction* High harvesting coefficients H_1 and H_2 may lead to the extinction of the species. This problem explores what happens.

(a) Explain the graphs in Figure 2.6.3.

(b) Set H_1 at a fixed positive value in system (9). Show that as $H_2 \to c^-$, the equilibrium point inside the population quadrant approaches the point $(0, (a + H_1)/b)$ on the y-axis and that if $H_2 = c$, all points on the y-axis are equilibrium points of system (9).

(c) Let $H_2 = c$ in system (9). Show that $dy/dx = (kxy)((a + H_1)x - bxy)^{-1}$. Separate the variables, solve, and find x as a function of y and the initial data x_0, y_0, where $x_0 > 0$, $y_0 > 0$.

(d) What do you think happens to the harvested species as time increases for the case $H_2 = c$? Give an informal explanation, but with reasons for your conclusions. As an aid, set $a = b = c = k = 1$, $H_1 = 0.1$, and plot a direction field and orbits.

Properties of the Predator-Prey System.

9. *Linearizing to Estimate the Periods of Cycles* To linearize a planar system $x' = f(x, y)$, $y' = g(x, y)$ about an equilibrium point (x_0, y_0), expand each rate function in a Taylor series about (x_0, y_0) and discard all terms higher than first-order. The result is the *linearization* of the original system at (x_0, y_0). Apply this process to the predator-prey rate equations $x' = -ax + bxy$, $y' = cy - kxy$ in order to estimate the periods of the population cycles near the equilibrium point $(c/d, a/b)$ inside the population quadrant.

(a) Show that the linearized rate equations about the equilibrium point $(c/k, a/b)$ form the linear system $x' = bcy/k - ac/k$, $y' = -akx/b + ac/b$.

(b) Show that, for certain values of ω, A, and B, the functions $x = c/k + A\cos\omega t$, $y = a/b + B\sin\omega t$ solve the linearized system in **(a)**. What are the values of A/B and ω?

 (c) Set $a = b = c = k = 1$ and plot orbits and component graphs of the nonlinear and the linearized systems, using the common initial data $x_0 = 1$, $y_0 = 1, 1.1, 1.3, 1.5, 1.9$. Plot over $0 \le t \le 20$. Explain the graphs and compare the periods of the cycles of the two systems using common initial points.

Answer 9: The xy-system is $x' = f(x, y) = -ax + bxy$, $y' = g(x, y) = cy - kxy$. Let $x_1 = c/k$ and $y_1 = a/b$, so (x_1, y_1) is the equilibrium point.

(a) To linearize we first find the partial derivatives of $f(x, y)$ and $g(x, y)$ with respect to both x and y. These are $\partial f/\partial x = -a + by$, $\partial f/\partial y = bx$, $\partial g/\partial x = -ky$, and $\partial g/\partial y = c - kx$. So with $x_1 = c/k$ and $y_1 = a/b$, we have $x' = \partial f/\partial x(x_1, y_1)(x - x_1) + \partial f/\partial y(x_1, y_1)(y - y_1) = (-a + b(a/b))(x - c/k) + b(c/k)(y - a/b) = bcy/k - ac/k$ and $y' = \partial g/\partial x(x_1, y_1)(x - x_1) + \partial g/\partial y(x_1, y_1)(y - y_1) = -k(a/b)(x - c/k) + (c - k(c/k))(y - a/b) = -akx/b + ac/b$. The linearized system is $x' = bcy/k - ac/k$, $y' = -akx/b + ac/b$.

(b) To show $x = c/k + A\cos\omega t$, $y = a/b + B\sin\omega t$ solves the linearized system, first solve for x'. We see that $x' = -A\omega\sin\omega t$. Substituting the value for y into the linearized system gives $x' = bc/k(a/b + B\sin\omega t) - ac/k = bcB/k\sin\omega t$. We get $-A\omega\sin\omega t = x' = bcB/k\sin\omega t$ so, $-A\omega = bcB/k$. Solving for ω gives $\omega = -Bbc/Ak$. Now $y' = B\omega\cos\omega t$. Substituting the value for x into the linearized system gives $y' = -ak(c/k + A\cos\omega t)/b + ac/b = -ak(A\cos\omega t)/b$. We get $B\omega\cos\omega t = -ak(A\cos\omega t)/b$, so $B\omega = akA/b$. Solving for ω gives $\omega = -Aak/Bb$. By setting the two expressions for ω equal to each other we get $-Bbc/Ak = -Aak/Bb$, so $A^2/B^2 = b^2c/ak^2$. So $A/B = \pm b/k\sqrt{c/a}$. Substituting this back in to either of the expressions for ω gives $\omega = \pm\sqrt{ac}$.

(c) See Graph 1 for plots of the predator-prey orbits and component graphs of the nonlinear Lotka–Volterra system, $x' = (-1 + y)x$, $y' = (1 - x)y$. Graph 2 shows the orbits and component graphs for the corresponding linearized system. Note from the x-component graphs near $t = 20$ how the period of the predator-prey cycles of the nonlinear system increases with amplitude, but the period of the linearized cycles is fixed at $2\pi/\sqrt{ac}$ ($= 2\pi$ if $a = c = 1$). The initial points used for the graphs in both figures are $x_0 = 1$, $y_0 = 1, 1.1, 1.3, 1.5$, and 1.9.

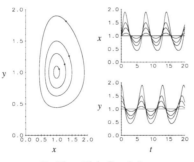

Problem 9(c), Graph 1.

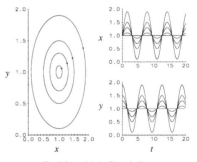

Problem 9(c), Graph 2.

10. Suppose that system (9) models a predator-prey system and that you are the harvester. Suppose we require $0 < x_m \le x(t) \le x_M$ and $0 \le y_m \le y(t) \le y_M$ for all t, where the positive population bounds x_m, x_M, y_m, y_M are given. Suppose also that $x(0) = c/k$, $y(0) = a/b$.

- Describe how you would choose the positive coefficients H_1 and H_2 to maintain each species within the prescribed bounds. Justify your arguments.

- High harvesting rates can be maintained if the harvesting season is short. Construct and justify your own strategy for maximizing the yield by imposing a limit on the harvest season, while still maintaining the populations within reasonable bounds.

Orbits of the Lotka–Volterra System Are Cycles.

11. Follow the steps below to show that for $0 < K \le K_0 = c^c(ke)^{-c}a^a(be)^{-a}$ the graph of the orbit defined by the equation $x^c e^{-kx} y^a e^{-by} = K$ is a simple closed curve (i.e., a cycle) in the interior of the population quadrant. The only exception occurs for $K = K_0$ when the graph is the equilibrium point $(c/k, a/b)$.

(a) Show that $f(x) = x^c e^{-kx}$ is defined for $x \ge 0$, rises from the value of 0 at $x = 0$ to its maximum value of $M_1 = c^c(ke)^{-c}$ at $x = c/k$, falls as x increases beyond c/k, and $f(x) \to 0$ as $x \to +\infty$. Show that the function $g(y) = y^a e^{-by}$ rises from the value of 0 at $y = 0$ to its maximum value $M_2 = a^a(be)^{-a}$ at $y = a/b$, and then decays to 0 as $y \to +\infty$.

(b) Show that no nonnegative values of x and y satisfy the equation $x^c e^{-kx} y^a e^{-by} = K$ if $K > M_1 M_2$. Show that equation $x^c e^{-kx} y^a e^{-by} = K$ has the unique solution $x = c/k$, $y = a/b$ if $K = M_1 M_2$.

(c) Suppose that γ is any positive number, $\gamma < M_1$. Show that $f(x) = \gamma$ has two solutions, x_1 and x_2, where $x_1 < c/k < x_2$. Show that the equation $g(y) = \gamma M_2/f(x)$ has no solution y if $x > x_2$ or $x < x_1$; exactly one solution, $y = a/b$, if $x = x_2$ or if $x = x_1$; and two solutions, $y_1(x) < a/b$ and $y_2(x) > a/b$, if $x_1 < x < x_2$. Show that $y_1(x) \to a/b$ and $y_2(x) \to a/b$ if $x \to x_1$ or $x \to x_2$.

(d) Explain why the equation $x^c e^{-kx} y^a e^{-by} = K$ defines a cycle inside the population quadrant if $0 < K < M_1 M_2$.

Answer 11:

(a) Let $f(x) = x^c e^{-kx}$, $g(y) = y^a e^{-by}$; according to text formula (6), each orbit is defined by $f(x)g(y) = K$, where $K_0 = (x_0^c e^{-kx_0})(y_0^a e^{-by_0})$ and (x_0, y_0) is a point on the orbit. We have that $f(0) = g(0) = 0$. Moreover, $f(x)$ and $g(y)$ are positive for $x > 0$ and $y > 0$, and tend to 0 as $x \to \infty$, $y \to \infty$, respectively. To show, for example, that $f(x) = x^c/e^{kx} \to \infty$ as $x \to \infty$, apply l'Hôpital's Rule, taking just enough derivatives of numerator and of denominator (say n) so that the power $c - n$ of x in the numerator is zero or negative. Then $\lim_{x \to \infty} f(x) = \lim_{x \to \infty} c(c-1)\cdots(c-n+1)x^{c-n}/(k^n e^{kx}) = 0$. Since $f'(x) = cx^{c-1}e^{-kx} - kx^c e^{-kx} = x^{c-1}e^{-kx}(c - kx)$, we see that $f(x)$ increases with x to the maximum value of $M_1 = f(c/k) = c^c(ke)^{-c}$ at $x = c/k$, and then decreases as x increases beyond c/k. Similar results hold for $g(y)$, whose maximum value is $g(a/b) = a^a(be)^{-a} = M_2$ at $y = a/b$.

(b) If $K > M_1 M_2 = [\max f(x)] \cdot [\max g(y)]$, then the equation $K = f(x)g(y)$ holds for no positive values of x and y. If $K = M_1 M_2$ then the equation has the unique solution $x = c/k$, $y = a/b$, given that $M_1 = f(c/k)$ for $x = c/k$, and $M_2 = g(a/b)$ for $y = a/b$ from part **(a)**.

(c) Since $f(x)$ is continuous, strictly increasing for $0 \le x \le c/k$, and strictly decreasing for $x \ge c/k$, then given any number $\triangle$, $0 < \triangle < M_1$, there are precisely two values of x, x_1 and x_2, $x_1 < c/k < x_2$, such that $f(x) = \triangle$. If $0 < x < x_1$, or $x > x_2$, then $0 < f(x) < \triangle$, and there is no positive y for which $f(x)g(y) = \triangle \cdot M_2$. In other words, the graph of $f(x)g(y) = \triangle \cdot M_2$ is contained entirely in the vertical strip $x_1 \le x \le x_2$. If $x = x_1$ or x_2, then $f(x) = \triangle$ and $f(x)g(y) = \triangle \cdot M_2$ has the unique y-solution $y = a/b$, since $g(a/b) = M_2$, and $g(y) < M_2$ for all other $y > 0$. On the other hand, for each x, $x_1 < x < x_2$, we have $f(x) > \triangle$, and the equation $f(x)g(y) = \triangle \cdot M_2$ has two solutions, $y_1(x) < a/b$ and $y_2(x) > a/b$ because of the way $g(y)$ rises to its maximum value M_2 and then falls as y increases through a/b. The same analysis shows that $y_1(x)$ and $y_2(x) \to a/b$ if $x \to x_1$ or $x \to x_2$, respectively. Similar results are obtained with the roles of f and g, x and y interchanged. The analogues of x_1 and x_2 are y_1 and y_2.

(d) The orbit defined by $f(x)g(y) = K$, $0 < K < M_1 M_2$ is intersected by each horizontal line $y = y_0$, $y_1 < y_0 < y_2$, exactly twice, and by each vertical line $x = x_0$, $x_1 < x_0 < x_2$ also twice. This is because the equation $f(x) = y$ has exactly two solutions, x_1 and x_2, such that $x_1 < x < x_2$ and the equation $g(y) = yM_2/f(x)$ has exactly two solutions, y_1 and y_2, such that $y_1(x) < a/b < y_2(x)$ if $x_1 < x < x_2$. Segments of the four lines $x = x_1$, $x = x_2$, $y = y_1$, $y = y_2$ form a rectangle just touching the four extreme points of the orbit. The orbit is a simple closed curve inside the rectangle, $x_1 \le x \le x_2$, $y_1 \le y \le y_2$.

Background Material: Occam's Razor

A guiding principle of modeling is Occam's Razor:

> **Occam's Razor.** What can be accounted for by fewer assumptions is explained in vain by more.

William of Occam (1285–1349) was an English theologian and philosopher who applied the Razor to arguments of every kind. The principle is called the Razor because Occam used it so often and so sharply.

2.7 Extension of Solutions: Long-Term Behavior

Comments:

Solutions of $y' = f(t, y)$ and the corresponding solution curves cannot just "die" without some reason. A solution can only stop by time T if the solution curve is about to enter a region where the conditions on f given in the Extension Principle fail to hold, or if the solution blows up as time nears the value T (i.e., the solution has a finite escape time). The Bounded Input-Bounded Output Theorem gives conditions that ensure that the solution $y(t)$ of a first-order linear IVP is bounded, that is that $|y(t)| \leq$ some constant for all $t \geq 0$. This is of considerable importance in the applications because it ensures that a solution won't become dangerously large in magnitude. See Problem 14 and the background material after the Problems for a discussion of the proof the theorem.

PROBLEMS _____

Maximally Extended Solutions. Find a formula for the maximally extended solution of each IVP below. Describe the t-interval on which the maximally extended solution is defined. Sketch the graph of this solution.

   1. $y' = y/t$, $y(1) = 1$ **2.** $y' - y = e^t \sec^2 t$, $y(0) = 1$

   3. $yy' = -t$, $y(0) = 1$ **4.** $2y' + y^3 = 0$, $y(0) = -1$

Answer 1: Using the Method of Separable Variables, the ODE can be written as $y'/y = 1/t$. Antidifferentiating both sides, we get $\ln|y| = \ln|t| + C$, where C is a constant. So we have $y = Ct$, Using $y(1) = 1$, we have $C = 1$. This means that the maximally extended solution to this IVP is $y = t, t > 0$ since y' is undefined at the origin, and so the maximal interval of existence of this solution is the positive $t-$axis. See the figure.

Answer 3: Writing the ODE $yy' = -t$ as $(y^2)'/2 = -t$ and then integrating, we have $y^2/2 = -t^2/2 + C$, so $y = \pm\sqrt{2C - t^2}$. Using the initial condition $y(0) = 1$, we find $C = 1/2$ and we take the plus sign. So the maximally extended solution satisfying $y(0) = 1$ is $y = \sqrt{1 - t^2}$ and the maximal interval of existence of this solution is $-1 < t < 1$. At $t = \pm 1$, the solution's slope would be infinite. See the figure for the graph of the maximally extended solution.

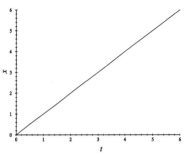

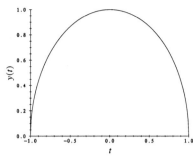

Problem 1.   Problem 3.

Investigating the Extension Theorem.

5. *Solution Formula for Example 2.7.2* Show that the solution $y = y(t)$ of the IVP, $y' = -t^2/[(y + 2)(y - 3)]$, $y(0) = 0$, satisfies the equation $2y^3 - 3y^2 - 36y = -2t^3$. [*Hint*: write the ODE as $(y^2 - y - 6)y' = [y^3/3 - y^2/2 - 6y]' = -t^2$ and apply the Antiderivative Theorem.] Show that $t = -(22)^{1/3}$ if $y = -2$ and that $t = (81/2)^{1/3}$ if $y = 3$. Explain why the solution $y(t)$ can't be extended beyond the interval $-(22)^{1/3} < t < (81/2)^{1/3}$.

   Answer 5: Separating variables in $y' = -t^2/[(y + 2)(y - 3)]$, we have $(y^2 - y - 6)\,dy = -t^2\,dt$, which integrates to $y^3/3 - y^2/2 - 6y = -t^3/3 + K$. Using the initial data $y(0) = 0$, we find $K = 0$. Multiplying through by 6, $2y^3 - 3y^2 - 36y = -2t^3$. If $y = -2$, then $-2t^3 = 44$ so $t = -(22)^{1/3}$. If $y = 3$, then $-2t^3 = -81$ so $t = (81/2)^{1/3}$. At $y = -2$ or $y = 3$, y' becomes infinite, so the solution can't be extended beyond the interval $-(22)^{1/3} < t < (81/2)^{1/3}$.

6. Plot a slope field for $y' = 3t^2/(3y^2 - 4)$, $|t| \le 2$, $|y| \le 2$. Then plot the maximally extended solution through $(0, 0)$. Find the t-interval on which this maximally extended solution is defined. [*Hint*: find the exact t-interval by first separating variables and then integrating each side of the ODE and then using the initial data.]

7. Find all maximally extended solutions of the ODE $y' = 2ty^2$. Sketch representative solution curves. [*Hint*: consult Example 1.3.3.]

   Answer 7: Using the Method of Separable Variables, the ODE can be written as $y'/y^2 = 2t$. Antidifferentiating both sides, we get $-1/y = t^2 + C$, $y = -1/(t^2 + C)$, C any constant. For each C the maximal interval of existence of a solution is:
   Case $C < 0$: $-\sqrt{-C} < t < \sqrt{-C}$ or $t > \sqrt{-C}$ or $t < \sqrt{-C}$
   Case $C > 0$: The whole t-axis.
   Case $C = 0$: $t > 0$ or $t < 0$.

   The figure shows solution curves (reading from top to bottom) for $C = -1, -4, 1, 1/4$. The respective initial values of y_0 with $t_0 = 0$ are $1, 1/4, -1, -4$ and the respective maximal intervals are $|t| < 1$, $|t| < 2$, and all t. See figure.

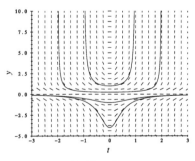

Problem 7.

8. *Long-Term Behavior*   Consider the IVPs $y' = -y\sin t + ct\cos t$, $y(-4) = -6$, $c = 0.7, 0.8, \cdots, 1.3$. Graph the seven solution curves for $|t| \leq 4$, and then for $-4 \leq t \leq 20$. Make conjectures about the long-term behavior of the solutions if $|c - 1| \leq 0.3$.

Finite Escape Time.

9. Do you expect finite escape times if the rate function involves y^2?

   (a) Find the solution of the IVP $y' = 1 - y^2$, $y(0) = y_0$, where $y_0 \neq -1$.

   (b) Find the finite escape time T_e as a function of y_0, where $y_0 < -1$.

   Answer 9:

   (a) Separating variables in the ODE, $y' = 1 - y^2$, and integrating, we obtain the general solution

$$\ln |(1 + y)/(1 - y)|^{1/2} = t + C, \quad \text{where } C \text{ is an arbitrary constant}$$

   Exponentiating, squaring, dropping the absolute value signs, and redefining the constant of integration we have

$$\frac{1 + y}{1 - y} = ce^{2t}, \quad \text{where } c \text{ is an arbitrary constant} \tag{i}$$

   Imposing the initial condition $y(0) = y_0$, we see that $c = (1 + y_0)/(1 - y_0)$. Solving (i) for y, we obtain

$$y = \frac{c - e^{-2t}}{e^{-2t} + c}$$

   (b) Now if $y_0 < -1$, then we have that $-1 < c < 0$, and $y(t)$ escapes to $-\infty$ as $t \to T_e^-$, where $e^{-2T_e} + c = 0$. Solving $e^{-2T_e} + c = 0$ for the escape time T_e,

$$T_e = (-1/2)\ln(-c) = (-1/2)\ln\frac{1 + y_0}{y_0 - 1}$$

10. Suppose that $y(t)$ is a positive solution of $y' = f(t, y)$ and that, for some constant $c > 0$, f satisfies the inequality $f(t, y) \geq cy^2$, for all $t \geq 0$ and $y > 0$. Show that $y(t)$ escapes to infinity in finite time. [*Hint*: suppose $y(0) > 0$. Write $y' - cy^2 \geq 0$ as $y^{-2}y' - c \geq 0$ for some interval $0 \leq t \leq T$. Then $(-y^{-1} - ct)' \geq 0$; what does this say about $-y^{-1} - ct$? Explain why $y(t) \geq y_0(1 - cy_0t)^{-1}$ and why this implies a finite escape time.]

Investigating BIBO.

11. *Hypotheses of BIBO*   Show that the hypothesis $p(t) \geq p_0 > 0$ in BIBO can't be weakened to the condition $p(t) > 0$. [*Hint*: show that the ODE $y' + (t + 1)^{-2}y = e^{1/(t+1)}$, $t \geq 0$, where $p(t) = (t + 1)^{-2}$ and $q(t) = e^{1/(t+1)}$, has unbounded solutions even though $p(t) > 0$ and $|q(t)| \leq e$ for $t \geq 0$.]

   Answer 11:   An integrating factor for the ODE, $y' + (t + 1)^{-2}y = e^{1/(t+1)}$, is the function $e^{-1/(t+1)}$. Solutions are given by $y = (C + t)e^{1/(t+1)}$, and so $y \to \infty$ as $t \to +\infty$. Note that $p(t) = (t + 1)^{-2} > 0$ but that there is no positive constant p_0 such that $p(t) \geq p_0$ for all $t \geq 0$. Thus, for this specific ODE, $|q(t)|$ is bounded and $p(t) > 0$ and yet the BIBO estimate fails. So, the condition $p(t) \geq p_0 > 0$ cannot in general be weakened to $p(t) > 0$.

12. *Intervals of BIBO*   Show that the equation $y' + y = t$ has unbounded solutions on the interval $0 \leq t$, thus showing that the hypothesis $|q(t)| \leq M$ can't be dropped from BIBO.

13. *Failure of BIBO for a Nonlinear ODE*   The following example shows how badly things may go wrong in the presence of a nonlinearity even though all solutions of the undriven ODE tend to 0 as $t \to +\infty$ and the driving term is bounded.

   (a) If $y(t)$ solves the separable ODE $y' = -y/(1 + y^2)$, show that $y(t) \to 0$ as $t \to +\infty$.

   (b) Show that every solution $y(t)$ of the driven ODE $y' = -y/(1 + y^2) + 1$ satisfies the inequality $y'(t) \geq 1/2$, for all $t \geq 0$. [*Hint*: find the minimum value of $-y/(1 + y^2) + 1$.]

(c) Integrating the inequality $y'(t) \geq 1/2$, we have $y(t) \geq t/2 + C$, where C is a constant. Show that if $y(t)$ solves the ODE in **(b)**, then $y(t)$ becomes unbounded as $t \to +\infty$.

Answer 13:

(a) The variables in the ODE $y' = -y/(1 + y^2)$ may be separated: $-(y^{-1} + y)y' = 1$. Integration gives $t = C - \ln|y| - y^2/2$. This equation can't be solved for y in terms of elementary functions of t. So, to show that $y \to 0$ as $t \to \infty$, we must argue indirectly. We have that $t + \ln|y| + y^2/2 = C$, or $e^t|y|e^{y^2/2} = e^C = K$, then we have $|y| = \dfrac{K}{e^t e^{y^2/2}}$. As $t \to \infty$, we have $\dfrac{K}{e^t e^{y^2/2}} \to \dfrac{K}{e^\infty e^{y^2/2}} = \dfrac{K}{\infty} = 0$, since $e^{y^2/2} \geq e^0 = 1$. Every solution $y = y(t)$ has the property that $y(t) \to 0$ as $t \to +\infty$.

(b) To find the extreme values of the rate function $r(y) = -y(1 + y^2)^{-1} + 1$, differentiate and equate to 0: $(y^2 - 1)/(y^2 + 1) = 0$. Solving for y, we obtain $y = \pm 1$. Since $r(1) = 1/2$ while $r(-1) = 3/2$ and $\lim_{y \to \pm\infty} r(y) = -1/\infty + 1 = 1$, we see that the minimum value of the rate function is $1/2$, and $y'(t) \geq 1/2$.

(c) Integrating each side of the inequality $y'(t) \geq 1/2$ from 0 to t, we have $y(t) - y_0 \geq t/2$ or $y(t) \geq y_0 + t/2$, and so $y(t) \to \infty$ as $t \to \infty$. We interpret this situation as one where there is bounded input [the function $+1$ on the right-hand side of $y' = -y(1 + y^2)^{-1} + 1$], while all solutions of the undriven equation $y' = -y(1 + y^2)^{-1}$ tend toward the equilibrium $y = 0$ (see **(a)**). This is the setting for the Bound Input–Bounded Output Theorem, except that the equation here is nonlinear and the Theorem need not apply. In this case, we see that the conclusion does not hold; the bounded input actually stimulates the production of an unbounded output.

Verification of BIBO Inequality (9).

14. The following steps show how to verify inequality (9) for the solution $y(t)$ of the IVP, $y' + p(t)y = q(t)$, $y(0) = y_0$, $t \geq 0$, in the special case that $p(t) = p_0 > 0$, for all $t \geq 0$. Assume that q is continuous and that $|q(t)| \leq M$, for all $t \geq 0$.

(a) Show that $|y(t)| \leq |y_0 e^{-p_0 t}| + |e^{-p_0 t} \int_0^t e^{p_0 s} q(s)\, ds|$. [*Hint*: use formula (8) in this section and the Triangle Inequality. The Triangle Inequality asserts that $|A + B| \leq |A| + |B|$ for any constants or functions A and B.]

(b) Then show that $|y| \leq |y_0|e^{-p_0 t} + e^{-p_0 t} \int_0^t e^{p_0 s} M\, ds \leq |y_0|e^{-p_0 t} + e^{-p_0 t}(M/p_0)|1 - e^{-p_0 t}|$. [*Hint*: the exponential terms in **(a)** are all positive.]

(c) Derive inequality (9) from the inequality in **(b)**.

15. *Estimate (9) Can't Be Improved* Show that inequality (9) of BIBO can't be improved. [*Hint*: find the solution $y(t)$ of the IVP $y' + p_0 y = M$, $y(0) = y_0$, where p_0, M, and y_0 are positive constants. What happens to $y(t)$ as $t \to +\infty$?]

Answer 15: The solution $y(t)$ of the IVP, $y' + p_0 y = M$, $y(0) = y_0$, is $y = y_0 e^{-p_0 t} + M(1 - e^{-p_0 t})/p_0$; that is, if y_0, M and p_0 are positive, then y precisely equals the right-hand side of the inequality given in problem 14(b). We have shown that for the specific case $p(t) = p_0$, $q(t) = M$, the inequality can't be improved. So, it is not in general possible to replace the right-hand side by a smaller quantity.

Modeling Problems.

16. *Restocking* A population $y(t)$ is replenished from time to time with new stock. Suppose that the rate law is $y'(t) = -r(t)y(t) + R(t)$, where all we know about $r(t)$ and $R(t)$ is that for $t \geq 0$, $0 < r_0 \leq r(t)$ and $|R(t)| \leq R_0$ for some positive constants r_0 and R_0. If $y(0) = y_0 > 0$, use BIBO to find an upper bound for $y(t)$ for all $t \geq 0$.

17. *Salt Solution* A vat contains 100 gal of brine in which initially 5 lb of salt are dissolved. More brine runs into the vat at a rate of $r(t)$ gal/min with a salt concentration of $c(t)$ lb/gal. The solution in the tank is thoroughly mixed and runs out of the tank at a rate of $r(t)$ gal/min. For safety, the salt concentration in the tank must never exceed 0.1 lb/gal. Suppose for $t \geq 0$ that $0 < r_0 \leq r(t) \leq r_1$ and $0 \leq c(t) \leq c_0$, where r_0, r_1, and c_0 are positive constants. What conditions must r_0, r_1, and c_0 satisfy to ensure safe operation? [*Hint*: first show that $S' = rc - rS/100$, where $S(t)$ is the amount of salt in the tank at time

$t \geq 0$. Then use BIBO.]

Answer 17: We are given that the vat has a 100 gal capacity, that r gal of brine run into the vat per minute, and the salt concentration in the inflow stream is $c(t)$ lbs/gal. The brine exits the tank at the rate of r gal/min. Let $S(t)$ be the lbs of salt dissolved in the brine in the tank at time t. We are given that $S(0) = 5$. Then $S'(t) = $ Rate in $-$ Rate out $= rc(t) - rS/100$, $S(0) = 5$, where $0 < r_0 \leq r \leq r_1$ and $0 \leq c(t) \leq c_0$. The Bounded Input–Bounded Output Principle applies to the ODE, $S' + rS/100 = rc$. Identifying $|y_0| = 5$, $p_0 = r_0/100$, and $M = r_1c_0$, we have that $|S(t)| \leq 5 + 100r_1c_0/r_0$. Since $c(t) = S(t)/100$, the safety rule that $c(t)$ must never exceed 0.1 lb/gal implies that we must be sure that $S(t) \leq 10$ at all times, which we can guarantee by setting $100r_1c_0/r_0 \leq 5$, that is, we require that $20r_1c_0 \leq r_0$ to ensure that the concentration of salt in the tank never exceeds 0.1 lbs/gal.

18. *Controlling a Reaction* Suppose that for the reaction of Example 2.7.3 you can control the initial concentration $y(0)$ of chemical A, but that it is harder to control the reaction rate constant k. If K is the maximal allowable concentration $y(t)$ of A and if $y(0) < K/10$, what is the smallest allowable value of k?

A Harvested Logistic Population. The problems below deal with the harvested logistic equation $y' = y - y^2/12 - H(t)$.

19. If $H(t) = 4$ tons/yr, describe what eventually happens to the fish population. Explain your reasoning. [*Hint*: show that $y' \leq -1$ by using calculus to find the maximum value of $y - y^2/12 - 4$.]

Answer 19: Substituting $H(t) = 4$ into the logistic equation gives

$$y' = y - y^2/12 - 4$$

where $y' < 0$ for all t since $y' < 0$ if $y = 0$ and the roots of the quadratic rate are complex conjugates, so y' never changes sign. The value of y' is actually no larger than -1. For, $y - y^2/12 - 4 = g(y)$ has its extreme value where $g'(y) = 1 - y/6 = 0$; this occurs at $y = 6$ and $y' = 6 - 6^2/12 - 4 = -1$. The fish population is always declining and lies below the line $y = -t + y_0$ passing through the point $t_0 = 0$, $y(0) = y_0$. Since there is no restocking of the population, then the fish will eventually die off no matter what the initial condition.

20. *Discontinuous Harvesting Term* Suppose that fishermen are allowed to fish at the rate of 4 tons/yr for the first 5 years, but not at all for the next 5 years. Use a numerical solver and describe what happens to the fish population over a 10 year span. Find the smallest initial fish population you can that will ensure that the population does not become extinct during the 10 year span. [*Hint*: compare your results with the Chapter 2 opening figure.]

Background: Proof of Theorem 2.7.2

Here is the proof of Theorem 2.7.2 that was omitted from the text. Suppose that $p(t) \geq p_0$ where $p_0 \neq 0$. Then by properties of integrals and exponentials, we have for all positive values of t that

$$p_0 t \leq \int_0^t p(s)\,ds$$

$$e^{-\int_0^t p(s)\,ds} \leq e^{-p_0 t}$$

Since the solution of the IVP $y' + p(t)y = q(t)$, $y(0) = y_0$ is given by

$$y(t) = e^{-\int_0^t p(s)\,ds}\left\{ y_0 + \int_0^t e^{\int_0^s p(r)\,dr} q(s)\,ds \right\}$$

$$= e^{-\int_0^t p(s)\,ds} y_0 + \int_0^t e^{-\int_s^t p(r)\,dr} q(s)\,ds$$

we have [using the triangle inequality and the fact that $|\int_s^t p(r)\,dr| \le \int_s^t |p(r)|\,dr$ if $t \ge s$]:

$$|y(t)| \le e^{-p_0 t}|y_0| + \int_0^t e^{-p_0(t-s)} M\,ds$$

$$\le e^{-p_0 t}|y_0| + \frac{M}{p_0} e^{-p_0 t}\left[e^{p_0 t} - 1\right]$$

$$\le e^{-p_0 t}|y_0| + \frac{M}{p_0}\left[1 - e^{-p_0 t}\right]$$

$$\le |y_0| + \frac{M}{p_0} \quad \text{(for } t \ge 0)$$

This proves Theorem 2.7.2.

2.8 Qualitative Analysis: State Lines, Sign Analysis

Comments:

The solution curves of a first-order autonomous ODE $y' = f(y)$ are easy to visualize through analysis of the fixed sign of y' between pairs of adjacent equilibrium lines. The corresponding state line encodes the needed information in a striking way. The state line allows us to visualize the sensitivity of solutions to changes in the initial data. (See also the background materal following the problems.) We like to emphasize the long-term behavior of solutions of the ODE $y' = f(y)$: every solution either tends to $+\infty$, or to $-\infty$, or to an equilibrium solution as t increases to $+\infty$. This question of the long-term behavior of solutions of autonomous ODEs is a hot issue today, because it is directly related to chaos. We come back to the question in Chapter 9 for a system of autonomous ODEs in three state variables. We explain in Section 9.2 just what the long-term behavior of a bounded solution of a planar autonomous system can be (chaos cannot occur in dimension 2, just as Theorem 2.8.1 says it can't occur in dimension 1). However, chaos can occur if an autonomous system has 3 state variables (see the Lorenz system of Section 9.4). What we just said applies to autonomous ODEs with one state variable, but chaos *can* occur in discrete autonomous dynamical systems in one state variable. As we note in the WEB SPOTLIGHT ON CHAOS IN NUMERICS, chaotic wandering *is* possible with a single state variable if we are dealing with certain discrete one-dimensional transformations.

PROBLEMS

Sign Analysis. Find all equilibrium solutions for each of the ODEs below. For every bounded and nonconstant solution $y(t)$, find the limit of $y(t)$ as $t \to +\infty$ and then as $t \to -\infty$.

1. $y' = y(y-1)(y-2)$ **2.** $y' = (1-y)(y+1)^2$ **3.** $y' = (y+1)(2-y)(1+y^2)$

4. $y' = |y-1|(y-3)$ **5.** $y' = \sin(y/2)$ **6.** $y' = 3y - ye^{y^2}$

Answer 1: The equilibrium solutions of $y' = y(y-1)(y-2)$ are $y = 0$, $y = 1$, and $y = 2$. Solutions between $y = 1$ and $y = 2$ tend to $y = 1$ as $t \to +\infty$ and to $y = 2$ as $t \to -\infty$ because $y' < 0$. Solutions between $y = 0$ and $y = 1$ tend to $y = 1$ as $t \to +\infty$ and to $y = 0$ as $t \to -\infty$ because $y' > 0$.

Answer 3: The equilibrium solutions of $y' = (y+1)(2-y)(1+y^2)$ are at $y = -1$ and $y = 2$. As $t \to +\infty$, solutions bounded by $y = -1$ and $y = 2$ will tend to $y = 2$. As $t \to -\infty$, these solutions tend to $y = -1$; $y' > 0$ in the region between $y = -1$ and $y = 2$.

Answer 5: The equilibrium solutions of $y' = \sin(y/2)$ are $y = 2k\pi$, $k = 0, \pm1, \pm2, \ldots$. The solutions between $y = 0$ and $y = 2\pi$ tend to $y = 2\pi$ as $t \to +\infty$ because the rate function is positive in that band and to $y = -2\pi$ as $t \to -\infty$. The solutions between $y = -2\pi$ and $y = 0$ tend to $y = -2\pi$ as $t \to +\infty$ because the rate function is negative in the band and to $y = 2\pi$ as $t \to -\infty$. The behavior as $t \to \infty$ alternates in this fashion in adjacent bands between equilibrium solutions.

For each ODE in Problems 1–6 sketch representative solution curves above, below, and between the equilibrium lines. Sketch the state lines.

7. Problem 1   **8.** Problem 2   **9.** Problem 3

10. Problem 4   **11.** Problem 5   **12.** Problem 6

Answer 7: See the figure for solution curves and for the state line.

Answer 9: See the figure for solution curves and for the state line.

Answer 11: See the figure for solution curves and for the state line.

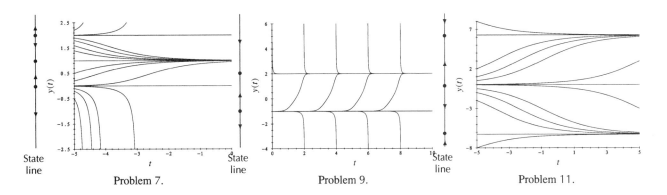

Problem 7.   Problem 9.   Problem 11.

Match the State Line with an ODE. Match each state line with one of the ODEs. Identify each equilibrium point as an attractor (A), a repeller (R), or an attractor/repeller (AR).

13. ⟶ y **A.** $y' = -y$

14. ⟶ y **B.** $y' = y^2(2 - y)$

15. ⟶ y **C.** $y' = y^2$

16. ⟶ y **D.** $y' = -(y - 1)(y - 2)(y - 3)$

17. ⟶ y **E.** $y' = 1$

18. ⟶ y **F.** $y' = -(y - 1)(y - 2)$

Answer 13: The correct answer is E. There are no equilibrium points for this ODE and $y' > 0$.

Answer 15: The correct answer is C. We see that the equilibrium point $y = 0$ is an attractor/repeller because y' is positive for all $y \neq 0$.

Answer 17: The correct answer is B. We see that the equilibrium point at $y = 0$ for ODE B is an attractor/repeller and the equilibrium point at $y = 2$ is an attractor because $y' > 0$ if $y < 0$ or $0 < y < 2$; while $y' < 0$ if $y > 2$.

Given a State Line, Find an ODE. Find a function $f(y)$ of the indicated type such that the autonomous ODE $y' = f(y)$ has a state line like that shown. Note that f is not unique.

19. $f(y)$ a polynomial of degree 2   

20. $f(y)$ a polynomial of degree 4   

21. $f(y)$ a polynomial of degree 3

A R A
-5 0 5 → y

22. $f(y)$ a polynomial of degree 5

AR A R A
-1 0 1 2 → y

23. $f(y)$ a trigonometric function

$\cdots$ $-\pi$ 0 π 2π $\cdots$ → y

Answer 19: $f(y) = -(y+1)(y-1) = 1 - y^2$ is one possible answer, as is $Cf(y)$ for any positive constant C; $f(y) < 0$ if $|y| > 1$ and $f(y) > 0$ if $|y| < 1$.

Answer 21: $f(y) = -y(y+5)(y-5) = -y^3 + 25y$ is one possible answer, as is $Cf(y)$ for any positive constant C; $f(y) > 0$ if $y < -5$ or $0 < y < 5$, while $f(y) < 0$ if $-5 < y < 0$ or $y > 5$.

Answer 23: $f(y) = \sin y$ is one possible answer, as is $Cf(y)$ for any positive constant C; $f(y) > 0$ if $2k\pi < y < (2k+1)\pi$ for any integer k, while $f(y) < 0$ if $(2k+1)\pi < y < 2k\pi$.

Given an ODE, Find Its State Line. Sketch the state line for each ODE. Label each equilibrium point as attractor (A), repeller (R), or attractor/repeller (AR).

24. $y' = y^2(y-1)^3$ **25.** $y' = \cos 2y$ **26.** $y' = 4e^y - 8$ **27.** $y' = y^2 \sin y$

Answer 25:

R A R A
$\cdots -\pi/4$ $\pi/4$ $3\pi/4$ $5\pi/4\cdots$ → y

Since $\cos 2y = 0$ at $y = (2k+1)\pi/4$, k any integer, while $\cos 2y$ changes sign every time y goes through a zero of $\cos 2y$, and since $\cos 2y > 0$ for $-\pi/4 < y < \pi/4$, the stateline is as shown.

Answer 27:

A R A R
$\cdots$ $-\pi$ 0 π 2π $\cdots$ → y

Since $y' = 0$ at $y = k\pi$, k any integer, while y' changes sign if y goes through a zero of $y^2 \sin y$, and since $y^2 \sin y > 0$ for $0 < y < \pi$, the stateline is as shown.

Equilibrium Solutions.

28. Show that the ODE $y' = y - ay^2 - b$, where a and b are positive constants, has equilibrium solutions if and only if $4ab \leq 1$. If equilibrium solutions exist, show that they are positive. [*Hint*: write $y - ay^2 - b$ as $-a(y^2 - y/a + b/a)$ and use the quadratic formula.]

Translation Property. Suppose that $f(y)$ is continuously differentiable on the entire y-axis.

29. Show directly that if $y(t)$ is any solution of the ODE $y' = f(y)$, then so is $z(t) = y(t - a)$, for any constant a. [*Hint*: $dz(t)/dt = y'(t-a) = f(y(t-a))$.]

Answer 29: Let $y(t)$ be a solution of the autonomous ODE $y' = f(y)$ on the interval $t_1 < t < t_2$. Put $z(t) = y(t-a)$, where a is any constant, and note that $z(t)$ lives on the interval $t_1 + a < t < t_2 + a$. Now by the Chain Rule

$$z'(t) = y'(t-a) \cdot 1 = y'(t-a) = f(y(t-a)) = f(z(t)),$$

for all $t_1 + a < t < t_2 + a$, and so $z(t)$ solves the ODE $z' = f(z)$ on the interval $t_1 + a < t < t_2 + a$.

30. Let $y(t)$ be the solution of the IVP $y' = f(y)$, $y(t_0) = y_0$, on some interval about t_0. For any constant a suppose that $z(t)$ is the solution of the IVP $z' = f(z)$, $z(t_0 + a) = y_0$ on some interval about $t_0 + a$. Show that $z(t + a) = y(t)$. [*Hint*: show that $y(t)$ and $z(t)$ solve the same IVP.] Verify this property directly for the IVP $y' = y^2$, $y(t_0) = y_0$.

Using the Translation Property.

31. Consider the IVP $y' = 3y^{2/3}$, $y(t_0) = 0$.

(a) Does the IVP satisfy the conditions of the Existence and Uniqueness Theorem for any value of t_0? Give reasons.

(b) Show that $y_1 = 0$ and $y_2 = t^3$ are solutions of the IVP $y' = 3y^{2/3}$, $y(0) = 0$.

(c) Find three solutions of the IVP $y' = 3y^{2/3}$, $y(t_0) = 0$, $t_0 \neq 0$ [*Hint*: show that $y_3(t)$ defined as 0 for $t \leq 0$, and as t^3, for $t \geq 0$, is a solution of the IVP in **(b)**. Then apply the translation property to get $y_1(t)$, $y_2(t)$, and $y_3(t)$.]

Answer 31:

(a) The rate function $f(y) = 3y^{2/3}$ is continuous for all values of y, including $y = 0$. However, $df/dy = 2y^{-1/3}$ is discontinuous at $y = 0$ (it isn't even defined there). So the conditions of the Existence and Uniqueness Theorem are satisfied everywhere in the ty-plane except along the line $y = 0$ (i.e., on the t-axis). The conditions are not satisfied for any value of t_0 if we stipulate that $y(t_0) = 0$.

(b) If we let $y_1(t) = 0$, all t, then $y_1'(t) = 0$, all t, and so $y_1(t) = 0$ is a solution of the IVP, $y' = 3y^{2/3}$, $y(0) = 0$. Now, suppose $y_2(t) = t^3$. Then $y_2' = 3t^2 = 3y_2^{2/3}$. Since $y_2(0) = 0$, $y_2(t)$ is another solution of the same IVP.

(c) By the translation property since $y = t^3$ is a solution of the ODE, $y' = 3y^{2/3}$, so is $y = (t - t_0)^3$ for any t_0. So the IVP, $y' = 3y^{2/3}$, $y(t_0) = 0$, has solutions $y = 0$ and $y = (t - t_0)^3$. There are other solutions. It can be shown that for any $a \leq t_0$ and $b \geq t_0$, the function

$$y(t) = \begin{cases} (t-a)^3, & t \leq a \\ 0, & a < t \leq b \\ (t-b)^3, & t > b \end{cases}$$

is a solution of this IVP. Note $y(t)$ is continuous for all t since $\lim_{t \to a^-} y(t) = \lim_{t \to b^+} y(t) = 0$. To show that each of these solutions is differentiable at the point $t = a$, proceed as follows:

$$\frac{y(t) - y(a)}{t - a} = \frac{(t-a)^3}{t - a} = (t-a)^2 \to 0 \text{ if } t \to a^-$$

so $\lim_{t \to a^-} y'(t) = 0$; the right-hand derivative $\lim_{t \to a^+} y'(t) = 0$ since $y(t) = 0$, $a < t \leq b$. So $y'(a)$ exists and is 0; similarly for $y'(b)$. We have constructed infinitely many solutions of the IVP.

Modeling Problems: Logistic Growth. Problems 32–35 involve logistic populations.

32. *Logistic Change*   Consider the logistic ODE $y' = (1 - y/20)y$.

    (a) Solve the IVPs when $y(0) = 5, 10, 20, 30$. [*Hint*: see text formula (9).]

    (b) Graph the solutions in **(a)** on the interval $0 \leq t \leq 10$. Highlight the carrying capacity.

    (c) Sketch the state line and identify the type of each equilibrium point (i.e., A, R, or AR).

33. *Logistic Growth*   A colony of bacteria grows according to the logistic law, with a carrying capacity of 5×10^8 individuals and natural growth coefficient $r = 0.01$ day^{-1}. What is the population after 2 days if the initial population is 1×10^8?

    Answer 33:   The logistic model is $y' = r(1 - y/K)y$, where $K = 5 \times 10^8$ and $r = 0.01$/day. The solution is given by formula (9) in the text: $y(t) = y_0 K/(y_0 + (K - y_0)e^{-0.01t})$. Set $t = 2$, $y_0 = 10^8$ and obtain $y(2) \approx 1.016 \times 10^8$ individuals.

34. *A Harvested Logistic Population: Sign Analysis.*   The ODE $y' = 3(1 - y/12)y - 8$ models population changes for a harvested, logistically changing species.

    (a) Find the equilibrium levels, and discuss the fate of the species if $y(0) = 2, 4, 6, 8$, or 10.

    (b) Highlight the equilibrium levels and sketch the solution curve of each IVP in **(a)** on the interval $0 \leq t \leq 5$.

    (c) Sketch the state line and identify the type of each equilibrium point (i.e., A, R, or AR).

35. *Harvested Logistic Population*   An IVP for a logistic population harvested at a constant rate is given by $y' = y(1 - y/10) - 9/10$, $y(0) = y_0$.

    (a) Sketch the equilibrium lines and some representative solution curves.

    (b) Sketch the state line and identify the type of each equilibrium point (i.e., A, R, or AR).

    (c) Find the solution formula for this IVP. [*Hint*: see Example 2.8.5.]

    Answer 35:

    (a) The rate function for the ODE, $y' = y(1 - y/10) - 9/10$, is the quadratic, $-y^2/10 + y - 9/10$, which factors to yield $y' = -(y - 1)(y - 9)/10$. The equilibrium populations are $y = 1$ and $y = 9$

because y' is zero if $y = 1, 9$. From the factorization of the rate function in the IVP we see that the population declines if $y > 9$ or if $0 < y < 1$ because y' is negative in those regions; $y(t)$ increases if $1 < y < 9$.

The population curves in the figure show the expected behavior; note the horizontal lines at the equilibrium population levels of $y = 1$ and $y = 9$.

(b) Repeller at $y = 1$, attractor at $y = 9$:

$$\xleftarrow{\qquad} \overset{\text{R}}{\underset{1}{\bullet}} \xrightarrow{\qquad} \overset{\text{A}}{\underset{9}{\bullet}} \xleftarrow{\qquad} y$$

(c) Separating variables in the ODE and using partial fractions,

$$\frac{1}{y-1} \cdot \frac{1}{y-9} \, dy = -\frac{1}{10} \, dt, \quad y \neq 1, \, 9$$

$$\frac{1}{8} \left(\frac{1}{y-9} - \frac{1}{y-1} \right) dy = -\frac{1}{10} \, dt$$

$$\ln \left| \frac{y-9}{y-1} \right|^{1/8} = -\frac{1}{10} t + k, \quad k \text{ a constant}$$

$$\left| \frac{y-9}{y-1} \right|^{1/8} = e^k e^{-t/10}$$

$$\frac{y-9}{y-1} = C e^{-4t/5}, \quad C = \pm e^{8k}$$

Solving for y and substituting initial conditions $y(0) = y_0$, we obtain

$$y = \frac{9 - C e^{-4t/5}}{1 - C e^{-4t/5}}, \quad C = \frac{y_0 - 9}{y_0 - 1}, \quad y_0 \neq 1$$

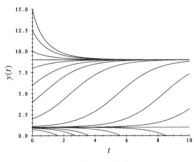

Problem 35(a).

Classification of Equilibrium Solutions.

☞ The notation $f_y(y_0)$ is shorthand for the partial derivative $\partial f / \partial y$ evaluated at y_0. **36.** Let $f(y)$ be continuously differentiable on the y-axis, and let y_0 be an equilibrium point of the autonomous ODE $y' = f(y)$. Explain why it is reasonable to classify y_0 as an attractor or a repeller according to whether $f_y(y_0)$ is negative or positive. Demonstrate the validity of this conjecture for the ODE in Figure 2.8.2. [*Hint:* see Problem 24 in Section 2.4.]

Background: Sign Analysis and Long-Term Behavior

Here is a proof of part of Theorem 2.8.1. Suppose that $y_1 < y_2$ are two consecutive zeros of $f(y)$; that is, $f(y_1) = f(y_2) = 0$, but $f(y)$ does not vanish in the interval $y_1 < y < y_2$. Then since $f(y)$ is continuous, $f(y)$ must be of one sign in the strip $y_1 < y < y_2$, say positive. Note that the lines $y = y_1$ and $y = y_2$ are solution curves for the ODE. Now, because f is positive, the solution that passes through the point (t_0, y_0), $y_1 < y_0 < y_2$, must be increasing as t increases, if the solution curve remains in the strip $y_1 < y < y_2$. To leave this strip as t increases, the solution curve would have to cross the equilibrium solution curve $y = y_2$, but the Uniqueness Theorem implies that this cannot happen. The Extension Theorem shows that our solution curve is defined for all $t \geq t_0$. Similarly, we see that for backward time our solution curve falls for decreasing t, but it never crosses the equilibrium solution $y = y_1$. The Extension Theorem shows again that the solution is defined for all $t \leq t_0$. If follows that

$$\lim_{t \to +\infty} y(t) = b \leq y_2 \quad \text{and} \quad \lim_{t \to -\infty} y(t) = a \geq y_1$$

both exist. We now show that $a = y_1$ and $b = y_2$. By shifting this solution curve in t we can completely fill up the strip $a < y < b$ with the solution curves of the ODE. Now suppose that $b < y_2$, contrary to our assertion. Then, as noted above, the solution curve defined by the IVP $y' = f(y)$, $y(t_0) = b$, can be extended to the entire t-axis. Extending the solution forward, the solution must cross the line $y = b$ because $f(b) > 0$, and so must intersect a solution curve in the strip $a < y < b$, in contradiction to the Uniqueness Theorem. It must be that $b = y_2$. Similarly, it can be shown that $a = y_1$.

2.9 Bifurcations: A Harvested Logistic Model

Comments:

Information about how the solution of an IVP changes as the data change is critical for a real understanding of the IVP, particularly if the IVP models some real phenomenon. The idea of sensitivity is fairly easy to grasp, even though we never give a precise definition. The informal explanation of the solution of an IVP being a continuous function of all the data is also easy to understand, but it must be emphasized that the time interval about t_0 usually must be small. Note that small changes in data may evolve over time into huge variations. This is good background for a later discussion on the Lorenz system (Section 9.4), where we speak of chaos.

Bifurcations are new to the introductory ODE syllabus, but they are an important part of any course that introduces the ideas of sensitivity to data changes and is serious about modeling physical systems. Although most changes in the data do not lead to drastic changes in solution behavior (at least not over a short span of time), there are those occasional data ranges where something completely new occurs, although the consequences may only become evident with time's advance. The term "bifurcation" is now applied to this kind of event, but it is not a precisely defined term. We introduce this idea via examples and suggestive definitions. We introduce the saddle-node bifurcation in the context of a harvested/restocked and logistically changing population. We do the important task of rescaling so that the only remaining parameter in the model ODE is the harvesting/restocking rate. This is a good example to see just why rescaling is often done before computing. In reading this material, you might think of four stages after the scaling has been done. First, find the equation of the equilibrium solutions in terms of the parameter c, and plot and interpret the graphs in the cy-plane (the bifurcation diagram), clearly locating the bifurcation point. Next, show graphs of solutions before, at, and beyond bifurcation. Then replot the bifurcation diagram, but this time introducing solid or dashed arcs depending upon whether an equilibrium solution is an attractor or a repellor. Finally, plot the rate function $f = f(y, c)$ in a yf-plane for various values of c below, at, and above the bifurcation value, and show that at the bifurcation value c^* the graph of $f = f(y, c^*)$ is tangent to $f = 0$. You may want to complete these stages in a different order.

PROBLEMS

Saddle-Node Bifurcations. Explain why there is a saddle-node bifurcation at some value of the parameter c for each ODE below. In each case sketch the saddle-node bifurcation diagram, using solid arcs for attracting equilibria and dashed arcs for repelling equilibria. [*Hint*: see the discussion about Figure 2.9.4.]

   1. $y' = c - y^2$ **2.** $y' = c - 2y + y^2$ **3.** $y' = c + 2y + y^2$

Answer 1: Since $f(y, c) = c - y^2$, if $c > 0$ the equilibrium solutions are at $y_{1,2} = \mp\sqrt{c}$. If $c = 0$, $y = 0$ is the only equilibrium solution, while if $c < 0$ there are no real equilibrium solutions since $\sqrt{c}$ is complex. So $c = 0$ is the critical value. As c increases through 0, the complex conjugate numbers $\pm\sqrt{c}$ move toward each other in the complex plane, meet when $c = 0$, and split apart into a pair of real equilibrium solutions. The solution $y_2 = \sqrt{c}$, $c > 0$, is an attractor since if $y < \sqrt{c}$ but near $\sqrt{c}$, then $f = c - y^2$ is positive and solutions of $y' = c - y^2$ rise up toward $y = \sqrt{c}$, while if $y > \sqrt{c}$, then $c - y^2$ is negative and solutions fall toward the equilibrium. A similar argument shows that $y_1 = -\sqrt{c}$ is a repeller. See the figure for the bifurcation diagram.

    NOTE: You could also use the following method to deal with such kind of problems:

1. First find the critical point c; for this specific problem c = 0.
2. Plot the ODE for three values of c: $c < 0$, $c = 0$, and for $c > 0$.
3. From the 3 plots you can see the equilibrium points and their attractor or repeller properties.
4. To draw the bifurcation diagram, plot the graph for $y = \pm\sqrt{c}$. Then, based on point 3. above, decide whether the equilibrium attracts or repels.

Answer 3: The zeros of $f(y, c) = c + 2y + y^2$ are $y_{1,2} = -1 \mp \sqrt{1 - c}$, and are real and distinct for $c < 1$, real and identical for $c = 1$, and complex conjugates for $c > 1$. For large y, $f(y, c)$ is positive, but for $y < y_2$ and near y_2, $f(y, c)$ is negative. So $y = y_2 = -1 + \sqrt{1 - c}$ is a repeller. A similar argument shows that $y = y_1 = -1 - \sqrt{1 - c}$ is an attractor. See the figure for the bifurcation diagram.

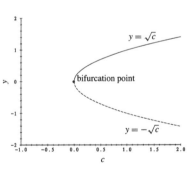

Problem 1.

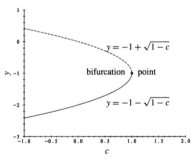

Problem 3.

Solution Curves of Saddle-Node Bifurcations. Plot state lines and several solution curves of the ODEs in Problems 1, 2, 3 for each of several values of c above, at, and below the saddle-node bifurcation value. Describe the behavior of the curves in each case as t increases.

   4. Problem 1   **5.** Problem 2   **6.** Problem 3

Answer 5: Graphs 1, 2, and 3 show, respectively, the rising and falling solution curves toward the attracting equilibrium $y = 0$ (where $c = 0$) and the falling and rising curves away from the repelling equilibrium $y = 2$, the curves rising toward and away from the single equilibrium at $y = 1$ ($c = 1$), and the rising curves where there is no real equilibrium at all ($c = 2$).

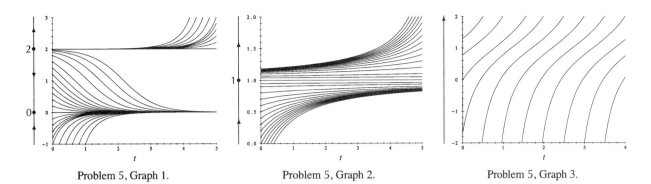

Problem 5, Graph 1.   Problem 5, Graph 2.   Problem 5, Graph 3.

Pitchfork Bifurcations. Explain why there is a pitchfork bifurcation at some value of the parameter c for each ODE. In each case sketch the pitchfork bifurcation diagram, using solid arcs for attracting equilibria, dashed arcs for repelling equilibria. [*Hint*: see the discussion about Figure 2.9.8.]

7. $y' = (c - 2y^2)y$ **8.** $y' = -(c + y^2)y$ **9.** $y' = (c - y^4)y$

Answer 7: The zeros of $f = (c - 2y^2)y$ are $y_1 = 0$, $y_{2,3} = \mp\sqrt{c/2}$, and the latter two are real for $c > 0$, merge with y_1 for $c = 0$, and are complex for $c < 0$. For $c > 0$, y_2 and y_3 are attractors since $f(y, c)$ is negative for y near y_2 [y_3] if $y > y_2$ [$y > y_3$] and $f(y, c)$ is positive if $y < y_2$ [$y < y_3$]. For $c > 0$, y_1 is a repeller since $f(y, c)$ is positive for y near 0, $y > 0$, and $f(y, c)$ is negative for $y < 0$. For $c < 0$, y_1 is an attractor since $f(y, c)$ is negative for $y > 0$ and $f(y, c)$ is positive for $y < 0$. See the figure.

Answer 9: The five zeros of $f = (c - y^4)y$ are $y_1 = 0$, $y_{2,3} = \mp c^{1/4}$, and $y_{4,5} = \mp i|c|^{1/4}$. The only zeros of interest here are y_1, y_2, y_3 because $y_{4,5}$ are complex, except if $c = 0$ in which case all five coincide at $y = 0$. The zeros $y_{2,3}$ are real for $c > 0$ where they are both attractors because $f(y, c)$ is negative for y near y_2 or y_3 with $y > y_2$ or $y > y_3$, but $f(y, c)$ is positive for $y < y_2$ or $y < y_3$. The zero y_1 is a repeller [attractor] for $c > 0$ [$c < 0$] because $f(y, c)$ is positive [negative] for y near 0, $y > 0$ and negative [positive] for $y < 0$. See Fig. 3(c).

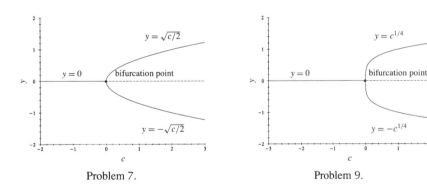

Problem 7.   Problem 9.

Solution Curves of Pitchfork Bifurcations. Plot state lines and several solution curves of the ODEs in Problems 7, 8, 9 for values of c above, at, and below the pitchfork bifurcation value. Describe the behavior of the curves in each case as t increases.

10. Problem 7   **11.** Problem 8   **12.** Problem 9

Answer 11: See Graphs 1, 2, and 3 for solution curves for $c = -1$, $c = 0$ (the bifurcation value), and

$c = 1$, respectively. At $c = -1$ there are three equilibrium solutions, $y_1 = 0$ (a repeller) and $y_{2,3} = \mp 1$ (attractors). At $c = 1$ there is a single attractor ($y = 0$).

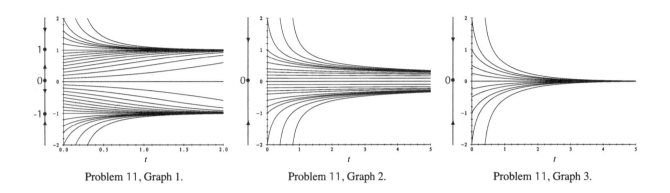

Problem 11, Graph 1.   Problem 11, Graph 2.   Problem 11, Graph 3.

☞ Yet another
kind of bifurcation.

Transcritical Bifurcations. In a *transcritical bifurcation*, as the parameter c in the rate function for the ODE $y' = f(y, c)$ is changed, a pair of equilibrium solutions, one an attractor and the other a repeller, merge and then separate, exchanging their attracting or repelling properties in the process. Explain why each of the following ODEs has a transcritical bifurcation. Draw a bifurcation diagram (solid arcs for attracting equilibria, dashed arcs for repelling equilibria). Then plot state lines and several solution curves for the ODEs for values of c above, at, and below the bifurcation values.

13. $y' = cy - y^2$ **14.** $y' = cy + 10y^2$

Answer 13: Here $f(y, c) = cy - y^2 = y(c - y)$, so $y_1 = 0$ and $y = c$ are the two equilibrium curves and the transcritical bifurcation occurs for $c = 0$. See Graphs 1–3 for solution curves for $c = -1$ ($y_1 = 0$ is an attractor and $y_2 = -1$ is a repeller), $c = 0$ at the transcritical bifurcation value ($y = 0$ attracts on one side and repels on the other), $c = 1$ ($y_1 = 0$ is a repeller and $y_2 = 1$ is an attractor). Graph 4 shows the bifurcation diagram in the cy-plane: the two lines are $y = 0$, $y = c$. Observe how the attractor switches to a repeller, and vice versa as c crosses the bifurcation value 0.

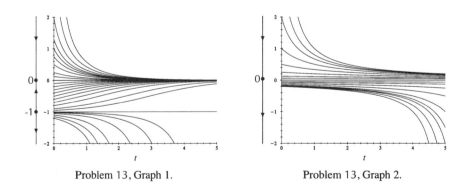

Problem 13, Graph 1.   Problem 13, Graph 2.

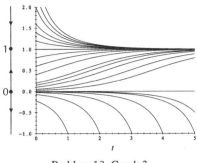

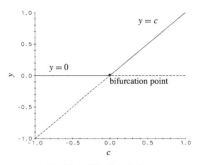

Problem 13, Graph 3.   Problem 13, Graph 4.

NOTE: The critical point is $c = 0$, since:
for $c = 0$, $y_1 = y_2 = 0$, i.e. we have only 1 equilibrium point.
for $c < 0$, we have 2 equilibrium points.
for $c > 0$, we have 2 equilibrium points.

From the plots we can see that:
For $c = 1$, $y = 0$ is repeller (R), and $y = 1$ is attractor (A).
For $c = -1$, $y = 0$ is (A), and $y = 1$ is (R).
Since the equilibrium points exchange their attracting or repeller properties, it follows that the ODE has a transcritical bifurcation.

Modeling Problems.

15. *Too Late to Save a Population from Extinction?*   Referring to the text and the discussion of the harvested population model ODE, $P' = r(1 - P/K)P - H$, $H > rK/4$, explain why the following is true: if the value of P is near 0, then restricting the harvest rate to slightly below the critical harvesting rate $rK/4$ will not save the population from extinction. What if you ban harvesting altogether in this case? Can you save the species?

Answer 15:   If the harvesting/restocking rate is, say, Q_1, slightly below the tangent bifurcation value of $rK/4$, then the rate function has two zeros, $P_{1,2} = K/2 \mp (K/2)(1 - 4Q_1/Kr)^{1/2}$, which are very close together. This means that in the tP-plane the two equilibrium solutions $y = P_1$, $y = P_2$, where $P_1 < P_2$, bound a very narrow band, with $y = P_1$ a repeller and $y = P_2$ an attractor. So if the population level is below the band, extinction is inevitable. Even if the level lies within the band, a small disturbance may send it below. Banning harvesting altogether would help in this situation since the model for subsequent times would be logistic and the population levels would then rise toward equilibrium. But any time population levels get too low, random events may have drastic effects and it is doubtful that even an outright ban on harvesting would have much effect.

16. *How Many Hunting Licenses Should Be Issued?*   The duck population near a hunting lodge is modeled by the ODE, $P' = (1 - P/1000)P - H$, where H is the harvesting rate.

(a) How many licenses can be issued per year so that the duck population has a chance of survival? Each hunter is allowed to shoot up to 20 ducks per year.

(b) Suppose N licenses are issued, where N is less than the maximal number found in **(a)**. What values of the initial duck population lead to total extinction of the species? Explain.

Spotlight on Approximate Numerical Solutions

Comments:

How does a numerical solver approximate the solution of the first-order IVP $y' = f(t, y)$, $y(t_0) = y_0$? The underlying principle of all solvers is simple enough: approximate the rate function at a point, multiply by a time increment, and repeat. The differences in the various methods have to do with how they approximate the rate function and how the step size is adjusted. In addition, many solvers have adjustable controls on allowable absolute and relative errors, minimum and maximum internal step size (for adaptive methods), number of points plotted, and so on. This section is a brief introduction to a complex and fascinating area of numerical analysis—the search for the best ODE solver. Of course, no such thing exists since for every numerical solver yet invented there is some IVP that breaks it. Nevertheless, there are some basic notions that can be conveyed about solvers. We focus on one-step algorithms such as the Euler Method and Runge-Kutta Method, and use examples to illustrate what they can do. Actual numerical estimates are given below only for the Euler and RK4 methods. See the background material following the Problems for more on round-off error and for the extension of the Euler and RK4 methods to planar systems.

PROBLEMS

🖳 **Comparison of Numerical Methods.**

1. Use the indicated method to estimate $y(1)$ if $y' = -y$, $y(0) = 1$, $h = 0.1$. Plot the broken-line approximation.

    (a) Euler's Method   **(b)** RK4

    Answer 1:

    (a) Euler's algorithm, $y_n = y_{n-1} - hy_{n-1}$, $1 \le n \le 10$, gives $y(1) \approx y_{10} = 0.348678$ (first six decimals). See the figure.

    (b) Fourth-order Runge-Kutta, $y_n = y_{n-1} + h(k_1 + 2k_2 + 3k_3 + k_4)/6$, where $1 \le n \le 10$, $k_1 = -y_{n-1}$, $k_2 = -(y_{n-1} + hk_1/2)$, $k_3 = -(y_{n-1} + hk_2/2)$, $k_4 = -(y_{n-1} + hk_3)$, gives $y(1) \approx y_{10} = 0.367879$ (first six decimals). See the figure for **(a)**.

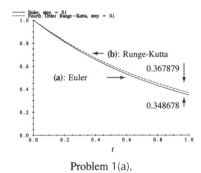

Problem 1(a).

2. Use the indicated method to estimate $y(1)$ if $y' = -y$, $y(0) = 1$, $h = 0.01, 0.001, 0.0001$, and plot the approximating polygons. Zoom in on part of plot if necessary to separate the polygons.

    (a) Euler's Method   **(b)** RK4

3. Use the indicated method to estimate $y(1)$ for the IVP $y' = -y^3 + t^2$, $y(0) = 0$, using the step sizes $h = 0.1, 0.01, 0.001$. Plot the broken-line approximation.

    (a) Euler's Method   **(b)** RK4

    Answer 3:

    (a) The first six decimals of the respective Euler approximations to $y(1)$ for the three values of h are 0.283786, 0.325058, and 0.329236. See the figure for the Euler polygons.

(b) The first six decimals of the respective RK4 approximations to $y(1)$ are 0.329699, 0.329700, and 0.329700. The RK4 polygons are very close together (Graph 1), but zooming in on the end segment of the polygons separates the $h = 0.001$ polygon from the other two (Graph 2).

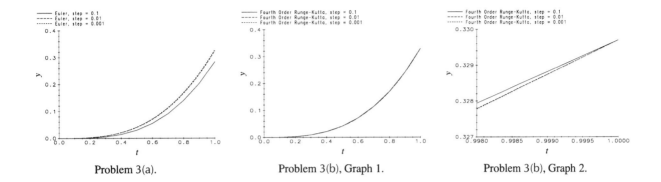

| Problem 3(a). | Problem 3(b), Graph 1. | Problem 3(b), Graph 2. |

4. *Convergence of Euler Approximations*   For each of the following IVPs, show that the Euler approximation $y_N(T) \to y(T)$ as $N \to \infty$ ($T > 0$ is fixed).

   (a) $y' = 2y$, $y(0) = 1$ **(b)** $y' = -y$, $y(0) = 1$

5. *RK4 and Simpson's Formula*   Simpson's formula for approximating an integral is

$$\int_a^b f(t)\,dt \approx \frac{b-a}{6}\left[f(a) + 4f\left(\frac{a+b}{2}\right) + f(b) \right]$$

Show that RK4 for the IVP $y' = f(t)$, $y(t_0) = y_0$, gives Simpson's formula at each step.

Answer 5:   We shall interpret a as t_j and b as t_{j+1} in an RK4 process for solving $y' = f(t)$, $y(t_0) = y_0$ over an interval containing a, b, $a < b$. Suppose y_j has been calculated as an estimate for $y(t_0 + jh)$, where $y'(t) = f(t)$, $y(t_0) = y_0$, $t_0 \le t \le T$. Then $y_{j+1} = y_j + \int_{t_j}^{t_{j+1}} f(t)dt \approx y_j + (h/6)[f(t_j) + 4f(t_j + h/2) + f(t_{j+1})]$ by Simpson's formula, where $h = t_{j+1} - t_j$. Fourth-order Runge-Kutta gives $y_{j+1} = y_j + (h/6)[k_1 + 2k_2 + 2k_3 + k_4]$, where $k_1 = f(t_j)$, $k_2 = f(t_j + h/2) = k_3$, $k_4 = f(t_j + h) = f(t_{j+1})$. So, Simpson's method and fourth-order Runge-Kutta yield identical values for y_{j+1}, once y_j has been calculated.

Background: Effect of Round-off Error

Suppose Euler's Method is used to estimate $y(3)$, where $y(t)$ solves the IVP,

$$y' = y, \qquad y(0) = 0$$

The figure below shows the error in approximating $y(3)$ as a function of the total number of steps.

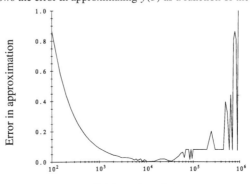

Total number of steps.

Background: Euler's Method and RK4 for Systems

Numerical methods for approximating solutions of a first-order ODE can be extended to approximate solutions of a system of first-order ODEs. We extend Euler's Method and RK4 to the IVP

$$x' = f(t, x, y), \qquad x(t_0) = x_0$$
$$y' = g(t, x, y), \qquad y(t_0) = y_0 \tag{i}$$

Euler's Method for IVP (i) has a geometric interpretation. Calculate $f(t_0, x_0, y_0)$ and $g(t_0, x_0, y_0)$ and move away from the point (x_0, y_0) in the xy-plane along the field vector $(f(t_0, x_0, y_0), g(t_0, x_0, y_0))$ of the system in IVP (i) as time runs from t_0 to $t_0 + h$ to obtain an Euler approximation (x_1, y_1) to the point $(x(t_0 + h), y(t_0 + h))$, where $x = x(t), y = y(t)$ is the exact solution of (i).

The nth step of the *Euler algorithm* for IVP (i) is given by

$$t_n = t_{n-1} + h$$
$$x_n = x_{n-1} + hf(t_{n-1}, x_{n-1}, y_{n-1})$$
$$y_n = y_{n-1} + hg(t_{n-1}, x_{n-1}, y_{n-1})$$

As before, this is a first-order algorithm.

The *RK4* algorithm for IVP (i) is given by

$$t_n = t_{n-1} + h$$
$$x_n = x_{n-1} + \frac{h}{6}(k_1 + 2k_2 + 2k_3 + k_4)$$
$$y_n = y_{n-1} + \frac{h}{6}(p_1 + 2p_2 + 2p_3 + p_4)$$

where h is the fixed step size and

$$k_1 = f(t_{n-1}, x_{n-1}, y_{n-1}) \qquad\qquad p_1 = g(t_{n-1}, x_{n-1}, y_{n-1})$$

$$k_2 = f\left(t_{n-1} + \frac{h}{2}, x_{n-1} + \frac{h}{2}k_1, y_{n-1} + \frac{h}{2}p_1\right) \qquad p_2 = g\left(t_{n-1} + \frac{h}{2}, x_{n-1} + \frac{h}{2}k_1, y_{n-1} + \frac{h}{2}p_1\right)$$

$$k_3 = f\left(t_{n-1} + \frac{h}{2}, x_{n-1} + \frac{h}{2}k_2, y_{n-1} + \frac{h}{2}p_2\right) \qquad p_3 = g\left(t_{n-1} + \frac{h}{2}, x_{n-1} + \frac{h}{2}k_2, y_{n-1} + \frac{h}{2}p_2\right)$$

$$k_4 = f(t_{n-1} + h, x_{n-1} + hk_3, y_{n-1} + hp_3) \qquad p_4 = g(t_{n-1} + h, x_{n-1} + hk_3, y_{n-1} + hp_3)$$

As the abbreviation suggests, RK4 is a fourth-order method. Euler's Method and RK4 can be extended to systems with any number of first-order ODEs.

Spotlight on Computer Implementation

Comments:

Given any numerical algorithm for solving ODEs, there will be some IVP for which the algorithm works poorly, if at all. This is a fact of numerical algorithms, so we address here a few of the things that can go wrong when using a solver. Problems with numerical algorithms show how important it is not to place unquestioning faith in their reliability. See also the tips before the answer to Problem 4.

PROBLEMS

Warning! Most of the problems here have to do with difficulties commonly encountered when using a numerical solver to approximate the solution of an IVP. The specific solve-time intervals and step sizes given in the problems are based on experience with one solver. Other solvers may behave differently, so be prepared to change the intervals or step sizes if necessary. It is even possible that your solver will not generate any bad approximate solutions.

Reversibility of an Approximation Method. If a numerical solver is used to solve an IVP forward in time, and then backward, it may return a number that is nowhere near the initial value. The following problems look at this unhappy fact of computational life.

☞ Dynamical systems are reversible, but solutions from a numerical solver may not be!

1. Assume that $f(t, y)$ satisfies the conditions of the Existence and Uniqueness Theorem and that $y = y(t)$, $t_0 \le t \le t_1$, solves the forward IVP, $y' = f(t, y)$, $y(t_0) = y_0$. Show that the same function $y(t)$ defines a solution $z = y(t)$ of the *backward* IVP $z' = f(t, z)$, $z(t_1) = y(t_1)$, $t_0 \le t \le t_1$, and that therefore $z(t_0)$ must be y_0.

   Answer 1: Since the solutions $z(t)$ and $y(t)$ coincide at $t = t_1$, the Existence and Uniqueness Theorem implies that $z(t) = y(t)$ for all t where these solutions are defined and, subsequently, $z(t_0) = y(t_0) = y_0$.

2. *A Good Solver May Do Bad Things* Use RK4 with step size $h = 0.1$ to find and plot an approximate solution of the IVP $y' = 3y \sin y - t$, $y(0) = 0.4$, over the interval $0 \le t \le 8$. Now, with the value for $y(8)$ just found, use RK4 to solve the *backward* IVP $z' = 3z \sin z - t$, $z(8) = y(8)$, from t-initial $= 8$ to t-final $= 0$. Plot the solution. How close is $z(0)$ to 0.4? Explain any significant difference.

3. *More on Reversibility* In Problems 1 and 2, if $f(t, y)$ satisfies the conditions of the Existence and Uniqueness Theorem, then the IVP, $y' = f(t, y)$, $y(t_0) = y_0$, $t_0 \le t \le t_1$, is reversible in theory, but experience with numerical solvers shows that actual practice with the nonlinear ODE of Problem 2 is different. Here, you will experiment with a linear IVP and show that the practical difficulties of running an IVP backward can sometimes be resolved (but only in part) by shortening the time step.

   (a) Find all solutions of the driven linear ODE $y' + 2y = \cos t$. Show that as $t \to +\infty$, all solutions approach the particular solution $y_d = 0.4 \cos t + 0.2 \sin t$.

   (b) Plot solution curves in the rectangle $0 \le t \le 20$, $|y| \le 1.5$ using as initial points $(0, 0.4)$, $(0, \pm 1.5)$, and several other points on the top and bottom sides of the rectangle. Observe that solutions converge to the solution $y_d(t)$ as t increases.

   (c) Use RK4 with step size $h = 0.1$ to solve the IVP with $y(0) = 0.4$ and plot an approximation to $y_d(t)$ for $0 \le t \le 20$.

☞ Good solver, bad behavior.

   (d) Using the value for $y_d(20)$ found in part **(c)**, solve the IVP $z' + 2z = \cos t$, $z(20) = y_d(20)$, backward from t-initial $= 20$ to t-final $= 0$, again using RK4 with $h = 0.1$. What happens? How would you explain the difficulty? [*Hint*: exact solutions of the ODE converge on $y_d(t)$ in forward time but diverge from $y_d(t)$ in backward time. The approximate nature of computed solutions suggests that the computed point $(20, y_d(20))$ is not quite on the solution curve of $y = y_d(t)$.]

   (e) Repeat **(d)**, but with $h = 0.01, 0.001$. Any improvement?

   Answer 3:

   (a) To solve $y' + 2y = \cos t$, use the integrating factor e^{2t} and a table of integrals (or integration by parts). We see that $(ye^{2t})' = e^{2t} \cos t$. Integrating, $ye^{2t} = \int e^{2t} \cos t \, dt + C$, so $y = Ce^{-2t} + 0.4 \cos t + 0.2 \sin t$, where C is an arbitrary constant. As $t \to +\infty$, $y(t) \to y_d(t) = 0.4 \cos t + 0.2 \sin t$.

   (b) See the figure for solution curves with initial points $(0, \pm 1.5)$, $(0, 0.4)$, $(2.5, 1.5)$, $(5, -1.5)$, $(7.5, 1.5)$, $(10, -1.5)$, $(12.5, 1.5)$, $(15, -1.5)$, and $(17.5, 1.5)$. As t increases, all of these solution curves converge to the graph of the particular solution y_p found in part **(a)**.

   (c) The figure shows the forward solution curve of $y' + 2y = \cos t$, $y(0) = 0.4$ as approximated by RK4 with step size $h = 0.1$; the computed approximation to $y(20)$ is $0.34582020\ldots$.

   (d) The lower curve in the figure shows what happens when the solver tries to go backward from the computed approximation to $y(20)$ with a step size $h = 0.1$. The approximation of $y(20)$ is of course not exact, so when we solve backwards from this approximation we get a very different solution curve.

(e) See Graphs 1 and 2, corresponding to $h = 0.01, 0.001$. The problem with going backward in time is that the forward solutions converge very rapidly to the particular solution $y_d = 0.4 \cos t + 0.2 \sin t$, but numerical solvers can only give approximate answers. Thus, $y(20) \approx y_d(20)$ can only be computed approximately, no matter how small a step size is used. In effect, when the backward solutions are attempted, they start at the different computed approximations corresponding to 0.01 and 0.001. While solving backwards, small errors occur. Due to sensitivity, these cause the failure to accurately solve backwards. The computed values of the approximations at $t = 20$ are printed on the graphs.

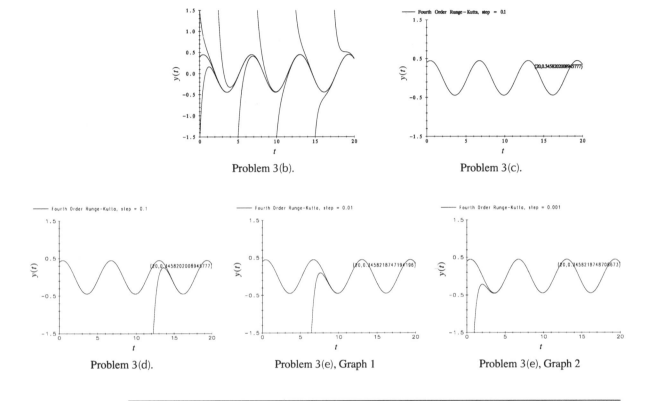

Problem 3(b).   Problem 3(c).

Problem 3(d).   Problem 3(e), Graph 1   Problem 3(e), Graph 2

Tips on using numerical solvers

In Section 2.7 we talked about extending a solution by taking the final values of the state variables and using them as initial data for the extended solution. There are at least two reasons why this procedure is not recommended when using numerical solvers:

- The solver's user interface may report numerical data with less precision than the solver itself uses for calculations (i.e., the solver may report and accept data with 4 digits of precision but use 15 digits of precision internally. Look at the ODE $y' = 3y \sin y + t$ some of whose solution curves appear in Figure 2.4.3. Imagine that $t_{final} = -6$ and that you wish to extend your solution forward in time. Can you see the trouble you will have? Problems 2, 3, and 11 address similar difficulties with other ODEs.

- Some solvers are adaptive in that they make decisions that affect the solver engine at every time step. These decisions are based on the recent history of the solution process. This is especially true of solvers based on multistep methods like the one used to produce the figures in this text. If the solution process is restarted at one point, then the history is lost and the solver is left to its own devices on how to start the solution process. Should this happen at a sensitive solution point like the one mentioned above, then some surprises may be in store for the user.

Investigating Approximation Methods.

    4. *High Rates: Finite Escape Times*   Consider the IVP $y' = y^3$, $y(0) = 1$.

    (a) Plot an Euler Solution of the IVP in the rectangle $0 \le t \le 1/2$, $0 \le y \le 20$. Use $h = 0.05$.

    (b) Use separation of variables to find a formula for the maximally extended solution of the IVP. On what t-interval is the solution defined?

    (c) Plot the exact solution found in **(b)** and the Euler Solution found in **(a)** in the same rectangle. Explain what you see near $t = 1/2$.

    5. *Long-Term Behavior*   The problems below show how shortening the step size may affect the long-term behavior of approximate solutions.

    (a) Use Euler's Method with step size $h = 0.1$ to plot an approximate solution of the IVP $y' = 1 + ty \cos y$, $y(0) = 1$, $0 \le t \le 20$. Repeat with $h = 0.01$. Describe what you see. [Our solver generated wild graphs like those shown in the figure. Yours may do something else. If you don't see any obviously bad behavior, try extending the solve-time interval beyond $t = 30$.]

    (b) Repeat **(a)** using RK4.

Answer 5:

(a) Using Euler's Method for the IVP, $y' = 1 + ty \cos y$, $y(0) = 1$, we have that $y_n = y_{n-1} + 0.1(1 + t_{n-1}y_{n-1} \cos y_{n-1})$, $n = 1, 2, \ldots, 300$, with $y_0 = 1$. See the figure. The Euler Solution (solid) for $h = 0.1$ becomes unstable at $t \approx 20$. The Euler Solution seems to be all right if h is shortened by an order of magnitude to $h = 0.01$ (dashed), but you should see what happens on an interval such as $0 \le t \le 50$.

(b) Use RK4 to obtain that $y_n = y_{n-1} + (k_1 + 2k_2 + 2k_3 + k_4)/60$ where $k_1 = 1 + t_{n-1}y_{n-1} \cos y_{n-1}$, $k_2 = 1 + (t_{n-1} + 0.05)(y_{n-1} + 0.05k_1) \cos(y_{n-1} + 0.05k_1)$, $k_3 = 1 + (t_{n-1} + 0.05k_2)(y_{n-1} + 0.05k_2) \cos(y_{n-1} + 0.05k_2)$ and $k_4 = 1 + t_n(y_{n-1} + 0.1k_3) \cos(y_{n-1} + 0.1k_3)$. See the figure. The RK4 solution goes wild if $h = 0.1$ (solid), but seems to behave nicely if $h = 0.01$ (dashed). You see the hazards of using one-step methods with a large step size over too long a span of time.

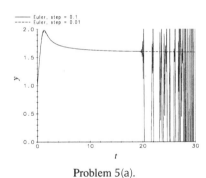

Problem 5(a).

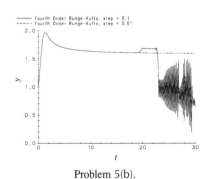

Problem 5(b).

    6. *Qualitative Behavior of Numerical Approximations*   The problems below show that some approximate solutions do not always describe the qualitative properties of exact solutions very accurately.

    (a) Use separation of variables to solve the IVP $y' = (1 - y)y$, $y(0) = y_0$. What happens to the solution as $t \to +\infty$ if $y_0 > 0$? Use a grapher to plot this solution over the interval $0 \le t \le 8$ when $y_0 = 2$. Then apply your best numerical solver to the same IVP. How does the approximate solution curve compare to the solution your solver drew?

(b) Now, solve the IVP in **(a)** using Euler's Method with $h = 0.75$ and $y_0 = 2$, and plot the Euler Solution on $0 \leq t \leq 8$. Does the Euler Solution curve behave qualitatively the same as the solution in **(a)**? Explain any differences.

(c) Repeat **(b)**, but with $h = 1.5$ and $y_0 = 1.4$.

(d) Repeat **(b)**, but with $h = 2.5$, $y_0 = 1.3$, and $0 \leq t \leq 50$.

7. *Sign Analysis and Approximate Solutions* Consider the IVP $y' = -y(1-y)^2$, $y(0) = 1.5$.

(a) Show by sign analysis (Section 2.8) that solution curves above the equilibrium line $y = 1$ fall toward the line with increasing time, while solution curves below the line fall away.

(b) Graph the approximate solutions given by Euler's method with step size $h = 0.1, 0.5, 1, 1.5$, and 2. Comment on any noteworthy behavior.

Answer 7:

(a) The rate function is $-y(1-y)^2$. Since $(1-y)^2$ is positive for $y \neq 1$, y' is negative for $y > 0$, so for $y_0 > 1$, solution curves of the ODE approach the equilibrium solution $y = 1$ from above, whereas for $y_0 < 1$ solutions diverge from the equilibrium $y = 1$ and fall toward the equilibrium $y = 0$. So, $y = 1$ is an attractor/repeller, attracting solutions from above and repelling solutions from below.

(b) From top to bottom the Euler Solutions in the figure correspond to $h = 0.1, 0.5, 1, 1.5$, and 2. The $h = 0.1$ curve is almost identical to the actual solution curve. For $h = 0.1, 0.5$, and 1, the Euler Solution approaches the correct equilibrium solution $y = 1$ in the limit because no segment of the Euler Solution ever dips below $y = 1$. But for $h = 1.5$ and 2, the Euler Solution overshoots the attractor/repeller equilibrium $y = 1$, drops into the region $0 < y < 1$, where y' is negative, and heads for the stable equilibrium $y = 0$. Note how the Euler Solution for $h = 2$ oscillates about $y = 0$. In a situation like this, a good adaptive solver would reduce the step size as the solution curve approaches $y = 1$.

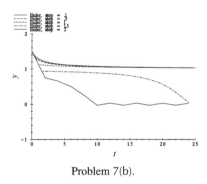

Problem 7(b).

8. *Sampling Rates and Aliasing* All the solutions of the autonomous linear system of ODEs, $x' = y$, $y' = -4x$, are given by the formulas $x = A\sin(2t+\phi)$, $y = 2A\cos(2t+\phi)$, where A and ϕ are arbitrary constants. In this problem we show that if we sample a computed solution only at relatively long time intervals, the resulting graphs may be quite misleading.

(a) Plot the tx- and ty- graphs and the xy-orbits of the solutions with initial points $x_0 = 0$, $y_0 = 1$, and then with $x_0 = 0$, $y_0 = 2$. Describe these graphs.

(b) Use RK4 for systems to solve the IVP $x' = y$, $x(0) = 0$, $y' = -4x$, $y(0) = 1$, over the time interval $0 \leq t \leq 100$ using step size $h = 0.1$. Plot $x(t)$ in the tx-plane.

(c) Use the step size $h = 0.1$ of **(b)** and plot only every fifth computed point of $x(t)$. Repeat, but plot only every tenth point, and then every twentieth point. Explain the differences in these plots from the plot in **(b)**. Each of these graphs is an *alias* for the true graph. An alias disguises the identity of somebody or something, so that's why we use the term here.

Convergent Euler Sequences. Consider the logistic IVP $y' = r(1 - y)y$, $y(0) = y_0 > 0$, where r is a positive constant. The Euler iterates are generated by $y_n = y_{n-1} + rh(1 - y_{n-1})$, $n = 1, 2, \ldots$, with $h > 0$. The Euler Solution depends on the product parameter $a = rh$, rather than on r or h separately. For each of the parameter intervals $0 < a \leq 1$ and $1 < a \leq 2$, the Euler Solutions exhibit qualitatively different properties in the long term. Perform the following simulations.

9. Let $r = 10$, $h = 0.05$ (so $a = rh = 0.5$). Construct and plot Euler Solutions for each of the initial values $y_0 = 0.3, 2.0$. Compare the Euler Solutions to the exact solution curve over $0 \leq t \leq 1$ (for $y_0 = 0.3$) and over $0 \leq t \leq 0.5$ (for $y_0 = 2.0$). Explain qualitative differences.

   Answer 9: See Graphs 1 and 2 for the Euler Solutions (solid) of $y' = 10y(1 - y)$, $y_0 = 0.3, 2.0$. In each case $h = 0.05$, and so $a = rh = 0.5$. The Euler Solution follows the true solution (dashed) quite closely. The Euler Solution of the IVP with $y_0 = 2$ reaches the equilibrium state $y = 1$ in one step and remains there forever.

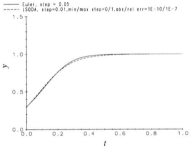
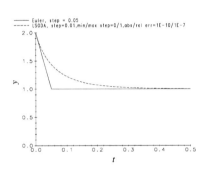

Problem 9, Graph 1.   Problem 9, Graph 2.

10. Let $r = 100$, $h = 0.015$ (so $a = rh = 1.5$). Construct and plot Euler Solutions for each of the initial values $y_0 = 0.5, 1.5$. Compare the Euler Solutions to the exact solution curve over $0 \leq t \leq 0.1$ (for $y_0 = 0.5$) and over $0 \leq t \leq 0.15$ (for $y_0 = 1.5$). Explain any differences.

11. *Runge-Kutta vs. Euler* Use Euler's Method and then RK4 for each IVP and time span in Problems 2, 3(d), 3(e), and 4(a). Describe and compare the graphical results. Which method seems to track the true solution the best?

   Answer 11: Note: Replace "figures 1–4" as printed in the first printing by "Problems 2, 3(d), 3(e), and 4(a)." Later printings are correct. Graphs 1 and 2, respectively, show Euler and RK4 approximations to the solution of the IVP, $y' = 3y \sin y - t$, $y(0) = 0.4, 0 \leq t \leq 8$, (see Problem 2) where the approximation goes forward from $t = 0$ to $t = 8$ and then backward from $t = 8$ to $t = 0$. In each case the step size is $h = 0.1$. There is more confidence in the forward RK4 approximation then in the forward Euler approximation because the former is a fourth order method and the latter is only first order. However, each method has accuracy problems as time goes backward from $t = 8$ (using the approximation to $y(8)$ computed by the method). It seems that there may be another solution $y = Y(t)$ to the ODE (where $Y(0) \approx 6.3$) with the property that $Y(8) \approx y(8)$ for each method. Since the computed value of $y(8)$ is only approximate, it seems that that value may be closer to $Y(8)$, and so when t goes backward the approximation ends up near $Y(0)$ instead of near $y(0)$.

   Graphs 3 and 4 show the respective Euler and RK4 approximations to the solution of the IVP of Problems 3(d) and 3(e); $y' = -2y + \cos t$, $y(0) = 0.4$, as t goes forward from 0 to 20 and then back to $t = 0$ from the approximation to $y(20)$ calculated by the method. The three step sizes are $h = 0.1$, 0.01, and 0.001. The forward solutions for both methods and all three step sizes seem to give good results, but the backward solution starting at $t = 20$ and using the forward approximation to $y(20)$ as the starting value is a different matter. The three Euler solutions going backward soon diverge from the forward solution, although the short-dashed backward curve (corresponding to $h = 0.001$) does better over a somewhat longer time span than the solid backward curve (corresponding to $h = 0.1$). The three backward RK4 approximations do better than the Euler approximations over a longer time span but even with $h = 0.001$ the backward RK4 curve (short dashes) ends up nowhere near $y(0) = 0.4$. The reason all the backward approximations go bad here is explained in the answer to Problem 3(e).

Finally, Graph 5 shows that RK4 with $h = 0.05$ does pretty well tracking the true solution to $y' = y^3$, $y(0) = 1$ when it escapes to infinity as $t \to 0.5$, but the Euler approximation seems to miss the escape as $t \to 0.5$; note the sharp upward turn at 0.55. The text explains why the Euler approximation is so bad. You should also try $h = 0.01$.

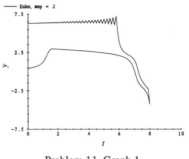

Problem 11, Graph 1

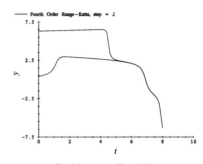

Problem 11, Graph 2

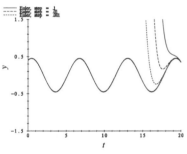

Problem 11, Graph 3

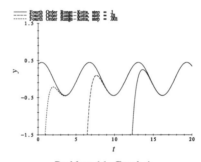

Problem 11, Graph 4

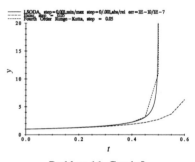

Problem 11, Graph 5

Spotlight on Steady States: Linear ODEs

Comments:

Steady-states are important in the applications because they represent how a system behaves in the long-term after local and transient behavior has disappeared.

PROBLEMS

Long-Term Behavior. Plot several solutions for each of the ODEs. Describe the long-term behavior of the solutions, and give reasons why the behavior is expected. If there is a steady state, plot it as well as you can (and identify it on your graph). If the steady state is periodic, find its period.

1. $y' + 2y = 4$
2. $y' + y = \text{SqWave}(t, 1, 1/2),\ 0 \le t \le 10$
3. $y' = 2y + 4$
4. $y' + y = 3\cos 4t,\ 0 \le t \le 10$
5. $y' - 3y = -1$
6. $y' + y = 3\,\text{SqWave}(t, 3, 3/4),\ 0 \le t \le 20$
7. $2y' = 5y - 10$
8. $y' + 0.75y = 3\,\text{SqWave}(t, 1, 1/2),\ 0 \le t \le 20$
9. $y' = 3\,\text{SqWave}(t, 2, 1),\ 0 \le t \le 5$
10. $y' + 0.2y = \text{SqWave}(t, 1, 1/2),\ 0 \le t \le 25$
11. $y' + y = \sin 3t - 2\cos 2t,\ 0 \le t \le 25$
12. $y' + 0.5y = \text{TRWave}(t, 2, 1, 2, 1),\ 0 \le t \le 15$

Answer 1:   Here, $p_0 = 2 > 0$, and $q_0 = 4$. Applying Theorem 1, the ODE has a unique steady-state solution $y = q_0/p_0 = 4/2 = 2$, which attracts all other solutions as $t \to \infty$ (see the figure).

Answer 3:   Here, $p_0 = -2 < 0$ and Theorem 1, does not apply. Using the integrating factor e^{-2t}, the solution for this ODE is $y = -2 + ce^{2t}$, c any constant. Since $y \to \infty$ as $t \to \infty$ if $c \ne 0$, the ODE has no steady-state solution.

Answer 5:   Here, $p_0 = -3 < 0$, so Theorem 1 does not apply. The general solution is $y = Ce^{3t} + 1/3$, C any constant. If $C \ne 0$, the solution moves away from the equilibrium $y = 1/3$ as t increases. See the figure.

Answer 7:   Here, $p_0 = -5/2 < 0$, so Theorem 2 does not apply. To study the long-term behavior of the ODE, we need to solve it. Using the integrating factor $e^{-5/2t}$, the solution for this ODE is $y = 4 + ce^{5/2t}$. Since $y \to \infty$ as $t \to \infty$ (if $c \ne 0$), the ODE does not have a steady-state solution. See the figure.

Answer 9:   Since $p_0 = 0$, Theorem 2 does not apply. The solution of the ODE is $y = \int_0^t 3\,\text{SqWave}(s, 2, 1)\,ds$ C, C any constant. The function $q(t)$ is a nonnegative, piecewise-continuous periodic function of period 2 where $\int_0^2 (\cdot)\,ds = 3$. As $t \to \infty$, the integral $\to \infty$, i.e. $y \to \infty$, which means that the ODE has no steady-state solution (see the figure).

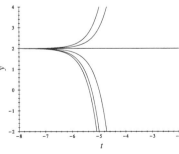

Problem 1.

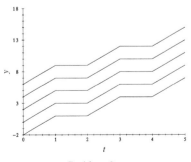

Problem 3.

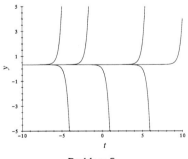

Problem 5.

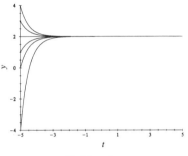

Problem 7.

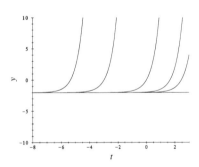

Problem 9.

Answer 11:   Here, $p_0 = 1 > 0$, and $q(t) = \sin 3t - 2\cos 2t$ is a continuous periodic function of period 2π. Applying Theorem 2, the ODE has a unique periodic solution $y_d(t) = A\sin 3t + B\cos 3t + C\sin 2t + D\cos 2t$. Inserting y_d in the ODE, we get $A = 1/10$, $B = -3/10$, $C = -4/5$, and $D = -2/5$. So $y_d(t) = 0.1\sin 3t - 0.3\cos 3t - .8\sin 2t - .4\cos 2t$ is a steady-state solution, and it attracts all other solutions as $t \to \infty$. See the figure.

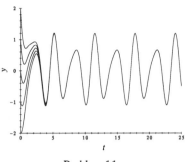

Problem 11.

💻 **Modeling Problems: Sensitivity to Input Frequency in *RC* Channels.**

13. *Will the Message Get Through?*   Plot the solutions of $V_O' + V_O = \text{SqWave}(t, T, T/2)$, $V_O(0) = 0$, where the period T of the square wave is 0.1, 10, 50 milliseconds. How closely does the message received match the message sent in each case? [*Hint*: see Example 3.]

Answer 13:   The IVP is $V_O' + V_O = \text{SqWave}(t, T, T/2)$, $V_O(0) = 0$. See Graphs 1, 2, 3, in which T is, respectively, 0.1, 10, and 50 milliseconds. The message received improves as the period T increases, i.e., as the frequency decreases.

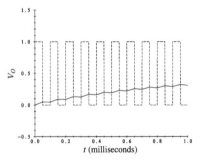

Problem 13, Graph 1.

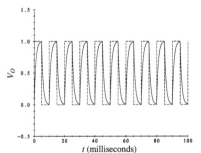

Problem 13, Graph 2.

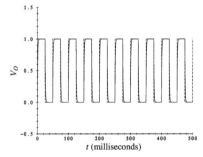

Problem 13, Graph 3.

14. In Example 3 we saw that the solution $V_O(t)$ of the IVP, $V_O' + V_O = \text{SqWave}(t, T, T/2)$, $V_O(0) = 0$, is quite sensitive to changes in the value of the period T. Explore the sensitivity of $V_O' + V_0 = q(t)$ using a numerical solver and the following input functions of period T: $q(t) = \sin(2\pi t/T)$, $\cos(2\pi t/T)$, $\text{SqWave}(t, T, T/10)$, and $\sin(2\pi t/T) + \sin(4\pi t/T)$. What happens as T ranges from 1.0 to 25.0 milliseconds? In each case find values of T for which the output voltage most closely mimics the input voltage.

 15. *S.O.S.*   The abbreviation S.O.S. is the universal code for "Send help!" In International Morse Code, S.O.S. is represented by three short pulses, three long pulses, and then another three short pulses. Use combinations of square waves to code a message $V_S(t)$ of repeated S.O.S. appeals.

- Simulate sending your message down the coaxial cable modeled by $V_O + V'_O = V_S(t)$, where $V_S(t)$ is your message.

- Adjust periods (frequencies) so that your message is understood at the receiving end.

- Now use the coaxial cable circuit modeled by $RCV'_O + V_O = V_S(t)$ and choose various values of the positive constant RC. Does your appeal for help get through? What values of RC work best?

Describe what you did and why. Attach graphs of the messages sent and received.

Answer 15: Group Project

Periodic Forced Oscillations. Theorem 2 and related matters are treated in Problems 16–19.

16. Plot the solution of the IVP, $y' + 0.5y = \text{SqWave}(t, 4, 2)$, $y(0) = 2$, over the interval $0 \le t \le 24$. Even though $y_0 = 2$ in this IVP is not the correct value to generate the periodic forced oscillation, one can see the periodic oscillation in the graph of the solution for large t. Why is this? Would this still be true for any value of $y(0)$?

17. *Proof of Theorem 2* Show that there is exactly one value y_0 for which the IVP, $y' + p_0 y = q(t)$, $y(0) = y_0$, has a periodic solution of period T for any constant $p_0 \ne 0$ and for any periodic and continuous function $q(t)$ with period T. [*Hint*: use a solution formula from Section 2.2, and the fact that a nonconstant continuous function $y(t)$ has period T if and only if $y(t + T) = y(t)$ for all t.]

Answer 17: We see that the unique solution of the IVP, $y' + p_0 y = q(t)$, $y(0) = y_0$, is given by $y(t) = e^{-p_0 t} y_0 + e^{-p_0 t} \int_0^t e^{p_0 s} q(s) ds$. This IVP has a periodic solution $y(t)$ with period T if and only if $y(T) = y(0) = y_0$. Applying this condition to the solution formula, we see that y_0 must satisfy the equation

$$y_0 = e^{-p_0 T} y_0 + e^{-p_0 T} \int_0^T e^{p_0 s} q(s) ds \tag{i}$$

which has a unique solution, $\bar{y}_0$, for y_0 because $p_0 \ne 0$ (hence, $e^{-p_0 T} \ne 1$). We shall show that the solution $\bar{y}(t)$ defined by

$$\bar{y}(t) = e^{-p_0 t} \bar{y}_0 + e^{-p_0 t} \int_0^t e^{p_0 s} q(s) ds$$

has period T by showing that for all t, $\bar{y}(t + T) = \bar{y}(t)$:

$$\bar{y}(t + T) = e^{-p_0(t+T)} \bar{y}_0 + e^{-p_0(t+T)} \int_0^{t+T} e^{p_0 s} q(s) ds$$

$$= e^{-p_0 t} \left[e^{-p_0 T} \bar{y}_0 + e^{-p_0 T} \int_0^T e^{p_0 s} q(s) ds \right] + e^{-p_0 t} \int_T^{t+T} e^{p_0(s-T)} q(s) ds$$

$$= e^{-p_0 t} \bar{y}_0 + e^{-p_0 t} \int_0^t e^{p_0 u} q(u) du \quad (\text{let } u = s - T \text{ and use } q(u+T) = q(u))$$

$$= \bar{y}(t)$$

where we have also used the fact that $y_0 = \bar{y}_0$ solves equation (i). Also, the uniqueness of $\bar{y}_0$ implies that this IVP has the unique periodic solution $\bar{y}(t)$.

18. Replace p_0 in the IVP (4) with a continuous periodic function $p(t)$ of the same period T as $q(t)$. If $\int_0^T p(s) ds \ne 0$, show that there is a unique periodic forced oscillation of period T. [*Hint*: proceed as in Problem 17. Use these facts: if $P_0(t) = \int_0^t p(s) ds$, then $P_0(t + T) = P_0(t) + P_0(T)$, and $\int_T^{t+T} e^{P_0(s)} q(s) ds = e^{P_0(T)} \int_0^t e^{P_0(s)} q(s) ds$ for any t.]

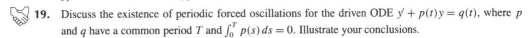

19. Discuss the existence of periodic forced oscillations for the driven ODE $y' + p(t)y = q(t)$, where p and q have a common period T and $\int_0^T p(s) ds = 0$. Illustrate your conclusions.

Answer 19: We know that in order for a continuous function $y(t)$ to be periodic with period T, it is necessary that $y(0) = y(T)$. As in problem 18, this implies that

$$y_0 = e^{-P_0(T)} y_0 + e^{-P_0(T)} \int_0^T e^{P_0(s)} q(s) ds$$

where $P_0(T) = \int_0^T p(v) dv$. Now if $P_0(T) = 0$, the only way this equation can hold is if the condition

$$\int_0^T e^{P_0(s)} q(s) \, ds = 0 \tag{i}$$

is satisfied, and in that case the equation above holds for all y_0. It follows that *every* solution

$$y = y_0 + e^{-P_0(t)} \int_0^t e^{P_0(s)} q(s) ds$$

of the ODE in this case is periodic with period T. If the integral condition (i) is not satisfied, then no solution of the ODE is periodic with period T. An example of the first case is $y' + (\cos t)y = \cos t$, where $T = 2\pi$. An example of the second case is $y' + (\cos t)y = 2 + \cos t$, where again $T = 2\pi$. Note that

$$\int_0^{2\pi} e^{\sin s}(2 + \cos s) ds > 0$$

because the integrand is positive.

Spotlight on Modeling: Cold Medication II

Comments:

The model that represents taking a fast-dissolving cold pill every four or six hours leads naturally to a system of linear ODEs with a periodic pulse input function such as $12 \cdot \text{SqWave}(t, 6, 1/2)$ of amplitude 12 and period 6 hours that is "on" only for the first half-hour of each period. Functions like this appear throughout the applications and we introduce them here in a natural setting. The system is most easily solved by using a numerical solver with piecewise continuous functions in its predefined list.

PROBLEMS

Cold Medication: Decongestant. Most cold pills contain a decongestant as well as an antihistamine. The form of the rate equations for the flow of decongestant is the same as that for antihistamine, but the rate constants are different. The values of the decongestant rate constants for one brand of cold pills now on the market are known: $k_1 \approx 1.386$ (hour)$^{-1}$ for the passage from the GI tract into the blood, and $k_2 \approx 0.1386$ (hour)$^{-1}$ for clearance from the blood. The respective amounts of decongestant in the GI tract and the blood are denoted by $x(t)$, $y(t)$. Use Examples 1–2 as guides.

One Dose.

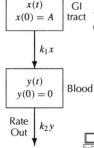

Flow of medication

1. *A Single Dose of Decongestant* Suppose that $x(0) = A$, $y(0) = 0$. Construct a labeled boxes and arrows diagram and the corresponding system of IVPs for the flow of decongestant.

   Answer 1: The boxes and arrows diagram is similar to the one given in Example 1, but the "Rate in" is zero and the initial amount in the GI-tract is A. So, $x'(t) = -k_1 x$, $x(0) = A$, and $y'(t) = k_1 x - k_2 y$, $y(0) = 0$, where the rate constants k_1 and k_2 have different values for the decongestant than for the antihistamine. The boxes and arrows diagram is shown in the margin.

2. Solve the IVPs of Problem 1. Then set $A = 1$, plot $x(t)$ and $y(t)$ over a six-hour time span, and describe what happens to the decongestant levels.

3. *Sensitivity* Plot decongestant levels in the blood if $A = 1$. Keep the value of k_1 fixed at 1.386 but set $k_2 = 0.01386, 0.06386, 0.1386, 0.6386, 1.386$. Describe what you see.

Answer 3:   See the figure. The larger the value of k_2, the sooner the antihistamine level peaks, the lower the peak level, and the more rapid the decline.

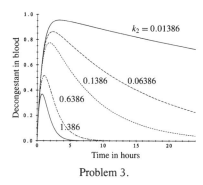

Problem 3.

Repeated Doses.

4. *One Dose Every Six Hours*   Suppose the decongestant is released into the GI tract at the constant rate of 12 units/hour for $1/2$ hour, and then repeated every 6 hours. Construct the model system of IVPs and plot the decongestant levels in the GI tract and blood over 48 hours (use $k_1 = 1.386$, $k_2 = 0.1386$). Compare and contrast your plots with Figure 2.

5. *Repeated Doses*   From data in Problem 4, plot decongestant levels in the GI tract and in the blood over a 5-day period. Repeat with $k_2 = 0.06386$, and 0.01386. The coefficient k_2 for the old and sick may be much smaller than that for the young and healthy. What advice would you give your grandmother when she comes down with a cold? Justify your answer.

Answer 5:   See Graphs 1, 2, and 3, where the respective values of k_2 are $0.1386, 0.06386$, and 0.01386. If the lowest value of k_2 is that of someone who is old, sick, or both, the rising decongestant level in Graph 3 is alarming.

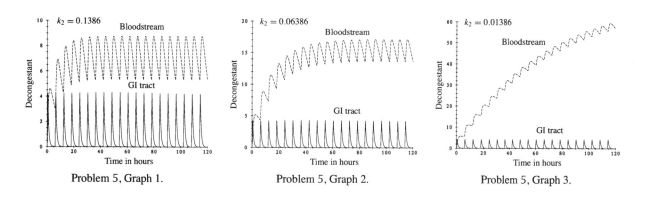

Problem 5, Graph 1.   Problem 5, Graph 2.   Problem 5, Graph 3.

Spotlight on Change of Variables: Pursuit Models

Comments:

There are hundreds of changes of variables for dealing with complicated (often nonlinear) ODEs; see, for example, *Handbook of Exact Solutions for Ordinary Differential Equations*, A.D. Polyanin, V.F. Zaitsev; CRC Press Inc., 1995, Boca Raton. See also D. Zwillinger's *Handbook of Differential*

Equations, 2nd ed., 1992; Academic Press, San Diego. In this section, we consider a few variable changes that reduce some simple (but nonlinear) first-order ODEs to forms that can be solved by methods of this chapter. Bernoulli ODEs (Problems 23–26), Riccati ODEs (Problems 27–31), and ODEs involving a function that is homogeneous of order 0 (this Spotlight) are all transformed to simpler ODEs by a variable change.

PROBLEMS

Solution Formulas: ODEs with Homogeneous Functions of Order Zero. Verify that each ODE can be written as $y' = f(x, y)$, where f is homogeneous of order zero. For each ODE find a formula that defines solution curves implicitly. [*Hint*: change variables from x and y to x and z by setting $y = xz$. Then solve as in Example 3.]

☞ A Computer
Algebra System
(CAS) is useful here.

1. $y' = (y + x)/x$ **2.** $y' = x/y + y/x$

3. $(x - y) dx + (x - 4y) dy = 0$ **4.** $(x^2 - xy - y^2) dx - xy dy = 0$

5. $(x^2 - 2y^2) dx + xy dy = 0$ **6.** $x^2 y' = 4x^2 + 7xy + 2y^2$

Answer 1: The ODE is $y' = (y + x)/x$. The rate function is homogeneous of order 0 because

$$f(kx, ky) = \frac{kx + ky}{kx} = \frac{x + y}{x} = f(x, y)$$

If $y = xz$, then $y' = xz' + z = (xz + x)/x = z + 1$, and $xz' = 1$. The solutions are $z = \ln|x| + C$. Using $y = xz$, $y = x \ln|x| + Cx$, C any constant, where the domain is restricted to $x > 0$, or to $x < 0$.

Answer 3: The ODE may be written as $y' = (y - x)/(x - 4y)$, and we have that

$$f(kx, ky) = \frac{ky - kx}{kx - 4ky} = \frac{y - x}{x - 4y} = f(x, y)$$

and so the rate function is homogeneous of order 0. After the variable change $y = xz$,

$$y' = xz' + z = (xz - x)/(x - 4xz) = (z - 1)/(1 - 4z)$$

$$xz' = (-1 + 4z^2)/(1 - 4z)$$

which separates to $\left[(1 - 4z)/(-1 + 4z^2)\right] dz = (1/x) dx$. Integral tables or partial fraction techniques yield the implicit solutions

$$\frac{1}{4} \ln\left|\frac{2z - 1}{2z + 1}\right| - \frac{1}{2} \ln|4z^2 - 1| = \ln|x| + C$$

In x, y variables, we have $(1/4) \ln|(2y/x - 1)/(2y/x + 1)| - (1/2) \ln|4y^2/x^2 - 1| = \ln|x| + C$. After a certain amount of fiddling with the logarithm terms, this simplifies to $(2y - x)(2y + x)^3 = C$, where C is a constant.

Answer 5: The ODE may be written as $y' = (2y^2 - x^2)/(xy)$, where $x, y > 0$, and we have that the rate function is homogeneous of order 0 because

$$f(kx, ky) = \frac{2k^2 y^2 - k^2 x^2}{k^2 xy} = \frac{2y^2 - x^2}{xy} = f(x, y)$$

Let $y = xz$ to obtain $xz' + z = (2x^2 z^2 - x^2)/(x^2 z) = (2z^2 - 1)/z$, or $xz' = (z^2 - 1)/z$. In separated form, we have $z \, dz/(z^2 - 1) = dx/x$, which integrates to $(\ln|z^2 - 1|)/2 = \ln|x| + K$, or replacing z by y/x, $y^2 - x^2 = Cx^4$. So the only solution for the ODE is $y = +x(1 + Cx^2)^{1/2}$, where C is a constant and x satisfies $1 + Cx^2 > 0$.

⊡ **Orbits: ODEs with Homogeneous Functions of Order Zero.** Write each ODE in Problems 1–2 in the form $N(x, y) dy/dx + M(x, y) = 0$. Use the equivalent system $dx/dt = N(x, y)$, $dy/dt = -M(x, y)$ and plot orbits in the xy-plane. For each initial point (x_0, y_0) chosen, solve both forward and backward in time and then repeat with the initial point $(-x_0, -y_0)$. Plot enough orbits to give a complete orbital portrait.

7. Problem 1   **8.** Problem 3   **9.** Problem 4

10. Problem 5   **11.** Problem 6   **12.** Problem 2

Answer 7: See the figure.

Answer 9: See the figure.

Answer 11: See the figure.

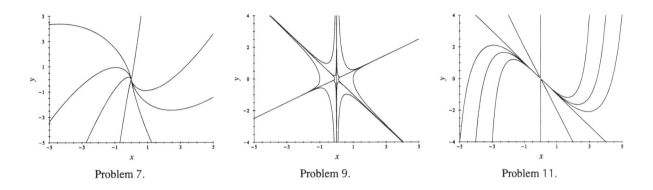

Problem 7.   Problem 9.   Problem 11.

13. *Orbital Symmetries*   Suppose that Γ is an orbit for $y' = f(x, y)$, where f is homogeneous of order 0. Show that $k\Gamma$ is also an orbit for any nonzero constant k. (This property explains the symmetries visible in the orbital graphs for Problems 7–12.)

Answer 13: If k is any nonzero constant, then

$$\frac{dy}{dx} = \frac{d(ky)}{d(kx)} = f(kx, ky) = f(x, y)$$

So if Γ is the orbit through the point (x, y), $k\Gamma$ is the orbit through (kx, ky).

Variation of Constants.

14. Suppose that $p(t)$ and $q(t)$ are continuous functions on the entire t-axis. The general solution of the undriven linear ODE $y' + p(t)y = 0$ is $y_u = Ce^{-P(t)}$, where C is an arbitrary constant and $P(t)$ is any (fixed) antiderivative of $p(t)$. Thinking of C as a function $C(t)$ instead of a constant, derive the general solution formula given in Section 2.1 for the driven linear ODE $y' + p(t)y = q(t)$. [*Hint*: make the substitution $y = C(t)e^{-P(t)}$ and find $C(t)$.]

Changing Variables: From Nonseparable to Separable ODEs.   Nonseparable ODEs may become separable by changing a variable. For each of the following cases, demonstrate this process and solve the new ODE. Then find the solution of the original ODE. [*Hint*: let $z = x + y$ in Problem 15, $z = 2x + y$ in Problem 16, $z = x + 2y$ in Problem 17.]

15. $dy/dx = \cos(x + y)$ **16.** $(2x + y + 1)\,dx + (4x + 2y + 3)\,dy = 0$

17. $(x + 2y - 1)\,dx + 3(x + 2y)\,dy = 0$ **18.** $e^{-y}(y' + 1) = xe^x$

Answer 15: Let $z = x + y$, in the ODE, $dy/dx = \cos(x + y)$. Then $dz/dx = 1 + dy/dx = 1 + \cos z$, so $(1 + \cos z)^{-1}\,dz = dx$. Using a table of integrals, we have $\tan(z/2) = x + C$, so $\tan[(x + y)/2] = x + C$. Then $x + y = 2\arctan(x + C) + n\pi$, so $y = -x + 2\arctan(x + C) + n\pi$, where C is any constant and n is any integer.

Answer 17:   Let $z = x + 2y$, so $dz/dx = 1 + 2dy/dx$. Then the ODE, $(x + 2y - 1)\,dx + 3(x + 2y)\,dy = 0$, becomes $(z - 1) + 3z(dz/dx - 1)/2 = 0$ or $3z\,dz/(z + 2) = dx$, which integrates to $3(z - 2\ln|z + 2|) = x + C$. In terms of x and y, $3(x + 2y - 2\ln|x + 2y + 2|) = x + C$, where C is any constant and $|x + 2y + 2| \neq 0$.

19. Suppose that $y' = f(rt + sy)$ where r and s are constants, $s \neq 0$, and f is a continuously differentiable function of a single variable.

(a) Show that the dependent variable change $z = rt + sy$ reduces the given ODE to an equivalent ODE that is separable.

(b) Find the general solution of the ODE $y' = (4t - y)^2$. [*Hint*: after making a variable change, use the integral table on the inside back cover.]

Answer 19:

(a) If $z = rt + sy$, then $y' = (z' - r)/s$ and the ODE $y' = f(rt + sy)$ becomes $(z' - r)/s = f(z)$, which is separable:

$$\frac{dz}{r + sf(z)} = dt$$

(b) Applying the technique of **(a)** to the ODE $y' = (4t - y)^2$ and set $z = 4t - y$ to get the separated ODE

$$\frac{dz}{4 - z^2} = dt$$

Use partial fractions to get

$$\frac{1}{4 - z^2} = \frac{1}{(2+z)(2-z)} = \frac{1/4}{z+2} - \frac{1/4}{z-2}$$

Integrate both sides of the ODE

$$\left(\frac{1/4}{z+2} + \frac{1/4}{z-2}\right) dz = dt$$

to get

$$\frac{1}{4}\ln|z+2| - \frac{1}{4}\ln|z-2| = t + C, \quad \ln\left|\frac{z+2}{z-2}\right| = 4t + C$$

Exponentiate:

$$\frac{z+2}{z-2} = Ke^{4t}, \quad K \text{ an arbitrary constant}$$

Solve for z:

$$z = 2 \cdot \frac{1 + Ke^{4t}}{Ke^{4t} - 1}, \quad K \text{ an arbitrary constant}$$

Since $y = 4t - z$,

$$y = 4t - 2\frac{1 + Ke^{4t}}{Ke^{4t} - 1}, \quad K \text{ an arbitrary constant}$$

Changing Variables: From Nonlinear to Linear. A change of variable converts each of the nonlinear ODEs in Problems 20–31 to a linear ODE solvable by the techniques of Sections 2.1 and 2.2.

20. Use the variable change $y = z^{1/2}$ to solve the IVP $yy'' + (y')^2 = 1$, $y(0) = 1$, $y'(0) = 0$.

☞ Alternative Derivation of Solution Formula (9) in Section 2.8.

21. The change of dependent variable $z = 1/y$ transforms the nonlinear ODE $y' = ry(1 - y/K)$ into the linear ODE $z' = -rz + r/K$.

(a) Show that the ODE for z is as claimed.

(b) Solve the ODE for z. Then show that $y = 1/z$ is given by solution formula (9) in Section 2.8 if $y(0) = y_0$.

Answer 21:

(a) Make the change of variable $z = 1/y$ in the logistic equation $y' = r(1 - y/K)y$. Since $y = 1/z$, $y' = -z^{-2}z'$, so

$$-z^{-2}z' = r\left(1 - \frac{1}{Kz}\right)\left(\frac{1}{z}\right) \quad -z' = r\left(z - \frac{1}{K}\right) \quad z' = -rz + \frac{r}{K}$$

The ODE for z is linear.

(b) Multiply through by the integrating factor e^{rt}:

$$(ze^{rt})' = \frac{r}{K}e^{rt}$$

Integrate:

$$ze^{rt} = \left(\frac{r}{K}\right)\frac{e^{rt}}{r} + C \quad z = Ce^{-rt} + 1/K$$

Change back to y:

$$y = \frac{1}{Ce^{-rt} + 1/K}$$

Use the initial condition $y(0) = y_0$ to evaluate C:

$$y_0 = \frac{1}{C + 1/K}, \quad C = \frac{1}{y_0} - \frac{1}{K}$$

So

$$y = \frac{1}{(1/y_0 - 1/K)e^{-rt} + 1/K} = \frac{Ky_0}{(K - y_0)e^{-rt} + y_0}$$

which is formula (9) of Section 2.5.

22. Consider the nonlinear ODE $y'(t) = (a + by)[c(t) + g(t)y]$, where a and b are constants, $b \neq 0$, and $c(t)$ and $g(t)$ are continuous on a t-interval I.

    (a) Show that the variable change $y = (z^{-1} - a)/b$ converts the given ODE into the linear ODE $dz/dt = [ag(t) - bc(t)]z - g(t)$.

    (b) Find all solutions of the ODE $y' = (3 - y)(2t + ty)$.

Bernoulli's ODE. The ODE $dy/dt + p(t)y = q(t)y^b$ is *Bernoulli's ODE*.

23. Show that the change of variable $z = y^{1-b}$, where b is a constant, $b \neq 0, 1$, changes the Bernoulli ODE to the linear ODE $dz/dt + (1 - b)p(t)z = (1 - b)q(t)$.

24. Show that the logistic ODE $y' = r(1 - y/K)y$ is a Bernoulli ODE with $b = 2$, and solve it.

25. Find all solutions of $dy/dt + t^{-1}y = y^{-4}$, $t > 0$. Graph several solutions for $t > 0$, $|y| \leq 5$.

26. Find all solutions of $dy/dt - t^{-1}y = -y^{-1}/2$, $t > 0$, $y \neq 0$. Graph several solutions for $1 \leq t \leq 2$, $1 \leq y \leq 3$.

Answer 23: Let $b \neq 0, 1$ and divide the ODE $dz/dt + (1 - b)p(t)z = (1 - b)q(t)$ through by y^b. Then multiply by $1 - b$ and the ODE can be written as

$$\frac{d}{dt}(y^{1-b}) + (1 - b)p(t)y^{1-b} = (1 - b)q(t)$$

Now make the change of variables $z = y^{1-b}$. The ODE takes the form

$$\frac{dz}{dt} + (1 - b)p(t)z = (1 - b)q(t) \tag{i}$$

which is a linear ODE. If $y(t)$ solves the original ODE, then $z(t) = (y(t))^{1-b}$ solves the linear ODE (i). Conversely, if z is any solution to ODE (i), then $y = z^{1/(1-b)}$ is a solution of the original ODE, or perhaps several solutions after taking the root $1/(1 - b)$.

Answer 25:   We may transform the Bernoulli ODE, $y' + t^{-1}y = y^{-4}$, to a linear ODE using the variable change $z = y^5$. We obtain $z' + 5t^{-1}z = 5$, where an integrating factor is $e^{5\ln t} = t^5$, and the solutions are given by $z = 5t/6 + C/t^5$. Then $y = z^{1/5} = (5t/6 + C/t^5)^{1/5}$, where $t > 0$ and C is an arbitrary constant. See the figure for some solution curves. Solution curves have infinite slope on the t- and y-axes, so the curves cannot touch the axes.

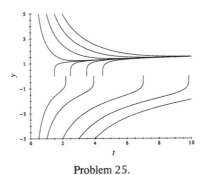

Problem 25.

Riccati's ODE.   The Riccati ODE is $dy/dt = a(t)y + b(t)y^2 + F(t)$. If $F(t) = 0$, the ODE is a special case of Bernoulli's ODE (Problems 23–26). Riccati's ODE may be reduced to a first-order linear ODE if one solution is known (Problem 27 below). Problems 28–31 contain examples.

27.   Let $g(t)$ be one solution of Riccati's ODE. Let $z = [y - g]^{-1}$. Show that $dz/dt + (a + 2bg)z = -b$, which is a first-order linear ODE in z. If $z(t)$ is the general solution of the linear ODE, show that the general solution $y(t)$ of the Riccati ODE is $y = g + 1/z$.

28.   Show that the ODE $dy/dt = (1 - 2t)y + ty^2 + t - 1$ has a solution $y = 1$. Let $z = (y - 1)^{-1}$ and show that $dz/dt = -z - t$. Then find the general solution $y(t)$ of the original ODE.

29.   Find all solutions of $dy/dt = e^{-t}y^2 + y - e^t$. [*Hint*: first show that $y = e^t$ is a solution.]

30.   Show that $y = t$ is a solution of $dy/dt = t^3(y - t)^2 + yt^{-1}$, $t > 0$, and then find all solutions.

31.   Show that the harvested logistic model ODE, $P' = rP - rP^2/K - H$, is a Riccati ODE. Solve it if $K = 1600, r = 0.01, H = 3$.

Answer 27:   Let the Riccati equation, $y' = ay + by^2 + F$, have a solution $y = g$. If $z = (y - g)^{-1}$, then $z' = -(y - g)^{-2}(y' - g') = -(y - g)^{-2}(ay + by^2 + F - ag - bg^2 - F) = -z^2(a/z + b(y^2 - g^2)) = -az - bz(y + g)$. Since $y - g = z^{-1}$, $y = g + z^{-1}$. So, $z' = -az - bz(2g + z^{-1})$, or $z' + (a + 2bg)z = -b$, which is linear in z. Once this has been solved for $z(t)$ by the integrating factor technique, we obtain y from $y = g + z^{-1}$.

Answer 29:   Direct calculation shows that the function $y = g(t) = e^t$ is a solution of the Riccati equation, $y' = e^{-t}y^2 + y - e^t$. So, the substitution $z = (y - e^t)^{-1}$ reduces the equation to the linear ODE, $z' + (1 + 2e^{-t}e^t)z = -e^{-t}$, or $z' + 3z = -e^{-t}$, whose solutions are given by $z = Ce^{-3t} - e^{-t}/2$. Solutions $y(t)$ are given by $y = e^t + 1/z = e^t + [Ce^{-3t} - e^{-t}/2]^{-1}$, where C is any constant and t is restricted to an interval for which $2Ce^{-3t} \neq e^{-t}$.

Answer 31:   The equation $P' = rP - (rP^2)/K - H$ is a Riccati equation with $a = r$, $b = -r/K$, $F = -H$. Let $K = 1600$, $r = 0.01$, $H = 3$. A direction calculation shows that $P(t) = g(t) = 1200$ is a solution. So, the change of variable $z = [P - 1200]^{-1}$ reduces the equation to the linear ODE, $z' + (0.01 - 0.015)z = 10^{-4}/16$, or $z' - 0.005z = 10^{-4}/16$, whose solutions are given by $z = Ce^{0.005t} - 1/800$. And so, $P = 1200 + 1/z = 1200 + 800(800Ce^{0.005t} - 1)^{-1}$, where C is any constant; t is restricted to an interval on which $800Ce^{0.005t} \neq 1$.

Changing Variables to Polar Coordinates.   Transform the ODE to polar coordinates and find the general solution of the transformed ODE. Plot a slope field and several solution curves of the given ODE in the rectangle $3 \le x \le 3$, $-3 \le y \le 3$. [*Hint*: see Example 4.]

32. $\dfrac{dy}{dx} = \dfrac{2y - x}{y + 2x}$

33. $\dfrac{dy}{dx} = \dfrac{2y + x}{-y + 2x}$

Answer 33: Following the initial steps described in Answer 33, take $A = 2$, $B = 4$, $C = 4$ and $D = -2$. The resulting transformed ODE is

$$\frac{dr}{d\theta} = 2r$$

whose solution is $r = Ke^{2\theta}$ for arbitrary constant K. See the figure for arcs of solution curves in the xy-plane. Look closely and you will see that arcs never cross the line $-y + 2x = 0$, because dy/dx is infinite there.

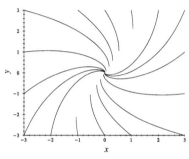

Problem 33.

Polar Coordinates.

 34. Find the coordinates of the endpoints of the solution curves in Example 4.

35. The path of a goose flying toward its nest was characterized by an ODE at the beginning of this Spotlight. Solve this ODE by changing the variables to polar coordinates. Convert this polar solution to cartesian coordinates to verify the solution of the goose problem given in the text of this Spotlight.

Answer 35: The flying goose model ODE is

$$y' = \frac{y - c(x^2 + y^2)^{1/2}}{x} = f(x, y)$$

where c is a positive constant. Now since $f(r\cos\theta, r\sin\theta) = (\sin\theta - c)/\cos\theta$ we see from (16) that the equivalent ODE in polar coordinates is

$$\frac{dr}{d\theta} = r\frac{1 - c\sin\theta}{-c\cos\theta} = r\left(\tan\theta - \frac{1}{c}\sec\theta\right)$$

which is linear. Use the fact that

$$\int \tan\theta \, d\theta = \ln\sec\theta, \quad \int \sec\theta = \ln(\sec\theta + \tan\theta)$$

to obtain the solution

$$r = k\sec\theta(\sec\theta + +\tan\theta)^{-1/c}, \quad k \text{ a constant}$$

Using the initial condition $r = a$ when $\theta = 0$, we finally have the solution of the goose IVP in polar form:

$$r = a(\cos\theta)^{(1-c)/c}(1 + \sin\theta)^{-1/c}$$

Recall that $c = w/b$, where w is the wind's speed and b is the bird's speed. Observe that

$$\text{if } c > 1, \quad \text{then } (1 - c)/c < 0 \text{ and } r \to +\infty \text{ as } \theta \to \pi/2$$

$$\text{if } c < 1, \quad \text{then } (1 - c)/c > 0 \text{ and } r \to 0 \text{ as } \theta \to \pi/2$$

Also note that if $c = 1$, then $r = a(1 + \sin\theta)^{-1}$ and $r \to a/2$ as $\theta \to \pi/2$. This gives the qualitative behavior of solutions which was observed from the solution curves of the original ODE obtained from numerically computed solution curves as well as a solution formula.

The cartesian coordinates for the solution curve are given parametrically by

$$x = r(\theta)\cos\theta = a\cos^{1/c}\theta(1 + \sin\theta)^{-1/c}$$

$$y = r(\theta)\sin\theta = a\cos^{1/c}(1 + \sin\theta)^{-1/c}\tan\theta$$

To eliminate the parameter θ between these two equations, observe that

$$y = x\tan\theta \quad \text{and} \quad (x/a)^c = \cos\theta(1 + \sin\theta)^{-1}$$

$$\left(\frac{x}{a}\right)^{-c} - \left(\frac{x}{a}\right)^c = \frac{1 + \sin\theta}{\cos\theta} - \frac{\cos\theta}{1 + \sin\theta} = 2\tan\theta$$

Therefore

$$y = \frac{x}{2}\left[\left(\frac{x}{a}\right)^{-c} - \left(\frac{x}{a}\right)^c\right]$$

Modeling Problems.

36. A ferryboat needs to sail directly across a river to a dock, but the current complicates things. The captain decides to aim the ferry toward the dock at all times. Will the ferry make it? Assume that the speed of the ferry is b and the speed of the river is w.

37. *Flight Path of a Goose, Wind from the Southeast*

(a) Set up the flight path problem for the goose flying at speed b, if the wind is blowing from the southeast at a speed of $w = b/\sqrt{2}$ and the goose starts at $x = a > 0$, $y = 0$. Solve the IVP in implicit form (don't attempt to solve for y in terms of x). [*Hint*: note that $x' = -b\cos\theta - b/2$, $y' = -b\sin\theta + b/2$. Use a table of integrals.]

(b) Set $b = 1$, $a = 1, 2, \ldots, 9$ and plot the paths. [*Hint*: use a numerical solver for the system that models the flight path.] Does the goose reach the nest? Does it overshoot?

Answer 37:

(a) The problem may be set up exactly as the pursuit problem in the text, except that the wind's velocity has component $-b/2$ along the x-axis and $b/2$ along the y-axis. So, $x' = -b\cos\theta - b/2$ and $y' = -b\sin\theta + b/2$. Divide to get

$$\frac{y'}{x'} = \frac{dy}{dx} = \frac{2\sin\theta - 1}{2\cos\theta + 1} = \frac{2y - (x^2 + y^2)^{1/2}}{2x + (x^2 + y^2)^{1/2}}$$

Since this quotient is a homogeneous function of order 0 in x and y, introduce $y = xz$ to obtain the separable ODE $xdz/dx + z = [2z - (1 + z^2)^{1/2}]/[2 + (1 + z^2)^{1/2}]$, or

$$\frac{-2 - (1 + z^2)^{1/2}}{(z + 1)(1 + z^2)^{1/2}}\,dz = \frac{1}{x}\,dx$$

Let $z = u - 1$ to reduce the left side to a form found in integral tables. After solving and returning to x, y variables and simplifying, we obtain $[\sqrt{2}(x^2 + y^2)^{1/2} + x - y]^2 = a^{\sqrt{2}}(3 + 2\sqrt{2})|x + y|^{2-\sqrt{2}}$, where the initial data are used. The goose follows the path defined by this expression if it heads towards its nest while the wind blows from the southeast.

(b) If $b = 1$, then the equivalent IVP for the flight path is $x' = -\cos\theta - 1/2 = -x(x^2 + y^2)^{-1/2} - 1/2$, $y' = -\sin\theta + 1/2 = -y(x^2 + y^2)^{-1/2} + 1/2$, $x(0) = a$, $y(0) = 0$. See Graph 1 for the flight paths if

$a = 1, 2, \ldots 9$. The goose always reaches the nest, but overshoots the mark a bit and has to head into the wind at the end so that it comes into the nest. See Graph 2 for a zoom on a portion of Fig. 11(b). Graph 2 shows that flight paths are tangent to the line $y = -x$ (dashed) at the origin.

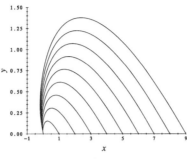

Problem 37(b), Graph 1.

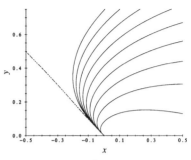

Problem 37(b), Graph 2.

38. *The Goose and a Moving Nest* On a windless day a goose sees its gosling aboard a raft in the middle of a river flowing at the rate of 8 yd/sec. When the raft is directly opposite the goose, the raft is 30 yd distant, and the goose instantly takes flight to save her gosling from going over a waterfall 60 yd downstream. If the goose flies directly toward the raft at the constant speed of 10 yd/sec, does she rescue her gosling before it tumbles over the falls? Follow the outline below for this problem.

At $t = 0$ place the raft and the goose in the xy-plane at the origin and at $(30, 0)$, respectively. Let the river flow in the positive y-direction. The parametric path $(x(t), y(t))$ followed by the goose in the xy-plane has the following properties: the goose's velocity vector at time t, $(x'(t), y'(t))$, always points toward the raft, and $((x')^2 + (y')^2)^{1/2} = 10$ yd/sec at all times. So, if the goose is at (x, y) at time t, then the raft is at $(0, 8t)$, and there is a factor $k > 0$ (which may depend on x, y, and t) such that $x' = k(-x)$, $y' = k(8t - y)$. Since $(x')^2 + (y')^2 = 100$, we find that $k = 10/(x^2 + (8t - y)^2)^{1/2}$, and we have the IVP

Gosling

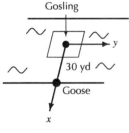

30 yd

Goose

x

☞ A nonautonomous planar system.

$$x' = \frac{-10x}{(x^2 + (8t - y)^2)^{1/2}}, \qquad x(0) = 30$$

$$y' = \frac{10(8t - y)}{(x^2 + (8t - y)^2)^{1/2}}, \qquad y(0) = 0$$

Since the rate functions in this system depend on t, we can't directly apply the technique used in this section. But the system is written in normal form, so a numerical solver can be used to solve and plot an orbit of this system to determine if the goose rescues the gosling. If the goose does reach the gosling in time, how long does it take?

39. In the goose model IVP described by text equations (1) and (2) (with $x(0) = 10$ miles replaced by $x(0) = 100$ meters) find the time it takes for the goose to reach the nest if in Figure 1, $a = 100$ meters, $b = 10$ meters/sec, and $w = 5$ meters/sec.

Answer 39: Use a numerical solver to solve the IVP

$$\frac{dx}{dt} = \frac{-10x}{(x^2 + y^2)^{1/2}} \quad x(0) = 100$$

$$\frac{dy}{dt} = \frac{-10y}{(x^2 + y^2)^{1/2}} + 5 \quad y(0) = 0$$

The time it takes for the goose to reach its nest in this case is about 13.3 sec.

40. What strategy should an eagle follow if it wants to fly along the x-axis to its nest at the origin from the point $(a, 0)$? See text Figure 1.

Spotlight on Continuity in the Data

Comments:

Exact values of parameters, rate functions, and initial conditions are almost never available in practice. So there is a real need to estimate the effects of the uncertaintity in the data, and that is what the concept of continuity does.

PROBLEMS

Continuity in the Data. Show long-term continuity or short-term continuity by using a solution formula for each IVP to obtain bounds.

1. *Long Term Continuity in the Data (Initial Values)* Find a bound for $|y_a(t) - y_A(t)|$ where $y_a(t)$ solves the IVP $y' = -y + e^{-t}$, $y(0) = a$, $t \geq 0$, while y_A solves the IVP with $y(0) = A$.

   Answer 1: Using the integrating factor e^t to solve $y' = -y + e^{-t}$, we see that $(ye^t)' = 1$, so integrating and using $y(0) = a$, we have $y_a(t) = ae^{-t} + te^{-t}$; similarly for $y(0) = A$, we have $y_A(t) = Ae^{-t} + te^{-t}$. Then

   $$|y_a(t) - y_A(t)| = |a - A|e^{-t} \leq |a - A| \quad t \geq 0$$

   since $e^{-t} \leq 1$ for $t \geq 0$. So you can make the difference between $|y_a(t) - y_A(t)|$ as small as desired by making $|a - A|$ small enough. So we have long-term continuity in the initial value.

2. *Long Term Continuity in the Data (Rate Function)* Find a bound for $|y_c(t) - y_C(t)|$ where $y_c(t)$ and $y_C(t)$ denote the respective solutions of the IVPs $y' = -y + c$, and $y' = -y + C$, $y(0) = 1$, $t \geq 0$.

3. *Long Term Continuity in the Data (Initial Value)* Find a bound for $|y_c(t) - y_C(t)|$ over an appropriate t-interval where $y_c(t)$ and $y_C(t)$ solves the respective IVPs $y' = -cy/t$, $y_C'(t) = -Cy/t$ $y(1) = 1$, $t \geq 1$.

   Answer 3: The solution of the IVP $y' = -y/t$, $y(1) = c$, is $y_c = c/t$, similarly $y_C = C/t$. Then

   $$|y_c(t) - y_C(t)| = |c/t - C/t| = |1/t||c - C| \leq |c - C| \quad t \geq 1$$

   since for $t \geq 1$, we have $|1/t| \leq 1$. You can make the difference $|y_c(t) - y_C(t)|$ as small as desired by making $|c - C|$ small enough, and we have long term continuity in C.

4. *Short Term Continuity* Show that the solution $y_{a,b}(t)$ of the IVP $y' = ay^2$, $y(0) = b$, $t \geq 0$, is continuous in the parameters a and b, for $a > 0$, $b > 0$, but only for $0 \leq t < 1/ab$. [*Hint:* see Example 1.]

Qualitative Analysis.

5. *Sensitivity of Solutions to Changes in Initial Data* Use your solver to do a sensitivity analysis of the IVP $y' = 3y \sin y + t$, $y(-6) = c$ as the parameter c changes. How close can you come to generating Figure 2.4.3 in Section 2.4?

   Answer 5: Compare your figure to ours.

Modeling Problem.

6. *Design Specifications for a Mixture Model* A salt solution runs into a tank, and mixes with the solution in the tank, and the mixture runs out. Under ideal operating conditions, suppose that

   Inflow rate: 10 gal/min   Volume of solution in tank: 100 gal
   Outflow rate: 10 gal/min   Initial amount of salt in tank: 15 lb
   Concentration of salt in inflow stream: 0.2 lb/gal

   Use this data to carry out the following analysis:

   (a) If $y(t)$ is the amount of salt in the tank at time t, show that $y(t)$ solves the first-order IVP, $y' + 0.1y = 2$, $y(0) = 15$.

   (b) Suppose that in reality the inflow salt concentration is not exactly 0.2 lb/gal and that we cannot be certain about the initial amount of salt in the tank. So the actual amount $z(t)$ of salt in the tank satisfies

 Use the Balance Law to get this ODE.

another IVP, $z' + 0.1z = m(t)$, $z(0) = c$, where $m(t)$ and c may be different from the respective values 2 and 15 in the ideal IVP. The design specifications require that for a prescribed positive constant K, $|y(t) - z(t)| \le K$, all $t \ge 0$. Find constraints on $|m(t) - 2|$ and $|c - 15|$ so that the design specifications are met? [*Hint*: see Example 1. Note that here $p(t) = p_0 = 0.1$.]

(c) Explain why this model displays long-term continuity in the data. [*Hint*: see the discussion following Example 1.]

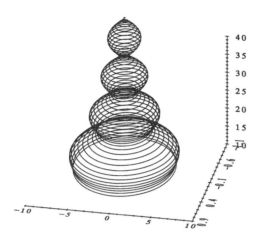

This time-state curve tracks the vertical
motion of a weight on a spring. It lives in the
three-dimensional space of position y,
velocity y', and time t. Check out
Example 3.1.6.

Second-Order
Differential Equations

The up-and-down motion of a body suspended by a spring may be modeled by a second-order
ODE (Section 3.1). Why second-order? Like most dynamical systems where forces are involved,
Newton's Second Law comes into play:

$$\text{mass} \times \text{acceleration of body} = \text{sum of forces acting on body}$$

Acceleration is the second derivative with respect to time of the state variable representing the
body's position. The proper use of Newton's Laws involves vectors, but the motion here is only
one-dimensional (along the local vertical under the point of suspension of the spring), and we can
get away with real-valued state variables. The vector treatment of Newton's Second Law is given in
Section 4.1, where we model pendulum motion.

All the basic facts about constant-coefficient second-order linear ODEs appear in the pages of
this chapter—and more besides. We have added some material that gives insight into some corners
that are often overlooked. The basic model in the chapter appears in the first section. We do the
standard linear spring here and in later sections hard and soft springs, and aging springs. The hard
and soft spring ODEs are nonlinear and provide us with a nice opportunity to compare solutions
of nonlinear ODEs and their linearizations (see especially Fig. 3.8.6). We take up the aging spring
model again in Chapter 11, where we show how to solve it using series methods, and in the WEB
SPOTLIGHT ON THE EXTENDED METHOD OF FROBENIUS, where the solution uses Bessel functions.

The geometry of solutions of second-order ODEs is a bit more complicated than for first-order
ODEs. Orbits and time state curves are introduced in Section 3.1 along with several examples. Also
in Section 3.1 is the Existence and Uniqueness Theorem for a general second-order initial value
problem. Sections 3.2–3.6 treat constant-coefficient linear second-order ODEs with and without
initial data. Direction fields are introduced in Section 3.3. The basic tool for finding solution formu-

las is the use of the operator $P(D) = D^2 + aD + b$ where a and b are real numbers. The emphasis of this chapter is on linear ODEs, because the solution set of a second-order linear ODE can be characterized in simple terms. General second-order linear ODEs are covered in Section 3.7, and some nonlinear second-order ODEs are treated in Section 3.8 (also in the direction fields problems of Section 3.3). In Section 3.4 we look at undriven linear ODEs with periodic solutions and the difficulty in tracking such solutions numerically.

3.1 Models of Springs

Comments:

This section is mostly introductory, presenting not much theory, but examples and comments about the spring model. Note that force models (typically) lead to second-order ODEs in a position variable and that second-order ODEs require two state variables, usually position and velocity. Second-order ODEs play a starring role in an ODE course because of the force laws. Spring models come up in this and subsequent chapters.

Whenever we introduce a new class of ODEs (second-order ODEs in this chapter), we need to assure ourselves that IVPs have unique solutions. As usual, this requires that the functions in the ODE satisfy certain continuity and smoothness conditions. That is the reason for Theorem 3.1.1. We also define the various graphs associated with a solution of a second-order linear ODE in y. We discuss and illustrate *solution curves* [y vs. t], *velocity curves* [y' vs. t], *orbits* [y' vs. y], and *time-state curves* [the parametrically-defined curve $t = t$, $y = y(t)$, $y' = y'(t)$ in tyy'-space]. It takes a while to absorb all of this, and so we have given lots of examples with pictures. Each graph gives its own kind of information about the behavior of a solution; if we don't have a solution formula, we need all the help we can get to determine the behavior of a solution.

Computer Tip

Here is how we coaxed our solver to plot Figure 3.1.4, a time-state curve and its orbit on the same set of axes.

Consider the first-order differential system with three state variables:

$$y' = v$$
$$v' = -65y - 0.4v$$
$$z' = f(z), \quad \text{where} \quad f(z) = \begin{cases} 0, & z = 0 \\ 1, & z \neq 0 \end{cases}$$

- Solve the system over $20 \leq t \leq 35$ under the conditions $y(20) = 9$, $v(20) = 0$, $z(20) = 20$, and plot the solution curve in xyz-space. [*Hint*: note that $z(t) = t$.]

- Repeat with $z(20) = 0$, and plot the orbit in xyz-space. [*Hint*: $z(t) = 0$, all t.]

PROBLEMS _____

Hooke's Law Spring. An undriven weight of mass m is suspended from an undamped Hooke's Law spring with spring constant k. Let y denote the position from its static deflection position, where "up" is the positive direction for y. The weight's static deflection is 24 in. [*Hint*: see Example 3.1.1.]

1. Show that $y(t)$ satisfies the ODE $y'' + 16y = 0$, where y is measured in inches. [*Hint*: g is 384 in/sec^2, the static deflection h satisfies the equation $kh = mg$, and $my'' + ky = 0$.]

2. For the model ODE in Problem 1, suppose that the weight is pushed up y_0 in and released from rest, where y_0 takes in turn each of the values 15, 10, 5, and 2 in. Plot the resulting orbits in yy'-state space, inserting arrowheads to indicate increasing time.

3. Plot the ty- and ty'-component graphs for each of the orbits in Problem 2. What can you say about the motion? Are any of these solutions periodic? If so, estimate the periods.

Answer 1:   The motion satisfies a form of ODE (5) of the text, $y'' + (c/m)y' + (k/m)y = F(t)/m$. Since there is no damping or driving force, $c = 0$ and $f = 0$, and the equation becomes $y'' + (k/m)y = 0$; so we need to find k/m. Since $kh = mg$, $k/m = g/h$, where $g = 384$ in/sec^2 (see Example 3.1.1), $k/m = g/h = (384$ in$/ \sec^2)/(24$ in$)= 16$ sec^{-2}. So the equation of motion is $y'' + 16y = 0$.

Answer 3:   See the figure for the plot in ty-space (top) and in ty'-space (bottom). The motion is periodic, and all nonconstant solutions seem to have period $T \approx 1.6$ sec (reading from the ty-graph).

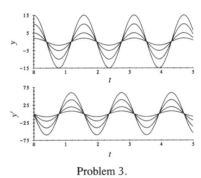

Problem 3.

Damped Hooke's Law Spring.   Consider an undriven 1-lb weight suspended from a Hooke's Law spring. Suppose that y denotes the deflection (in inches) of the weight from its static deflection position, where "up" is the positive direction for y. Assume the damping force is viscous, the static deflection is 24 inches, and the damping coefficient is $c = 2.60 \times 10^{-4}$ lb·sec /in.

4. Show that $y(t)$ satisfies the ODE $y'' + ay' + 16y = 0$, $a \approx 0.1$. [*Hint*: see Example 3.1.1.]

5. For the model ODE in Problem 4, push the weight up y_0 inches and release it from rest, where y_0 takes, in turn, the values 15, 10, and 2 in. Plot the resulting orbits in yy'-state space, inserting arrowheads to show the direction of increasing time. [*Hint*: see Example 3.1.5.]

6. Plot the ty- and ty'-component curves for each of the orbits in Problem 5. What can you say about the body's motion and the sequence of transit times through the t-axis?

Answer 5:   See the figure for the three orbits; the initial data is $y(0) = 15, 10, 2$ and $y'(0) = 0$. Adjacent loops of the innermost spiral are so close together that they seem to be on top of one another, but that only shows the limitations of computer graphics.

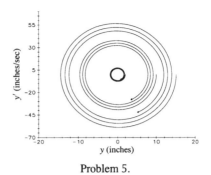

Problem 5.

![computer icon] **Plotting Curves.** Verify that each IVP satisfies the conditions of Theorem 3.1.1. Plot component curves, orbits, and time-state curves. Are there any periodic solutions? In Problems 7–10 discuss the motion of the spring as t increases through the given interval.

7. *Undamped Hooke's Law Spring* $y'' + y = 0$, $y(0) = 0, 0.5, 1.0, 1.5$, $y'(0) = 0$. Use the t-interval $0 \le t \le 20$ and the state variable ranges $|y| \le 2$, $|y'| \le 3$.

8. *Damped Hooke's Law Spring* $y'' + 0.1y' + 4y = 0$, $y(0) = 0, 0.5, 1.0, 1.5$, $y'(0) = 0$. Use the t-interval $0 \le t \le 40$ and the state variable ranges $|y| \le 2$, $|y'| \le 4$.

9. *Driven, Undamped Hooke's Law Spring* $y'' + y = \cos(1.1t)$, $y(0) = 5$, $y'(0) = 0$. Solve over the interval $0 \le t \le 200$.

10. *Driven and Damped Hooke's Law Spring* $y'' + 0.05y' + 25y = \sin(5.5t)$, $y(0) = 0$, $y'(0) = 0$. Use the t-interval $0 \le t \le 40$, and the state variable ranges $|y| \le 0.4$, $|y'| \le 2.1$.

11. *Going Backward in Time* $y'' - 0.1y' + 4y = 0$, $y(0) = 0, 1, 2$, $y'(0) = 0$. Use the t-interval $-100 \le t \le 0$ and the state variable ranges $|y| \le 2$, $|y'| \le 4$.

Answer 7: The ODE has the form $y'' + a(t)y' + b(t)y = f(t)$, where $a(t) = 0$, $b(t) = 1$ and $f(t) = 0$. Since the constant functions a, b are continuous everywhere, Theorem 3.1.1 is satisfied. The solutions $y = y(t)$ and velocities $v = y'(t)$ of the IVPs $y'' + y = 0$, $y(0) = 0.5, 1.0, 1.5$, $y'(0) = 0$ seem to be periodic with a common period T a little larger than 6. (As we shall see in the next section, the period is $2\pi \approx 6.28$.) There is a single equilibrium solution, $y = 0$ for all t, whose graph is a straight line. The nonconstant orbits seem to be simple closed curves (which again suggests that the solutions are periodic). The equilibrium solution corresponds to a point orbit at the origin. The time-state curves are circular helices, while the equilibrium time-state curve is a vertical line. See the figures for the solution and velocity curves (Graph 1), orbits (Graph 2) and time-state curves (Graph 3). The spring oscillates periodically forever; the model ODE has no damping term, so the motion is not realistic over a long span of time.

Answer 9: Theorem 3.1.1 applies because 1 (the coefficient of y) and $\cos(1.1t)$ are continuous for all t. See the figures for the graphs of the ty- and ty'-component curves, the orbit in the yy'-plane, and the time-state curve for the IVP, $y'' + y = \cos(1.1t)$, $y(0) = 5$, $y'(0) = 0$. From the component curves it appears that the period is about 62. The spring oscillates rapidly with rising and then falling amplitudes; this behavior repeats endlessly because there is no damping.

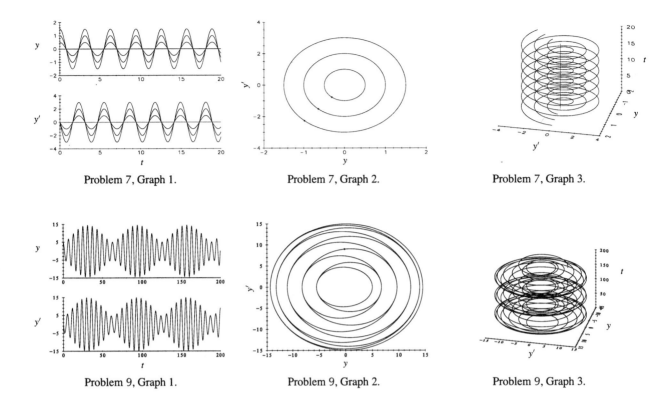

Problem 7, Graph 1.   Problem 7, Graph 2.   Problem 7, Graph 3.

Problem 9, Graph 1.   Problem 9, Graph 2.   Problem 9, Graph 3.

Answer 11:   The ODE has the form $y'' + a(t)y' + b(t)y = f(t)$, where $a(t) = -0.1$, $b(t) = 4$ and $f(t) = 0$. Since the constant functions are continuous everywhere, Theorem 3.1.1 is satisfied. Here the IVPs are $y'' - 0.1y' + 4y = 0$, $y(0) = 0, 1, 2$, $y'(0) = 0$, but we solve *backward* from $t = 0$ to $t = -100$. There is a single equilibrium solution, $y = 0$, all t. See Graph 1 for the ty and ty' curves, Graph 2 for the orbits and Graph 3 for time-state curves. Solutions tend to the equilibrium as t diminishes.

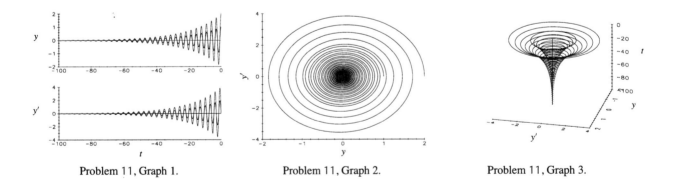

Problem 11, Graph 1.   Problem 11, Graph 2.   Problem 11, Graph 3.

🖳 **Concavity.**   The sign of the second derivative of $y(t)$ determines whether the graph of $y = y(t)$ is concave up ($y'' > 0$) or down ($y'' < 0$). For each of the ODEs determine the regions in the ty-plane where solution curves are concave down [up]. Then plot solutions that meet given initial conditions and verify that the solution curves are concave down [up] in the proper regions.

12. $y'' = -1$; $y(0) = 1$; $y'(0) = 2, 0, -2$; use the window $0 \leq t \leq 10$, $-27 \leq y \leq 3$.

13. $y'' = -y$; $y(0) = 1, 0, -1$; $y'(0) = 0$; use the window $0 \leq t \leq 20$, $|y| \leq 1$.

Answer 13: Solution curves are concave down when y is positive because then $y'' = -y$ is negative, but concave up when y is negative. See the figure for some ty-curves. Note how the arcs of solution curves above the t-axis are concave down, while those below the t-axis are concave up, and the points where $y = y(t)$ crosses the t-axis are the inflection points.

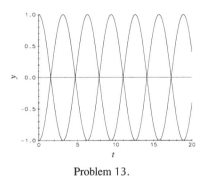

Problem 13.

Existence and Uniqueness.

14. Solve the IVP $y'' + 4y = 0$, $y(\pi/2) = 1$, $y'(\pi/2) = 0$. [*Hint*: see Example 3.1.3.]

15. Solve the IVP $y'' + 25y = 0$, $y(0) = 10$, $y'(0) = -10$. [*Hint*: see Example 3.1.3, but use $\cos 5t$ and $\sin 5t$.]

16. Find all solutions of the IVP $y'' + a(t)y' + b(t)y = 0$, $y(t_0) = 0$, $y'(t_0) = 0$, where t_0 is any (fixed) point on the t-axis, and the coefficients $a(t)$ and $b(t)$ are continuous for all t.

Answer 15: Try to find solutions of the form $y = A\cos 5t + B\sin 5t$, where A and B are to be determined. insert y into the ODE:

$$y'' + 25y = (-25A\cos 5t - 25B\sin 5t) + (25A\cos 5t + 25B\sin 5t) = 0, \quad \text{all } t, A, B$$

Now choose A, B so that $y(0) = 10$, $y'(0) = 10$. So $A = 10$, $B = -2$. The function $y(t) = 10\cos 5t - 2\sin 5t$ is a solution of the IVP, and is the only solution by Theorem 3.1.1.

17. Let $a(t)$ and $b(t)$ have derivatives of all orders on an interval I. Suppose that $y(t)$ is a solution of the ODE $y'' + a(t)y' + b(t)y = 0$ that lives on I. Show that $y(t)$ has derivatives of all orders on I.

Answer 17: Since $y(t)$ is a solution, $y(t)$ and $y'(t)$ are continuous on I by the definition of a solution. Since $y''(t) = -a(t)y'(t) - b(t)y(t)$ and a, b, y, and y' are continuous, then y'' exists and is continuous. Since a, b, y, and y' are then continuously differentiable, so is y''. That means y''' exists and is continuous. Proceeding in this fashion, we see that $y^{(n)}$ exists and is continuous for $n = 2, 3, 4, \ldots$.

Modeling Problems.

18. *Driven Hooke's Law Spring*   In a Hooke's Law spring model ODE, suppose that the spring constant is 1.01 newtons/m and the mass of the body is 1 kg. The body is acted on by viscous damping with a coefficient equal to 0.2 newton · sec/m, and driven by the periodic force (measured in newtons) described by ODE (1) with $F(t) = \text{SqWave}(t, 2\pi, \pi)$. Graph the response $y(t)$ of the system if it is at rest in its static deflection position $y = 0$ at $t = 0$. Interpret what you see.

3.2 Undriven Constant-Coefficient Linear Differential Equations

Comments:

We show how to find the general real-valued solution to the undriven, constant-coefficient ODE $y'' + ay' + by = 0$ in terms of the roots of the quadratic $r^2 + ar + b$, where a and b are real constants. We show that any two solutions of the ODE generate all the others if the two satisfy a certain condition, in which case the two solutions are a basic solution set. Add a pair of initial conditions to the ODE and there is exactly one solution. We like to do a lot of graphing of solution curves, especially with $y(t_0)$ fixed, but $y'(t_0)$ varying. This highlights once more that solution curves of normalized second-order ODEs require two items of initial data to determine a unique solution, that solution curves of a second-order ODE can intersect with impunity, and that one can imagine each point (t_0, y_0) in the ty-plane as a source of solution curves emerging from the point in all directions (except vertical), moving to the right if t increases from t_0, to the left if it decreases.

PROBLEMS

General Solution. Find the general solution of the ODEs below. [*Hint*: see Theorem 3.2.1.]

1.	$y'' - 4y = 0$	**2.**	$y'' = 0$	**3.**	$y'' - 4y' + 4y = 0$
4.	$y'' + y' - 2y = 0$	**5.**	$5y'' - 10y' = 0$	**6.**	$2y'' + 12y' + 18y = 0$
7.	$y'' + 2y = 0$	**8.**	$y'' + 4y' - y = 0$	**9.**	$y'' + 2y' + y = 0$
10.	$y'' + 2y' + 101y = 0$	**11.**	$y'' - 2y' + 101y = 0$	**12.**	$y'' + 4\pi^2 y = 0$

Answer 1: The ODE $y'' - 4y = 0$ has the characteristic polynomial $r^2 - 4$ which has roots $r_1 = 2$ and $r_2 = -2$. So the general solution is $y(t) = C_1 e^{2t} + C_2 e^{-2t}$, where C_1 and C_2 are arbitrary real numbers.

Answer 3: The ODE $y'' - 4y' + 4y = 0$ has the characteristic polynomial $r^2 - 4r + 4$ which has a double root, $r = 2$. Using this root of the characteristic polynomial, the general solution is $y(t) = C_1 t e^{2t} + C_2 e^{2t}$, where C_1 and C_2 are arbitrary real numbers.

Answer 5: The ODE $5y'' - 10y' = 0$ has the characteristic polynomial $5r^2 - 10r$ which has roots $r_1 = 0$ and $r_2 = 2$. So the general solution is $y(t) = C_1 + C_2 e^{2t}$, where C_1 and C_2 are arbitrary real numbers.

Answer 7: The ODE $y'' + 2y = 0$ has the characteristic polynomial $r^2 + 2$ which has roots $r_1 = -\sqrt{2}i$, $r_2 = \sqrt{2}i$. So the general solution is $y(t) = C_1 \cos\sqrt{2}t + C_2 \sin\sqrt{2}t$, where C_1 and C_2 are arbitrary real numbers.

Answer 9: The ODE $y'' + 2y' + y = 0$ has the characteristic polynomial $r^2 + 2r + 1$ which has a double root, $r = -1$. So the general solution is $y(t) = C_1 t e^{-t} + C_2 e^{-t}$, where C_1 and C_2 are arbitrary real numbers.

Answer 11: The ODE $y'' - 2y' + 101y = 0$ has the characteristic polynomial $r^2 - 2r + 101$, which has roots $r_1 = 1 - 10i, r_2 = 1 + 10i$. So the general solution is $y(t) = e^t(C_1 \cos 10t + C_3 \sin 10t)$, where C_1 and C_2 are arbitrary real numbers.

IVPs. Solve the IVPs below. Plot the solution curves.

13.	$y'' + y' = 0$, $y(0) = 1$, $y'(0) = 2$	**14.**	$y'' + 3y' + 2y = 0$, $y(0) = 0$, $y'(0) = 1$
15.	$y'' - 9y = 0$, $y(0) = 2$, $y'(0) = -1$	**16.**	$y'' - 4y' + 4y = 0$, $y(0) = 1, y'(0) = 1$
17.	$y'' + 6y = 0$, $y(0) = 1$, $y'(0) = -1$	**18.**	$y'' - 25y = 0$, $y(1) = 0, y'(1) = 1$

Answer 13: The ODE $y'' + y' = 0$ has characteristic polynomial $r^2 + r$ which has roots $r_1 = 0$ and $r_2 = -1$. So the general solution is $y(t) = C_1 + C_2 e^{-t}$. Using $y(0) = 1$ and $y'(0) = 2$ gives $C_1 + C_2 = 1$, $-C_2 = 2$, and so $C_1 = 3$ and $C_2 = -2$. So our final solution is $y(t) = 3 - 2e^{-t}$. See the figure.

Answer 15: The characteristic polynomial of $y'' - 9y = 0$ has roots ± 3. The general solution is $y(t) = C_1 e^{3t} + C_2 e^{-3t}$. Applying initial conditions $y(0) = 2$ and $y'(0) = -1$ gives $C_1 + C_2 = 2$ and $3C_1 - 3C_2 = -1$, so $C_1 = 5/6$ and $C_2 = 7/6$. The final solution is $y(t) = 5/6e^{3t} + 7/6e^{-3t}$. See the figure.

Answer 17: The characteristic polynomial of $y'' + 6y = 0$ has roots $-\sqrt{6}i$ and $+\sqrt{6}i$. The general solution is $y(t) = C_1 \cos \sqrt{6}t + C_2 \sin \sqrt{6}t$. Applying initial conditions $y(0) = 1$ and $y'(0) = -1$ gives $C_1 = 1$ and $C_2 = -1/\sqrt{6}$. The final solution is $y(t) = \cos \sqrt{6}t - (1/\sqrt{6}) \sin \sqrt{6}t$. See the figure.

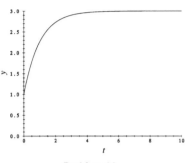

Problem 13.

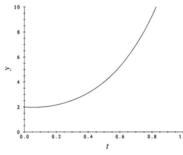

Problem 15.

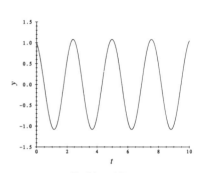

Problem 17.

Given a Solution, Find the ODE. Find an ODE of the form $y'' + ay' + by = 0$, where a and b are real constants, for which the given function is a solution, or else explain why no such ODE exists. [*Hint*: look at the exponents. For example, $e^{2t} - 7e^{3t}$ suggests that $r_1 = 2$ and $r_2 = 3$ are characteristic roots. So $(r - 2)(r - 3) = r^2 - 5r + 6$ is the characteristic polynomial and the ODE is $y'' - 5y' + 6y = 0$.]

19. $e^t - e^{-t}$ **20.** $e^t - te^t$ **21.** $e^{-t} + e^{-2t}$ **22.** $1 + e^{-3t}$

23. $e^{2t} + 10000e^{3t}$ **<u>24.</u>** $e^{\sqrt{2}t} + e^{-\sqrt{2}t}$ **25.** $e^{\pi t} - 3$ **26.** e^{t^2}

27. $t^2 e^{-t}$ **28.** $t + 2$ **29.** $e^{-5t} - e^{-2t}$ **30.** $3e^{4t} + 2te^{4t}$

<u>31.</u> $e^t \sin 2t$ **32.** $7e^{-t} \cos \pi t + e^{-t} \sin \pi t$ **33.** $8e^{-4t} \sin t$

Answer 19: In the case of $e^t - e^{-t}$ we see that the roots are $r_1 = 1$ and $r_2 = -1$. So the characteristic polynomial must have roots 1 and -1: $(r - r_1)(r - r_2) = (r - 1)(r + 1) = r^2 - 1$. The original ODE is $y'' - y = 0$.

Answer 21: In the case of $e^{-t} + e^{-2t}$ we see that the roots are $r_1 = -1$ and $r_2 = -2$: So $(r + 1)(r + 2) = r^2 + 3r + 2$. The original ODE is $y'' + 3y' + 2y = 0$.

Answer 23: In the case of $e^{2t} + 10000e^{3t}$ we see that the roots are $r_1 = 2$ and $r_2 = 3$. So the characteristic polynomial is $(r - 2)(r - 3) = r^2 - 5r + 6$. The ODE is $y'' - 5y' + 6y = 0$.

Answer 25: In the case of $e^{\pi t} - 3$ we see that the roots are $r_1 = \pi$ and $r_2 = 0$. So the characteristic polynomial is $(r - \pi)r = r^2 - r\pi$ and the original ODE is $y'' - \pi y' = 0$.

Answer 27: In the case of $t^2 e^{-t}$ we see that t^2 in the coefficient cannot occur in the solution of an undriven linear *second order* ODE with constant coefficients. (It could occur in the solution of a third or higher order ODE, but we aren't concerned with that here.)

Answer 29: Here, $r_1 = -5$, $r_2 = -2$ so the characteristic polynomial is $(r + 5)(r + 2) = r^2 + 7r + 10$. The ODE is $y'' + 7y' + 10y = 0$.

Answer 31: The characteristic roots are the complex conjugates $1 + 2i$ and $1 - 2i$. The characteristic polynomial is $(r - 1 - 2i)(r - 1 + 2i) = r^2 - 2r + 5$, and the ODE is $y'' - 2y' + 5y = 0$.

Answer 33: In this case a pair of complex roots $-4 \pm i$ must be involved. The characteristic polynomial is $(r + 4 - i)(r + 4 + i) = r^2 + 8r + 17$. The ODE is $y'' + 8y' + 17y = 0$.

⬜ **Solution Curve.** Find the solution for which $y(-1) = 2$, $y'(-1) = 0$. Graph the solution curve.

34. $y'' = 0$ **35.** $y'' + y' - 2y = 0$ **36.** $y'' - 4y' + 4y = 0$

37. $y'' = 1$ **38.** $y'' - 4y' + 5y = 0$

Answer 35:   Applying the initial conditions $y(-1) = 2$, $y'(-1) = 0$ to the general solution $y(t) = C_1 e^{-2t} + C_2 e^t$, we have $C_1 e^2 + C_2 e^{-1} = 2$ and $-2C_1 e^2 + C_2 e^{-1} = 0$. Solving we obtain $C_1 = (2/3)e^{-2}$, $C_2 = (4/3)e$. So $y = (2/3)e^{-2(t+1)} + (4/3)e^{t+1}$ is the solution of the IVP. See the figure.

Answer 37:   Integrating once, we have $y' = t + C_1$. Integrating again we have $y = t^2/2 + C_1 t + C_2$. Applying the initial conditions $y(-1) = 2$, $y'(-1) = 0$ to this general solution we have that $2 = 1/2 - C_1 + C_2$, $0 = -1 + C_1$. Therefore $C_1 = 1$ and $C_2 = 5/2$ and so $y = t^2/2 + t + 5/2$ solves the IVP. See the figure.

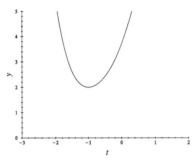
Problem 35.

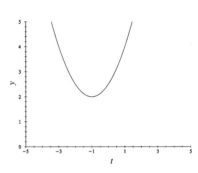
Problem 37.

IVPs, Solution Curves, Long-Term Behavior.   Find the real-valued solution. What happens as $t \to +\infty$? As $t \to -\infty$?

39. $y'' + 4y = 0$, $y(\pi) = 1$, $y'(\pi) = 0$

40. $y'' - y = 0$, $y(-1) = 0$, $y'(-1) = 1$

41. $y'' + 2y' + 2y = 0$, $y(0) = 1$, $y'(0) = 0$

42. $y'' + 4y' + 4y = 0$, $y(1) = 2$, $y'(1) = 0$

43. $y'' + y' - 6y = 0$, $y(-1) = 1$, $y'(-1) = -1$

Answer 39:   Since the characteristic polynomial for $y'' + 4y = 0$ is $r^2 + 4$, whose roots are $\pm 2i$, the general real-valued solution is $y = C_1 \cos 2t + C_2 \sin 2t$. The initial data $y(\pi) = 1$, $y'(\pi) = 0$ are used to find C_1 and C_2: $C_1 = 1$ and $C_2 = 0$. The solution is $y = \cos 2t$. As $t \to +\infty$, $y(t)$ oscillates within the bounds $|y| \leq 1$. The solution $y(t)$ is periodic with period π.

Answer 41:   Since the characteristic polynomial for $y'' + 2y' + 2y = 0$ is $r^2 + 2r + 2$, whose roots are $-1 \pm i$, the general real-valued solution is $y = e^{-t}(C_1 \cos t + C_2 \sin t)$. The initial data $y(0) = 1$, $y'(0) = 0$ are used to find C_1 and C_2: $C_1 = 1$ and $C_2 = 1$. The solution is $y = e^{-t}(\cos t + \sin t)$. As $t \to +\infty$, we have that $y(t) \to 0$ in the fashion of a decaying sinusoid. As $t \to -\infty$, the exponential factor $\to +\infty$, so the trig functions continue to oscillate with period 2π, but $e^{-t} \to +\infty$.

Answer 43:   The characteristic polynomial for $y'' + y' - 6y = 0$ is $r^2 + r - 6$ which has the roots -3 and 2. The general real-valued solution of the ODE is $y = C_1 e^{-3t} + C_2 e^{2t}$. The initial data $y(-1) = 1$, $y'(-1) = -1$ imply that $C_1 e^3 + C_2 e^{-2} = 1$, and $-3C_1 e^3 + 2C_2 e^{-2} = -1$. So $C_1 = 3e^{-3}/5$, $C_2 = 2e^2/5$ and $y = 3e^{-3(t+1)}/5 + 2e^{2(t+1)}/5$. As $t \to +\infty$ the second term in the solution formula tends to $+\infty$ and the first tends to 0, so $y \to +\infty$. As $t \to -\infty$, the situation is the other way around and $y \to +\infty$.

Wronskians and Basic Solution Sets.

44. Show that the Wronskian of every solution pair listed in (7) is never zero.

45. Suppose that $\{y_1, y_2\}$ is any basic solution set for ODE (6) and α, β, γ, δ are any constants such that $\alpha\delta - \beta\gamma \neq 0$. Show that if $z_1 = \alpha y_1 + \beta y_2$ and $z_2 = \gamma y_1 + \delta y_2$, then $\{z_1, z_2\}$ is also a basic solution set for ODE (6). [*Hint*: show that $W[z_1, z_2] = (\alpha\delta - \beta\gamma)W[y_1, y_2]$ by direct calculation.]

Answer 45: With $z_1 = \alpha y_1 + \beta y_2$, $z_2 = \gamma y_1 + \delta y_2$ we have

$$
\begin{aligned}
W[z_1, z_2] &= (\alpha y_1 + \beta y_2)(\gamma y_1' + \delta y_2') - (\alpha y_1' + \beta y_2')(\gamma y_1 + \delta y_2) \\
&= \alpha\delta y_1 y_2' + \beta\gamma y_2 y_1' - \alpha\delta y_1' y_2 - \beta\gamma y_2' y_1 \\
&= \alpha\delta(y_1 y_2' - y_1' y_2) - \beta\gamma(y_1 y_2' - y_1' y_2) \\
&= (\alpha\delta - \beta\gamma)W[y_1, y_2]
\end{aligned}
$$

Now $\{y_1, y_2\}$ is a basic solution set and $W[y_1, y_2] \neq 0$, for all t. So if $\alpha\delta - \beta\gamma \neq 0$, then $W[z_1, z_2] \neq 0$, for all t. Hence, $\{z_1, z_2\}$ is also a basic solution set.

46. *Basic Solution Sets* Find a basic solution set for the ODE $y'' - y' - 2y = 0$ other than the one given in Theorem 3.2.1.

47. Show that if $\{z_1, z_2\}$ is any basic solution set for ODE (6), then the general solution is $y = C_1 z_1 + C_2 z_2$, where C_1 and C_2 are arbitrary constants. [*Hint*: let $\{y_1, y_2\}$ be the basic solution set given in (7) and α, β, γ, δ be constants such that $z_1 = \alpha y_1 + \beta y_2$, $z_2 = \gamma y_1 + \delta y_2$. Conclude that $\alpha\delta - \beta\gamma \neq 0$ (see the Hint in Problem 45). Solve for y_1 and y_2 in terms of z_1 and z_2. Then apply Theorem 3.2.1.]

Answer 47: If $\{z_1, z_2\}$ is any given basic solution set and $\{y_1, y_2\}$ is the basic solution set given in (7), then there exist constants α, β, γ, δ such that

$$z_1 = \alpha y_1 + \beta y_2, \quad z_2 = \gamma y_1 + \delta y_2$$

with $\alpha\delta - \beta\gamma \neq 0$ (see the answer to Problem 45). Hence we can solve for y_1 and y_2 in terms of z_1 and z_2 to obtain

$$y_1 = \frac{\delta z_1 - \beta z_2}{\alpha\delta - \beta\gamma}, \quad y_2 = \frac{-\gamma z_1 + \alpha z_2}{\alpha\delta - \beta\gamma}$$

So for any constants C_1, C_2, we have

$$C_1 y_1 + C_2 y_@ = K_1 z_1 + K_2 z_2$$

for

$$K_1 = \frac{\delta C_1 - \gamma C_2}{\alpha\delta - \beta\gamma}, \quad K_2 = \frac{-\beta C_1 + \alpha C_2}{\alpha\delta - \beta\gamma}$$

Hence every solution of the ODE can be represented as $K_1 z_1 + K_2 z_2$, for suitable constants K_1, K_2. Conversely, since all the steps are reversible, for *any* constants K_1 and K_2 there are constants C_1 and C_2 such that $K_1 z_1 + K_2 z_2 = C_1 y_1 + C_2 y_2$. Therefore, $y = K_1 z_1 + K_2 z_2$, for arbitrary constants K_1 and K_2, is a general solution formula for the ODE.

Complex Characteristic Roots.

48. *Nodes and Complex-Conjugate Characteristic Roots* Suppose that the characteristic roots for the ODE $y'' + ay' + by = 0$ are $\alpha \pm \beta i$, $\beta \neq 0$.

(a) Show that the general solution of the ODE has the form $Ae^{\alpha t}\cos(\beta t - \delta)$, where A and δ are arbitrary real constants. [*Hint*: use the phase shift formula on the inside back cover.]

(b) Explain the nodes in Figure 3.2.1. What is the time span between successive nodes?

3.3 Visualizing Graphs of Solutions: Direction Fields

Comments:

In Section 3.2 we characterized the solution set of the second-order, constant coefficient, undriven, linear ODE, no matter whether the characteristic roots are real or complex. Now we generalize differentiation to complex-valued functions of a real variable. We show how to get real-valued solutions from complex-valued solutions when the ODE has real coefficients. In addition, you will visualize the solutions of a damped, undriven Hooke's Law model ODE, $y'' + ay' + by = 0$, in four ways: the ty- and ty'- curves, orbits in the yy'-state plane, and time0state curves in tyy'-space. Direction fields in the yy'-plane help you to visualize the behavior of orbits. You can even draw vector fields and orbits for some nonlinear ODEs of the form, $y'' = F(y, y')$.

PROBLEMS

Draw a Direction Field, Create a Portrait of Orbits. In Problems 1–8 draw a direction field for $y' = v$, $v' = F(y, v)$ in the rectangle $|y| \leq 5$, $|v| \leq 5$. Sketch a portrait of the orbits (including equilibrium points) in the rectangle. Do this by hand or with a numerical solver that has a direction field routine.

1. $F = 1$	**2.** $F = -2y$	**3.** $F = -4v$	**4.** $F = -y - v$
5. $F = 3y$	**6.** $F = -y - y^3$	**7.** $F = -5 \sin y$	**8.** $F = \cos y$

Answer 1: See the figure. There are no equilibrium points.

Answer 3: See the figure. All points on the y-axis are equilibrium points.

Answer 5: See the figure. There is a single equilibrium point at $(0, 0)$.

Answer 7: See the figure. The equilibrium points are $(n\pi, 0)$, $n = 0, \pm 1, \pm 2, \ldots$.

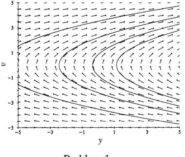

Problem 1.

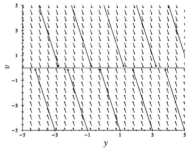

Problem 3.

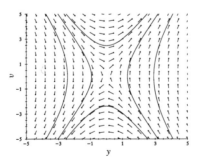

Problem 5.

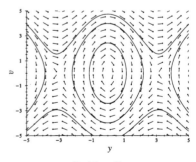

Problem 7.

Matching up a Direction Field with a System. In Problems 9–12, match the direction field with one of the eight systems $y' = v$, $v' = F(y, v)$, where F is given in (**a**)–(**h**). [*Hint*: see Examples 3.3.2–3.3.4. All direction fields here have $(0, 0)$ as an equilibrium point (and Problem 12 has two others near $(\pm 3, 0)$),]

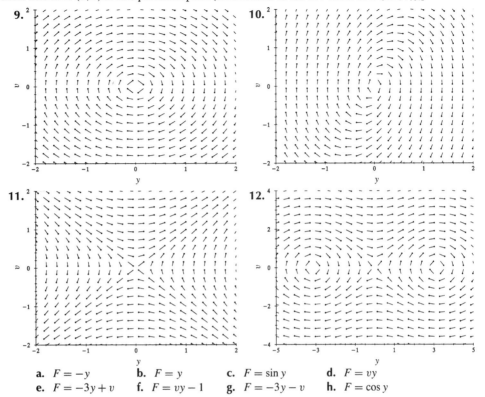

a. $F = -y$	**b.** $F = y$	**c.** $F = \sin y$	**d.** $F = vy$
e. $F = -3y + v$	**f.** $F = vy - 1$	**g.** $F = -3y - v$	**h.** $F = \cos y$

Answer 9: **a**; since $dv/dy = -y/v$, orbits are circles, $y^2 + v^2 = $ constant.

Answer 11: **b**; the system $y' = v$, $v' = y$ has orbits $y^2 - v^2 = c$ since $dy/dv = v/y$.

More Direction Fields and Orbital Portraits.. For each function F given below, plot a direction field of the system $y' = v$, $v' = F(y, v)$ on the given rectangle. Mark all equilibrium points. Plot enough orbits that a clear orbital portrait can be seen. Describe what each orbit does as t increases and as t decreases. [*Hint*: orbits move left to right for $v > 0$, the other way around for $v < 0$.]

13. $F = -4y - 5v$; $|y| \le 2$, $|v| \le 2$ **14.** $F = -4y + 5v$; $|y| \le 2$, $|v| \le 2$

15. $F = -2y - 3v$; $|y| \le 5$, $|v| \le 5$ **16.** $F = -3y - v$; $|y| \le 2$, $|v| \le 2$

17. $F = -100.25y - v$; $|y| \le 2$, $|v| \le 20$ **18.** $F = -y/16 - v/2$; $|y| \le 2$, $|v| \le 2$

19. $F = -10y - 0.2v - 9.8$; $|y| \le 5$, $|v| \le 5$

Answer 13: An equilibrium point at $(0, 0)$; nonconstant orbits shown tend to $(0, 0)$ tangent to the line $v = -y$ as $t \to +\infty$ and become unbounded as $t \to -\infty$.

Answer 15: An equilibrium point at $(0, 0)$; nonconstant orbits shown tend to $(0, 0)$ tangent to line $v = -y$ as $t \to +\infty$ and become unbounded as $t \to -\infty$.

Answer 17: An equilibrium point at $(0, 0)$; nonconstant orbits spiral clockwise toward $(0, 0)$ as $t \to +\infty$ and become unbounded as $t \to -\infty$.

Answer 19: An equilibrium point at $(-.98, 0)$; nonconstant orbits spiral clockwise toward this point as $t \to +\infty$, but become unbounded as $t \to -\infty$.

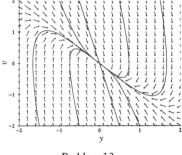

Problem 13.

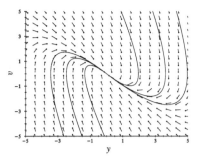

Problem 15.

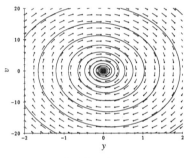

Problem 17.

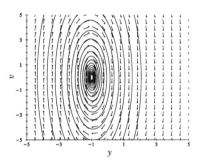

Problem 19.

General Real-Valued Solution.   Find all real-valued solutions of each ODE.

20. $y'' - 6y' + 10y = 0$ **<u>21.</u>** $y'' - 4y' + 5y = 0$ **22.** $y'' - 2y' + 5y = 0$

Answer 21:   The characteristic polynomial is $r^2 - 4r + 5$ whose roots are $2 \pm i$. So, $y = e^{2t}(C_1 \cos t + C_2 \sin t)$ gives the real-valued solutions, where C_1 and C_2 are arbitrary real constants.

General Solutions, Solution Curves, Orbits.   Find the general real-valued solution of each ODE. Plot solution curves in the ty-plane for $-1 \leq t \leq 5$, where $y(0) = 1$, $y'(0) = -6, -3, 0, 3, 6$. Plot the corresponding orbits and describe their behavior.

23. $y'' + y' = 0$ **24.** $y'' + 2y' + 65y = 0$ **25.** $y'' + 3y' + 2y = 0$

26. $y'' + 10y = 0$ **27.** $y'' - y/9 = 0$ **28.** $y'' - 3y'/4 + y/8 = 0$

Answer 23:   The characteristic polynomial for $y'' + y' = 0$ is $r^2 + r$, which factors to $(r + 1)r$. Thus, solutions of the ODE are given by e^{-t}, $e^{0 \cdot t}$, and the linear combinations of these functions. That is, the real-valued solutions are given by $y = C_1 e^{-t} + C_2$. From the figures below and the solution formulas we see that as $t \to +\infty$, $y(t) \to C_2$, while (if $C_1 \neq 0$), as $t \to -\infty$, $y(t) \to \pm\infty$ depending on the sign of C_1 (Graph 1). The nonconstant orbits (Graph 2) are slanted rays that approach the y-axis in the yy'-plane as $t \to +\infty$. The constant orbits are all the points on the $y' = 0$ line; these points are the equilibrium points of the equivalent system $y' = v$, $v' = -v$.

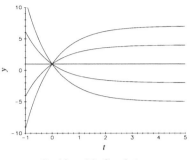

Problem 23, Graph 1.

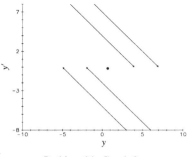

Problem 23, Graph 2.

Answer 25: The characteristic polynomial of the ODE $y'' + 3y' + 2y = 0$ is $r^2 + 3r + 2$ which factors to $(r+1)(r+2)$. So the real-valued solutions are given by $y = C_1 e^{-t} + C_2 e^{-2t}$. From the formula and Graph 1, as $t \to +\infty$ nonconstant solutions decay to 0, but as $t \to -\infty$, solutions become unbounded. The orbits (Graph 2) tend to the origin of the yy'-plane with definite tangents as $t \to +\infty$.

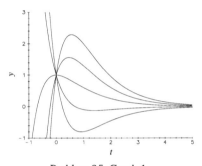

Problem 25, Graph 1.

Problem 25, Graph 2.

Answer 27: The characteristic polynomial of the ODE $y'' - y/9 = 0$ is $r^2 - 1/9$ which factors to $(r + 1/3)(r - 1/3)$. The real-valued solutions are given by $y = C_1 e^{-t/3} + C_2 e^{t/3}$. The solution curves (Graph 1) become unbounded as $t \to \pm\infty$ if C_1 and C_2 are nonzero. The time span used here is fairly short, so we only see arcs of orbits (in some cases) that would move out of the rectangle if the time span, both forward and backward, were longer (Graph 2). In Graph 3 we have enlarged the xy-rectangle and lengthened the time span, so we can really see what is going on.

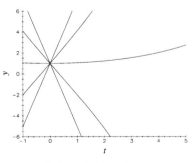

Problem 27, Graph 1.

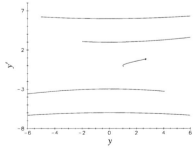

Problem 27, Graph 2.

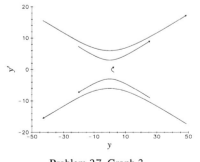

Problem 27, Graph 3.

Effect of Coefficients on Solutions of ODEs.

29. *Decaying Solutions*   Consider the ODE $y'' + ay' + by = 0$, where a and b are real constants.

(a) Show that if $a > 0$, $b > 0$, then every solution tends to zero as $t \to +\infty$. [*Hint*: treat the cases $a^2 > 4b$, $a^2 = 4b$ and $a^2 < 4b$ separately.]

(b) Show that if every solution tends to zero as $t \to +\infty$, then $a > 0$ and $b > 0$.

Answer 29:

(a) The roots of the characteristic polynomial $r^2 + ar + b$ are $-a/2 \pm (a^2/4 - b)^{1/2}$. If $(a/2)^2 > b$, so both roots r_1 and r_2 are negative since $((a/2)^2 - b)^{1/2} < a/2$. In this case, the general solution $y = C_1 e^{r_1 t} + C_2 e^{r_2 t} \to 0$ as $t \to +\infty$. If $b = (a/2)^2$, then there is a double root $r_1 = -a/2$ and the general solution is $y = C_1 e^{r_1 t} + C_2 t e^{r_1 t}$ and again $y \to 0$ as $t \to +\infty$ since r_1 is negative and $t e^{r_1 t} \to 0$ as $t \to +\infty$ (apply L'Hopital's Rule to $t/e^{-r_1 t}$). If $b > (a/2)^2$, then the roots $\alpha \pm i\beta$ are imaginary with real part $-a/2 = \alpha$, which is negative. The general solution $y = e^{\alpha t}(C_1 \cos \beta t + C_2 \sin \beta t) \to 0$ as $t \to +\infty$ because $e^{\alpha t} \to 0$, while $C_1 \cos \beta t + C_2 \sin \beta t$ oscillates and remains bounded.

(b) Suppose all solutions $y(t) \to 0$ as $t \to +\infty$. Then the real parts of r_1 and r_2 must both be negative as can be seen by inspecting the three formulas in Theorem 3.2.1. Let $r^2 + ar + b = (r - r_1)(r - r_2) = r^2 - (r_1 + r_2)r + r_1 r_2$. Now either r_1 and r_2 are both real, or else $r_1 = \alpha + i\beta$, $r_2 = \alpha - i\beta$, α and $\beta \neq 0$ real. In either case $a = -(r_1 + r_2)$, $b = r_1 r_2$. In the first case (r_1 and r_2 real), $a > 0$ and $b > 0$ since r_1 and r_2 are negative. In the second case (r_1 and r_2 imaginary), $a > 0$ and $b > 0$ since $a = -(r_1 + r_2) = -2\alpha$, where α is negative and $b = r_1 r_2 = \alpha^2 + \beta^2 > 0$.

30. *Positively Bounded Solutions*   A solution $y(t)$ of $y'' + ay' + by = 0$, where a and b are real numbers, is *positively bounded* if there is a positive constant M such that $|y(t)| \leq M$ for all $t \geq 0$.

(a) Show that all solutions are positively bounded if $a \geq 0$, $b \geq 0$, but not both a and b are zero.

(b) Show that if all solutions of the ODE are positively bounded, then $a \geq 0$, $b \geq 0$, but not both a and b are zero.

Real-Valued Solutions from Complex-Valued Solutions.
Let $y(t) = u(t) + iv(t)$ solve the ODE $y'' + ay' + by = 0$, where a and b are real numbers, and both $u(t)$ and $v(t)$ are *real-valued* functions.

31. Show that $u(t)$ and $v(t)$ are both real-valued solutions of the ODE.

Answer 31: We can show this as follows:

$$(u + iv)'' + a(u + iv)' + b(u + iv) = 0$$

or, collecting real and imaginary parts,

$$u'' + au' + bu + i(v'' + av' + bv) = 0 + 0i$$

But two complex quantities are equal if and only if the real parts match up and the imaginary parts match up. So $u'' + au' + bu = 0$, $v'' + av' + bv = 0$, and we are done.

32. If $e^{(\alpha + i\beta)t}$ solves the ODE for some real numbers α and β, then show that $u(t) = \mathrm{Re}\left[e^{(\alpha + i\beta)t}\right]$ and $v(t) = \mathrm{Im}\left[e^{(\alpha + i\beta)t}\right]$ are real-valued solutions. Show that $e^{(\alpha - i\beta)t}$ is also a solution.

33. Find an ODE with real coefficients that has $u(t) = \mathrm{Re}\left[e^{(-1 + 2i)t}\right]$ as a real-valued solution.

Answer 33:   The real part of $e^{(-1 + 2i)t}$ is $e^{-t} \cos 2t$. The roots of the characteristic polynomial of the ODE, $y'' + ay' + by = 0$, must be $r_1 = -1 + 2i$ and $r_2 = -1 - 2i$. So $(r - r_1)(r - r_2) = r^2 + 2r + 5$ is the characteristic polynomial and $a = 2$, $b = 5$; an ODE that $e^{-t} \cos 2t$ solves is $y'' + 2y' + 5 = 0$.

ODEs with Complex Coefficients.
Find the complex-valued general solution of the ODE. Give all the real-valued solutions (if any) of the ODE. [*Hint*: use *De Moivre's Formula* in the WEB SPOTLIGHT ON COMPLEX-VALUED FUNCTIONS to find the characteristic roots, if necessary.]

34. $y'' + iy' + 2y = 0$ <u>**35.**</u> $y'' + iy = 0$.

36. $y'' + (1 + i)y' + iy = 0$ **37.** $y'' + (-1 + 2i)y' - (1 + i)y = 0$

Answer 35: Let z be a root of $r^2 + i$. Then $z = r_0 e^{i\theta}$ and $z^2 = r_0^2 e^{2i\theta} = -i = e^{(3\pi/2 + 2n\pi)i}$. Matching magnitudes and angles, we have that $r_0 = 1$ and $2\theta = 3\pi/2 + 2n\pi$. The only two distinct angles correspond to $n = 0, 1$. So, $r_0 = 1$, $\theta_0 = 3\pi/4$, $\theta_1 = 7\pi/4$, and the two roots of $r^2 + i$ are $e^{3\pi i/4} = (-1 + i)/\sqrt{2}$ and $e^{7\pi i/4} = (1 - i)/\sqrt{2}$. The general solution is $y = k_1 e^{(-1+i)t/\sqrt{2}} + k_2 e^{(1-i)t/\sqrt{2}}$, where k_1 and k_2 are any complex constants. If $y(t)$ is a real solution, then y'' is real and $y'' = iy$ implies that $y = 0$ all t, is the only real solution.

Answer 37: The characteristic polynomial is $r^2 + (-1 + 2i)r - 1 - i$, which factors to $(r + i)(r - 1 + i)$. The functions e^{-it} and $e^{(1-i)t}$ are solutions. The general solution is $y = k_1 e^{-it} + k_2 e^{(1-i)t}$, k_1 and k_2 any real constants. If $y(t)$ is a real solution than the ODE implies that $2iy' - iy = 0$, so $y' - y/2 = 0$ and $y = ce^{t/2}$, c any real constant, are the only real solutions.

3.4 Periodic Solutions: Simple Harmonic Motion

Comments:

Periodic solutions play an important role in the study of ODEs, so we thought we should give them a section of their own. Readers probably have seen periodic functions before, but they may have forgotten the terminology associated with them. It is especially important to read the discussion of when the sum of two periodic functions is periodic (and how to find the period of the sum). The simple harmonic motion examples (Example 3.4.1) should be read carefully. Example 3.4.2 brings up the important point that when tracking a periodic solution (or an oscillating solution for that matter) with a numerical solver, decreasing the number of plotted points per unit time causes an interesting modulation effect on the output (aliasing). Not much is made of this phenomenon other than to say that when using a numerical solver it's important to try various numbers of plotted points before accepting the results. Incidentally, the aliasing effect is the reason stage-coach wheels in the movies often seem to rotate slowly or even rotate backwards. It all depends on the frame rate of the movie cameras.

PROBLEMS

Periodic Functions. Determine whether each function is periodic. If it is periodic, find its period, frequency, and circular frequency.

1. $3\cos 4t$ **2.** $3\cos 4t + 4\sin 4t$ **3.** $\cos 3t + \sin 5t$

4. $\sin t + \sin \pi t$ **5.** $\sin 2\pi t + \cos 5\pi t$ **6.** $\sin t + \cos \sqrt{2}t$

Answer 1: Periodic with period $2\pi/4 = \pi/2$, frequency $2/\pi$, and circular frequency 4.

Answer 3: $T_1 = 2\pi/3$ and $T_2 = 2\pi/5$ are the respective periods of the two summands. So the sum is periodic since $T_1/T_2 = 5/3$ which is a rational number. The period of the sum is 2π since $3T_1 = 5T_2 = 2\pi$. The frequency is $1/(2\pi)$ and the circular frequency is 1.

Answer 5: The respective periods of the two summands are $T_1 = 1$ and $T_2 = 2/5$. The sum is periodic since $T_1/T_2 = 5/2$, which is a rational number. Since $2T_1 = 5T_2 = 2$, the period is 2, the frequency is $1/2$ and the circular frequency is π.

Phase/Amplitude Form. Write each function in the form $A\sin(\omega t + \theta)$ for some phase angle θ, $-\pi \leq \theta < \pi$. Sketch the graph of $f = A\sin \omega t$ and $g = A\sin(\omega t + \theta)$. [*Hint*: see Example 3.4.1 and the paragraph following it.]

7. $-3\cos \pi t + 4\sin \pi t$ **8.** $5\sin 2t + 7\cos 2t$ **9.** $\cos t + 3\sin t$

10. $3\cos \pi t - 4\sin \pi t$ **11.** $2\sin 2t - 3\cos 2t$ **12.** $5\sin(7t + 7\pi/2)$

Answer 7: From Example 3.4.1, we have $-3\cos \pi t + 4\sin \pi t = \sqrt{4^2 + (-3)^2}\sin(\pi t + \theta) = 5\sin(\pi t + \theta)$, $\cos\theta = 4/5$ and $\sin\theta = -3/5$. The signs of $\cos\theta$ and $\sin\theta$ imply that $-\pi/2 < \theta < 0$ since θ is restricted to the interval $-\pi \le \theta < \pi$. In this case θ is negative, so the graph of $g = 5\sin(\pi t + \theta)$ is the same as the graph of $f = 5\sin \pi t$, but shifted $|\theta|/\pi$ units to the right. So $A = 5$ and $\theta = \arcsin(-3/5)$ here. See the figure for the graphs of f (dashed) and g (solid).

Answer 9: The example tells you that $\cos t + 3\sin t = \sqrt{3^2 + 1^2}\sin(t + \theta) = \sqrt{10}\sin(t + \theta)$ where $\cos\theta = 3/\sqrt{10}$ and $\sin\theta = 1/\sqrt{10}$. The value of θ lies between 0 and $\pi/2$ since $\sin\theta > 0$, $\cos\theta > 0$ and $-\pi \le \theta < \pi$. Since θ is positive, the graph of $g = \sqrt{10}\sin(t + \theta)$ is the same as the graph of $f = \sqrt{10}\sin t$ shifted $\theta/1 = \theta$ units to the left. Here $A = \sqrt{10}$, $\theta = \arcsin(1/\sqrt{10})$. See the figure for the graphs of f (dashed) and g (solid).

Answer 11: According to the example, $2\sin 2t - 3\cos 2t = \sqrt{2^2 + 3^2}\sin(2t + \theta) = \sqrt{13}\sin(2t + \theta)$, where $\cos\theta = 2/\sqrt{13}$ and $\sin\theta = -3/\sqrt{13}$. So θ is in the fourth quadrant, $-\pi/2 < \theta < 0$. Since θ is negative, the graph of $g = \sqrt{13}\sin(2t + \theta)$ is that of $f = \sqrt{13}\sin 2t$ shifted to the right by $|\theta|/2$; $A = \sqrt{13}$, $\theta = \arcsin(-3/\sqrt{13})$. See the figure for the graphs of f (dashed) and g (solid).

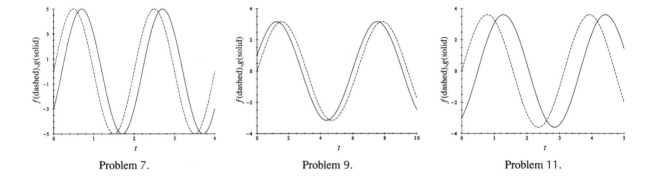

Problem 7.   Problem 9.   Problem 11.

Periodic Function Facts.

13. Show that if $f(t)$ has period T, then $f(\omega t)$ has period T/ω.

14. Suppose $f(t)$ and $g(t)$ have periods T and S, respectively, with T/S rational. Show that $h(t) = f(t) + g(t)$ is periodic with a period equal to the smallest value k such that $k = mT = nS$ for positive integers m and n. Illustrate this claim by graphing $f(t) = \sin(2t)$, $g(t) = \sin(5t)$, and $h(t) = f(t) + g(t)$ on the same plot. What are the periods of these functions?

Answer 13: If $f(t)$ has period T, then T is the smallest positive number such that for any time t, $f(t + T) = f(t)$. Let $g(t) = f(\omega t)$ and suppose that $g(t)$ has period T_1; that is, T_1 is the smallest positive number such that $g(t + T_1) = g(t)$. Then $f(\omega(t + T_1)) = f(\omega t)$ by the definition of $g(t)$, so $f(\omega t + \omega T_1) = f(\omega t)$. But the smallest positive value of T_1 for which this holds satisfies $\omega T_1 = T$ since T is the period of f. So, $T_1 = T/\omega$ is the period of $g(t) = f(\omega t)$.

Periodic Solutions of ODEs. Find a formula for all real-valued solutions of each ODE; find all periodic solutions (if any) and their periods. [*Hint*: for the driven ODEs see Example 3.4.3.]

15. $y'' + 25y = 0$ **16.** $y'' + \pi^2 y = 0$ **17.** $y'' + 25y = 5\cos 6t$

18. $y'' + y = \cos \pi t$ **19.** $y'' + \pi^2 y = 2\cos t$ **20.** $y'' - y = \cos t$

Answer 15: The roots of the characteristic polynomial $r^2 + 25$ are $\pm 5i$, so the general solution of $y'' + 25y = 0$ is $y = C_1 \sin 5t + C_2 \cos 5t$, C_1 and C_2 arbitrary constants. Every nonconstant solution is periodic with period $2\pi/5$.

Answer 17: Following the approach of Example 3.4.3, the general solution of the driven ODE, $y'' + 25y = 5\cos 6t$ is the sum of any one particular solution $y_d(t)$ of the driven ODE and the general solution $y_u(t)$ of the undriven ODE, $y'' + 25y = 0$. Guess a y_d of the form $A \sin 6t + B \cos 6t$. Insert this form for y_d into $y'' + 25y = 5 \cos 6t$:

$$y_d'' + 25y_d = -36A \sin 6t - 36B \cos 6t + 25A \sin 6t + 25B \cos 6t$$
$$= -11A \sin 6t - 11B \cos 6t = 0 \sin 6t + 5 \cos 6t$$

Matching the coefficients of like terms, we see that $A = 0$ and $B = -5/11$. So $y_d = -(5/11) \cos 6t$ is a particular solution of $y'' + 25y = 5 \cos 6t$.

Since the characteristic polynomial of $y'' + 25y$ is $r^2 + 25$, the characteristic roots are $\pm 5i$ and so the general solution of the undriven ODE is $y_u = C_1 \sin 5t + C_2 \cos 5t$, C_1 and C_2 arbitrary constants. The general real-valued solution of the driven ODE is

$$y(t) = y_d + y_u = -(5/11) \cos 6t + C_1 \sin 5t + C_2 \cos 5t.$$

The period of the first term in the solution formula is $T_1 = 2\pi/6 = \pi/3$ and the period of each of the other two terms is $T_2 = 2\pi/5$. Since $T_1/T_2 = 5/6$, a rational number and since $6T_1 = 5T_2 = 2\pi$ and 6 and 5 have no common divisors, the period of $y(t)$ is 2π, except that the solution y_d (corresponding to $C_1 = C_2 = 0$) has period $\pi/3$.

Answer 19: Following Example 3.4.3, we see that the driven ODE, $y'' + \pi^2 y = 2 \cos t$, has a solution y_d of the form $A \sin t + B \cos t$. Inserting y_d into the ODE, we obtain $(-A + \pi^2 A) \sin t + (-B + \pi^2 B) \cos t = 0 \sin t + 2 \cos t$. Matching coefficients of like terms, we see that $A = 0$ and $B = 2/(\pi^2 - 1)$. So $y_d = (2 \cos t)/(\pi^2 - 1)$. Since $\pm \pi i$ are the roots of the characteristic polynomial $r^2 + \pi^2$, the general solution of the undriven ODE, $y' + \pi^2 y = 0$ is $y_u = C_1 \sin \pi t + C_2 \cos \pi t$, C_1 and C_2 arbitrary constants. The general solution of $y'' + \pi^2 y = 2 \cos t$ is

$$y = y_d + y_u = (2 \cos t)/(\pi^2 - 1) + C_1 \sin \pi t + C_2 \cos \pi t$$

Since the first term is this formula has period $T_1 = 2\pi$ and the other two have period $T_2 = 2$ and the ratio $T_1/T_2 = \pi$, which is not a rational number, the general solution is not periodic, except that y_d (corresponding to $C_1 = C_2 = 0$) has period 2π.

21. Does the ODE $y'' + 4y = \sin 2t$ have any periodic solutions? [*Hint:* use the approach in Example 3.4.3, but with a guess of $y_d(t) = At \sin 2t + Bt \cos 2t$ for constants A and B.]

Answer 21: To determine whether the ODE $y'' + 4y = \sin 2t$ has any periodic solutions, we will find and examine the general solution. Note that the general solution of the undriven ODE is $C_1 \sin 2t + C_2 \cos 2t$. Now assume a particular solution of the form $y_d = At \sin 2t + Bt \cos 2t$. Calculating $y_d'' + 4y_d$ and equating it to $\sin 2t$, we obtain $4A \cos 2t - 4B \sin 2t = \sin 2t$. Equating the coefficients of like terms on either side of the equation, we see that $A = 0$ and $B = -1/4$. The general solution of the driven equation is $y(t) = C_1 \sin 2t + C_2 \cos 2t - (1/4)t \cos 2t$. There are no periodic solutions because of the factor t that is always present, although all solutions do oscillate with oscillation time π.

22. Show that the ODE $y'' + ay' + by = 0$, for real constants a and b, has a periodic solution if and only if $a = 0$ and $b > 0$.

23. *Aliasing, Amplitude Modulation* Duplicate Figures 3.4.2–3.4.3. Explain the modulation phenomenon in Figures 3.4.2 and 3.4.3. Formulate a prediction about how the modulation varies with the sampling rate, and verify your prediction experimentally. Repeat with a simple harmonic oscillator of your choice and using several different numbers of plotted points.

Answer 23: Group project.

Modeling Problem.

☞ See the Spotlight on Modeling: Vertical Motion.

24. *Gravity as a Source of Energy* It can be shown that the gravitational force exerted on a body at a distance r from the center of the earth arises solely from the mass $M(r)$ of that part of the earth closer to the earth's center than r. Hence, if a body of mass m is *inside* the earth at a distance r from the earth's center, the gravitational force **F** acting on the body is given by

$$\mathbf{F} = -\frac{GM(r)m}{r^2}\hat{\mathbf{r}} = -\frac{G\left(\frac{4}{3}\pi r^3 \rho\right)m}{r^2}\hat{\mathbf{r}} = -\frac{4\pi \rho Gm}{3}r\hat{\mathbf{r}} = -\frac{4\pi \rho Gm}{3}\mathbf{r}$$

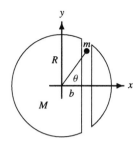

where **r** is the vector from the earth's center to the body's center, $\hat{\mathbf{r}}$ is the unit vector in the direction of **r**, ρ is the density of the earth (assumed to be constant), and G is the universal gravitational constant. Now assume that a tunnel is bored through the earth and an object of mass m is dropped into it. Neglect friction and air resistance.

(a) Show that if the hole is drilled through the center of the earth, then, if we use the notation indicated in the margin figure with $b = 0$, the equation of motion of the body is given by

$$my'' = -\frac{4\pi\rho Gm}{3}y$$

where y is measured from the center of the earth. Given that $\rho = 5.51 \times 10^3$ kg/m³, $G = 6.67 \times 10^{-11}$ Nm²/kg², and $R \approx 6.38 \times 10^6$ m, show that the body returns to the starting point in roughly 5070 sec (84.5 min). Calculate its speed at the center of the earth.

(b) Show that if the hole does not go through the center of the earth, then the period of oscillation is the same as in **(a)**, but the velocity when the body is midway through the hole will be different. This has been proposed for rapid transit systems of the future. [1]

3.5 Driven Linear ODEs: Undetermined Coefficients I

Comments

We do two things in this section: first, we present the idea of the Method of Undetermined Coefficients, and second, we take the reader through the algebraic intricacies of actually constructing a particular solution. A CAS would be helpful for the algebra. We have included many examples, because that is probably the only way to see how to carry out the undetermined coefficients approach. Although we do some numerical solving/plotting, the geometry of solution graphs is *not* the point of this section. The point is the construction of solution formulas. **Warning:** some problems involve a lot of crank work.

PROBLEMS

Using Operators.   Apply each operator to the function $y(t) = e^{-t}$.

   1. $D - 2$ **2.** $D + 3$ **3.** $D^2 + D - 6$

   4. $(D - 2)(D + 3)$ **5.** $(D + 3)(D - 2)$ **6.** $(D + 1)^2$

   Answer 1: $(D - 2)[e^{-t}] = D[e^{-t}] - 2e^{-t} = -e^{-t} - 2e^{-t} = -3e^{-t}$.

   Answer 3: $(D^2 + D - 6)[e^{-t}] = D^2[e^{-t}] + D[e^{-t}] - 6e^{-t} = e^{-t} - e^{-t} - 6e^{-t} = -6e^{-t}$.

   Answer 5:   Using the fact from Problem 1 that $(D - 2)[e^{-t}] = -3e^{-t}$,

$$(D + 3)(D - 2)[e^{-t}] = (D + 3)[(D - 2)e^{-t}] = (D + 3)[-3e^{-t}]$$
$$= 3e^{-t} - 9e^{-t} = -6e^{-t}$$

   which equals the results in Problems 3 and 4 since $D^2 + D - 6 = (D + 3)(D - 2) = (D - 2)(D + 3)$.

Finding Operators.   Find a polynomial operator $P(D) = D^2 + aD + b$, where a and b are real numbers, for which the indicated function $y(t)$ yields $P(D)[y] = 0$.

---

[1]Edwards, L. K., "High Speed Tube Transportation," *Scientific American*, 213 (Aug, 1965), 30–40.

7. $y(t) = 2e^{5t}$ **8.** $y(t) = e^{2t} + e^{-t}$ **9.** $y(t) = te^t$ **10.** $y(t) = \sin t$

Answer 7: For $y(t) = 2e^{5t}$, one of the roots of the characteristic polynomial $P(r) = r^2 + ar + b$ must be 5. There is no information about the other root r_2, so $P(r) = (r - 5)(r - r_2) = r^2 - (5 + r_2)r + 5r_2$ and the operator is $P(D) = D^2 - (5 + r_2)D + 5r_2$, where r_2 is any real number.

Answer 9: For $y(t) = te^t$, the characteristic polynomial $P(r) = r^2 + ar + b$ has the double root 1. So the polynomial operator is $(D - 1)^2 = D^2 - 2D + 1$.

Particular Solutions. For each function $f(t)$, find a function $y(t)$ such that $(D - 2)[y(t)] = f(t)$.

11. $f(t) = e^{-t}$ **12.** $f(t) = t - 1$ **13.** $f(t) = \sin t + 4$

Answer 11: We guess that $y(t) = Ae^{-t}$ for some value of A. Then

$$(D - 2)[Ae^{-t}] = -Ae^{-t} - 2Ae^{-t} = -3Ae^{-t} = e^{-t}$$

which holds if $A = -1/3$, so $y(t) = (-1/3)e^{-t}$.

Answer 13: We guess that $y(t) = A\cos t + B\sin t + C$ for some choice of A, B, C. Then

$$(D - 2)[y(t)] = -A\sin t + B\cos t - 2A\cos t - 2B\sin t - 2C$$

$$= (-A - 2B)\sin t + (B - 2A)\cos t - 2C = \sin t + 4$$

So $A = -1/5$, $B = -2/5$, $C = -2$, and $y(t) = -1/5\cos t - 2/5\sin t - 2$.

Polynomial-Exponentials. Show that each function below is a polynomial-exponential.

14. $\cos 2t - \sin t$ **15.** $t\sin^2 t$ **16.** $t^2 \sin 2t - (1 + t)\cos^2 t$

17. $\sin^3 t$ **18.** $(1 - t)e^{it}\cos 3t$ **19.** $(i + t - t^2)e^{(3+i)t}\sin^2 3t$

Answer 15: $t\sin^2 t = t[(e^{-it} - e^{it})i/2]^2 = t[2 - (e^{2it} + e^{-2it})]/4.$

Answer 17: $\sin^3 t = (-e^{it} + e^{-it})^3 i^3/8 = i(e^{3it} - 3e^{it} + 3e^{-it} - e^{-3it})/8.$

Answer 19: Since $\sin^2 3t = (1 - \cos 6t)/2 = (2 - e^{6it} - e^{-6it})/4$, we have $(i + t - t^2)e^{(3+i)t}\sin^2 3t = (i + t - t^2)e^{(3+i)t}/2 - (i + t - t^2)(e^{(3+7i)t} + e^{(3-5i)t})/4.$

☞ You may want to use a Computer Algebra System (CAS) for some of these IVPs.

Initial Value Problems. Solve each of the following initial value problems:

20. $y'' - 4y = 2 - 8t,\quad y(0) = 0,\quad y'(0) = 5$

21. $y'' + y = 10e^{2t},\quad y(0) = 0,\quad y'(0) = 0$

22. $y'' - 3y' + 2y = 8t^2 + 12e^{-t},\quad y(0) = 0,\quad y'(0) = 2$

23. $y'' + 2y' + y = 1 + 2te^t \quad y(0) = 0,\quad y'(0) = 0$

24. $y'' + y' = te^t \quad y(0) = 0,\quad y'(0) = 0$

25. $y'' - y' - 2y = 2e^t,\quad y(0) = 0,\quad y'(0) = 1$

26. $y'' + 3y' + 2y = t^2 e^{2t} + t \quad y(0) = 0,\quad y'(0) = 0$

27. $y'' - 4y = 1 + t + t^3 e^t \quad y(0) = 0,\quad y'(0) = 0$

28. $y'' - 9y = t^3 e^{-t} \quad y(0) = 0,\quad y'(0) = 0$

Answer 21: Using the notation and method of the procedure: Case 1 to find a particular solution for $y'' + y = 10e^{2t}$, we have $a = 0, b = 1, p(t) = 10, n = 0, s = 2$. The characteristic roots of $P(r) = r^2 + 1$ are $r_1 = i, r_2 = -i$. Then y_d has the form $y_d = A_0 e^{2t}$. Here it is easiest to insert y_d directly into the ODE and match coefficients : $(4A_0 + A_0)e^{2t} = 10e^{2t}$, so $A_0 = 2$ and $y_d = 2e^{2t}$

The general solution of $y'' + y = 0$ is $y = C_1\cos t + C_2\sin t$, so the general solution of the driven ODE is $y = 2e^{2t} + C_1\cos t + C_2\sin t$, C_1 and C_2 any constants.

Imposing the ICs, $y(0) = 0, y'(0) = 0$, we see that $0 = 2 + C_1$ and $0 = 4 + C_2$. So $C_1 = -2$ and $C_2 = -4$. The solution of the IVP is $y = 2e^{2t} - 2\cos t - 4\sin t$

Answer 23:   Here $P(D) = D^2 + 2D + 1$ and -1 is a double root of the characteristic polynomial. To find a particular solution y_d of $P(D)[y] = 1 + 2te^t$, we first find a solution y_{d_1} of $P(D)[y] = 1$ and then a solution y_{d_2} of $P(D)[y] = 2te^t$. By inspection we see that $y_{d_1} = 1$. For y_{d_2} we see that $a = 2$, $b = 1$, $p(t) = 2t$, $n = 1$, $s = 1$, $y_{d_2} = h(t)e^t$, where $h(t) = \sum_{k=0}^{1} A_k t^k = A_0 + A_1 t$. So $(D^2 + 4D + 4)[h(t)] = 4A_1 + 4A_0 + 4A_1 t = 2t$. So $A_1 = 1/2$ and $A_0 = -1/2$, and $y_{d_2} = (1/2)(t - 1)e^t$ and $y_d = y_{d_1} + y_{d_2} = 1 + (1/2)(t - 1)e^t$.

Since the general solution of $P(D)[y] = 0$ is $y = C_1 e^{-t} + C_2 t e^{-t}$, the general solution of $P(D)[y] = 1 + 2te^t$ is

$$y = 1 + (1/2)(t - 1)e^t + C_1 e^{-t} + C_2 t e^{-t}, \quad C_1 \text{ and } C_2 \text{ arbitrary constants}$$

Imposing the ICs, $y(0) = 0$ and $y'(0) = 0$, we have $0 = 1 - 1/2 + C_1$ and $0 = 1/2 - 1/2 - C_1 + C_2$. So, $C_1 = -1/2$, $C_2 = -1/2$ and the solution of the IVP is

$$y = 1 + (1/2)(t - 1)e^t - e^{-t}/2 - te^{-t}/2$$

Answer 25: $P(D) = D^2 - D - 2$ and the characteristic roots are $r_1 = -1$, $r_2 = 2$. To find a particular solution y_d of $P(D)[y] = 2e^t$, it is easiest to guess $y_d = A_0 e^t$ and insert y_d into the ODE: $(A_0 - A_0 - 2A_0)e^t = -2A_0 e^t = 2e^t$. So $A_0 = -1$ and $y_d = -e^t$.

Since the general solution of $P(D)[y] = 0$ is $y = C_1 e^{-t} + C_2 e^{2t}$, the general solution of $P(D)[y] = 2e^t$ is $y = -e^t + C_1 e^{-t} + C_2 e^{2t}$, where C_1 and C_2 are arbitrary constants.

Imposing the ICs $y(0) = 0$, $y'(0) = 1$, we see that $0 = -1 + C_1 + C_2$ and $1 = -1 - C_1 + 2C_2$. So $C_1 = 0$, $C_2 = 1$ and $y = -e^t + e^{2t}$ is the solution of the IVP.

Answer 27: $P(D) = D^2 - 4$ and $r_1 = -2$, $r_2 = 2$ are the roots of the characteristic polynomial. To find a particular solution y_d of $P(D)[y] = 1 + t + t^3 e^t$ we first find a particular solution y_{d_1} of $P(D)[y] = 1 + t$ and then a particular solution y_{d_2} of $P(D)[y] = t^3 e^t$. For y_{d_1}, we see that $y_{d_1} = A_0 + A_1 t$, where $P(D)[A_0 + A_1 t] = -4A_0 - 4A_1 t = 1 + t$. So $A_0 = -1/4 = A_1$ and $y_{d_1} = -1/4 - t/4$. For y_{d_2}, we have $a = 0$, $b = -4$, $p(t) = t^3$, $n = 3$, $s = 1$, $h(t) = A_0 + A_1 t + A_2 t^2 + A_3 t^3$ and $(D^2 + 2D - 3)[h(t)] = -3A_0 + 2A_1 + 2A_2 + (-3A_1 + 4A_2 + 6A_3)t + (-3A_2 + 6A_3)t^2 - 3A_3 t^3 = t^3 e^t$. So $-3A_3 = 1$, $-3A_2 + 6A_3 = 0$, $-3A_1 + 4A_2 + 6A_3 = 0$, $-3A_0 + 2A_1 + 2A_2 = 0$. We have $A_3 = -1/3$, $A_2 = -2/3$, $A_1 = -14/9$, $A_0 = -40/27$, and the solution $y_d = y_{d_1} + y_{d_2}$ is

$$y_d = -1/4 - t/4 - (40/27 + 14t/9 + 2t^2/3 + t^3/3)e^t$$

Since the general solution of $P(D)[y] = 0$ is $y = C_1 e^{-2t} + C_2 e^{2t}$, the general solution of $P(D)[y] = 1 + t + t^3 e^t$ is $y = C_1 e^{-2t} + C_2 e^{2t} - 1/4 - t/4 - (40/27 + 14t/9 + 2t^2/3 + t^3/3)e^t$, where C_1 and C_2 are arbitrary real numbers.

Imposing the initial conditions $y(0) = 0$ and $y'(0) = 0$, we have $0 = -1/4 - 40/27 + C_1 + C_2$, and $0 = -1/4 - 40/27 - 14/9 - 2C_1 + 2C_2$. So $C_1 = 19/432$ and $C_2 = 27/16$ and the solution of the IVP is

$$y = -1/4 - t/4 - (40/27 + 14t/9 + 2t^2/3 + t^3/3)e^t + (19/432)e^{-2t} + (27/16)e^{2t}$$

Operator Identities.   The problems below give some useful identities.

29. Show that if n is any positive integer, r_0 is any constant, and h is any n-times differentiable function, then $(D - r_0)^n[he^{r_0 t}] = e^{r_0 t}D^n[h]$. [*Hint:* do for $n = 1$ and iterate.]

Answer 29:   For $n = 1$, we have that $(D - r_0)[he^{r_0 t}] = (he^{r_0 t})' - r_0(he^{r_0 t}) = (h'e^{r_0 t} + r_0 h e^{r_0 t}) - r_0 h e^{r_0 t} = e^{r_0 t}h' = e^{r_0 t}D[h]$. For $n = 2$, we have that $(D - r_0)^2[he^{r_0 t}] = (D - r_0)[(D - r_0)[he^{r_0 t}]] = (D - r_0)[D[h]e^{r_0 t}] = e^{r_0 t}D[Dh] = e^{r_0 t}D^2[h]$, where we have used the result for $n = 1$ twice. Similarly, if $(D - r_0)^{n-1}[he^{r_0 t}] = e^{r_0 t}D^{n-1}[h]$, then $(D - r_0)^n[he^{r_0 t}] = (D - r_0)[(D - r_0)^{n-1}[he^{r_0 t}]] = (D - r_0)[e^{r_0 t}D^{n-1}[h]] = e^{r_0 t}D[D^{n-1}[h]] = e^{r_0 t}D^n[h]$ as claimed. The assertion has been proved by induction on n.

30. For any polynomial $p(t)$ of degree $n - 1$ (or less), show that $y = p(t)e^{r_0 t}$ is a solution for the nth-order ODE $(D - r_0)^n[y] = 0$ for any constant r_0.

 31. For any polynomial operator $P(D)$ of degree n, any function $h(t)$ with n continuous derivatives and any real number s show that $P(D)[he^{st}] = e^{st}P(D + s)[h]$. Use this fact to find a particular solution of the ODE $y''' - 2y'' + 2y' - y = t^2 e^t$.

Answer 31: Group Problem.

3.6 Driven Linear ODEs: Undetermined Coefficients II

Comments:

We continue to use the Procedure given in the previous section, but now with complex exponentials. This speeds up the construction of a particular solution whenever sinusoids are involved.

PROBLEMS

☞ You may want to use a Computer Algebra System (CAS) to find the solution formula for some of the ODEs in this problem set.

Complex Polynomial-Exponentials. Write each function as the real part of a polynomial-exponential. [*Hint:* express $\sin^2 t$ and $\cos^2 t$ in terms of $\cos 2t$. See the trig identities on the inside back cover.]

1. $\cos 2t - \sin t$ **2.** $t \sin^2 t$ **3.** $t^2 \sin 2t - (1 + t)\cos^2 t$

Answer 1: $\cos 2t - \sin t = \text{Re}[e^{2it} + ie^{it}]$ because $e^{2it} = \cos 2t + i\sin 2t$ and $ie^{it} = i\cos t - \sin t$

Answer 3: $t^2 \sin 2t - (1 + t)\cos^2 t = \text{Re}[-it^2 e^{2it} - (1 + t)(1 + e^{2it})/2]$ since $\sin 2t = \text{Re}[-ie^{2it}]$ and $\cos^2 t = (1 + \cos 2t)/2$, while $\cos 2t = \text{Re}[e^{2it}]$.

General Real-Valued Solution. Find the general real-valued solution formula for each ODE. Use the procedure in Section 3.5 to find particular solutions. See the Examples for more help.

4. $y'' - y = \sin 2t$ **5.** $y'' - y' - 2y = \cos t$

6. $y'' + 2y' + y = e^{-t}$ **7.** $(D + 1)(D - 3)[y] = t + e^{-t}$

Answer 5: Writing the ODE $y'' - y' - 2y = \cos t$ in operator form $(D^2 - D - 2)[y] = \cos t$, we notice that the two characteristic roots are $r_1 = 2$, $r_2 = -1$. Factoring the operator, the ODE becomes $(D - 2)(D + 1)[y] = \cos t$. The general solution for the undriven ODE is $y_u = C_1 e^{2t} + C_2 e^{-t}$. $P(D) = D^2 - D - 2$, $a = -1$, $b = -2$, $r_1 = 2$, $r_2 = -1$, $p(t)e^{st}$ is $1e^{it}$ because $\cos t = \text{Re}[e^{it}]$, $p(t) = 1$, $n = 0$, $s = i$; see Example 3.6.3. Case 1 applies since $s \neq r_1, r_2$. So $h(t) = A_0$ since $n = 0$, where $(D^2 + (2i - 1)D - 3 - i)[A_0] = 1$; $A_0 = -1/(3 + i) = (-3 + i)/10$. A complex particular solution is $(-3 + i)e^{it}/10$. The corresponding real particular solution is the real part, so $y_d = (-3/10)\cos t - (1/10)\sin t$. Hence the general solution for the ODE is given by $y = -(1/10)(\sin t + 3\cos t) + C_1 e^{2t} + C_2 e^{-t}$, C_1 and C_2 arbitrary constants, any t.

Answer 7: First, the general solution to the undriven ODE is $y_u = C_1 e^{-t} + C_2 e^{3t}$ since the roots of the characteristic polynomial are $r_1 = -1$ and $r_2 = 3$. $P(D) = D^2 - 2D - 3$, $a = -2$, $b = -3$, $r_1 = -1$, $r_2 = 3$, $p_1(t)e^{s_1 t} = t$ with $s_1 = 0$ and $p_1(t) = t$ and $n = 1$; $p_2(t)e^{s_2 t} = 1e^{-t}$, with $s_2 = -1$ and $p_2(t) = 1$, $n = 0$. Since $s_1 \neq r_1, r_2$, but $s_2 = r_1$, we see that the particular solution y_d has the form $A_1 t e^{-t} + A_0^* + A_1^* t$ where (Case 2 with $s_2 = -1$) $(D^2 - 4D)[A_1 t] = 1$ and (Case 2 with $s = 0$) $(D^2 - 2D - 3)[A_0^* + A_1^* t] = t$. Thus $-4A_1 = 1$ and $A_1 = -1/4$, while $-2A_1^* - 3A_0^* - 3A_1^* t = t$, $A_1^* = -1/3$ and $A_0^* = 2/9$: hence, $y_d = -te^{-t}/4 + 2/9 - t/3$. The general solution is $y = y_d + C_1 e^{-t} + C_2 e^{3t}$, where C_1 and C_2 arbitrary constants.

Finding General Solution Formulas. Find all real-valued solutions of each ODE.

8. $y'' - y' - 2y = 2\sin 2t$ **9.** $y'' - y' - 2y = t^2 + 4t$ **10.** $y'' - 2y' + y = -te^t$

11. $y'' - 2y' + y = 2e^t$ **12.** $y'' + 2y' + y = e^t \cos t$ **13.** $y'' + 4y' + 5y = e^{-t} + 15t$

14. $y'' + y' + y = \sin^2 t$ **15.** $y'' + 4y = e^{2it}$

Answer 9: From Problem 8, $y_u = c_1 e^{-t} + c_2 e^{2t}$ for the ODE $y'' - y' - 2y = (D^2 - D - 2)[y] = t^2 + 4t$, since the roots of the characteristic polynomial $r^2 - r - 2$ are $-1, 2$. Let's guess that $y_d = A + Bt + Ct^2$ and so $y_d'' - y_d' - 2y_d = 2C - B - 2A - 2Bt - 2Ct - 2Ct^2 = t^2 + 4t$. So $-2A - B + 2C = 0$, $-2B - 2C = 4$, $-2C = 1$. So $A = 1/4$, $B = -3/2$ and $C = -1/2$, and $y = y_d + y_u = (1 - 6t - 2t^2)/4 + c_1 e^{-t} + c_2 e^{2t}$.

Answer 11: The ODE $y'' - 2y' + y = (D^2 - 2D + 1)[y] = 2e^t$ has characteristic polynomial $r^2 - 2r + 1$ with a double root 1, so $y_u = (c_1 + c_2 t)e^t$ and (by Case 3 of the procedure) we look for y_d in the form $y_d = At^2 e^t$. Then $y_d'' - 2y_d' + y_d = 2Ae^t$, which equals $2e^t$ if $A = 1$. So $y = (c_1 + c_2 t + t^2)e^t$.

Answer 13: The ODE $y'' + 4y' + 5y = (D^2 + 4D + 5)[y] = e^{-t} + 15t$ has characteristic polynomial $r^2 + 4r + 5$ with complex roots $-2 \pm i$, so $y_u = e^{-2t}(c_1 \cos t + c_2 \sin t)$. We can guess y_d to have the form $y_d = Ae^{-t} + B + Ct$. We have $y_d'' + 4y_d' + 5y_d = 2Ae^{-t} + (5B + 4C) + 5Ct$, which equals $e^{-t} + 15t$ if $A = 1/2$, $B = -12/5$, $C = 3$. Then $y = e^{-2t}(c_1 \cos t + c_2 \sin t) + e^{-t}/2 - 12/5 + 3t$.

Answer 15: The ODE $y'' + 4y = e^{2it}$ does not have a real-valued solution. If a solution $y(t)$ were real-valued, then $y'' + 4y$ would also be a real-valued function and cannot be equal to e^{2it} for all t because e^{2it} has a nonzero imaginary part.

Plot solution graphs of the respective ODEs of Problems 8–15 with initial data $y(0) = 0$, $y'(0) = -1, 0, 1$. What is the long term behavior of each solution as $t \to +\infty$?

16. Problem 8   **17.** Problem 9   **18.** Problem 10   **19.** Problem 11

20. Problem 12   **21.** Problem 14   **22.** Problem 13   **23.** Problem 15

Answer 17: See the figure. All these solution curves tend to $+\infty$ as $t \to +\infty$.

Answer 19: See the figure. All these solution curves tend to $+\infty$ as $t \to +\infty$

Answer 21: See the figure. All these solution curves tend to a sinusoid of period about π and centered around $y = 0.5$. Checking the general solution formula for this ODE given in Answer 18, we see that in fact all solutions tend to $y_d = 1/2 + (3 \cos 2t - 2 \sin 2t)/26$, which is a sinusoid of period π centered at $y = .5$.

Answer 23: There are no real-valued solutions.

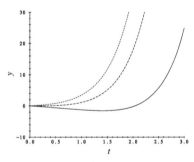

Problem 17.

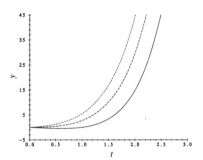

Problem 19.

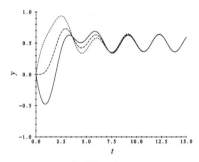

Problem 21.

Initial Value Problems. Find the unique solution of each IVP.

24. $y'' + 9y = 81t^2 + 14\cos 4t$, $y(0) = 0$, $y'(0) = 3$

25. $y'' - y = e^{-t}(2\sin t + 4\cos t)$, $y(0) = 1$, $y'(0) = 1$

26. $y'' + 2y' + 2y = 2t$, $y(1) = 1$, $y'(1) = 0$

Answer 25: The IVP is $y'' - y = e^{-t}(2\sin t + 4\cos t)$, $y(0) = y'(0) = 1$. The ODE has $P(D) = D^2 - 1$. Although we could use Case 1 of the procedure, in this case we find a particular solution y_d by the method of undetermined coefficients; y_d has the form

$$y_d = Ae^{-t}\sin t + Be^{-t}\cos t$$

Inserting y_d into the ODE, we have

$$(2B - A)e^{-t}\sin t - (2A + B)e^{-t}\cos t = 2e^{-t}\sin t + 4e^{-t}\cos t$$

Matching coefficients of like terms, $2B - A = 2$, $-2A - B = 4$, we have $A = -2$, $B = 0$, so $y_d = -2e^{-t}\sin t$ is a particular solution. The characteristic polynomial $r^2 - 1$ has roots ± 1, so the general solution of the undriven equation is $C_1 e^t + C_2 e^{-t}$. The general solution of the driven ODE is $y = -2e^{-t}\sin t + C_1 e^t + C_2 e^{-t}$. The initial conditions, $y(0) = 1$, $y'(0) = 1$ imply that

$$1 = C_1 + C_2, \quad 1 = C_1 - C_2 - 2$$

so $C_1 = 2$, $C_2 = -1$. The solution of the IVP is

$$y = -2e^{-t}\sin t + 2e^t - e^{-t}$$

27. $y'' + 25y = \sin 4t$, $y(0) = 0$, $y'(0) = 0$. Plot the component curves and the orbit, the latter for the rectangle $|y| \le 0.25$, $|y'| \le 1$. Any surprises?

Answer 27: First we must find the general real-valued solution of the ODE

$$y'' + 25y = \sin 4t \tag{i}$$

which models the vertical displacement y of an undamped Hooke's Law spring with a sinusoidal driving term. By Theorem 3.6.1 if we had a solution z_d of the ODE

$$z'' + 25z = e^{4it} \tag{ii}$$

then $y_d = \text{Im}[z_d]$ solves the ODE $P(D)[y] = f(t)$ because $\text{Im}[e^{4it}] = \sin 4t$.

Writing ODE $z'' + 25z = e^{4it}$ as $P(D)[z] = e^{4it}$, with $P(D) = D^2 + 25$ we can use Case 1 of the procedure to guess that z_d is of the form $z_d = A_0 e^{4it}$. Now $P(D)[A_0 e^{4it}] = A_0 P(4i)e^{4it}$, and since $P(4i) \ne 0$ we can take $A_0 = 1/P(4i)$, so

$$z_d = \frac{e^{4it}}{(4i)^2 + 25} = \frac{1}{9}e^{4it}$$

Since the imaginary part of $e^{4it}/9$ is $(\sin 4t)/9$, we see by Theorem 3.6.1 that a particular solution of the ODE $y'' + 25y = \sin 4t$ is $y_d = (\sin 4t)/9$. Since the characteristic roots of $D^2 + 25$ are $r_1 = 5i$, $r_2 = -5i$, we see from Theorem 3.2.1 that the general solution of $(D^2 + 25)[y] = 0$ is $C_1 \cos 5t + C_2 \sin 5t$ for arbitrary reals C_1 and C_2. Putting everything together, we see that the general, real-valued solution of $y'' + 25y = \sin 4t$ is

$$y = \frac{1}{9}\sin 4t + C_1 \sin 5t + C_2 \cos 5t \tag{iii}$$

where C_1 and C_2 are any real constants. To satisfy the initial conditions $y(0) = 0$, $y'(0) = 0$, we must choose $C_1 = -4/45$ and $C_2 = 0$. This solution appears in the figure.

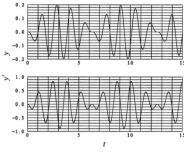
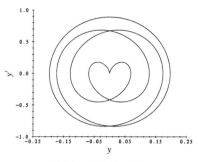

Problem 27, Graph 1   Problem 27, Graph 2

28. *Hearts and Eyes*   Find a solution formula for $y'' + 25y = \sin \omega t$, where $\omega \neq 5$. Plot the solution curve of the IVP with $y(0) = y'(0) = 0$, where $\omega = 4$. Plot the orbit for $0 \leq t \leq 20$ in the rectangle $|y| \leq 0.1$, $-0.5 \leq |y'| \leq 0.3$. Repeat with $\omega = 1$. Overlay the graphs.

Chapter 3's Opening Figure and the ODE of Example 3.1.2.   Find the solution formula for the indicated IVP using the procedure given in this section. Use your numerical solver to plot the time-state curve, orbit, and component curves over the indicated time span. What happens to the solution curve as $t \to +\infty$? Explain why the solution approaches a periodic solution. Find the formula for the periodic solution; what is the period? Each IVP models the motion of a damped, driven Hooke's Law spring; describe the long-term behavior of the spring.

29. *Chapter 3's Opening Figure* $y'' + 0.1y' + 64y = \sin 8.6t$, $y(13) = -0.75$, $y'(13) = 1.2$. Use the t-interval $13 \leq t \leq 54$.

30. *Example 3.1.2* $z'' + 0.05z' + 25z = 100 \sin 5.6t$, $z(0) = 0$, $z'(0) = 0$, $0 \leq t \leq 40$.

Answer 29:   The characteristic roots of $P(D) = D^2 + 0.1D + 64$ are $r_1 = -0.05 + 7.994i$, $r_2 = -0.05 - 7.994i$, where 7.994 is an approximation to the exact number $(\sqrt{255.99})/2$. The driving term is $\sin 8.6t = \mathrm{Im}[e^{8.6it}]$, so $s = 8.6i \neq r_1, r_2$ and we are in Case 1 of the procedure. A particular complex solution of $P(D)[y] = e^{8.6it}$ has the form $Ae^{8.6it}$ where $A \cdot ((8.6i)^2 + 0.86i + 64) = A(-9.96 + 0.86i) = 1$. So $A = 1/(-9.96 + 0.86i) \approx -0.0996 - 0.0086i$ and $Ae^{8.6it} \approx (-0.0996 - 0.0086i)e^{8.6it}$. The imaginary part of this is

$$y_d \approx \mathrm{Im}[Ae^{8.6it}] = -(0.0996 \sin 8.6t + 0.0086 \cos 8.6t)$$

The general solution of $P(D)[y] = \sin 8.6t$ is then $y \approx y_d + e^{-0.05t}(C_1 \cos 7.994t + C_2 \sin 7.994t)$, C_1 and C_2 arbitrary constants, any t.

Imposing the initial condition $y(13) = -0.75$ and $y'(13) = 1.2$, we could do the algebra (a lot of it) to find C_1 and C_2 such that

$$-0.75 = y_d(13) + e^{-0.65}(C_1 \cos 103.922 + C_2 \sin 103.922)$$

$$1.2 = y_d'(13) - .05e^{-0.65}(C_1 \cos 103.922 + C_2 \sin 103.922) +$$

$$7.994e^{-0.65}(C_2 \cos 103.922 - C_1 \sin 103.922)$$

but we don't. The figures show the time-state curve (Graph 1), the orbit (Graph 2), the ty-solution curve and the ty'-curve (Graph 3).

The particular solution y_d is periodic with period $\approx 2\pi/7.994 \approx 0.7860$. The general solution of the undriven equation involves sines and cosines whose amplitude have the exponential factors $e^{-0.05t}$ and so those terms $\to 0$ as $t \to +\infty$. That means all solutions tend to y_d as $t \to \infty$. The formula for y_d is given above. A damped, driven Hooke's Law spring modeled by the IVP of this problem eventually settles down to periodic motion of period $= 2\pi/7.994 \approx 0.7860$. See Graph 4 which shows the nearly-periodic solution curve over the range $160 \leq t \leq 163$.

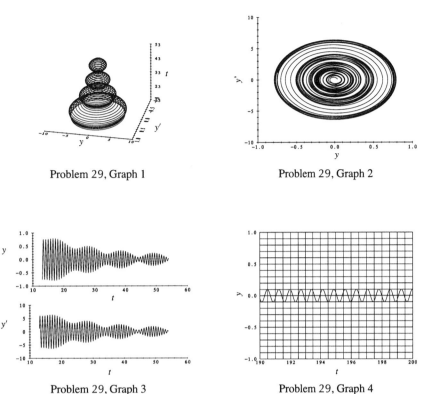

Problem 29, Graph 1   Problem 29, Graph 2

Problem 29, Graph 3   Problem 29, Graph 4

Undetermined Coefficients. Consider the ODE $P(D)[y] = f(t)$, where $P(D) = D^2 + aD + b$ and a and b are constants. Let r_1 and r_2 be the roots of $P(r)$, and carry out the indicated tasks.

31. Suppose $f(t) = Ae^{st}$, where A and s are real or complex constants. Show that a particular solution y_d of $P(D)[y] = Ae^{st}$ has the indicated form:

- If $s \neq r_1, r_2$, then $y_d = [A/P(s)]e^{st}$.

- If $s = r_1$, $r_2 \neq r_1$, then $y_d = [A/(r_1 - r_2)]te^{r_1 t}$.

- If $s = r_1$, $r_2 = r_1$, then $y_d = (A/2)t^2 e^{r_1 t}$.

[*Hint*: when s is a characteristic root, look for y_d in the form $y_d = h(t)e^{st}$ and apply the identity $P(D)[h(t)e^{st}] = e^{st}P(D+s)[h]$ to conclude that $P(D+s)[h] = A$. Solve for $h(t)$ using undetermined coefficients.]

Answer 31:

- If $y_d = [A/P(s)]e^{st}$ then $y'_d = [sA/P(s)]e^{st}$, and $y''_d = [s^2 A/P(s)]e^{st}$. Substituting these expressions into the ODE, we get

$$\left[Ae^{st}/P(s)\right](s^2 + as + b) = Ae^{st}$$

since $P(s) = s^2 + as + b$. So the given y_d is indeed a particular solution.

- If $y_d = [A/(r_1 - r_2)]te^{r_1 t}$ then $y'_d = [A/(r_1 - r_2)]e^{r_1 t} + [Atr_1/(r_1 - r_2)]e^{r_1 t}$. $y''_d = [A/(r_1 - r_2)]e^{r_1 t}(2r_1 + r_1^2 t)$. Substituting these expressions into the ODE, where $a = -(r_1 + r_2)$ and $b =$

$r_1 r_2$, we get

$$y_d'' - (r_1 + r_2)y_d + r_1 r_2 y_d = \left[2r_1 A/(r_1 - r_2)\right]e^{r_1 t} + \left[Atr_1^2/(r_1 - r_2)\right]e^{r_1 t} - \left[(r_1 + r_2)A/(r_1 - r_2)\right]e^{r_1 t}$$
$$- \left[(r_1 + r_2)Atr_1/(r_1 - r_2)\right]e^{r_1 t} + \left[Atr_1 r_2/(r_1 - r_2)\right]e^{r_1 t}$$
$$= 2r_1 Ae^{r_1 t}/(r_1 - r_2) + Atr_1^2 e^{r_1 t}/(r_1 - r_2) - \left[(r_1 + r_2)/(r_1 - r_2)\right](Ae^{r_1 t}$$
$$+ Atr_1 e^{r_1 t}) + \left[Atr_1 r_2/(r_1 - r_2)\right]e^{r_1 t}$$
$$= \frac{Ae^{r_1 t}\left[(2r_1 + tr_1^2) - (r_1 + r_2) - tr_1(r_1 + r_2) + tr_1 r_2\right]}{(r_1 - r_2)}$$
$$= \frac{Ae^{r_1 t}[(r_1 - r_2)]}{(r_1 - r_2)} = Ae^{r_1 t} = Ae^{st}.$$

The given expression is thus a particular solution.

- If $y_d = (A/2)t^2 e^{r_1 t}$ then $y_d' = Ate^{r_1 t} + (A/2)t^2 r_1 e^{r_1 t}$, and $y_d'' = Ae^{r_1 t} + 2Atr_1 e^{r_1 t} + (A/2)t^2 r_1^2 e^{r_1 t}$. Substituting these expressions into the ODE, we get

$$y_d'' - (r_1 + r_2)y_d + r_1 r_2 y_d = (Ae^{r_1 t} + 2Atr_1 e^{r_1 t} + (A/2)t^2 r_1^2 e^{r_1 t}) - (r_1 + r_2)(Ate^{r_1 t}$$
$$+ (A/2)t^2 r_1 e^{r_1 t}) + (r_1 r_2)(A/2)e^{r_1 t}$$
$$= Ae^{r_1 t}(1 + 2r_1 t + t^2 r_1^2/2 - tr_1 - tr_2 - t^2 r_1^2/2 - t^2 r_1 r_2/2 + t^2 r_1 r_2/2)$$
$$= Ae^{r_1 t}(1 + r_1 t - r_2 t) = Ae^{st} \qquad \text{(since } r_1 = r_2)$$

So y_d is a particular solution.

32. Find a particular solution of $(D^2 + 2D + 10)[y] = e^{-t}\cos 3t$.

 Periodic Solutions of a First-Order ODE.

33. Follow the outline below to show that the ODE $y' + ay = A\cos\omega t$, where a, A, and ω are nonzero constants, has a unique periodic solution.

(a) First, show that if $z_d(t)$ is a particular solution of the ODE $z' + az = Ae^{i\omega t}$, then $y_d(t) = \text{Re}[z_d(t)]$ is a particular solution of the given ODE $y' + ay = A\cos\omega t$.

(b) Use the operator formula $(D + a)[Ce^{i\omega t}] = C(i\omega + a)e^{i\omega t}$ to get the particular solution

$$z_d = \frac{A(a - i\omega)}{a^2 + \omega^2}e^{i\omega t}$$

of the ODE in **(a)**.

(c) Let ϕ be an angle such that $\cos\phi = a/\sqrt{a^2 + \omega^2}$ and $\sin\phi = \omega/\sqrt{a^2 + \omega^2}$. Show that

$$y_d = \frac{A}{\sqrt{a^2 + \omega^2}}\cos(\omega t - \phi)$$

is a particular solution of the ODE $y' + ay = A\cos\omega t$.

(d) Find the general solution of the ODE. How many periodic solutions are there?

(e) Now adapt the process described above to the ODE $y' + ay = A\sin\omega t$. Repeat **(a)**–**(d)**, suitably modified to handle the driving term $A\sin\omega t$.

Answer 33:

(a) If $z_d(t)$ is a solution of the ODE $z' + az = Ae^{i\omega t}$, then, since the coefficients in the operator $P(D) = D + a$ are real numbers, it follows from Theorem 3.6.1 that $y_d = \text{Re}[z_d]$ solves the same ODE but with the driving term replaced by $\text{Re}[Ae^{i\omega t}] = A\cos\omega t$.

(b) Comparing the formula $(D + a)[Ce^{i\omega t}] = C(i\omega + a)e^{i\omega t}$ with the ODE $(D + a)[z] = Ae^{i\omega t}$, we see that we can take $z_d = Ce^{i\omega t}$ if $C(i\omega + a) = A$. So, $z_d = (A/(i\omega + a))e^{i\omega t}$. After multiplying numerator and denominator of the coefficient fraction by $a - i\omega$, we obtain the desired formula for z_d.

(c) Writing

$$y_d = \text{Re}[z_d] = \frac{Aa}{a^2 + \omega^2} \cos \omega t + \frac{A\omega}{a^2 + \omega^2} \sin \omega t$$

and choosing an angle φ as indicated, we have

$$y_d = \frac{A}{\sqrt{a^2 + \omega^2}} \left\{ \frac{a}{\sqrt{a^2 + \omega^2}} \cos \omega t + \frac{\omega}{\sqrt{a^2 + \omega^2}} \sin \omega t \right\}$$

$$= \frac{A}{\sqrt{a^2 + \omega^2}} \{\cos \varphi \cos \omega t + \sin \varphi \sin \omega t\}$$

$$= \frac{A}{\sqrt{a^2 + \omega^2}} \cos(\omega t - \varphi)$$

where we have used the formula $\cos(\alpha - \beta) = \cos \alpha \cos \beta + \sin \alpha \sin \beta$.

(d) The general solution of the undriven ODE $y' + ay = 0$ is $y_u = Ce^{-at}$, where C is an arbitrary real number. So the general solution of $y' + ay = A \cos \omega t$ is $y = Ce^{-at} + y_d$. If $C \neq 0$, then the solution is not periodic. There is only one periodic solution of the given ODE (namely, when $C = 0$).

(e) We leave this to the group.

Modeling Problem. Refer to Newton's Law of Cooling on page 37.

See the Library entry *Newton's Law of Cooling II* under First order Equations.

34. Suppose that the temperature of the air surrounding a house varies sinusoidally. Show that the temperature variation in the house also varies sinusoidally with the same period but lags behind the temperature variation of the outside air by a fixed time. [*Hint*: see Problem 33.]

Higher-Order Constant-Coefficient Undriven Linear ODEs. Use Problems 37 and 38 to find all solutions of the ODEs in Problems 39–49. [*Hint*: operator identities (9) and (10) in Section 3.5 hold for *any* operator $P(D)$.]

35. Find all solutions of the ODE $(D + 1)^3[y] = 0$. [*Hint*: show that the ODE is equivalent to the linear cascade $(D + 1)[z] = 0$, $(D + 1)[w] = z$, $(D + 1)[y] = w$.]

36. Suppose that $P(r)$ is a cubic polynomial with real coefficients. Find all solutions of the ODE $P(D)[y] = 0$. [*Hint*: either all the roots of $P(r)$ are equal, or there is a real root r_1 that is different from the other two roots (the first case is treated in Problem 35). Write $P(r) = (r - r_1)Q(r)$ and observe that the ODE $(D - r_1)Q(D)[y] = 0$ is equivalent to the linear cascade $Q(D)[y] = w$, $(D - r_1)[w] = 0$.]

37. Show that the operator identities (9), (10) in Section 3.5 hold for any operator $P(D)$.

38. It is a fact that if the polynomials $Q(r)$ and $R(r)$ have no common roots, then a basic solution set for $Q(D)R(D)[y] = 0$ is the union of basic solution sets for $Q(D)[y] = 0$ and $R(D)[y] = 0$. Show this agrees with the result of Problem 36.

39. $y''' - 2y'' - y' + 2y = 0$

40. $y''' - y'' - y' + y = 0$

41. $y''' + 3y'' + 3y' + y = 0$

42. $y''' + y'/4 + 5y/4 = 0$

43. $y''' + y'' - 16y' - 16y = 17e^{3t} - 28te^{3t}$

44. $y''' + y'' = 12t^2$

45. $y''' + 6y'' + 11y' + 6y = 2e^{-3t} - te^{-t}$ [*Hint*: $r_1 = -3$ is a characteristic root.]

46. $y'''' + 2y'' + y = 3 + \cos 2t$

47. $y'''' - 10y''' + 38y'' - 64y' + 40y = 153e^{-t}$

48. $(D^2 + 1)(D^2 - 1)D[y] = 0$

49. $(D^2 - 9)(D^2 - 4)(D^2 - 1)[y] = e^t$

Answer 35: From the operator identities, the ODE $(D + 1)^3[y] = 0$ is equivalent to the cascade

$$(D + 1)[z] = 0$$

$$(D + 1)[w] = z$$

$$(D + 1)[y] = w$$

Solving the top ODE of the system we have $z = C_1 e^{-t}$ where C_1 is an arbitrary constant. Inserting $C_1 e^{-t}$ for z in the second equation of the system and solving, we get $w = (C_1 t + C_2)e^{-t}$, where C_2 is an arbitrary constant. Inserting $(C_1 t + C_2)e^{-t}$ for w in the last ODE of the system and solving for y we get $y(t) = C_1 e^{-t} + C_2 t e^{-t} + C_3 t^2 e^{-t}$, where C_1, C_2, and C_3 are arbitrary constants.

Answer 37: $P(D) = D^n + aD^{n-1} + bD^{n-2} + \cdots + c$ is an nth polynomial operator. From 3.5, identity (9), $P(D)[e^{st}p(s)]$

$$P(D)[e^1] = (D^n + aD^{n-1} + bD^{n-2} + \cdots + c)[e^{st}]$$
$$= D^n[e^{st}] + aD^{n-1}[e^{st}] + bD^{n-2}[e^{st}] + \cdots + c[e^{st}]$$
$$= s^n e^{st} + as^{n-1}e^{st} + bs^{n-2}e^{st} + \cdots + ce^{st}$$
$$= e^{st}(s^n + as^{n-1} + bs^{n-2} + \cdots + c) = e^{st}p(s)$$

From 3.5, identity (10), we see that $P(D)[h(t)e^{st}] = e^{st}P(D+s)[h]$. We know from section 3.5 that $D[h(t)e^{st}] = e^{st}(D+s)[h]$ implies that $D^n[he^{st}] = e^{st}(D+s)^n[h]$ Then, $P(D)[h(t)e^{st}] = D^n[he^{st}] + aD^{n-1}[he^{st}] + bD^{n-2}[he^{st}] + \cdots + c[he^{st}]$

$$= e^{st}P(D+s)[h]$$

, where $P(D+s) = (D+s)^n + a(D+s)^{n-1} + b(D+s)^{n-2} + \cdots + c$

Answer 39: The roots of the characteristic polynomial $r^3 - 2r^2 - r + 2$ are $r = 1, r = 2$ and $r = -1$. Hence, the general solution to the ODE $y''' - 2y'' - y' + 2y = 0$ is $C_1 e^t + C_2 e^{2t} + C_3 e^{-t}$, where C_1, C_2, C_3 are arbitrary constants.

Answer 41: The characteristic polynomial $r^3 + 3r^2 + 3r + 1$ has a three-time repeated root $r = -1$. Thus, the general solution to the ODE $y''' + 3y'' + 3y' + y = 0$ is $C_1 e^{-t} + C_2 t e^{-t} + C_3 t^2 e^{-t}$, where C_1, C_2, C_3 are arbitrary constants.

Answer 43: Recall from Theorem 3.5.2 that if $P(D) = D^2 + a(t)D + b(t)$, ($a(t)$ and $b(t)$ are continuous on a t-interval I), and $f(t)$ is continuous on I, then the general solution of the ODE $P(D)[y] = f$ can be written as $y = y_d(t) + y_u(t)$ where y_d is a particular solution of the driven ODE and y_u is the general solution of the undriven ODE $P(D)[y] = 0$.

To find the general solution to the undriven ODE $y''' + y'' - 16y' - 16y = 0$ we must first find the roots of the characteristic polynomial $r^3 + r^2 - 16r - 16$ which are $r = -4, r = -1$ and $r = 4$. So the general solution to this undriven ODE is

$$y_u = C_1 e^{-4t} + C_2 e^{-t} + C_3 e^{4t}$$

where C_1, C_2, C_3 are arbitrary constants.

Since the driving function is $17e^{3t} - 28te^{3t}$, a good guess for a particular solution to the driven ODE is $y_d = Ae^{3t} + Bte^{3t}$. Substituting this guess into the left-hand side of the ODE we have (after a lot of algebra)

$$y_d''' + y_d'' - 16y_d' - 16y_d = -28Ae^{3t} - 28Bte^{3t} + 17Be^{3t}$$
$$= 17e^{3t} - 28te^{3t}$$

The only way for this equality to hold for all t is if $B = 1$ and $A = 0$. Hence, a particular solution to the driven ODE is

$$y_d = te^{3t}.$$

Summing the general solution to the undriven ODE and the particular solution which we found to the driven ODE gives the *general* solution to the *driven* ODE:

$$y(t) = y_u + y_d = C_1 e^{-4t} + C_2 e^{-t} + C_3 e^{4t} + te^{3t}$$

Answer 45: Using Theorem 3.5.2 again, we must find the general solution of the undriven ODE $y''' + 6y'' + 11y' + 6y = 0$ and add it to a particular solution of the driven ODE.

Since the roots of the characteristic polynomial $r^3 + 6r^2 + 11r + 6$ are $r = -3, r = -2$ and $r = -1$, then the general solution to the undriven ODE is

$$y_u = C_1 e^{-3t} + C_2 e^{-2t} + C_3 e^{-t}$$

To find a particular solution to the driven ODE, follow the hint to make the educated guess $y_d = Ate^{-3t} + Bt^2 e^{-t} + Cte^{-t}$. Substituting this into the left-hand side of the ODE and setting it equal to the right-hand side give us (after a lot of algebra) $A = 1$, $B = -1/4$ and $C = 3/4$. Therefore, a particular solution to the driven ODE is

$$y_d = te^{-3t} - (t^2/4)e^{-t} + 3/4te^{-t}$$

Adding the general solution of the undriven ODE and the particular solution we have found for the driven ODE gives us

$$y(t) = C_1 e^{-3t} + C_2 e^{-2t} + C_3 e^{-t} + te^{-3t} - t^2/4e^{-t} + 3/4te^{-t}$$

which is the general solution to the driven ODE.

Answer 47: To find the general solution of the homogeneous ODE $y'''' - 10y''' + 38y'' - 64y' + 40y = 0$, first find the roots of the characteristic polynomial, $r^4 - 10r^3 + 38r^2 - 64r + 40$, which are $r = 2$ (repeated), and $r = 3 \pm i$. So the general solution is

$$y_u = C_1 e^{2t} + C_2 te^{2t} + C_3 e^{3t} \cos t + C_4 e^{3t} \sin t$$

Next, since the driving function is $153e^{-t}$, we should guess $y_d = Ae^{-t}$ as a particular solution to the driven ODE $y'''' - 10y''' + 38y'' - 64y' + 40y = 153e^{-t}$. Substituting our guess into the left side of the ODE and setting it equal to the right side gives us

$$y'''' - 10y''' + 38y'' - 64y' + 40y = Ae^{-t} + 10Ae^{-t} + 38Ae^{-t} + 64Ae^{-t} + 40Ae^{-t}$$
$$= 153Ae^{-t}$$
$$= 153e^{-t}$$

so that $A = 1$. Hence, $y_d = e^{-t}$ is a particular solution to the driven ODE.

Adding y_u and y_d we obtain the general solution to the driven ODE:

$$y(t) = C_1 e^{2t} + C_2 te^{2t} + C_3 e^{3t} \cos t + C_4 e^{3t} \sin t + e^{-t}$$

Answer 49: Group project.

3.7 Theory of Second-Order Linear Differential Equations

Comments:

The two main results in this section, the Existence and Uniqueness Theorem for Linear ODEs and the Method of Variation of Parameters, are mostly of theoretical value, but form a nice way to pull things together. If you intend to read about series methods in Chapter 11, then we suggest that you at least read the highlights of this section. We cover the Method of Variation of Parameters. Although it's nice to know the statement of the Method, it's a lot of work to do examples where the driving term is not a polynomial-exponential.

So we concentrate on constant-coefficient operators $P(D)$ and polynomial-exponential driving terms. This covers most of the important applications in a first ODE course.

Which of the two methods for finding y_d is to be preferred? It depends on whether you prefer to integrate (variation of parameters) or to solve linear algebraic equations (undetermined coefficients). The latter method only works when the driving function is polynomial expoentnial, the former always works but you may have to leave the answer in terms of unevaluated integrals.

PROBLEMS

Basic Solution Sets and IVPs. First verify that $\{y_1, y_2\}$ is a basic solution set for the given ODE, $t > 0$. Then use y_1 and y_2 to find all solutions of the IVP.

1. $t^2 y'' + ty' - y = 0$, $y(1) = 0$, $y'(1) = -1$; $y_1 = t^{-1}$, $y_2 = t$

2. $t^2 y'' - ty' + y = 0$, $y(1) = 1$, $y'(1) = 0$; $y_1 = t$, $y_2 = t \ln t$

3. $t^2 y'' + ty' - 4y = 0$, $y(0) = 0$, $y'(0) = 0$; $y_1 = t^2$, $y_2 = t^{-2}$

> **Answer 1:** Although the details aren't shown here, first check that $y_1 = t^{-1}$ and $y_2 = t$ are solutions of the ODE $t^2 y'' + ty' - y = 0$. Then $\{t^{-1}, t\}$ is a basic solution set because the Wronskian $W(t^{-1}, t) = t^{-1} \cdot 1 - (-t^{-2}) \cdot t = 2t^{-1} \neq 0$ for $t > 0$. The general solution of the ODE, then, is $y = c_1 t^{-1} + c_2 t$, where c_1 and c_2 are arbitrary constants. The initial conditions $y(1) = 0$, $y'(1) = -1$ imply that $0 = c_1 + c_2$, $-1 = -c_1 + c_2$, so $c_1 = 1/2$ and $c_2 = -1/2$. The solution of the IVP is $y = (t^{-1} - t)/2$, $t > 0$.

> **Answer 3:** First check that $y_1 = t^2$ and $y_2 = t^{-2}$ are solutions of the ODE, $t^2 y'' + ty' - 4y = 0$. Then $\{t^2, t^{-2}\}$ is a basic solution set because the Wronskian $W(t^2, t^{-2}) = t^2 \cdot (-2t^{-3}) - (2t) \cdot t^{-2} = -4t^{-1} \neq 0$ for $t > 0$. The general solution of the ODE is $y = c_1 t^2 + c_2 t^{-2}$, where c_1 and c_2 are arbitrary constants. The initial conditions $y(0) = 0$ and $y'(0) = 0$ are permissible here because $y = c_1 t^2$ does meet them for all t and all c_1, even though t^{-2} is not defined for $t = 0$. Theorem 3.7.1 is not applicable at $t = 0$, so there is no violation of the uniqueness part of the theorem.

Basic Solution Sets.

4. *Uniqueness Depends on Initial Data*   Consider the polynomial operator $P(D) = tD^2 + D$.

   (a) Find a basic solution set for $P(D)[y] = 0$ over $t > 0$. [*Hint*: write $P(D)[y]$ as $(ty')'$.]

   (b) Show that the IVP, $P(D)[y] = 2t$, $y(0) = a$, $y'(0) = b$, has a unique solution on any interval containing the origin, if a and b are chosen suitably. [*Hint*: $t^2/2$ is a particular solution.]

   (c) Why doesn't the result in **(b)** contradict Theorem 3.7.1?

5. Find a basic solution set for $P(D)[y] = ((1 - t^2)D^2 - 2tD)[y] = 0$. Find the general solution of $P(D)[y] = 0$ on the interval $|t| < 1$. [*Hint*: $(1 - t^2)D^2 - 2tD = D[(1 - t^2)D]$.]

> **Answer 5:** The general solution of $P(D)y = (1 - t^2)y'' - 2ty' = [(1 - t^2)y']' = 0$ can be found by integration: $(1 - t^2)y' = C_1$, so $y' = C_1(1 - t^2)^{-1}$. Integrating again,
>
> $$y = \frac{C_1}{2} \ln \frac{1+t}{1-t} + C_2, \quad |t| < 1$$
>
> where C_1 and C_2 are any constants. (Since C_1 is arbitrary, we can replace it by $2C_1$.) So a basic solution set for $P(D)[y] = 0$ is $\{\ln((1 + t)/(1 - t)), 1\}$.

☞ This is one of the few second-order linear ODEs with nonconstant coefficients for which we can find a simple solution formula.

Euler ODE.   For real constants p, q, the ODE $t^2 y'' + pty' + qy = 0$ is an *Euler ODE*.

6. Show that Euler's ODE has the solution $y = t^r$ if r solves the quadratic equation $Q(r) = r^2 + (p - 1)r + q = 0$. Show that $y_1 = t^{r_1}$ and $y_2 = t^{r_2}$ form a basic solution set if r_1 and r_2 are distinct real roots of $Q(r)$. If $r = \alpha + i\beta$ is a complex root, show that $y_1 = t^\alpha \cos(\beta \ln t)$, $y_2 = t^\alpha \sin(\beta \ln t)$ form a basic solution set for the Euler ODE. [*Hint*: write the complex-valued function $t^{\alpha+i\beta}$ as $e^{(\alpha+i\beta)\ln t}$.]

7. Use the result presented in Problem 6 to solve the IVP $t^2 y'' - 2ty' + 2y = 0$, $y(1) = 0$, $y'(1) = -1$. What is the largest interval on which this solution is defined?

> **Answer 7:** The polynomial $Q(r)$ for the Euler ODE, $t^2 y'' - 2ty' + 2y = 0$ is $Q = r^2 - 3r + 2 = (r - 1)(r - 2)$. The roots are 1 and 2, so a basic solution set is $y_1 = t$, $y_2 = t^2$. The general solution

of the ODE is $y = C_1 t + C_2 t^2$. Imposing the initial conditions $y(1) = 0$ and $y'(1) = -1$, we see that $0 = C_1 + C_2$ and $-1 = C_1 + 2C_2$. So, $C_1 = 1$ and $C_2 = -1$; the solution is $y = t - t^2$, all t.

8. Use the result in Problem 6 to solve the IVP, $t^2 y'' + 3ty' + 2y = 0$, $y(1) = 0$, $y'(1) = 1$.

9. Use the result in Problem 6 to solve the IVP, $t^2 y'' + 2ty' + 2y = 0$, $y(1) = 0$, $y'(1) = 1$.

Answer 9: The polynomial $Q(r)$ for this Euler ODE is $r^2 + r + 2$ with roots $r_{1,2} = -1/2 \pm \sqrt{7}i/2$. The general solution of the ODE is

$$t^{-1/2}\left[C_1 \cos\left((\sqrt{7}/2)\ln t\right) + C_2 \sin\left((\sqrt{7}/2)\ln t\right)\right]$$

where C_1 and C_2 are arbitrary constants. Imposing the initial conditions we have

$$0 = C_1, \quad 1 = \sqrt{7}C_2$$

and the solution of the IVP is

$$y = \frac{2}{\sqrt{7}} t^{-1/2} \sin(\sqrt{7}(\ln t)/2)$$

10. Searching for solutions on $t > 0$, show that the change of independent variable $t = e^s$ converts the Euler ODE into the constant-coefficient linear ODE

$$\frac{d^2 y}{ds^2} + (p - 1)\frac{dy}{ds} + qy = 0$$

11. Use Problem 10 to find a basic solution set for an Euler equation if $r^2 + (p - 1)r + q$ has a double root.

Answer 11: $t^2 y'' + py' + qy = 0$ is the Euler equation. Let $t = e^s$. We get $\frac{d^2 y}{ds^2} + (p - 1)\frac{dy}{ds} + 9y = 0$, as in Problem 10. If $r^2 + (p - 1)r + q$ has a double root, $r_1 = r_2$, then $y(s) = (C_1 s + C_2)e^{r_1 s} = (C_1 \ln t + C_2)t^{r_1}$ is the general solution since $s = \ln t$. Hence, $\{t^{r_1} \ln t, t^{r_1}\}$ is a basic solution set.

12. *Another Euler ODE* Show that the IVP $t^2 y'' - 2ty' + 2y = 0$, $y(0) = 0$, $y'(0) = 0$ has infinitely many solutions. Why does this fact not contradict the assertion of uniqueness in Theorem 3.7.1?

Solution Techniques. Here are some useful techniques in coming up with solution formulas.

13. *Reduction of Order Technique* Let $y = u(t)$ be a solution of the ODE $y'' + a(t)y' + b(t)y = 0$, where $a(t)$ and $b(t)$ are continuous on a t-interval I.

(a) Show that $y = u(t)z(t)$ solves that ODE if $z(t)$ solves the ODE $uz'' + (2u' + a(t)u)z' = 0$.

(b) Show that

$$z(t) = \int_{t_0}^{t} \left\{ \frac{1}{(u(s))^2} \exp\left[-\int^s a(r)\, dr\right] \right\} ds$$

is a solution of the z-equation in **(a)** if $u(t) \neq 0$ for any t in I.

(c) Show that the pair of solutions $\{u, uz\}$ constructed in **(a)** and **(b)** is a basic solution set for the given ODE.

(d) Find all solutions of $ty'' - (t + 2)y' + 2y = 0$, for $t > 0$, given that e^t is one solution.

Answer 13:

(a) Suppose $y = u(t)$ is a solution of $y'' + a(t)y' + b(t)y = 0$. Let $z(t)$ satisfy $uz'' + (2u' + au)z' = 0$. Then if $y = uz$ is substituted into $y'' + ay' + by = 0$, we have $(u''z + 2u'z' + uz'') + a(u'z + uz') + buz = uz'' + z'(2u' + au) + z(u'' + au' + bu) = 0$ since $u'' + au' + bu = 0$ and $uz'' + (2u' + au)z' = 0$ by the hypothesis. So $y = u(t)z(t)$ is a solution of $y'' + a(t)y' + b(t)y = 0$.

(b) Let $v = z'$. Then the ODE $uz'' + (2u' + a(t)u)z' = 0$ from **(a)** may be written in terms of v and normalized as $v' + (2u'/u + a)v = 0$, which is linear. Its integrating factor is $u^2(t)\exp\int^t a(r)\, dr$.

So $v(t) = [u(t)]^{-2}\exp[-\int^t a(r)\,dr]$ is one solution of the ODE for v. We find z by an integration: $z(t) = \int_{t_0}^t v(s)\,ds$.

(c) We have that $W[u, uz] = u(uz)' - (uz)u' = u^2 z' = u^2 v = \exp[-\int^t a(r)\,dr] > 0$ where we have used $z' = v$ and v from part **(b)**, $v(t) = [u(t)]^{-2}\exp[-\int^t a(r)\,dr]$. Since the Wronskian is not 0 and $\{u, uz\}$ is a set of solutions of the normal ODE $y'' + ay' + by = 0$, the set $\{u, uz\}$ is a basic solution set.

(d) Using the preceding results, we see that if $u = e^t$ is one solution of $ty'' - (t+2)y' + 2y = 0$ then a second solution is given by $y = uz = e^t z$, where $z = \int^t e^{-2s}\exp(\int^s(1 + 2/r)\,dr)\,ds$ and we have normalized the ODE to $y'' - (1 + 2/t)y' + 2t^{-1}y = 0$. Carrying out the integration, we have that $z = \int^t e^{-2s}(s^2 e^s)\,ds = \int^t s^2 e^{-s}\,ds = -t^2 e^{-t} - 2e^{-t}(t+1) = -e^{-t}(t^2 + 2t + 2)$. So, $uz = -(t^2 + 2t + 2)$. Since by part **(c)** $\{e^t, -(t^2 + 2t + 2)\}$ forms a basic solution set, we obtain all solutions from u and uz in the form $y = c_1 e^t + c_2(2 + 2t + t^2)$, where c_1 and c_2 are arbitrary constants.

14. *Wronskian Reduction of Order*   Suppose that $P(D) = D^2 + a(t)D + b(t)$, where $a(t)$, $b(t)$ are continuous on the t-interval I. Suppose that $W(t) = W[y_1, y_2](t)$ is the Wronskian of a pair of solutions $\{y_1, y_2\}$ of $P(D)[y] = 0$.

(a) Show that $W(t)$ satisfies the first-order linear equation $W' = -a(t)W$, so W is given by *Abel's Formula*

$$W(t) = W(t_0)\exp\left[-\int_{t_0}^t a(s)\,ds\right] \quad \text{for } t_0, t \text{ in } I$$

[*Hint*: differentiate W directly.]

(b) Use Abel's Formula to show that if $u(t) \neq 0$ is a solution of $P(D)[y] = 0$, then a second solution $v(t)$ of the ODE can be found by solving the following first-order linear ODE for v:

$$u(t)v' - u'(t)v = \exp\left[-\int^t a(s)\,ds\right]$$

Show that the pair $\{u(t), v(t)\}$ is a fundamental set for $P(D)[y] = 0$.

(c) Given that e^t is a solution of $ty'' - (t+2)y' + 2y = 0$, find a second solution v such that $\{e^t, v\}$ is a fundamental set. [*Hint*: remember to normalize the ODE first.] Show that by an appropriate choice of the limits of integration, the solution v can be made identical to the solution uz in Problems 13**(a)** and 13**(b)**.

The Method of Variation of Parameters.

15. If $W(t) = W[y_1, y_2](t)$ is the Wronskian of the basic solution set $\{y_1, y_2\}$, show directly that the functions $C_1(t)$ and $C_2(t)$ below satisfy the condition $C_1'y_1 + C_2'y_2 = 0$:

$$C_1(t) = \int_{t_0}^t \frac{-y_2(s)f(s)}{W(s)}\,ds \quad \text{and} \quad C_2(t) = \int_{t_0}^t \frac{y_1(s)f(s)}{W(s)}\,ds$$

Show directly that $y_d = C_1 y_1 + C_2 y_2$ solves the IVP, $P(D)[y] = f$, $y(t_0) = 0$, $y'(t_0) = 0$.

Answer 15:   Let

$$C_1(t) = \int_{t_0}^t \frac{-y_2(s)f(s)}{W(s)}\,ds, \quad \text{and} \quad C_2(t) = \int_{t_0}^t \frac{y_1(s)f(s)}{W(s)}\,ds$$

We need to show that $C_1'y_1 + C_2'y_2 = 0$ and that $y_d(t)$ actually solves the IVP. Using the Fundamental Theorem of Calculus,

$$C_1'y_1 + C_2'y_2 = \frac{-y_2(t)f(t)}{W(t)} \cdot y_1(t) + \frac{y_1(t)f(t)}{W(t)} \cdot y_2(t) = 0$$

Thus, by the work done above, in order to show that $y_d(t)$ is a solution of the ODE $P(D)[y] = f$ we need only show that $C_1' y_1' + C_2' y_2' = f$. Verifying this,

$$C_1' y_1' + C_2' y_2' = \frac{-y_2(t)f(t)}{W(t)} \cdot y_1'(t) + \frac{y_1(t)f(t)}{W(t)} \cdot y_2'(t)$$

$$= \frac{f(t)}{W(t)} \left[y_1 y_2' - y_2 y_1' \right] = \frac{f(t)}{W(t)} \cdot W(t) = f(t)$$

Finally, we need to show that $y_d(t)$ satisfies the initial conditions $y_d(t_0) = 0$ and $y_d'(t_0) = 0$:

$$y_d(t_0) = C_1(t_0) y_1(t_0) + C_2(t_0) y_2(t_0)$$

$$= \int_{t_0}^{t_0} \frac{-y_2(s)f(s)}{W(s)} \, ds \cdot y_1(t_0) + \int_{t_0}^{t_0} \frac{y_1(s)f(s)}{W(s)} \, ds \cdot y_2(t_0)$$

$$= 0 \cdot y_1(t_0) + 0 \cdot y_2(t_0) = 0$$

$$y_d'(t_0) = C_1(t_0) y_1'(t_0) + C_2(t_0) y_2'(t_0) + C_1'(t_0) y_1(t_0) + C_2'(t_0) y_2(t_0)$$

$$= 0 \cdot y_1'(t_0) + 0 \cdot y_2'(t_0) + 0 = 0$$

where we have used the fact that $C_1'(t) y_1(t) + C_2'(t) y_2(t) = 0$ for all t.

16. Show that the solution of the IVP, $P(D)[y] = f$, $y(t_0) = y_0$, $y'(t_0) = v_0$ is

$$y = [C_1(t) y_1(t) + C_2(t) y_2(t)] + [c_1 y_1(t) + c_2 y_2(t)]$$

where $\{y_1, y_2\}$ is a basic solution set fo $P(D)[y] = 0$, and the functions $C_1(t)$ and $C_2(t)$ are given in Problem 15 and the constants c_1 and c_2 are

$$c_1 = \frac{y_0 y_2'(t_0) - v_0 y_2(t_0)}{W[y_1, y_2](t_0)}, \qquad c_2 = \frac{v_0 y_1(t_0) - y_0 y_1'(t_0)}{W[y_1, y_2](t_0)}$$

17. *Green's Kernel*   Show that $y_d(t)$, given in formula (8), may be written in the form

☞ Green's kernel
$K(t, s)$ looks like it
depends on the basic
solution set $\{y_1, y_2\}$,
but it is exactly the
same for every basic
solution set. [*Hint:*
use Problem 45 in
Section 3.2 to show
this.]

$$y_d(t) = \int_{t_0}^{t} K(t, s) f(s) \, ds$$

where $K(t, s)$ is the *kernel function* (also called *Green's kernel*):

$$K(t, s) = \frac{y_1(s) y_2(t) - y_1(t) y_2(s)}{y_1(s) y_2'(s) - y_2(s) y_1'(s)} = \frac{y_1(s) y_2(t) - y_1(t) y_2(s)}{W[y_1, y_2](s)}$$

Find K for the operators $P(D) = D^2 + 1$, and $P(D) = D^2 - 2/t^2$.

Answer 17:   From

$$C_1(t) = \int_{t_0}^{t} \frac{-y_2(s) f(s)}{W(s)} \, ds, \quad \text{and} \quad C_2(t) = \int_{t_0}^{t} \frac{y_1(s) f(s)}{W(s)} \, ds$$

we see that

$$y_d(t) = C_1(t) y_1(t) + C_2(t) y_2(t)$$

$$= \int_{t_0}^{t} \frac{-y_2(s) f(s)}{W(s)} \, ds \cdot y_1(t) + \int_{t_0}^{t} \frac{y_1(s) f(s)}{W(s)} \, ds \cdot y_2(t)$$

$$= \int_{t_0}^{t} \frac{-y_2(s) y_1(t) f(s)}{W(s)} + \frac{y_1(s) y_2(t) f(s)}{W(s)} \, ds$$

$$= \int_{t_0}^{t} \frac{y_1(s) y_2(t) - y_1(t) y_2(s)}{W[y_1, y_2](s)} f(s) \, ds$$

$$= \int_{t_0}^{t} K(t, s) f(s) \, ds$$

That is, the particular solution $y_d(t)$ given in formula (8) may be written in the form

$$y_d(t) = \int_{t_0}^{t} K(t, s) f(s) \, ds$$

18. Use the Method of Variation of Parameters to solve the IVP $y'' + y = \tan t$, $y(0) = 0$, $y'(0) = 0$. Find the general solution of $y'' + y = \tan t$. Find the solution of the IVP $y'' + y = \tan t$, $y(0) = y_0$, $y'(0) = v_0$.

19. *A Driven Euler ODE*   Show that the Euler ODE, $t^2 y'' - 4ty' + 6y = 0$ has the basic solution set $\{t^2, t^3\}$. Find the general solution of the driven ODE, $t^2 y'' - 4ty' + 6y = t^4 e^t$, $t > 0$, by the Method of Variation of Parameters. [*Hint:* normalize before using the method.]

Answer 19:   Direct calculation shows that t^2 and t^3 form a basic solution set for the undriven ODE $t^2 y'' - 4ty' + 6y = 0$. The Wronskian of this basic solution set is

$$W(t) = t^2 \cdot (t^3)' - t^3 \cdot (t^2)' = t^4 \neq 0 \quad \text{for } t > 0$$

The normalized and driven ODE is $y'' - 4y'/t + 6y/t^2 = t^2 e^t$, $t > 0$. Then by (11) in the text

$$C_1(t) = \int^{t} \frac{-s^3 \cdot s^2 e^s}{s^4} \, ds = \int^{t} -s e^s \, ds = e^t(t - 1)$$

$$C_2(t) = \int^{t} \frac{s^2 \cdot s^2 e^s}{s^4} \, ds = \int^{t} e^s \, ds = e^t$$

where we have not specified a lower limit of integration since we are looking for the general solution. So the general solution of the driven ODE $t^2 y'' - 4ty' + 6y = e^t/t$ is given by

$$y_d(t) = C_1(t)t^2 + C_2(t)t^3 + c_1 t^2 + c_2 t^3$$

with $C_1(t), C_2(t)$ given above and c_1, c_2 arbitrary constants.

20. *Method of Variation of Parameters vs. the Method of Undetermined Coefficients*   It is known that the ODE $y'' + 3y' + 2y = 4 \sin 2t$ has a unique periodic solution. Use the Method of Variation of Parameters to find it. Then use the Method of Undetermined Coefficients to find it. Which method is easier? What happens to the solutions as $t \to +\infty$?

21. *Engineering Function Inputs*   Use the Method of Variation of Parameters to solve the IVP, $y'' + 0.4y' + y = 20 \, \text{SqWave}(t, 2\pi, \pi) - 10$, $y(0) = y_0$, $y'(0) = v_0$, leaving any awkward integrals unevaluated. Then use your numerical solver to plot solutions for the following pairs of values of (y_0, v_0): $(0, 0)$, $(40, 0)$, $(-40, 0)$. Looking at your plots, describe the long-term behavior of all solutions as $t \to +\infty$.

Answer 21:   The undriven equation $y'' + 0.4y' + y = 0$ has characteristic polynomial $r^2 + 0.4r + 1$ with roots $-0.2 \pm i\sqrt{0.96}$. Thus, a basic solution set is

$$\{y_1, y_2\} = \{e^{-0.2t} \cos(\sqrt{0.96}t), \, e^{-0.2t} \sin(\sqrt{0.96}t)\}$$

The Wronskian of this basic solution set is

$$W(t) = e^{-0.2t} \cos(\sqrt{0.96}t) \cdot \left[e^{-0.2t} \sin(\sqrt{0.96}t) \right]'$$

$$\qquad\qquad - e^{-0.2t} \sin(\sqrt{0.96}t) \cdot \left[e^{-0.2t} \cos(\sqrt{0.96}t) \right]'$$

$$= \sqrt{0.96} e^{-0.4t}$$

Then

$$C_1(t) = \int_0^t \frac{-e^{-0.2s} \sin(\sqrt{0.96}s) \cdot \left[20\,\mathrm{SqWave}(s, 2\pi, \pi) - 10\right]}{\sqrt{0.96}e^{-0.4s}}\,ds,$$

$$= \int_0^t \frac{-1}{\sqrt{0.96}} e^{0.2s} \sin(\sqrt{0.96}s) \cdot \left[20\,\mathrm{SqWave}(s, 2\pi, \pi) - 10\right]\,ds,$$

$$C_2(t) = \int_0^t \frac{e^{-0.2s} \cos(\sqrt{0.96}s) \cdot \left[20\,\mathrm{SqWave}(s, 2\pi, \pi) - 10\right]}{\sqrt{0.96}e^{-0.4s}}\,ds,$$

$$= \int_0^t \frac{1}{\sqrt{0.96}} e^{0.2s} \cos(\sqrt{0.96}s) \cdot \left[20\,\mathrm{SqWave}(s, 2\pi, \pi) - 10\right]\,ds$$

The solution of the IVP is given by

$$y(t) = \left[C_1(t)y_1(t) + C_2(t)y_2(t)\right] + \left[c_1 y_1(t) + c_2 y_2(t)\right]$$

where $y_1(t)$ and $y_2(t)$, $C_1(t)$ and $C_2(t)$ are as given above, and c_1 and c_2 are given in Problem 16. See the figure, where the initial data $y_0 = 0, \pm 40$ and $v_0 = 0$ are used. As $t \to +\infty$, all solutions approach a periodic solution.

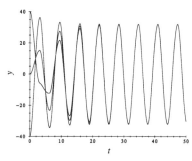

Problem 21.

22. *Engineering Function Inputs* Repeat Problem 21 for $y'' + y = \mathrm{TRWave}(t, 2\pi, \pi, 2\pi, 10)$. What happens to the solutions as $t \to +\infty$?

Properties of Zeros of a Basic Solution Set. Suppose that $a(t)$ and $b(t)$ are continuous on $-\infty < t < \infty$, and suppose that $\{y_1(t), y_2(t)\}$ is a basic solution set for the ODE $y'' + a(t)y' + b(t)y = 0$.

23. *Interlacing of Zeros of a Basic Solution Pair* Show that between any two consecutive zeros of one solution of a basic solution set, there is precisely one zero of the other solution. Why are the margin graphs *not* solution curves of a basic solution set?

Answer 23: First note that if $y_1(\alpha) = 0$, where $y_1(t)$ is a nonconstant solution of the ODE, then $y_1'(\alpha) \neq 0$. For, if $y_1'(\alpha) = 0$, then the IVP consisting of the ODE and the condition $y(\alpha) = 0$, $y'(\alpha) = 0$ has two solutions, $y_1(t)$ and $y = 0$ for all t, but this violates uniqueness. Hence, any nonconstant solution curve that touches the t-axis must cross it and then have a fixed sign as t increases until the curve touches the axis again and the sign reverses.

Now suppose that α and β, $\alpha < \beta$, are consecutive zeros of $y_2(t)$. We want to show that $y_1(t)$ has exactly one zero, say γ, between α and β. First, $y_1(t)$ could not have a zero at α or β, because then the Wronskian $W[y_1, y_2]$ would have value 0 at that point and so everywhere on the interval I, thus violating Abel's Theorem. So if $y_2(t)$ has a zero γ between α and β, we must have $\alpha < \gamma < \beta$. Now $W[y_1, y_2](\alpha) = y_1(\alpha)y_2'(\alpha)$ and $W[y_1, y_2](\beta) = y_1(\beta)y_2'(\beta)$ since $y_2(\alpha) = y_2(\beta) = 0$. Since W is never zero and W is a continuous function of t, W has a fixed sign, which we assume to be positive. Since $y_2(t)$ has a fixed sign for $\alpha < t < \beta$, we will take that sign to be positive. (The other three alternatives for the signs of W and y_2 can be treated similarly.) This means that the solution curve $y = y_2(t)$ cuts across the t axis with a positive slope ($y_2'(\alpha) > 0$), and similarly $y_2'(\beta) < 0$. But that implies (because $y_1(\alpha)$ and $y_2'(\alpha)$ have a common sign, as do $y_1(\beta)$ and $y_2'(\beta)$), that $y_1(\alpha) > 0$ and $y_1(\beta) < 0$. Then since $y_1(t)$ is continuous, it must have a zero γ somewhere between α and β in order to change signs like that.

Could $y_1(t)$ have had more than one zero between α and β? No, because we can interchange the roles of y_1 and y_2 in the above reasoning and so show that y_2 has to have at least one zero between consecutive zeros of $y_1(t)$. The curves in the margin figure do not have this property, and so cannot be the solution curves of a basic solution set.

3.8 Nonlinear Second-Order Differential Equations

Comments:

Now we survey properties of orbits of autonomous second-order ODEs of the form $y'' = F(y, y')$ that may be nonlinear. If F has certain special forms we can find solution formulas. In any case, we can use direction fields and numerical solvers to get some idea of orbital behavior. Nonlinear models of hard and soft springs are introduced and then approximated by linear models.

PROBLEMS

Direction Fields and Orbital Portraits. For each function F given below plot a direction field of the system $y' = v$, $v' = F(y, v)$ on the given rectangle. Mark all equilibrium points. Plot enough orbits that a clear orbital portrait can be seen. Describe what each orbit does as t increases and as t decreases. Any periodic orbits?

1. $F = vy$; $|y| \le 5$, $|v| \le 4$
2. $F = vy - 1$; $|y| \le 2$, $|v| \le 2$
3. $F = v\cos y$; $|y| \le 5$, $|v| \le 4$
4. $F = y\cos v$; $|y| \le 10$, $|v| \le 5$
5. $F = y\sin v$; $|y| \le 10$, $|v| \le 5$
6. $F = y\sin y$; $|y| \le 10$, $|v| \le 5$
7. $F = -10\sin y - v$; $|y| \le 5$, $|v| \le 25$
8. $F = y\cos y$ $|y| \le 10$, $|v| \le 5$
9. $F = y - y^2$; $-1 \le y \le 3$, $|v| \le 3$
10. $F = y - y^4$; $-1 \le y \le 3$, $|v| \le 3$
11. $F = yv - y^3$; $|y| \le 4$, $-10 \le v \le 5$

Answer 1: All points on the y-axis are equilibrium points. There are no periodic orbits. Since $dv/dy = y$, we have $v = y^2 + C$; all nonconstant orbits are arcs of these parabolas. As the direction field and the orbits in the figure suggest, some orbits tend to equilibrium points as t increases and decreases. Others to infinity as t changes in one direction and to an equilibrium point as t changes in the reverse sense. Still other orbits tend to infinity as t increases and as t decreases.

Answer 3: All points on the y-axis are equilibrium points; there are no periodic orbits since the ODE is autonomous and there are no orbits that are simple closed curves. Since $dv/dy = \cos y$, we have $v = \sin y + C$, so orbits are arcs on these sine curves. If $C > 1$, then $v = \sin y + C$ is a vertically displaced sine curve that is traced out from left to right as t increases, and the other way around if $C < -1$. If $|C| \le 1$ each orbit is an arc of $v = \sin y + C$ that tends to an equilibrium point as $t \to +\infty$ and to an equilibrium point as $t \to -\infty$. See the figure.

Answer 5: The y-axis is a line of equilibrium points; there are no periodic orbits. The direction field and orbits in the figure suggest that some orbits tend to the horizontal line orbit $v = \pi$ or to an equilibrium point t increases or decreases. The orbits at the bottom of the figure approach the line orbit $v = -\pi$ as t increases, but then turn away.

Answer 7: The equilibrium points are $(\pm n\pi, 0)$, $n = 0, 1, 2, \ldots$, The orbits and direction field suggest that each nonconstant orbit tends to an equilibrium point as t increases and becomes unbounded as t

decreases. Orbits are symmetric about $(0,0)$.

Answer 9:   The two equilibrium points are $(0,0)$ and $(1,0)$. There is a nest of periodic orbits enclosing $(1,0)$ in the figure. Other orbits seem to enter the screen as t increases and then leave. Since $dv/dy = (y - y^2)/v$, $v^2 = y^2 - 2y^3/3 + C$. The curve $v^2 = y^2 - 2y^3/3$ consists of the equilibrium point $(0,0)$, an orbit that tends to $(0,0)$ as t increases and as t decreases, another that tends to $(0,0)$ as t increases, and a third that does so as t decreases. Orbits are symmetric about the y-axis.

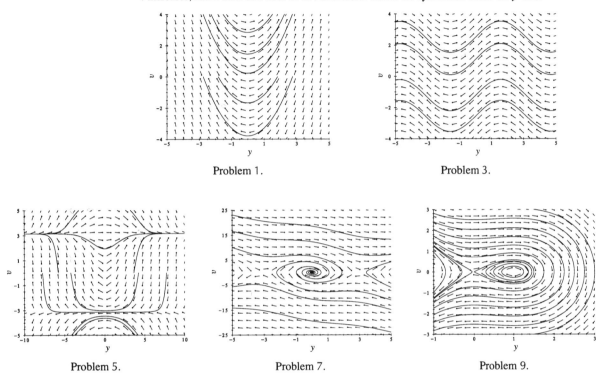

Problem 1.   Problem 3.

Problem 5.   Problem 7.   Problem 9.

Answer 11:   The origin is the single equilibrium point. All other orbits seem to be periodic.

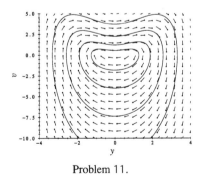

Problem 11.

Nonlinear Springs.   A body of mass m is attached to a wall by means of a spring and moves back and forth along a straight line on a horizontal frictionless surface. Suppose that y measures the displacement of the body along the spring's axis from the equilibrium position; $y < 0$ corresponds to a stretched spring, $y > 0$

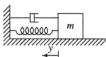

to a compressed spring. The margin figure shows the body with viscous damping forces represented by a damper. The downward gravitational force on the body is exactly canceled by an upward force exerted by the supporting surface, so the mass moves only horizontally, unaffected by gravity. Under these assumptions, Newton's Second Law implies that

$$my'' = S(y) - cy'$$

where $S(y)$ is the *spring force* and $c > 0$ is the damping constant. For each model ODE below first write the equivalent system as in IVP (2). Plot the orbits through the given initial points (at $t_0 = 0$) in the yv-plane, and plot the corresponding solution curves in the ty-plane. Interpret your graphs in terms of the motion of a spring, and discuss how realistic each orbit is. Estimate the periods of periodic solutions; does the period increase, decrease, or stay the same as the magnitude of the initial velocity v_0 increases?

12. *Undamped Hard Spring* $\quad y'' = -0.2y - 0.02y^3, \ 0 \le t \le 25; \ y_0 = 0, \ v_0 = 0, 1, 3, 9.$

13. *Undamped Soft Spring* $\quad y'' = -0.2y + 0.02y^3, \ 0 \le t \le 30; \ y_0 = 0, \ v_0 = 0, 0.4, 0.9; \ y_0 = -5,$
$v_0 = 1.49, 1.51; \ y_0 = 5, v_0 = -1.49, -1.51.$

14. *Damped Soft Spring* $\quad y'' = -y + 0.1y^3 - 0.1y', \ 0 \le t \le 40; \ y_0 = 0, \ v_0 = 0, 2.44, 2.46.$

15. *Alternate Model for Undamped Soft Spring* $\quad y'' = -y + 0.1y|y|, \ |t| \le 20; \ y_0 = 0,$
$v_0 = 2, 5, -6, 6.$

Answer 13: In this case the system is $y' = v, \ v' = -0.2y + 0.02y^3, \ 0 \le t \le 30$. See the figures for orbits and solution curves with the given initial points. The point at $(0,0)$ in Graph 1 and the straight line $y = 0$ in Graph 2 correspond to the spring at rest in static equilibrium. The initial points $y_0 = 0$, $v_0 = 0.4, 0.9$ give periodic solutions with respective periods of approximately 14.5, 19 and amplitudes of approximately 0.9, 2.4. Here the periods seem to *increase* with increasing amplitude and initial velocity, but that is expected because the restoring force weakens as the spring stretches farther and farther. The remaining solutions, plotted along with their orbits, correspond to the spring stretching or compressing far beyond the limits of validity of the model. See the curves in Graph 2 that "escape" as t increases.

Answer 15: $y' = v, v' = -y + 0.1y|y|, |t| \le 20$

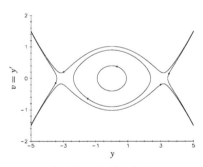

Problem 13, Graph 1.

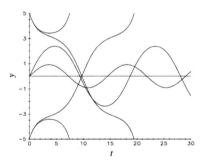

Problem 13, Graph 2.

Linearization. Linearize each ODE at the static equilibrium $y = 0$, $y' = 0$. Find the general solution of the linearized ODE and describe what happens to solutions as $t \to +\infty$.

16. Problem 12   **17.** Problem 13   **18.** Problem 14   **19.** Problem 15

Answer 17: Following Example 3.8.4, we rewrite $y'' = -.2y + 0.02y^3$ as $y'' = F(y, y')$. Notice first that $F(0,0) = 0$. Taking partial derivatives, $F_y = -.2 + 0.06y^2$ and $F_{y'} = 0$. So $F_y(0,0) = -.2$ and $F_{y'}(0,0) = 0$. The linear approximation of F at $(0,0)$ is $G(y, y') = -0.2y$. Thus, the linear approximation to the original ODE is $y'' = -0.2y$; the general solution of this ODE is $y = C_1 \cos \beta t + C_2 \sin \beta t$ (as in Problem 16), $\beta = \sqrt{0.2}$. All solutions are periodic of period $2\pi/\beta$.

Answer 19: Following Example 3.8.4, we rewrite $y'' = -y + 0.1y|y|$ as $y'' = F(y, y')$. Notice first that $F(0, 0) = 0$. Taking partial derivatives, $F_y = -1 \pm 0.2y$ and $F_{y'} = 0$. So $F_y(0, 0) = -1$ and $F_{y'}(0, 0) = 0$. The linear approximation of F at $(0, 0)$ is $G(y, y') = -y$. Thus, the linear approximation to the original ODE is $y'' = -y$, whose general solution is $y = C_1 \cos t + C_2 \sin t$. All nonconstant solutions are periodic of period 2π.

Compare a Nonlinear ODE and Its Linearization. Plot orbits of the original and of the linearized system near $y_0 = 0$, $y_0' = 0$, using the two initial data points $(0, .5)$ and $(.5, 0)$: overlay the orbits and explain the difference between the two state portraits. Using the same data sets, plot ty-solution curves of the two systems, overlay the graphs, and explain what you see. [*Hint*: see Example 3.8.4.]

   20. Problem 16   **21.** Problem 17   **22.** Problem 18   **23.** Problem 19

Answer 21: Graphs 1 and 2 show the orbits of the nonlinear system $y' = v$, $v' = -0.2y + 0.02y^3$, and of the linearized system $y' = v$, $v' = -0.2y$, respectively. The only discernible difference is that the nonlinear y-amplitude of the larger orbit is slightly larger than the linearized y-amplitude. Graph 3 shows the corresponding y-solution curves of the nonlinear system (solid) and the linearized system (dashed). The time lags are now apparent, as is the difference in y-amplitudes.

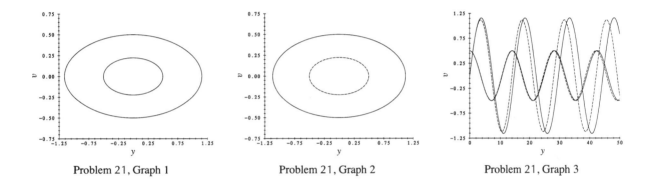

Problem 21, Graph 1   Problem 21, Graph 2   Problem 21, Graph 3

Answer 23: Graphs 1, 2, and 3 show no detectable differences between the orbits or between the y-solution curves of the nonlinear system $y' = v$, $v' = -y + 0.1y|y|$ and the linearized system $y' = v$, $v' = y$.

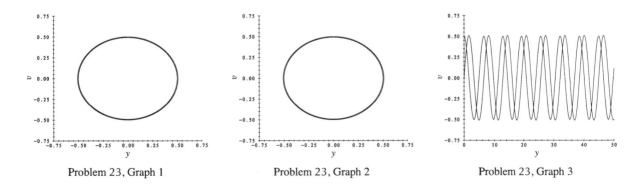

Problem 23, Graph 1   Problem 23, Graph 2   Problem 23, Graph 3

Fundamental Theorem, Plotting Curves, Equilibrium Solutions. Verify that each IVP satisfies the conditions of Theorem 3.8.1 throughout tyy'-space. Plot representative component curves, orbits, and time-state curves. Find all equilibrium solutions. Make conjectures about the long-term behavior of solutions.

24. *Vertical Oscillations of a Damped Hard Spring* $y'' + 0.2y' + 10y + 0.2y^3 = -9.8$, $y(0) = -4$, 0, $y'(0) = 0$; $0 \le t \le 40$, $-5 \le y \le 3$, $|y'| \le 9$.

25. *Nonlinear ODE* $y'' + (0.1y^2 - 0.08)y' + y^3 = 0$, $y(0) = 0$, $y'(0) = 0.5$; $0 \le t \le 100$, $-1.9 \le y \le 2.1$, $|y'| \le 2.5$.

Answer 25:   The ODE can be rewritten as $y'' = F(t, y, y') = -(0.1y^2 - 0.08)y' - y^3$. The function $F = -(0.1y^2 - 0.08)y' - y^3$ is continuous, as are $\partial F/\partial y = -0.2yy' - 3y^2$ and $\partial F/\partial y' = -0.1y^2 + 0.08$, so Theorem 3.8.1 is satisfied. The solution of the nonlinear IVP $y'' + (0.1y^2 - 0.08)y' + y^3 = 0$, $y(0) = 0$, $y'(0) = 0.5$ seems to oscillate with a slowly growing amplitude [Graph 1]. The orbit [Graph 2] suggests that as $t \to +\infty$ the orbit spirals to the orbit of a slightly tilted, periodic "racetrack" orbit. Turning back to the solution curve in Graph 1, we see from the solution curve for large t that the period of the racetrack orbit is about 4.5. The time-state curve is shown in Graph 3.

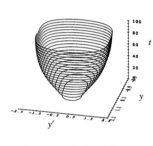

Problem 25, Graph 1.   Problem 25, Graph 2.   Problem 25, Graph 3.

Modeling Problems.

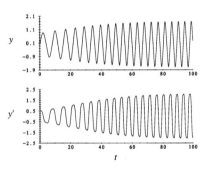

26. *Modeling a Magnetically Driven Spring*   As in the margin figure, a magnet is suspended by a spring above a fixed iron plate. Suppose that z is the vertical displacement upward of the magnet from its rest position at distance b from the plate. Assume that the force of magnetic attraction is inversely proportional to the square of the distance between the magnet and the plate, and that the spring force is directly proportional to the displacement z. Find the ODE for the magnet's motion. Is the ODE linear?

27. *Effect of Gravity on Spring Motion*   Does gravity affect the vertical motion of a block on a hard spring? On a soft spring? Contrast your conclusions with those drawn for the Hooke's Law spring as described in Section 3.1, and give reasons.

Answer 27:   We will account for gravity by simply changing variables to shift the equilibrium position of the spring from h in the old variable y to 0 in the new variable z. Substituting $y = z - h$ in the hard/soft spring models, $my'' + cy' + ky \pm jy^3 = -mg + f(t)$ yields $mz'' + cz' + kz \pm j(z^3 - 3z^2h + 3zh^2) = f(t)$, where we have used the fact that the equilibrium displacement h is the value of y at which the gravitational force is exactly balanced by the spring force, so h satisfies $k(-h) \pm j(-h)^3 = -mg$. Since our transformed ODE for motion about the equilibrium changes by a nonzero term $\pm j(-3z^2h + 3zh^2) = \pm 3zhj(-z + h)$ when gravity is introduced, gravity *does* affect the motion of hard and soft springs more than just by changing the position of static equilibrium, and it does so by changing the form of the ODE of the motion near the point of static equilibrium. In the Hooke's Law spring, this substitution does not change the equation of motion since $j = 0$ under Hooke's Law, so the Hooke's Law spring's motion is not affected by gravity.

 More Solver Techniques.   Some solvers may not be able to overlay plots from two different ODEs on the same set of axes. But there are ways to coax your solver into being more cooperative.

28. *One Way to Generate Figures 3.8.6 and 3.8.7* Follow the outline below to trick your solver into reproducing these figures. Let's consider the IVPs

$$y'' = (1-c)(-10y + 0.2y^3 - 0.2y' - 9.8) + c[-9.4(y+1) - 0.2y']$$

$$y(0) = -1, \quad y'(0) = 5, 13.27$$

where c is a constant.

(a) Explain why plotting solutions of these IVPs for $c = 0$ and $c = 1$ generates the graphs in Figures 3.8.6 and 3.8.7.

(b) Show that the IVPs in **(a)** can be converted to the equivalent IVPs for a first-order system:

$$y' = v, \qquad\qquad\qquad\qquad\qquad\qquad\qquad y(0) = -1$$
$$v' = (1-c)(-10y + 0.2y^3 - 0.2v - 9.8) + c[-9.4(y+1) - 0.2y'], \qquad v(0) = 5, 13.27$$

For $c = 0$ and $c = 1$ graph the orbits of these IVPs.

29. *Another Way to Generate Figures 3.8.6 and 3.8.7.* Suppose that $c = c(t)$ is another state variable and consider the following IVPs for a first-order system with three state variables:

$$y' = v, \qquad\qquad\qquad\qquad\qquad\qquad\qquad y(0) = -1$$
$$v' = (1-c)(-10y + 0.2y^3 - 0.2v - 9.8) + c[-9.4(y+1) - 0.2y'], \qquad v(0) = 5, 13.27$$
$$c' = 0, \qquad\qquad\qquad\qquad\qquad\qquad\qquad c(0) = c_0$$

Show that if these IVPs are solved for $c_0 = 0$ and $c_0 = 1$ and y plotted against t in both cases, then Figures 3.8.6 and 3.8.7 result. Graph the orbits of these IVPs in the yv-plane.

Answer 29: If $C = 0$, then the system of ODEs is that of Figure 3.8.7. If $c = 1$, the system is that of Figure 3.8.6. The respective orbits are shown in Graphs 1 and 2. The initial data for each graph is $y(0) = -1$, $y'(0) = 5, 13.27$.

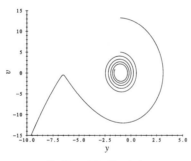

Problem 29, Graph 1

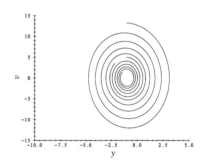

Problem 29, Graph 2

30. Plot the solution curves of $y'' = -y + cy^3$, $y(0) = 1$, $y'(0) = 0$, for $c = 0, -0.1, 0.1, 0 \le t \le 50$ on the same screen. Interpret in terms of particular types of springs.

Solution Behavior.

31. Consider the nonlinear and nonautonomous IVP

$$t[y''y + (y')^2] + y'y = 1, \qquad y(1) = 1, \quad y'(1) = v_0$$

Use the screen size $0 \le t \le 3$, $|y| \le 5$ and plot several solution curves using several values of v_0 with $|v_0| \le 10$. What happens to the curves as $t \to 0^+$? Why would you expect some kind of bad behavior? What happens to solutions as t increases far beyond $t = 3$? [*Hint*: use your solver to extend t.]

Answer 31: See the figure for graphs of several solution curves of $t(y''y + (y')^2) + y'y = 1$ with initial points $y(1) = 1$, $|y'(1)| = |v_0| \le 10$. As $t \to 0^+$, some solution curves of the IVPs in normal form, $y'' = 1/(ty) - y'/t - (y')^2/y$ tend to $+\infty$. Other solution curves simply stop when they hit the line $y = 0$. Neither result is surprising because y'' is undefined if $t = 0$ or if $y = 0$. As t increases (or decreases) some of these solutions seem to become unbounded, but others are bounded below by $y = 0$.

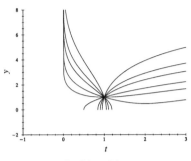

Problem 31.

Concavity. The sign of the second derivative of $y(t)$ determines whether the graph of $y = y(t)$ is concave up ($y'' > 0$) or down ($y'' < 0$). Determine the regions in the ty-plane where the solution curves of the ODE are concave down. Then plot solutions that meet given initial conditions and verify that the solution curves are concave down in the proper regions. See also the Problem Set in the WEB SPOTLIGHT ON OPTICAL ILLUSIONS.

32. $y'' = -y + t$, $y(0) = 0$, $y'(0) = -1, 0, 1$ 33. $y'' = -4y + \sin t$, $y(0) = y'(0) = 0$

Answer 33: The concavity of the solution curve changes when it crosses the curve $-4y + \sin t = 0$. Above that sine curve the solution curve is concave down since $y'' < 0$, below the concavity is up. The figure shows the solution curve (solid) and the curve $-4y + \sin t = 0$ (dashed).

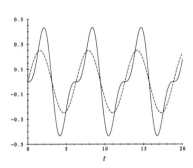

Problem 33.

Incompatible Functions. Not every pair of functions can be solutions to a second-order ODE in normal form. Suppose the functions $F(t, y, y')$, $\partial F/\partial y$, and $\partial F/\partial y'$ are continuous for all t, y, y'.

34. Explain why $y_1 = \sin t$ and $y_2 = t$, $-\infty < t < \infty$, can't *both* be solutions of the same ODE $y'' = F(t, y, y')$, for *any* F. [*Hint*: note that $y_1(0) = y_2(0) = 0$, $y_1'(0) = y_2'(0) = 1$.]

35. Repeat Problem 34, but with $y_1 = e^t$ and $y_2 = 1 + t + t^3/3$.

Answer 35: Both functions y_1 and y_2 have value 1 and derivative 1 at $t = 0$. If y_1 and y_2 were solutions of the same ODE $y'' = F(t, y, y')$, where F, $\partial F/\partial y$, and $\partial F/\partial y'$ are continuous for all t, y, y', that would violate uniqueness. So they cannot both be solutions of $y'' = F(y, y', t)$.

Properties of Orbits. Suppose that $y' = v$, $v' = F(y, v)$, where $F(y, v)$, $F_y(y, v)$ and $F_v(y, v)$ are continuous for all y, v.

36. As t increases, why do all orbits above the y-axis move to the right and orbits below the axis move to the left?

37. Why do nonconstant orbits that cross the y-axis always do so perpendicular to the axis?

38. Suppose that $F(y, v) = vG(y, v)$, where the functions G, G_y, and G_v are continuous for all y, v. Why don't any orbits cross the y-axis?

39. The parabola $v = 1 - y^2$, $-\infty < y < \infty$, is *not* an orbit of any system $y' = v$, $v' = F(y, v)$. Why not?

40. The circle $(x - 1)^2 + (y - 1)^2 = 16$ is not an orbit of any system $y' = v$, $v' = F(y, v)$. Why not?

Answer 37: The nonconstant orbits that cross the y-axis have $(0, F(y, v))$ as the vector tangent to the orbit. Since $y' = v = 0$, the vector is upward or downward. Therefore, orbits that cross the y-axis are perpendicular to the axis.

Answer 39: The parabola $v = 1 - y^2$, $-\infty < y < \infty$ cuts the y-axis at a slant. From Problem 37, orbits that cross the y-axis do so perpendicular to the axis. Thus, $v = 1 - y^2$ is not an orbit of the system $y' = v$, $v' = F(y, v)$

Orbits of ODEs: $y'' = F(y)$.

41. *Closed Curve Orbits of Harmonic Oscillator* Show that the nonconstant orbits of the general harmonic oscillator ODE $y'' + \omega^2 y = 0$, where ω is a positive constant, are ellipses.

42. *Closed Curve Orbits of Undamped Hard Spring* Show that the nonequilibrium orbits of the ODE $y'' = -4y - y^3$ are simple closed curves. [*Hint*: show that the nonconstant orbits are given by $y' = v = \pm(C - 4y^2 - y^4/2)^{1/2}$, where C is any positive constant. Then show that each orbit is bounded, symmetric about the y-axis and cuts the y-axis at two points.]

43. *Orbits of Undamped Soft Spring*

(a) Find an equation for the orbits of the ODE $y'' = -10y + .2y^3 - 9.8$.

(b) Plot a variety of orbits in the rectangle $|y| \le 12.5$, $|v| \le 25$. Compare your graphs with those of the *damped* soft spring shown in Figure 3.8.2. Explain the differences in terms of the behaviors of damped and undamped soft springs.

Answer 41: Multiply the ODE $y'' + \omega^2 y = 0$ by y' to obtain $y'y'' + \omega^2 yy' = 0$. This can be written as $((y')^2)'/2 + (\omega^2 y^2)'/2 = 0$, which integrates to $(y')^2/2 + (\omega^2 y^2)/2 = C$, where C is any non-negative constant. The graph in the yy'-plane is an ellipse if $C > 0$ since ω^2 is positive.

Answer 43:

(a) Since the ODE $y'' = -10y + 0.2y^3 - 9.8$ has the form given in Theorem 3.8.2, an equation for the orbits can be found by multiplying through by y' and integrating to obtain $(y')^2/2 = -5y^2 + 0.05y^4 - 9.8y + C$. Solving for y', we obtain the orbital equations

$$y' = \pm(-10y^2 + 0.1y^4 - 19.6y + K)^{(1/2)}$$

where K is an arbitrary constant.

(b) This undamped soft spring has some periodic orbits enclosing the static equilibrium point at $y = -1$, $y' = 0$; the damped soft spring model of Example 3.8.1 and Figure 3.8.2 has no periodic orbits. Aside from that, the orbits look similar. That is not surprising because for large values of y, the dominant term, $0.2y^3$, is the same in both ODEs. See the figure.

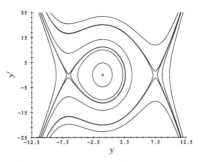

Problem 43(b).

Orbits of ODEs: $y'' = F(y)y'$.

44. Suppose that $F(y)$ and $F'(y)$ are continuous. Describe a method for finding all orbits of the ODE. $y'' = F(y)y'$. [*Hint*: suppose that $G(y)$ is an antiderivative of $F(y)$. Then antidifferentiate each side of $y'' = F(y)y'$.]

45. Find the solution formula for the IVP $y'' = 2yy'$, $y(0) = 1$, $y'(0) = 1$. Plot the solution curve and orbit of this IVP. What is the largest t-interval on which the solution of this IVP is defined?

Answer 45: To solve the IVP $y'' = 2y'y$, $y(0) = 1$, $y'(0) = 1$ we use the hint in Problem 44 to see that $y'' = (y^2)'$. Integrating and applying the initial conditions, we see that $y' = y^2$, so $y'/y^2 = 1$. Integrating again and applying the initial conditions, we obtain $-1/y + 1 = t$. Solving for y, we have $y = 1/(1 - t)$. The largest t-interval for this IVP is $t < 1$. See Graph 1 and Graph 2 for the solution graph and orbit, respectively.

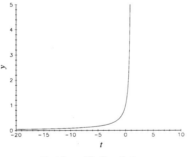

Problem 45, Graph 1.

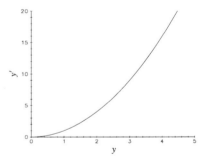

Problem 45, Graph 2.

Rescaling, Qualitative Behavior.

 46. Rescale the planar autonomous system $dx/dt = f(x, y), dy/dt = g(x, y)$ by the change of independent variable $t = t(s)$ where $dt/ds > 0$. Show that both systems have exactly the same orbits.

Spotlight on Modeling: Vertical Motion

Comments

The section has a proof of the existence of the escape velocity from a massive body. Besides being a nice example of the use of the reduction of order technique, it also provides a striking example of what the computer cannot do i.e., prove the existence of an escape velocity.

PROBLEMS

Reduction of Order. Solve the following ODEs by reduction of order. Plot solution curves $y = y(t)$ in the ty-plane using various sets of initial data. [*Hint*: if y does not appear explicitly, set $y' = v$, $y'' = v'$. If t does not appear explicitly, set $y' = v$ and $y'' = v\,dv/dy$.]

1. $ty'' - y' = 3t^2$ **2.** $y'' - y = 0$ **3.** $yy'' + (y')^2 = 1$ **4.** $y'' + 2ty' = 2t$

Answer 1: Let $v = y'$. The ODE $ty'' - y' = 3t^2$ becomes $tv' - v = 3t^2$, so $v' - t^{-1}v = 3t$, which is a first-order linear equation in v and has solutions $v = 3t^2 + Ct$. So, $y = \int^t (3s^2 + Cs)ds = t^3 + Ct^2/2 + C_2$ or $y = t^3 + C_1 t^2 + C_2$, where C_1 and C_2 are any constants. See the figure.

Answer 3: Let $v = y'$, $v\,dv/dy = y''$. The ODE $yy'' + (y')^2 = 1$ becomes $yv\,dv/dy + v^2 = 1$, which separates to $v(1 - v^2)^{-1}dv = y^{-1}dy$. Integrating, $-(\ln|1 - v^2|)/2 = \ln|y| + K$, or $\ln(y^2|1 - v^2|) = -2K$. So, $y^2(1 - (y')^2) = C_1$, or $y' = \pm y^{-1}(-C_1 + y^2)^{1/2}$. So, $y(y^2 - C_1)^{-1/2}dy = \pm dt$. Integrating, $(y^2 - C_1)^{1/2} = \pm t + C_2$. Solving for y^2, we get $y^2 = C_1 + (C_2 \pm t)^2$. So, we have the four solution formulas, $y = \pm[C_1 + (C_2 \pm t)^2]^{1/2}$, where C_1 and C_2 are arbitrary constants and t is such that $C_1 + (C_2 \pm t)^2 \geq 0$. See the figure.

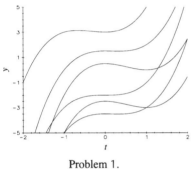

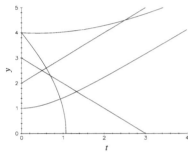

Problem 1.   Problem 3.

Reduction of Order: IVP. Solve the given initial value problems.

5. $2yy'' + (y')^2 = 0$, $y(0) = 1$, $y'(0) = -1$

6. $y'' = y'(1 + 4/y^2)$, $y(0) = 4$, $y'(0) = 3$

7. $y'' = -g - y'$, $y(0) = h$, $y'(0) = 0$, g, h are positive constants

Reduction of Order Methods.

8. The ODE $y'' = F(t, y')$ can be reduced to a first-order ODE via the substitution $v = y'$.

(a) Show that the second-order ODE $y'' = F(t, y')$ (where the derivatives occur with respect to t) is transformed to the equivalent system, $y' = v$, $v' = F(t, v)$. [*Hint*: introduce a new state variable $v = y'$.]

(b) Find the general solution of the ODE $y'' = y' - t$.

9. In the second-order autonomous ODE $y'' = f(y)$, assume that $f(y)$ is a continuously differentiable function on the y-axis. Let $F(y)$ be an antiderivative of $f(y)$.

(a) Show that $y(t)$ is a solution of $y'' = f(y)$ if and only if $(y')^2$ is equal to $2F(y) + C$, where C is a constant. [*Hint*: multiply the ODE $y'' = f(y)$ through by y' and apply the Antiderivative Theorem.]

(b) Derive formula (13) by using the result in (a).

Answer 5: Let $y' = v$, $y'' = v\,dv/dy$. The ODE $2yy'' + (y')^2 = 0$ becomes $2yv\,dv/dy + v^2 = 0$. So $v = 0$ for all t, or $2y\,dv/dy + v = 0$. If $v = 0$ for all t, then y is a constant, but the initial condition $y'(0) = -1$ implies that y is nonconstant. So, v is not always 0. Separating variables, we have $2v^{-1}dv + y^{-1}dy = 0$. Integrating, $\ln|v^2 y| = K$, so $v^2 y = \pm e^K$. So, $v = y' = K_1 y^{-1/2}$. Separating variables, $y^{1/2}dy = K_1 \, dt$. Integrating, $2y^{3/2}/3 = K_1 t + K_2$ or $y = (C_1 t + C_2)^{2/3}$, where C_1 and C_2 are constants to be determined by the initial data $y(0) = 1$, $y'(0) = -1$. So, $C_2 = 1$, $C_1 = -3/2$. The solution is $y = (-3t/2 + 1)^{2/3}$.

Answer 7: Let $y' = v$, $y'' = v'$. The ODE $y'' = -g - y'$ becomes $v' = -g - v$ or $(g + v)^{-1}dv = -dt$. Integrating, $\ln|g + v| = -t + K$, so $g + v = C_1 e^{-t}$. Since $v = y'$, we have $y' = C_1 e^{-t} - g$. Integrating, $y = -C_1 e^{-t} - gt + C_2$. Since $y = h$ and $y' = 0$, at $t = 0$, we have $y = -ge^{-t} - gt + h + g$.

Answer 9:

(a) Multiply $y'' = f(y)$ through by y' to get $y'y'' = f(y)y'$. Let $dF/dy = f(y)$. Then integrating we obtain $(y')^2/2 = F(y) + C$, where C is a constant.

(b) IVP (8) is

$$my'' = -\frac{mMG}{(y + R)^2}, \quad y(0) = 0, \quad y'(0) = v_0 > 0$$

Using the integration method in **(a)** and the fact that

$$\frac{d}{dy}\left(\frac{MG}{y + R}\right) = -\frac{MG}{(y + R)^2}y'$$

we obtain

$$\frac{1}{2}v^2 = \frac{MG}{y + R} + C$$

The initial conditions $y = 0$, $y' = v_0$ imply that $v_0^2/2 = MG/R + C$, which gives formula (13).

Reduction of Order by Varied Parameters.

10. Suppose that $z(t)$ is a solution of the linear *undriven* ODE $z'' + a(t)z' + b(t)z = 0$, where $a(t)$ and $b(t)$ are continuous on an interval I. The function $z(t)$ can be used to find solutions of the *driven* ODE $y'' + a(t)y' + b(t)y = f(t)$, where $f(t)$ is also continuous on I.

    (a) Show that $y(t) = u(t)z(t)$ is a solution on I of the ODE $y'' + a(t)y' + b(t)y = f(t)$ if $u'(t)$ solves the first-order linear ODE $zu'' + (2z' + az)u' = f$. [The function $u(t)$ is said to be a *varied parameter*.]

    (b) If $z(t) \neq 0$ on I, put the linear ODE for u' in **(a)** into normal linear form by dividing by z. Find $u'(t)$ by using the integrating factor $z^2 e^{A(t)}$, where $A(t)$ is an antiderivative of $a(t)$.

    (c) Find another solution of $tz'' - (t + 2)z' + 2z = 0$ for $t > 0$, given that $z = e^t$ is one solution.

11. Using the method described in Problem 10, find a solution of the ODE $t^2 y'' + 4ty' + 2y = \sin t$, $t > 0$, where it is known that $z(t) = t^{-2}$ solves the ODE $t^2 y'' + 4ty' + 2y = 0$. [*Hint*: first put the ODE in normal linear form by dividing by t^2.]

    Answer 11: The normalized ODE is $y'' + 4t^{-1}y' + 2t^{-2}y = t^{-2}\sin t$, $t > 0$. Following the technique described in Problem 10 and using the fact that $z = t^{-2}$ is a solution of $z'' + 4t^{-1}z' + 2t^{-2}z = 0$ (check by direct substitution), we see that $y = t^{-2}u(t)$, where $u(t)$ solves $t^{-2}u'' + (-4t^{-3} + 4t^{-3})u' = t^{-2}\sin t$, solves the normalized ODE. This last equation simplifies to $u'' = \sin t$. After two integrations, we have $u = -\sin t + C_1 t + C_2$, where C_1 and C_2 are any constants. So $y = -t^{-2}\sin t + C_1 t^{-1} + C_2 t^{-2}$, $t > 0$, gives a family of solutions of the ODE.

Modeling Problems: Newtonian Damping.

12. *Projectile Motion: Newtonian Damping* A spherical projectile weighing 100 lb is observed to have a limiting velocity of magnitude 400 ft/sec.

    (a) Show that the velocity v of the projectile undergoing either upward or downward vertical motion and acted on by Newtonian air resistance is given by $v' = -g - (gc/w)v|v|$, where the magnitude of c is 1/1600 and $w = mg$ is the weight of the projectile, where m is its mass.

    (b) Model the projectile motion as a system in the state variables y and v, where $v = y'$.

    (c) If the projectile is shot straight upward from the ground with an initial velocity of 500 ft/sec, what is its velocity when it hits the ground? [*Hint*: use a numerical solver to solve the system of **(b)** with suitable initial conditions. Use the solver to plot the solution $v = v(t)$ and $y = y(t)$ as a parametric curve in the vy-plane. Estimate the impact velocity from the graph.]

(d) *Longer to Fall?* Does the projectile in **(c)** take longer to rise or to fall? Explain. Repeat for $v_0 = 100$, $200, \ldots, 1000$ ft/sec. [*Hint*: use a numerical solver and graph $y(t)$.]

13. *The Newtonian Skydiver's Velocity* In Example 1 it was shown that the skydiver's velocity $v(t)$ satisfies the IVP $v' = -g + cv^2/m$, $v(0) = 0$. Separate variables and use a formula in the Integral Table on the inside back cover and algebraic manipulation to derive the solution formula

$$v(t) = \left(\frac{mg}{c}\right)^{1/2} \frac{e^{-At} - 1}{e^{-At} + 1}, \quad \text{where } A = 2\left(\frac{gc}{m}\right)^{1/2}$$

Answer 13: Observe that the ODE $v' = -g + (c/m)v^2$ is separable. Separating the variables, we obtain

$$\int \frac{1}{-g + (c/m)v^2}\, dv = \int dt$$

Using a formula in the integral table on the inside back cover, we have

$$\frac{1}{2}\left(\frac{m}{gc}\right)^{1/2} \ln\left|\frac{v - (mg/c)^{1/2}}{v + (mg/c)^{1/2}}\right| = t + C$$

Plugging in the initial condition $v(0) = 0$ gives $C = 0$. Setting $A = 2(gc/m)^{1/2}$ and taking exponentials of both sides we obtain

$$\left|\frac{v - (mg/c)^{1/2}}{v + (mg/c)^{1/2}}\right| = e^{At}$$

Note that the quantity between the absolute value bars is negative for all $t \geq 0$, and so changing signs and dropping the absolute value, we solve for v to obtain

$$v(t) = \left(\frac{mg}{c}\right)^{1/2} \frac{e^{-At} - 1}{e^{-At} + 1}$$

14. *Skydiver: Newtonian Damping* A skydiver and equipment weigh 240 lb. [Note: a body of mass m has weight $= mg$.] In free fall the skydiver reaches a limiting velocity of 250 ft/sec. Some time after the parachute opens, the skydiver reaches the limiting velocity of 17 ft/sec. Suppose that later the skydiver jumps out of an airplane at 10,000 feet. Use $g = 32.2$ ft/sec^2 and answer the following questions about the second jump:

(a) How much time must elapse before the free-falling skydiver falls at a speed of 100 ft/sec? [*Hint*: use the data to determine the coefficient c in IVP (1); then use formula (3).]

(b) When falling at 100 ft/sec, the skydiver pulls the rip cord and the chute opens instantaneously. How much longer does it take for the speed of the descent to drop to 25 ft/sec?

(c) How long does the jump last? What is the velocity on landing?

15. *Newtonian vs. Viscous Damping* Design an experiment to determine whether Newtonian or viscous damping better describes the motion of the skydiver of Problem 14. Carry out a computer simulation and discuss the results. [*Hint*: compare with the viscous damping model in Section 1.4 and with Example 1. The damping constant can be determined from the terminal velocity.]

Answer 15: Group Project.

Modeling Problems: Viscous Damping.

16. *Scaling the Whiffle Ball IVP: Viscous Damping* The velocity of a vertically moving whiffle ball subject to viscous damping is given by the ODE $v' = -g - (c/m)v$, where g, c, and m are positive constants. Rescale the state and time variables so that the scaled ODE is free of these constants. Solve

the scaled ODE and draw a conclusion about the limiting velocity for the original ODE. [*Hint*: let $t = aT$, $v = bV$; see also Example 2.]

Modeling Problems: Escape Velocities.

17. *Projectile Motion*   Suppose that a projectile of mass m is launched vertically from ground level with initial velocity v_0 that is less than the earth's escape velocity (i.e., $v_0^2 < 2MG/R$). What is the maximum height achieved by the projectile? Neglect air resistance.

Answer 17:   From formula (13)

$$v^2 = \left(v_0^2 - \frac{2MG}{R} \right) + \frac{2MG}{y + R}$$

At the top point y_{max} of the projectile's path, $v = 0$:

$$0 = v_0^2 - \frac{2MG}{R} + \frac{2MG}{y_{max} + R}$$

$$y_{max} = \frac{2MG}{2MG/R - v_0^2} - R = \frac{2MGR}{2MG - v_0^2 R} - R = \frac{v_0^2 R^2}{2MG - v_0^2 R} > 0$$

18. *A Falling Body*   A body is dropped from rest at a height of y_0 units above the ground. Neglect air resistance and find the velocity of the body when it hits the ground.

19. *Escape Velocity in an Inverse Cube Universe*   Suppose that on a planet in another universe the magnitude of the force of gravity obeys the Inverse Cube Law:

$$|\mathbf{F}| = \frac{\tilde{G}mM}{(y + R)^3}$$

where m is the mass of the object, y its location above the surface of the planet, M the mass of the planet, $\tilde{G}$ a new universal constant, and R the radius of the planet.

(a) What is the escape velocity $\tilde{v}_0$ from this planet?

(b) What is the ratio of $\tilde{v}_0$ to the Inverse Square Law escape velocity $(2MG/R)^{1/2}$?

Answer 19:

(a) The mass of the object times its acceleration equals the net force acting on the object. The equation of motion is $mz'' = -mM\tilde{G}/(z + R)^3$. Let $v = z'$, $v\, dv/dz = z''$ and separate the z and v variables to obtain $v\, dv = -M\tilde{G}(z + R)^{-3} dz$, which integrates to $v^2 = M\tilde{G}(z + R)^{-2} + v_0^2 - M\tilde{G}R^{-2}$ where it is assumed that at $t = 0$, $v = v_0$ and $z = 0$. So, v remains positive for all z if $v_0 > (M\tilde{G})^{1/2}/R$, which is the escape velocity.

(b) The ratio of the two escape velocities is given by

$$\frac{(M\tilde{G})^{1/2}/R}{(2MG/R)^{1/2}} = \left(\frac{\tilde{G}}{2GR} \right)^{1/2}$$

20. *Escape to Infinity*   Let $y(t)$ be the solution of IVP (8) with $v_0^2 > 2MG/R$. Show that $y(t) \to +\infty$, as $t \to +\infty$. [*Hint*: if $A = (v_0^2 - 2MG/R)^{1/2}$, then show that $dy/dt > A$, for all $t \geq 0$. Hence, $(y(t) - At)' > 0$ for all $t > 0$.]

Scaling.

 21. *Scaling and Escape Velocity*   Rescale IVP (8) by defining $z = y/R$ and $t = s(R/MG)^{1/2}$, and derive a new IVP using the new variables z and s. Let $z'(0) = a$, and use a as a scaled "velocity" parameter. Use a numerical solver to find as good a lower bound for the initial escape velocity a as you can.

Answer 21:   Group project.

Spotlight on Modeling: Shock Absorbers

Comments:

Drive on a bumpy road. Do the car's shock absorbers vibrate at maximal amplitude at high speeds? This section does the modeling and answers the question.

PROBLEMS

1. *Calculating the Damping Constant*   In a spring/damper system where the spring force follows Hooke's Law, the spring and damping constants must be determined before the model ODE can be used. The spring constant k can be determined experimentally as follows: hang a weight of mass m on the spring and observe the distance h that it descends until equilibrium is reached; then from Section 3.1, $k = mg/h$. Design an experiment that will measure the damping constant c.

   Answer 1:   The ODE $y'' + (c/m)y' + (k/m)y = 0$ has the characteristic polynomial $r^2 + (c/m)r + (k/m) = 0$ and the characteristic roots

   $$r_1, r_2 = -\frac{c}{2m} \pm \frac{1}{2m}\sqrt{c^2 - 4km}$$

   Suppose that the spring is underdamped so that oscillatory behavior results; i.e., assume that $c^2 < 4km$ (which can always be arranging by taking m large enough). Then the natural circular frequency of the spring is $\beta = \sqrt{4km - c^2}/(2m)$ and oscillations are damped by the factor $e^{-ct/(2m)}$. The period of the oscillations is

   $$\frac{2\pi}{\beta} = \frac{4m\pi}{\sqrt{4km - c^2}} = T$$

   Using suitable lab equipment T can be measured, so we assume that T is known. Solving for c, $16m^2\pi^2 = T^2(4km - c^2)$ and $c^2 = 4km - (16m^2\pi^2)/T^2$. Now with this m determine k via the static deflection h: $kh = mg$. Then eliminate m from the equation for c:

   $$c^2 = 4\frac{k^2h}{g} - \frac{16k^2h^2\pi^2}{T^2g^2} = \frac{4k^2h}{g^2T^2}(gT^2 - 4\pi^2h)$$

   So $c = (2k/gT)h^{1/2}(gT^2 - 4\pi^2h)^{1/2}$.

2. *Steady-State Response*   Find the general solution of ODE (2). Show that as $t \to +\infty$, all solutions decay to a periodic function with the same frequency as the driving term.

 3. *Effect of Speed on Steady-State Response*   Reproduce Figure 2. Convince yourself graphically that the amplitude of the steady-state response is greater the more closely the driving frequency matches the natural frequency of ODE (2).

   Answer 3:   Group project.

Spotlight on Einstein's Field Equations

PROBLEMS

Einstein's Field Equations of General Relativity.

 1. *Einstein's Field Equations of General Relativity*
   For each of the three cases, $k = 0, +1, -1$, decide whether the universe is born with a "big bang" [i.e., at some time t_1, as $t \to t_1^+$, $R(t) \to 0^+$, and $R'(t) \to +\infty$]. In each case, what happens to the

universe as t increases? Does the universe die in a big crunch [i.e., for some t_2, as $t \to t_2^-$, $R(t) \to 0^+$, and $R'(t) \to -\infty$]. Address the following points:

- **Reduction of Order** Multiply the ODE in (1) by R' and rewrite as $[R(R')^2 + kc^2 R]' = 0$. Show that $R(R')^2 + kc^2 R = R_0 v_0^2 + kc^2 R_0 = C_0$.

- Set $k = 0$ (a Euclidean or flat universe) and show that $R^{3/2} = R_0^{3/2} \pm 1.5 C_0^{1/2}(t - t_0)$. Show that the minus sign leads to the eventual collapse of the universe (the "big crunch") at some time $T > t_0$, while the plus sign leads to perpetual expansion. What happens in both cases as $R_0 \to 0^+$? See Graph 1.

- Set $k = 1$ (a spherical universe). Solve the ODE, $R(R')^2 + c^2 R = C_0$, in terms of a parameter u, by setting $R = C_0 c^{-2} \sin^2 u$, rewriting the ODE with u as the dependent variable, and solving to find t in terms of u. Explain why $t = t(u)$, $R = R(u)$ are the parametric equations of a cycloid in the tR-plane. Interpret each arch of the cycloid in terms of a big bang and big crunch cosmology. See Graph 2.

- Set $k = -1$ (a pseudospherical universe). Follow the steps of the spherical case, but set $R = C_0 c^{-2} \sinh^2 u$. Interpret the solution in cosmological terms.

- Rewrite the original ODE as a system of ODEs in dimensionless variables: $dx/ds = 2y$, $dy/ds = -x^{-1}(k + y^2)$, where $x(s_0) = 1$, $y(s_0) = v_0/c$, $x = R/R_0$, $y = R'/c$, $s = ct/2R_0$. Now use an ODE solver/graphics package to plot solutions $x = x(s)$ and $y = y(s)$ as functions of s for $k = 0$, $+1, -1$. Interpret the graphs in cosmological terms. Discuss the advantage of scaling the original variables as indicated before computing.

Answer 1: Group project. Here is an outline of one way to address the first four points:

- Multiply the differential equation by R' to obtain $2RR'R'' + (R')^3 + kc^2 R' = 0$, or $(R(R')^2 + kc^2 R)' = 0$. Let $R = R_0$, $R' = v_0$ at $t = t_0$. Then $R(R')^2 + kc^2 R = C_0 = R_0 v_0^2 + kc^2 R_0$.

- If $k = 0$, separate variables to obtain $R^{1/2} dR = \pm C_0^{1/2} dt$, which integrates to $R^{3/2} = R_0^{3/2} \pm 3C_0^{1/2}(t - t_0)/2$. Since $R = 0$ at $t = t_1 = 2R_0^{3/2} C_0^{-1/2}/3$, set $T = t - t_1$ to obtain $R = AT^{2/3}$, where $A = (3/2)^{2/3} C_0$. The plus sign in $R^{1/2} dR = \pm C_0^{1/2} dt$ carries down in the preceding algebra to the minus sign in the definition of t_1. At $t = t_1$, $R = 0$ and the universe is "created" in a "big bang", the latter term being used since $dR/dT = 2AT^{-1/3}/3$, which is infinite at $T = 0$, that is, at $t = t_1$. Observe that the universe subsequently expands like $T^{2/3}$ as T increases. On the other hand, if the minus sign is used in the differential equation for R, then $t_1 = t_0 + 2R_0^{3/2} C_0^{-1/2}/3$ and for $T = t - t_1 < 0$ we have that $R = AT^{2/3}$ decreases to 0 as T increases to 0. The universe expires in a "big crunch" since $|dR/dT|$ tends to infinity as $T \to 0$. Graph 1 shows the universe expanding (solid curve) from $R = 0.001$ at $t = 0.001$ (right after a big bang) if $R' = R^{-1/2}$. The dashed curve shows a future big crunch (dashed curve) if $R(0) = 5$ and $R' = -R^{-1/2}$.

- Let $k = 1$. The ODE $R(R')^2 + c^2 R = C_0$ is solved parametrically by expressing both R and t in terms of a parameter u. Let $R = C_0 c^{-2} \sin^2 u = C_0(1 - \cos 2u)/2c^2$. Replacing R' by $(dR/du)(du/dt)$ in the ODE, and separating variables, we have $du/dt = \pm c^3 (2C_0 \sin^2 u)^{-1}$, or $dt/du = \pm 2C_0 \sin^2 u/c^3$. So, $t = \pm C_0(2u - \sin 2u)/(2c^3)$, where we assume $t = 0$ and $R = 0$ at $u = 0$. Since we require $t \geq 0$ here, we have for $u \geq 0$

$$t = \frac{C_0}{2c^3}(2u - \sin 2u), \qquad R = \frac{C_0}{2c^2}(1 - \cos 2u)$$

which are the parametric equations of a cycloid of period $T = \pi C_0/c^3$ since $u = \pi$ corresponds to $t = \pi C_0/c^3$ and $R = 0$ at $u = n\pi$, $n = 0, 1, 2, \ldots$. Note that

$$\left| \frac{dR}{dt} \right| = \left| \frac{dR/du}{dt/du} \right| = \left| c \frac{2 \sin 2u}{2 - 2 \cos 2u} \right| \to \infty \quad \text{as} \quad u \to n\pi$$

This model leads to periodic expansion and collapse, that is, to alternating episodes of big bang and big crunch. Graph 2 shows the alternating bangs and crunches of a universe modeled by the spherical geometry corresponding to $k = 1$.

- The ODE here is $R(R')^2 - c^2 R = C_0$ since $k = -1$. Following the steps of 7(c) but using $R = C_0 c^{-2} \sinh^2 u = C_0(\cosh 2u - 1)/2c^2$ instead of $C_0 c^{-2} \sin^2 u$ and using appropriate indentities for hyperbolic functions, we have $t = C_0(\sinh 2u - 2u)/(2c^3)$, $R = C_0(\cosh 2u - 1)/(2c^2)$. As u decreases to 0 both t and R tend to 0. Creation begins from nothing and, since R and $t \to +\infty$ as $u \to +\infty$, the universe expands without bound. Graph 3 shows the future and the past of a pseudospherical universe ($k = -1$) which is at $R = 1$ when $t = 1$ with $R' = 1, 2, 3$. Note that each of these three universes starts from nothing sometime in the past, but not all start with a bang (e.g., $R'(0) \neq \infty$ if $R'(1) = 1$). In each of the three universes, $R(t)$ always increases as t increases, so each is an expanding universe.

All of these models ignore "pressure", which may be acceptable on the large scale of an essentially empty universe, but cannot be correct for small R. The scales in the figures are arbitrary.

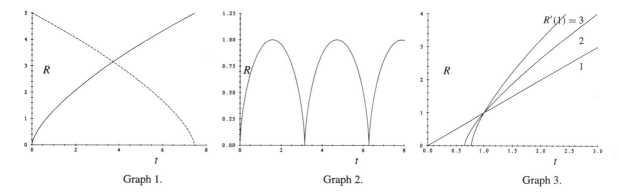

Graph 1.   Graph 2.   Graph 3.

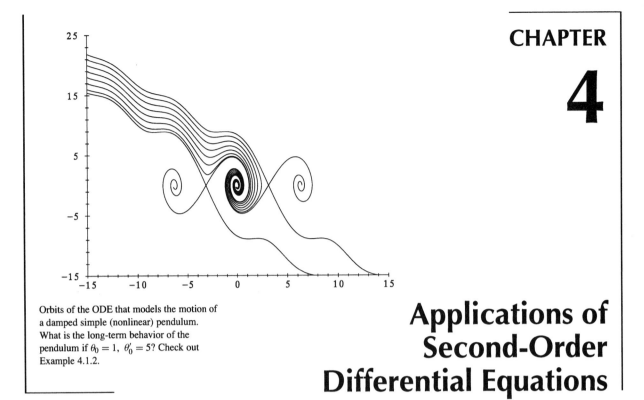

Orbits of the ODE that models the motion of
a damped simple (nonlinear) pendulum.
What is the long-term behavior of the
pendulum if $\theta_0 = 1$, $\theta_0' = 5$? Check out
Example 4.1.2.

CHAPTER 4

Applications of Second-Order Differential Equations

There are two main applications in this chapter: the pendulum and electrical circuits. The pendulum model reappears in later chapters. In Section 4.1, we derive the model ODE of the simple pendulum starting from Newton's Laws of Motion. In Section 4.4, we use Kirchhoff's Voltage and Current Laws to derive ODEs the models for current and charge in electrical circuits.

The material in Section 4.2 is a general treatment of beats and resonance for linear second-order ODEs with constant coefficients. This section can be covered independently of the applications sections. Engineering students may already have seen frequency response modeling in a systems or lumped parameters course, so the material of Section 4.3 may not come as a big surprise to them. Our approach builds on the properties of linear ODEs developed in Chapter 3.

4.1 Newton's Laws: The Pendulum

Comments:

The approach to vectors in this section is traditional and nonaxiomatic. Vectors are thought of as arrows, but are denoted by boldface type. The dot product is introduced and some of its properties are listed (Problem 1). Unit coordinate vectors $\mathbf{i}$, $\mathbf{j}$, $\mathbf{k}$, $\hat{\mathbf{x}}$, $\hat{\mathbf{y}}$, $\hat{\mathbf{z}}$, $\hat{\mathbf{r}}$, $\hat{\boldsymbol{\theta}}$ are used. The main application of the vector approach is the derivation of the equation of motion of the simple pendulum. We have opted to leave the derivation of various properties of the solutions and orbits of the simple undamped pendulum (both nonlinear and linearized) to the problem set (Problems 11–15).

PROBLEMS

Properties of the Dot Product. Let $\mathbf{u} = u_1\mathbf{i} + u_2\mathbf{j} + u_3\mathbf{k}$ and $\mathbf{v} = v_1\mathbf{i} + v_2\mathbf{j} + v_3\mathbf{k}$.

1. Show that $\|\mathbf{u}\|^2 = u_1^2 + u_2^2 + u_3^2$ for any vector $\mathbf{u}$.

   Answer 1: Let $\mathbf{u} = (u_1, u_2, u_3)$. The value of $\|\mathbf{u}\|^2$ is the square of the distance from the origin $(0, 0, 0)$ to the point (u_1, u_2, u_3). So, $\|\mathbf{u}\|^2 = \left(\sqrt{(u_1 - 0)^2 + (u_2 - 0)^2 + (u_3 - 0)^2}\right)^2 = u_1{}^2 + u_2{}^2 + u_3{}^2$ for any vector $\mathbf{u}$.

2. Show that $\mathbf{u} \cdot \mathbf{v} = u_1 v_1 + u_2 v_2 + u_3 v_3$ for any vectors $\mathbf{u}$ and $\mathbf{v}$. [*Hint*: imagine $\mathbf{u}$ and $\mathbf{v}$ to have their feet at the origin, then use the Law of Cosines on the inside back cover, the definition of $\mathbf{u} \cdot \mathbf{v}$, and Problem 1.]

3. *Symmetry* Show that $\mathbf{u} \cdot \mathbf{v} = \mathbf{v} \cdot \mathbf{u}$ for all $\mathbf{u}, \mathbf{v}$.

   Answer 3: Let $\mathbf{u} = (u_1, u_2, u_3)$ and $\mathbf{v} = (v_1, v_2, v_3)$. Then $\mathbf{u} \cdot \mathbf{v} = u_1 v_1 + u_2 v_2 + u_3 v_3 = v_1 u_1 + v_2 u_2 + v_3 u_3 = \mathbf{v} \cdot \mathbf{u}$.

4. *Bilinearity* Show that $(\alpha\mathbf{u} + \beta\mathbf{w}) \cdot \mathbf{v} = \alpha\mathbf{u} \cdot \mathbf{v} + \beta\mathbf{w} \cdot \mathbf{v}$ for all scalars α, β and vectors $\mathbf{u}, \mathbf{v}$, and $\mathbf{w}$.

5. *Positive Definiteness* Show that $\mathbf{u} \cdot \mathbf{u} \geq 0$ for all $\mathbf{u}$ and that $\mathbf{u} \cdot \mathbf{u} = 0$ if and only if $\mathbf{u} = \mathbf{0}$.

   Answer 5: Let $\mathbf{u} = (u_1, u_2, u_3)$. Then $\mathbf{u} \cdot \mathbf{u} = u_1^2 + u_2^2 + u_3^2 \geq 0$ for all $\mathbf{u}$. Since a sum of squares of real numbers is 0 if and only if each number is 0, then $\mathbf{u} \cdot \mathbf{u} = u_1^2 + u_2^2 + u_3^2 = 0$ if and only if $\mathbf{u} = \mathbf{0}$.

6. *Cauchy–Schwarz Inequality* Show that $|\mathbf{u} \cdot \mathbf{v}| \leq \|\mathbf{u}\| \|\mathbf{v}\|$ for all vectors $\mathbf{u}$ and $\mathbf{v}$.

Modeling Problems.

<u>7.</u> *Vector Addition* An airplane flies from point P in space. It flies 20 mi due south, turns left $90°$ and goes into a climb 8 mi long at an angle of $10°$ with the horizontal, turns left again, and flies horizontally 42 mi due north. Let $\mathbf{i}, \mathbf{j}$, and $\mathbf{k}$ point north, west, and upward, respectively, from P. What is the final position of the airplane relative to P?

   Answer 7: Since we are looking for the position relative to P, let $P = (0, 0, 0)$. Then the final position is given by the sum of the vectors $\mathbf{u}_1 = -20\mathbf{i}$, $\mathbf{u}_2 = -8\cos(10°)\mathbf{j} + 8\sin(10°)\mathbf{k} = -7.88\mathbf{j} + 1.39\mathbf{k}$, and $\mathbf{u}_3 = 42\mathbf{i}$, where the axis $\mathbf{i}$ is positive to the north, $\mathbf{j}$ is positive to the west, and $\mathbf{k}$ is positive upwards. The final position is approximately $22\mathbf{i} - 7.88\mathbf{j} + 1.39\mathbf{k}$, which translates to 22 miles North, 7.88 miles East, and 1.39 miles above P.

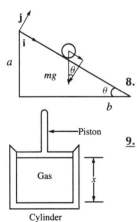

8. *Inclined Plane* A ball is released from rest at the top of an inclined plane and rolls without friction down the plane until it hits the bottom (see the margin figure). How long does it take the ball to reach the bottom? [*Hint*: set up a fixed frame $\{\mathbf{i}, \mathbf{j}\}$ as shown and apply Newton's laws in each component of the frame. There are two forces acting on the ball: gravity and the reaction force of the plane.]

<u>9.</u> *Ideal Gas Law* According to the *Ideal Gas Law*, the pressure P, volume V, temperature T, and the number of moles (1 mole $= 6.02 \times 10^{23}$ molecules) n of a gas in a closed container satisfy the equation $PV = nRT$, where R is a universal constant. Suppose that a cylinder contains an ideal gas with a piston of mass m on top (see the margin figure). Assume that the temperature is constant and that the only forces acting on the piston are gravity and gas pressure. Find the ODE for the position of the piston measured from the bottom of the cylinder. (P is defined to be the magnitude of the gas pressure per unit area of the piston.)

   Answer 9: Applying Newton's Second Law (in the vertical direction) to the piston, we have $mx'' =$ force of gravity $+$ force due to gas pressure. The magnitude of the force due to gas pressure is $P \cdot A$ where A is the area of the piston. But $P = nRT/V = nRT/(x \cdot A)$, so the force due to gas pressure is $P \cdot A = nRT/x$. Since the force of gravity is directed downward while the force due to gas pressure is directed upward, $mx'' = -mg + nRT/x$.

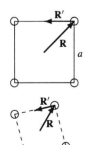

10. *Creeping Bugs* Four bugs sit at the corners of a square table whose sides are of length a. See the margin figure. The bugs begin to move at the same instant, each crawling at the same constant speed directly toward the bug on its right. Find the path of each bug. Do the bugs ever meet? If so, when? [*Hint*: let the vector $\mathbf{R}(t) = r(t)\hat{\mathbf{r}}$ point from the center of the table to one of the bugs, where $r(0) = a/\sqrt{2}$ and $\theta(0) = 0$. Explain why the velocity vector $\mathbf{R}'(t)$ makes a $135°$ angle with $\mathbf{R}(t)$ for $t \geq 0$.]

Behavior of Pendulum.

11. *Linearized Pendulum*   Consider a simple linearized and undriven pendulum described by $\theta'' + c\theta'/m + g\theta/L = 0$.

(a) If $c = 0$, find the period T of the pendulum in terms of L and g.

(b) If $c = 0$, find the length of a pendulum whose period is exactly 1 sec.

(c) If $c = 0$ and the pendulum is 1 m long and swings with an amplitude of 1 rad, compute the angular velocity of the pendulum at its lowest point. Find the accelerations θ'' at $\theta = \pm 1$.

(d) Suppose that $c > 0$. Show that if $c^2 < 4g/L$ then the pendulum oscillates with declining amplitudes about $\theta = 0$, $\theta' = 0$.

(e) Now set the pendulum in a barrel of molasses. What happens? Explain.

Answer 11:

(a) The equation of motion is $\theta'' + g\theta/L = 0$. The general solution is $\theta(t) = C_1 \cos(t\sqrt{g/L}) + C_2 \sin(t\sqrt{g/L})$, where C_1 and C_2 are arbitrary constants. Since the sum of two periodic functions with period T is also a periodic function of period T, we see that the period of the pendulum is $T = 2\pi\sqrt{L/g}$. Note that it is independent of the mass of the bob.

(b) Set $2\pi\sqrt{L/g} = 1$ and obtain $L = g/(4\pi^2)$. In mks units where $g = 9.8 \; m/s^2$, then $L \approx \frac{1}{4}$ meter.

(c) From (a) we have $\theta = C_1 \cos(t\sqrt{g/L}) + C_2 \sin(t\sqrt{g/L})$. Let the pendulum be at its lowest point $\theta = 0$ at $t = 0$. But $\theta(0) = 0$ forces $C_1 = 0$, so $\theta = C_2 \sin(t\sqrt{g/L})$. Since the pendulum swings with amplitude 1 radian, $C_2 = 1$, so $\theta = \sin(t\sqrt{g/L})$. Then $\theta' = \sqrt{g/L}\cos(t\sqrt{g/L})$ and at $t = 0$, $\theta' = \sqrt{g/L} = \sqrt{9.8/1} \approx 3.13$ rad/sec. Every time the pendulum passes through the lowest point, $\theta' = \sqrt{g/L} \approx 3.13$ rad/sec or $-\sqrt{g/L} \approx -3.13$ rad/sec.

Since $\theta = 0$ at $t = 0$, say that $\theta = 1$ first at $t = (\pi/2)\sqrt{g/L}$ and $\theta = -1$ when $t = (3\pi/2)\sqrt{g/L}$. Now $\theta'' = -g/L \sin(t\sqrt{g/L}) = -g/L = -9.8$ rad/(sec)^2 when $t = (\pi/2)\sqrt{g/L}$. Depending on whether $\theta = +1$ or -1, the acceleration is -9.8 ot $+9.8$ rad/sec^2, respectively.

(d) if $c > 0$, the roots of the characteristic polynomial are $r_{1,2} = \frac{-(c/m)\pm\sqrt{(c^2/m^2)-4(g/L)}}{2}$. The solutions oscillate with declining amplitude if and only if the roots have the form $\alpha \pm i\beta$, where α is real and negative and $\beta \neq 0$. Here $\alpha = -c/(2m) < 0$, and if $c^2/m^2 - 4(g/L) < 0$ then $\beta = (1/2)\sqrt{4g/L - c^2/m^2}$ is real. So $c^2/m^2 < 4g/L$ reduces to $c^2 < 4g/L$ if $m = 1$. The fact that $m = 1$ was mistakenly omitted in the problem statement.

(e) In molasses there is a damping force acting on the pendulum. Hence the damping coefficient $c > 0$ and the behavior of the pendulum is described by (d).

12. *Undriven Simple Pendulum and Its Linearization*   The ODE for the simple undriven pendulum is $mL\theta'' + cL\theta' + mg\sin\theta = 0$.

(a) *Undamped*   Set $m = 1$, $c = 0$, $g/L = 10$ and plot a detailed portrait of the orbits of the system in θ, v variables, $\theta' = v$, $v' = -10\sin\theta$. Use the screen size $|\theta| \leq 15$, $|\theta'| \leq 15$. [*Hint*: see Figure 4.1.1.]

(b) *Damped*   Set $c = 1$, so that the ODE becomes $\theta'' + \theta' + 10\sin\theta = 0$. Plot some orbits for the equivalent system $\theta' = v$, $v' = -10\sin\theta - v$ with screen size $|\theta| \leq 15$, $|\theta'| \leq 15$. Compare this portrait with those in (a) and the chapter's opening figure. Explain the differences.

(c) Now repeat (a) and (b) but with the linearized ODE $mL\theta'' + cL\theta' + mg\theta = 0$. Use $m = 1$, $g/L = 10$, $c = 0$ (first), and $c = 1$ (second). Compare your graphs with those obtained earlier for the nonlinear ODEs and explain any differences.

13. *Variable-Length Pendulum*   Show that if the length $L(t)$ of a pendulum is a function of time, then the equation of motion of the pendulum is $mL\theta'' + (2mL' + cL)\theta' + mg\sin\theta = F$. [*Hint*: use $\mathbf{R}' = (L\hat{\mathbf{r}})'$, but don't assume that L is constant. Compare with ODE (11).]

Answer 13:   If $L = L(t)$, then equations (8) in the text must be altered to take that fact into account. We have that

$$\mathbf{v} = \mathbf{R}' = (L\hat{\mathbf{r}})' = L'\hat{\mathbf{r}} + L\hat{\mathbf{r}}' = L'\hat{\mathbf{r}} + L\hat{\boldsymbol{\theta}}\theta'$$

and

$$\mathbf{a} = \mathbf{R}'' = (L'\hat{\mathbf{r}} + L\theta'\hat{\boldsymbol{\theta}})' = L''\hat{\mathbf{r}} + L'\hat{\mathbf{r}}' + L'\theta'\hat{\boldsymbol{\theta}} + L\theta''\hat{\boldsymbol{\theta}} + L\theta'\hat{\boldsymbol{\theta}}' = L''\hat{\mathbf{r}} + 2L'\theta'\hat{\boldsymbol{\theta}} + L\theta''\hat{\boldsymbol{\theta}} - L'(\theta')^2\hat{\mathbf{r}}$$

Equating the $\hat{\boldsymbol{\theta}}$ components in Newton's Second Law $m\mathbf{a} = \mathbf{F}$ we have

$$m(2L'\theta' + L\theta'') = -cL\theta' - mg\sin\theta + F(t)$$

or

$$mL\theta'' + (2mL' + cL)\theta' + mg\sin\theta = F(t)$$

where $F(t)$ is the $\hat{\boldsymbol{\theta}}$ component of the external force $\mathbf{F}$.

Closed Orbits of an Undriven, Undamped Simple Pendulum of Constant Length L.

14. *Orbits of Undriven, Undamped Simple Pendulum*   A formula for the orbits of the undriven, undamped simple pendulum may be derived directly from the ODE $mL\theta'' + mg\sin\theta = 0$.

☞ This technique
is also used in
Theorem 3.8.2.

(a) Multiply each side of the ODE by θ' to get $mL\theta'\theta'' + mg\theta'\sin\theta = \left[\frac{1}{2}mL(\theta')^2 - mg\cos\theta\right]' = 0$. Show that $(1/2)mL(\theta')^2 - mg\cos\theta = c$, where c is a constant, is the equation of an orbit. Let $g/L = 10$, and reproduce Figure 4.1.1 by plotting orbits for several values of c.

(b) *Conservation of Energy*   The *kinetic energy* (KE) of the pendulum is $(1/2)m(L\theta')^2$ and the *potential energy* (PE) is $mgL(1 - \cos\theta)$. Show that, given θ and θ' at time 0, the total energy $E(t) = KE + PE$ at time t is the same as $E(0)$. Explain why $E(t) = E(0)$ is the equation of an orbit. [*Hint*: show that $E'(t) = 0$, for all t.]

(c) *Periodic, Separatrix, Tumbling Orbits for Figure 4.1.1*   Consider the particular simple pendulum ODE, $\theta'' + 10\sin\theta = 0$, whose orbital equations by **(a)** are $(\theta')^2/2 - 10\cos\theta = c$ for various constants c. Show that if $\theta(0) = 0$ and $\theta'(0) = v_0$, then $c = v_0^2/2 - 10$. Show that if $0 < v_0^2 < 40$, then the orbit is periodic, but if $v_0^2 > 40$, then the orbit is tumbling. If $v_0^2 = 40$, explain why the corresponding orbit is a separatrix.

 15.   The ODEs

$$mL\theta'' + mg\sin\theta = 0 \quad \text{and} \quad mL\theta'' + mg\theta = 0$$

model the motion of an undriven, undamped simple pendulum and of an undriven, undamped linearized pendulum, respectively. Orbital portraits of each ODE show a region of closed orbits encircling the origin. These closed orbits (or *cycles*) correspond to periodic solutions. Find and compare the periods of the cycles for the simple pendulum and for the linearized pendulum. Follow the outline below in proving the existence of cycles and in studying the periods. [*Hint*: see also Problem 14.]

- *Linearized Pendulum*   The equation of the linearized pendulum is $mL\theta'' + mg\theta = 0$. Show that all nonconstant solutions are periodic of period $T = 2\pi\sqrt{L/g}$. Show that the corresponding orbits in the $\theta\theta'$-state space are elliptical cycles. Choose a value for L, plot a portrait of cycles in the state space, plot component graphs, and verify graphically the formula for T.

- *Closed Orbits of the Simple Pendulum*   Suppose that the simple pendulum modeled by $mL\theta'' + mg\sin\theta = 0$ is released from rest when $\theta = \theta_0$, where $0 < \theta_0 < \pi$. Show that the subsequent motion is periodic. [*Hint*: use Problem 14 to show that orbits are described by the relation $(\theta')^2 = (2g/L)(\cos\theta - \cos\theta_0)$, and use symmetries in this relation to show that the orbits are closed, and so represent periodic solutions.]

- *Periods of the Simple Pendulum*   Let T be the period of the orbit with $\theta(0) = \theta_0 > 0$, $\theta'(0) = 0$. Show that T is given by

$$T = 4\sqrt{\frac{L}{2g}} \int_0^{\theta_0} \frac{d\theta}{\sqrt{\cos\theta - \cos\theta_0}}$$

[*Hint*: since $\theta(t)$ initially decreases as t increases, $\theta' = -(2g/L)^{1/2}(\cos\theta - \cos\theta_0)^{1/2}$. Show that θ continues to decrease until the time $t = t_1$ for which $\theta(t_1) = -\theta_0$.]

- *Elliptic Integrals and the Periods of the Simple Pendulum* Show that the change of variables $k = \sin(\theta_0/2)$, $\sin\phi = (1/k)\sin(\theta/2)$ gives

$$T = 4\sqrt{\frac{L}{g}} \int_0^{\pi/2} \frac{d\phi}{\sqrt{1 - k^2\sin^2\phi}}$$

The integral is an *elliptic integral of the first kind* Its approximate values have been tabulated[1] for various values of k. For example, if $\theta_0 = 2\pi/3$, then $k = \sqrt{3}/2$, and the value of the integral is about 2.157. The corresponding period is about $8.628\sqrt{L/g}$, quite different from the period of the linearized pendulum, which is $2\pi\sqrt{L/g} \approx 6.282\sqrt{L/g}$. Why would you expect the period of the nonlinear pendulum to be greater than the period of the linearized pendulum? Choose various values for L and use an ODE solver to verify the above estimate for the periods.

- *Asymptotic Values of the Periods of the Simple Pendulum* Explain why $T \to 2\pi\sqrt{L/g}$ as $\theta_0 \to 0$. It is known that $T \to \infty$ as $\theta_0 \to \pi$, although a complete mathematical proof of this fact is not given here. Why are these results expected on physical grounds?

Answer 15:   This problem can be approached as follows. The solutions of the linearized pendulum ODE, $mL\theta'' + mg\theta = 0$, have the form $\theta = A\cos(t\sqrt{g/L} - \delta)$, where A and δ are arbitrary constants; the period is $2\pi\sqrt{L/g}$. Since $\theta' = -A\sqrt{g/L}\sin(t\sqrt{g/L} - \delta)$, we have that $\theta^2 + L(\theta')^2/g = A^2$, which for each $A > 0$ is the equation of an ellipse in the $\theta\theta'$-plane. The major axis lies along the θ-axis if $L/g > 1$, along the θ'-axis if $L/g < 1$, while the orbit is a circle if $L = g$.

If $\theta(0) = \theta_0$ and $\theta'(0) = 0$, then the orbit of the nonlinear pendulum ODE, $mL\theta'' + mg\sin\theta = 0$, is defined by

$$\frac{1}{2}mL(\theta')^2 - mg\cos\theta = -mg\cos\theta_0$$

The motion is periodic if the graph of the orbit in the $\theta\theta'$-plane is a simple closed curve not passing through any equilibrium points. The equation of the orbit is $L(\theta')^2 = 2g(\cos\theta - \cos\theta_0)$ and has the property that if (θ, θ') is a point on the orbit, then the points $(-\theta, \theta')$, $(\theta, -\theta')$, and $(-\theta, -\theta')$ also are on the orbit. We only need to show that the orbit in the first quadrant is an arc reaching from the positive θ'-axis to the positive θ-axis, then reflect this arc through the axes to obtain the simple closed curve which is the full orbit. None of the equilibrium points $\theta = n\pi$, $\theta' = 0$ lie on the given orbit since $0 < \theta_0 < \pi$. The arc of the orbit in the first quadrant falls steadily from the point $\theta = 0$, $\theta' = [2g(1 - \cos\theta_0)/L]^{1/2}$ to the point $\theta = \theta_0$, $\theta' = 0$, as we see directly from the equation of the orbit. After the reflections, we obtain the simple, closed orbit corresponding to a periodic solution.

In the first quadrant, we have that $\theta' = \sqrt{2g/L}\sqrt{\cos\theta - \cos\theta_0}$. Separating variables and integrating from $t = 0$ to $t = T/4$, where T is the period once around the full orbit, we have that $4\int_0^{\theta_0}(\cos\theta - \cos\theta_0)^{-1/2}d\theta = T(2g/L)^{1/2}$, where we have used the fact that the arc in the first quadrant is traversed in one fourth of the period. Note that we have initialized time at $t_0 = 0$, then set $\theta = 0$ at $t_0 = 0$ and $\theta = \theta_0$ at $t = T/4$. The integral may be transformed into the elliptic integral given in the problem by the changes in constants and variables, $k = \sin(\theta_0/2)$, $\sin\phi = (1/k)\sin(\theta/2)$. The function $\sin\phi$ is defined by the latter equation since $\sin(\theta_0/2) \geq \sin(\theta/2) > 0$ if $0 \leq \theta \leq \theta_0 < \pi$. To change from the dummy variable of integration θ to the variable ϕ, note that

$$\cos\theta - \cos\theta_0 = 2(\sin^2(\theta_0/2) - \sin^2(\theta/2)) = 2k^2(1 - \sin^2\phi) = 2k^2\cos^2\phi$$

by a double angle formula of trigonometry. Next we have that

$$\cos\phi\, d\phi = [(2k)^{-1}\cos(\theta/2)]d\theta$$
$$= (2k)^{-1}[1 - \sin^2(\theta/2)]^{1/2}d\theta = (2k)^{-1}(1 - k^2\sin^2\phi)^{1/2}d\theta \quad \text{or}$$
$$d\theta = 2k\cos\phi(1 - k^2\sin^2\phi)^{-1/2}d\phi$$

---

[1] M. Abramowitz and I. A. Stegun, eds., *Handbook of Mathematical Functions* (Washington, D.C.: National Bureau of Standards, 1964); Dover (reprint), New York, 1965.

So

$$(\cos\theta - \cos\theta_0)^{-1/2}\,d\theta = (2k^2\cos^2\phi)^{-1/2}\,d\theta = 2^{1/2}(1 - k^2\sin^2\phi)^{-1/2}\,d\phi$$

Since $\phi = 0$ when $\theta = 0$ and $\phi = \pi/2$ when $\theta = \theta_0$, the limits on the transformed integral are as given and the θ integral transforms to the ϕ integral.

We expect the nonlinear pendulum motion modeled by $mL\theta'' = -mg\sin\theta$, $\theta(0) = \theta_0$, $\theta'(0) = 0$ to have a longer period than the motion of the linearized pendulum modeled by $mL\theta'' = -mg\theta$, $\theta(0) = \theta_0$, $\theta'(0) = 0$, since the magnitude of the restoring force of the nonlinear pendulum is less than that of the linearized pendulum, $|-mg\sin\theta| < |-mg\theta|$ with equality only at $\theta = 0$. So the nonlinear pendulum moves more slowly and has a longer period than the linearized pendulum, given identical initial conditions.

As $\theta_0 \to 0$, $k = \sin\theta_0/2 \to 0$ and so the period T given by the integral tends to $4\sqrt{L/g}\int_0^{\pi/2}\,d\phi = 2\pi\sqrt{L/g}$. This is expected since for small $|\theta|$, $\sin\theta \approx \theta$, and the period of the nonlinear pendulum motion for small $|\theta_0|$, $\theta_0' = 0$, should be close to $2\pi\sqrt{L/g}$, which is the period of the linear pendulum motion regardless of initial data. At the other extreme as $\theta_0 \to \pi$, one expects the period to tend to ∞ since $\theta_0 = \pi$ corresponds to the pendulum standing at rest vertically upwards. With a slight displacement to $\pi - \varepsilon$, the pendulum would slowly move downward, gathering speed as it goes, then slow down as it climbs back up towards the top to come to rest at $-\pi + \varepsilon$ before it reverses direction. The smaller the value of ε, the longer the period.

4.2 Beats and Resonance

Comments

In this section we see how second-order, constant coefficient linear ODEs respond to a sinusoidal driving force. ODEs of this kind often appear in the sciences and engineering. The notions of beats, resonance, free and forced oscillations (both in the undamped and damped cases) all arise in this setting. Take a pair of tuning forks tuned to slightly different frequencies and listen to the beats.

PROBLEMS

Free Oscillations. Identify whether solutions are overdamped, critically damped, or underdamped. If underdamped, what is the oscillation time? Justify your answers.

1. $y'' + y' + y = 0$ **2.** $y'' + 2y' + y = 0$ 3. $4y'' + 4y' + y = 0$

4. $y'' + 3y' + y = 0$ 5. $y'' + 4y' + 4y = 0$ 6. $y'' + 4y' - 3y = 0$

Answer 1: The solutions of $y'' + 2cy' + \omega_0^2 y = 0$, where $c \geq 0$ and $\omega_0 > 0$ are overdamped if $c > \omega_0$, critically damped if $c = \omega_0$, underdamped if $c < \omega_0$. In this case $c = \frac{1}{2}$, $\omega_0 = 1$, so the system is underdamped. Solutions have the form $y = (C_1\cos\beta t + C_2\sin\beta t)e^{\alpha t}$ where $\alpha = -1/2$, $\beta = \sqrt{3}/2$; so oscillation time is $2\pi/\beta = 4\pi/\sqrt{3}$.

Answer 3: $c = \frac{1}{2}$, $\omega_0 = \frac{1}{2}$, so the system is critically damped. (See Answer 1)

Answer 5: $c = 2$, $\omega_0 = 2$, so the system is critically damped. (See Answer 1)

Graphing Damped Oscillations. Use the two sets of initial data $y(0) = 0$, $y'(0) = 1$ and $y(0) = 1$, $y'(0) = 0$ and graph the damped oscillations of the given ODE. Identify whether solutions are overdamped, critically damped, or underdamped.

7. $y'' + 20y' + y = 0$ **8.** $y'' + y' + 10y = 0$ 9. $y'' + 8y' + 16y = 0$

Answer 7: Overdamped since $c = 10$, $\omega_0 = 1$. See the figure.

Answer 9: Critically damped since $c = 4$, $\omega_0 = 4$. See the figure.

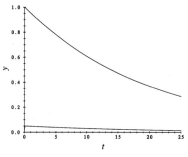

Problem 7.   Problem 9.

Beats and Resonance.   Find the solutions of the IVPs. Any resonance or beats?

10. $y'' + y = \sin t$, $y(0) = 0$, $y'(0) = 0$

11. $y'' + y = \sin 1.1t$, $y(0) = 0$, $y'(0) = 0$

12. $y'' + 9y = 5\cos 2t$, $y(0) = 0$, $y'(0) = 0$

13. $y'' + 4y = \cos 3t$, $y(0) = 1$, $y'(0) = -1$

14. $y'' + y = \cos t$, $y(0) = 1$, $y'(0) = 0$

15. $y'' + 4y = \sin(2t + 1)$, $y(0) = 0$, $y'(0) = 0$

> **Answer 11:**   For the ODE $y'' + y = \sin 1.1t$, we have $\omega_0 = 1$ and $\omega = 1.1$, so the solution exhibits beats. Following the technique of beats where $\omega \neq \omega_0$ on page 271 of the text (with $F_0 = 1$, $\omega_0 = 1$, $\omega = 1.1$, $z_d = A_d e^{1.1it}$, $y_d = \text{Im}[z_d]$), we get the particular solution $y_d = -(1/0.21)\sin 1.1t$. The general solution of the ODE is $y = C_1 \cos t + C_2 \sin t + y_d$, where C_1 and C_2 are arbitrary constants. The initial conditions $y(0) = 0$, $y'(0) = 0$ imply that $C_1 = 0$ and $C_2 = 1.1/0.21$, so the solution of the initial value problem is $(1/0.21)(1.1\sin t - \sin 1.1t)$.

> **Answer 13:**   For the ODE $y'' + 4y = \cos 3t$, we have $F_0 = 1$, $\omega_0 = 2$, and $\omega = 3$, so the solution exhibits beats. Use the technique given for beats on page 271 and in Answer 11, but with $y_d = \text{Re}[z_d]$, so the solution is $y(t) = (6/5)\cos 2t - (\sin 2t)/2 - (1/5)\cos 3t$.

> **Answer 15:**   For the ODE $y'' + 4y = \sin(2t + 1)$, we have $F_0 = 1$, $\omega_0 = \omega = 2$, so resonance occurs. Follow the technique of Pure Resonance on page 270 of the text but with $e^{i\omega_0 t}$ replaced by $e^{i(2t+1)}$, $z_d = A_d t e^{i(2t+1)}$, $y_d = \text{Im}[z_d]$. The general solution is $y(t) = C_1 \cos 2t + C_2 \sin 2t - (1/4)t\cos(2t + 1)$. Since $y(0) = 0$ and $y'(0) = 0$, $C_1 = 0$ and $C_2 = (1/8)\cos 1$, so the solution is
>
> $$y(t) = (1/8)\cos 1 \sin 2t - (1/4)t\cos(2t + 1)$$

Visualizing Beats and Resonance.   Use a numerical solver to study the solution curves of the ODE $y'' + 25y = \sin \omega t$ for initial data $(0,0)$, other data sets $(y(0), y'(0))$ of your choice, and a time span that shows at least 3 "humps" of the graph. Plot the results on separate graphs. Compare these graphs and record your observations. [*Hint*: see Figures 4.2.5 and 4.2.6.]

16. $\omega = 5.6$ **17.** $\omega = 5.2$ **18.** $\omega = 5.0$ **19.** $\omega = \pi$

> **Answer 17:**   Now the input frequency 5.2 is close to the natural frequency 5.0, and the beats are pronounced, but the beat frequency of $(5.2 - 5.0)/2 = 0.1$ is very small (Graph 1). The beats are distinct, but not as sharp for the respective initial data $(0, 10)$ and $(4, 0)$ for Graphs 2 and 3 below.

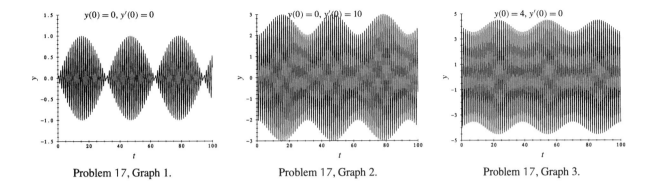

Problem 17, Graph 1. Problem 17, Graph 2. Problem 17, Graph 3.

Answer 19: The natural frequency $\omega_0 = 5$ and the input frequency $\omega = \pi$ are not rationally related. Graph 1 below shows what the superposition of the two frequencies looks like with the initial data $(0, 0)$. For respective initial data $(0, 10)$ and $(4, 0)$ of Graphs 2 and 3 below the two frequencies are still there, but are not nearly so obvious.

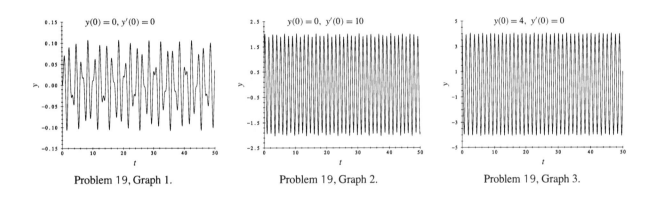

Problem 19, Graph 1. Problem 19, Graph 2. Problem 19, Graph 3.

Investigating Beats.

20. *Beats* For $\omega \neq \omega_0$ show that the solution of the IVP $y'' + \omega_0^2 y = A \sin \omega t$, $y(0) = 0$, $y'(0) = -A/(\omega_0 + \omega)$, exhibits the phenomenon of beats similar to that in Figure 4.2.5. [*Hint*: use the trigonometric identity $\sin \alpha - \sin \beta = 2 \sin((\alpha - \beta)/2) \cos((\alpha + \beta)/2)$.]

21. Explore the phenomenon of beats for a lightly damped and driven oscillator. Begin your study with the ODE for the damped and driven Hooke's Law spring of Example 3.1.2. Then replace the driving function in that example first by $100 \, \text{SqWave}(t, 5.6, 2.8)$ and then by $100 \, \text{TRWave}(t, 50, 5.6)$. Do you still get beats? What happens as you vary the damping constant and the driving period?

Answer 21: Group project. [Example 3.1.3 (*not* Example 3.1.2) and $100 \, \text{TRWave}(t, 5.6, 1.4, 2.8, 1)$ (*not* $100 \, \text{TRWave}(t, 50, 5.6)$); these misprints in the first printing will be corrected in later printings.]

22. *Detecting a Periodic Forced Oscillation* Solutions of the ODE $y'' + \omega_0^2 y = F_0 \cos \omega t$, where F_0, ω_0, ω ($\omega_0 \neq \omega$) are positive constants, are superpositions of sinusoids of period $2\pi/\omega_0$ and of period $2\pi/\omega$. Given the graph of a solution, is it possible to detect the periodic forced oscillation (i.e., the unique solution of period $2\pi/\omega$)? Does the periodic forced oscillation have a phase angle? Are the nonconstant solutions periodic, and if they are, what is their period? In addressing these questions, start out by solving graphically the two IVPs $y'' + 4y = \cos \omega t$, $y(0) = 0$, $y'(0) = 0$, for $\omega = 3, \pi$. You should also find solution formulas.

Archimedes' Buoyancy Principle.

 Thus, a
wooden ship with
dimensions identical
to those of an iron
ship would float
higher on the water.

> Archimedes' Principle. The buoyant force acting on a body wholly or partially immersed in a fluid equals the weight of the fluid displaced. Hence, at equilibrium, the mass of the body equals the mass of the displaced water.

23. Ignoring friction, show that a flat wooden block of face L^2 square feet and thickness h feet floating half-merged in water will act as a harmonic oscillator of period $2\pi\sqrt{h/2g}$ if it is depressed slightly. [*Hint*: show that if x is a vertical coordinate from the midpoint of the block to the surface of the water, then $(L^2 h\rho/2)x'' + L^2\rho g x = 0$, where ρ is the water density.]

   Answer 23: Suppose that $x(t)$ is the distance from the midpoint of the block at time t to the surface of the water. Since the block is floating in equilibrium half-submerged, the mass of the block must equal the mass, $(hL^2/2)\rho$, of the displaced water. Now suppose the block is pushed down a distance $x < h/2$ into the water. Then the buoyant force acting on it must be $L^2 x\rho g$, the weight of the extra water displaced. By Newton's Second Law, $(hL^2/2)\rho x'' = -L^2 x\rho g$, or $x'' + 2gx/h = 0$. The solutions are $x = C_1 \cos t\sqrt{2g/h} + C_2 \sin t\sqrt{2g/h}$ of period $T = 2\pi\sqrt{h/2g}$, and so the block oscillates vertically with period $2\pi\sqrt{h/2g}$.

24. *Solid Spheres and Soft Springs* Replace the block in Problem 20 by a buoyant solid sphere of radius R floating half-submerged, and show that if the initial displacement is small, the sphere will act as a harmonic oscillator of period $2\pi\sqrt{2R/3g}$. Show that if the displacement is large, the nonlinearities become important in the ODE modeling the motion, and so can't be ignored. Show that, in this case, the buoyancy ODE has the form of a soft-spring ODE of Section 3.8.

25. Release a hollow ball under water and describe what happens. Model the dynamics and explain. [*Hint*: use an approach similar to that for Problem 23.]

   Answer 25: Group project.

4.3 Frequency Response Modeling

Comments

Frequency response modeling is a meat-and-potatoes item on the engineering menu. Others are less accustomed to seeing the damped resonance phenomena couched in the engineering language of gain, phase shift, and Bodé plots used here, but once you get used to the new terms they seem natural and down-to-earth. The only real problem that we see is not the new terminology, but the need for care in the algebraic manipulation. We also touch on one of the most important engineering problems, the problem of parameter identification: how can you determine the coefficients of an ODE from a knowledge of the input and output?

PROBLEMS

Periodic Forced Oscillations. Find the forced oscillation y_d for each ODE.

1. $y'' + 2y' + 7y = 10\sin 2t$ 2. $y'' + 2y' + 7y = 20\sin t$

   Answer 1: You can use either the methods of Section 3.6 or formulas (11) and (12) in the text to find the forced oscialltion y_d. Using (11) and (12) with $c = 1$, $\omega_0 = \sqrt{7}$, $\omega = 2$, $F_0 = 10$, we see that the forced oscillation of periods $2\pi/\omega = \pi$ is $y_d = 10M(2)\sin(2t + \varphi(2))$ where $M(2) = 1/5$ and $\varphi(2) = -\cot^{-1}(-3/4)$.

Periodic Response, Gain, and Phase.

3. Consider the IVP $y'' + 0.5y' + 16y = 100 \sin t$, $y(0) = 0$, $y'(0) = 0$. Find the periodic response component y_p and the transient component y_{tr} of the unique solution of the IVP. On a single pair of axes, plot the solution graph of the IVP along with the periodic and transient components of the solution of the IVP.

Answer 3: Guessing a particular solution of the form $A \sin t + B \cos t$ for the ODE $y'' + 0.5y' + 16y = 100 \sin t$, we see that $A = 60/9.01$, $B = -2/9.01$, so $y_p = (60 \sin t - 2 \cos t)/9.01$ is the unique periodic response. The roots of the characteristic polynomial $r^2 + 0.5r + 16$ are $-1/4 \pm i\sqrt{255/16}$, so the transients are given by $y_{tr} = C_1 e^{-t/4} \cos \beta t + C_2 e^{-t/4} \sin \beta t$, where $\beta = \sqrt{255/16} \approx 3.992$. Applying the initial conditions $y(0) = y'(0) = 0$ to $y = y_p + y_{tr}$, we obtain $C_1 \approx 0.222$ and $C_2 \approx -1.654$. So,

$$y_{tr} = 0.222 e^{-t/4} \cos(3.992t) - 1.654 e^{-t/4} \sin(3.992t)$$

The figure below shows the solution curve (solid) of the IVP, along with the periodic component y_p (long dashes) and transient component y_{tr} (short dashes) of the solution.

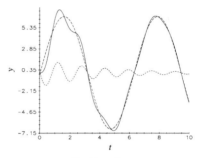

Problem 3.

4. *Periodic Forced Oscillation* Let c, ω_0^2, F_0, and ω be positive constants. Find the unique periodic solution of the ODE $y'' + 2cy' + \omega_0^2 y = F_0 \cos \omega t$.

Gain and Phase Shift.

5. Use the solution of the IVP

$$y'' + 2y' + 4y = 3 \sin 2t, \qquad y(0) = 0.5, \quad y'(0) = 0.5$$

to estimate the gain $M(2)$ and phase shift $\varphi(2)$ for the steady-state forced response. [*Hint*: see Figure 4.3.4 and Example 4.3.4.] Compare these estimates with the exact values calculated from formulas (11). Find the system transfer function $H(i\omega)$.

Answer 5: The figure shows the input $3 \sin 2t$ (dashed) of circular frequency $\omega = 2$ and the output $y(t)$ (solid) of the IVP, $y'' + 2y' + 4y = 3 \sin 2t$, $y(0) = 0.5$, $y'(0) = 0.5$. $M(2)$ may be estimated from the plot by dividing the amplitude of the steady-state response by 3, the input amplitude: $M(2) \approx 1/4$. The phase shift may be estimated by finding how far to the right the output is shifted. The magnitude of this time shift is $|\varphi(\omega)/\omega| = |\varphi(2)/2|$ here. Multiplying this value by $\omega = 2$ gives us $-\varphi(2)$ since we're looking at how far to the *right* the output is shifted (a "lag time"). Using formula (11) with $2c = 2$ and $\omega = \omega_0 = 2$, we obtain $M(2) = 0.25$ and $\varphi(2) = -\pi/2$. The transfer function is $H(r) = 1/P(r) = 1/(r^2 + 2r + 4)$. In this case $H(i\omega) = H(2i) = -i/4$.

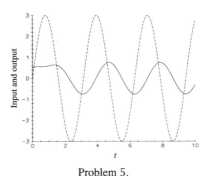

Problem 5.

6. Consider the ODE $y'' + y' + 2y = F_0 \sin \omega t$.

(a) Use an ODE solver to find enough values of the gain $M(\omega)$ and the phase shift $\varphi(\omega)$ to make rough sketches of M and φ vs. ω.

(b) Use formulas (11) to plot the graphs of $M(\omega)$ and $\varphi(\omega)$. Compare with plots in (a).

Parameter Identification.

7. Estimate the parameters c and ω_0 of the system whose Bodé plots are shown in Figure 4.3.3. [*Hint:* pick a frequency ω and find M and φ from the Bodé plots. The equations in (11) can then be solved for c and ω_0. The exact values are $c = 0.1$ and $\omega_0 = 1$. How close did you get to these values using the data you read off the graphs in Figure 4.3.3?]

Answer 7: To find c and ω_0 from Fig. 4.3.3, pick a particular frequency ω_0 and find the corresponding values of $M(\omega_0)$ and $\varphi(\omega_0)$ from the Bodé plot. Formulas (11) then provide us with two equations in the unknowns $2c$ and ω_0^2, allowing us to solve for approximate values for $2c$ and ω_0^2. Answers may vary significantly, due to the exponential frequency scale which amplifies small errors in reading values off the plot.

8. *Damped Undriven Hooke's Law Spring*   In a spring system modeled by $y'' + 2cy' + \omega_0^2 y = 0$ the static deflection is 0.0127 m, while damped vibrations decay from amplitude 0.01016 m to amplitude 0.00254 m in 20 cycles. Assume a mass of 1 kg.

(a) Find the damping constant $2c$. [*Hint:* first find ω_0, and then find c such that amplitudes decay by a factor of 4 over a $20T$ time span, where $T = 2\pi / \sqrt{\omega_0^2 - c^2}$.]

(b) Find the resonant frequency ω_r and calculate the maximal amplitude of the steady-state response of the system to a driving force $A \cos \omega_r t$.

 Gain and Phase Shift.

9. For the IVP $y'' + 0.2y' + y = \cos \omega t$, $y(0) = 0$, $y'(0) = 0$, do some frequency response modeling by performing the tasks below:

- Let $\omega = 0.5$. Plot the solution of the IVP over a time sufficiently long that the transient part of the solution has become negligible and the steady-state response is clearly visible. From the graph, determine the amplitude of the steady-state response, compare it to the amplitude of the driving function, and then determine the gain M.

- Now determine the phase shift φ between the driving function and the steady-state response. [*Hint:* see Figure 4.3.4.]

- Explain why the amplitude M and the phase shift φ do not depend on initial conditions. Confirm this fact by repeating the above steps for several sets of initial conditions.

- Determine the amplitude M and the phase shift φ for various other values of ω for the IVP. For instance, choose $\omega = 0.25, 0.75, 1, 1.5, 2$, and 3. Try some values of ω very close to 1. Sketch the graphs of M vs. ω and φ vs. ω.

- Solve the ODE analytically and determine the steady-state response. Find M and φ as functions of ω. Compare these results with those found above.

Answer 9: Group project.

 10. The steady-state response of the ODE $y'' + 2cy' + \omega_0^2 y = F_0 \sin \omega t$, with $0 < c \leq \omega_0$, can be characterized via an amplitude magnification and a phase shift applied to the sinusoidal driving term $f(t)$. See formulas (11). Explore this connection by performing the tasks below:

- Verify that the general features of the graphs of $M(\omega)$ and $\varphi(\omega)$, for fixed positive constants ω_0 and c with $\omega_0^2 > 2c^2$, are similar to the plots in Figure 4.3.3. What is the value ω_r (the resonant frequency) where $M(\omega)$ achieves a maximum? What are $M(0)$, $M(\omega_r)$? What is the value ω_0 where $\varphi(\omega)$ changes inflection?

- What happens to the frequency ω_r and the maximum value $M(\omega_r)$ determined above for fixed $\omega_0 > 0$ as $c \to 0$? Interpret this observation for a vibrating mechanical system generated by ODE (1) with very small damping constant c and a sinusoidal driving term with frequency close to a natural (i.e., undamped) frequency of the system.

 Periodic Forced Oscillations.

11. The ODE is $y'' + ay' + by = f(t)$, where a and b are constants and $f(t)$ is a periodic continuous function with period T. Follow the outline below to show that if the ODE $y'' + ay' + by = 0$ does *not* have a periodic solution with period T, then the driven ODE $y'' + ay' + by = f(t)$ has a unique periodic solution of period T.

☞ This problem proves the existence of a periodic forced oscillation; it uses the Method of Variation of Parameters (Theorem 3.7.4).

- Show that a solution $y(t)$ of the driven ODE is periodic with period T if and only if $y(0) = y(T)$ and $y'(0) = y'(T)$. [*Hint*: to show that $y(t)$ repeats in the time interval $T \leq t \leq 2T$, replace t by $s + T$ and start the clock at $s = 0$.]

- Suppose that y_1, y_2 are solutions of the undriven ODE $y'' + ay' + by = 0$ with $y_1(0) = 1$, $y_1'(0) = 0$; $y_2(0) = 0$, $y_2'(0) = 1$. Suppose that $y_d(t)$ is the particular solution of the driven ODE $y'' + ay' + by = f(t)$ given by $y_d = C_1(t)y_1(t) + C_2(t)y_2(t)$ with

$$C_1(t) = \int_0^t \frac{-y_2(s)f(s)}{W(s)}\,ds, \qquad C_2(t) = \int_0^t \frac{y_1(s)f(s)}{W(s)}\,ds$$

where $W = y_1y_2' - y_1'y_2$ is the Wronskian of y_1 and y_2. (These formulas are given in Theorem 3.7.4.) Show that $y = y_0 y_1 + y_0' y_2 + y_d$ is the solution of the IVP

$$y'' + ay' + by = f(t), \qquad y(0) = y_0, \quad y'(0) = y_0'$$

- From the above, we know that the conditions $y(0) = y(T)$ and $y'(0) = y'(T)$ applied to the solution $y(t)$ above guarantee that $y(t)$ is periodic, and vice versa. Show that these conditions on y are equivalent to the matrix equation

$$(I - M(T)) \begin{bmatrix} y_0 \\ y_0' \end{bmatrix} = \begin{bmatrix} y_d(T) \\ y_d'(T) \end{bmatrix}$$

where I is the matrix $\begin{bmatrix} 1 & 0 \\ 0 & 1 \end{bmatrix}$ and the matrix $M(T)$ is $\begin{bmatrix} y_1(T) & y_2(T) \\ y_1'(T) & y_2'(T) \end{bmatrix}$.

☞ A matrix is not invertible if and only if its determinant is zero.

- Suppose that the matrix $I - M(T)$ is *not* invertible. Then there exists a nonzero vector v with components α and β such that $(I - M(t))v = 0$, so

$$\begin{bmatrix} \alpha \\ \beta \end{bmatrix} = M(T) \begin{bmatrix} \alpha \\ \beta \end{bmatrix}$$

Show that $z(t) = \alpha y_1(t) + \beta y_2(t)$ is a nontrivial periodic solution of the undriven ODE $y'' + ay' + by = 0$, contradicting the condition given in the opening statement of this problem.

- The contradiction to our hypothesis establishes that $I - M(T)$ *is* invertible. Show that the ODE $y'' + ay' + by = f(t)$ has a unique periodic solution with period T.

- This result holds when $f(t)$ is a piecewise continuous periodic function, provided we allow the second derivative of a solution to be only piecewise continuous. So the ODE $y'' + 4y' + 3y =$ SqWave$(t, 4, 2)$ has a unique periodic solution of period $T = 4$. Without actually calculating the values of y_0 and y_0' necessary to generate the periodic solution, produce a graph approximating this periodic solution. How do you know your graph stays close to the periodic solution for large t?

Answer 11:   Group project.

4.4   Electrical Circuits

Comments

Electrical circuits provide a very nice application of second-order constant coefficient linear ODEs, an application of immense importance in today's technological world of communication channels. We start from scratch here and talk about charge carriers, voltage, currents, the whole bit. The going is heavy at times, but no one says that electricity is easy to understand. When all of the characters in the electrical circuit story have been introduced, we turn to the simple RLC circuit and solve the corresponding model ODE for the current in the circuit—no new solution techniques here, just the new language of circuits. The solution calculations are neither easier nor harder than before.

PROBLEMS

Find the Current.   Suppose that a simple RLC circuit is charged up by a battery so that the charge on the capacitor is $q_0 = 10^{-3}$ coulomb, while $I_0 = 0$. The battery is then removed. In each case find the charge $q(t)$ on the capacitor and the current $I(t)$ in the circuit for $t \geq 0$. [*Hint*: use IVP (5) with $I_0 = 0$, $E(t) = 0$, $t_0 = 0$, $q_0 = 10^{-3}$, and find the current. Then use formula (1) for the charge.]

   1. $L = 0.3$ H, $R = 15$ Ω, $C = 3 \times 10^{-2}$ F   **2.** $L = 1$ H, $R = 1000$ Ω, $C = 4 \times 10^{-4}$ F

   3. $L = 2.5$ H, $R = 500$ Ω, $C = 10^{-6}$ F

      Answer 1:   The roots of the characterstic polynomial $r^2 + 50r + 1000/9$ are approximately -47.7 and -2.33; so $I(0) = I_0' = -1/9$. $I(t) \approx .0024(e^{-47.7t} - e^{-2.33t})$, $q(t) \approx .00005 + .001e^{-2.33t} - .00005e^{-47.7t}$

      Answer 3:   The roots of the characteristic polynomial $r^2 + 200r + 4 \cdot 10^5$ are approximately $-100 \pm 624i$; so $I(0) = I_0' = -400$.

$$I(t) \approx -.64e^{-100t} \sin 624t, \quad q(t) \approx .00001 + .00016e^{-100t} \sin 624t + .00099e^{-100t} \cos 624t$$

Simple Circuit.

   4.   *Damped Oscillation*   Look at IVP (5) for current in an RLC circuit.

      (a) Solve IVP (5) if $R = 20$ Ω, $L = 10$ H, $C = 0.05$ F, $E(t) = 12$ volts, $q_0 = 0.6$ coulomb, and $I_0 = 1$ ampere.

      (b) Explain why the solution of IVP (5) in this case is a damped oscillation and plot its graph.

      (c) To what value must the resistance be increased in order to reach the overdamped case?

   5.   Find the charge $q(t)$ on the capacitor of the simple RLC circuit in the margin. Assume zero initial charge and current, $R = 20$ Ω, $L = 10$ H, $C = 0.01$ F, and $E(t) = 30 \cos 2t$ volts.

Answer 5: The IVP to be solved is $10q'' + 20q' + 100q = 30\cos 2t$, where $q(0) = 0$ and $q'(0) = I(0) = 0$. We first find a particular solution of the form $q_d = A\sin 2t + B\cos 2t$. Inserting q_d into the ODE and matching coefficients of like terms to find A and B, we see that $6A - 4B = 0$, $4A + 6B = 3$, and so $A = 3/13$, $B = 9/26$ and $q_d = (3/13)\sin 2t + (9/26)\cos 2t$. Since the roots of the characteristic polynomial $10r^2 + 20r + 100$ are $-1 \pm 3i$, the general solution of the charge ODE is $q(t) = e^{-t}[C_1\cos 3t + C_2\sin 3t] + q_d(t)$. The constants C_1 and C_2 may be found from the initial data, $q(0) = q'(0) = 0$: $0 = C_1 + 9/26$, $0 = -C_1 + 3C_2 + 6/13$, and so $C_1 = -9/26$, $C_2 = -7/26$. The solution is $q(t) = (1/26)[-e^{-t}(9\cos 3t + 7\sin 3t) + 6\sin 2t + 9\cos 2t]$.

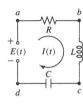

6. A simple RLC circuit has a capacitor with $C = 0.25$ F, a resistor with $R = 7 \times 10^4$ Ω, and an inductor with $L = 2.0$ H. The initial charge on the capacitor is zero, as is the initial current. If an impressed voltage of 60 volts is connected to the circuit, determine the charge on the capacitor for $t > 0$. Estimate the charge when $t = 0.1$ sec. [*Hint*: first solve the current ODE for $I(t)$, then integrate to get $q(t)$. Use IVP (5).]

7. *Resonance* Let's look at ODE (3) for the charge in an RLC circuit.

   (a) Show that if $R = 0$ in the simple RLC circuit and the impressed voltage is of the form $E_0\cos\omega t$, the charge on the capacitor will become unbounded as $t \to \infty$ if $\omega = 1/\sqrt{LC}$. This is the phenomenon of resonance. [*Hint*: look at the form of particular solutions.]

   (b) Show that the charge will always be bounded, no matter what the choice of ω, provided that there is some resistance in the circuit.

   Answer 7:
   (a) The ODE is $Lq'' + q/C = E_0\cos\omega t$, or $q'' + q/LC = (E_0/L)\cos\omega t$, where $\omega = 1/\sqrt{LC}$. The general solution of the undriven ODE is $C_1\cos\omega t + C_2\sin\omega t$. The frequency ω of the impressed voltage matches the natural frequency $1/\sqrt{LC}$ of the oscillations of the undriven system. There is a particular solution $At\cos\omega t + Bt\sin\omega t$. Matching coefficients, we have $A = 0$, $B = E_0/(2\omega L)$. So, $q = C_1\cos\omega t + (C_2 + E_0 t/2\omega L)\sin\omega t$ and $q(t)$ undergoes unbounded oscillations as $t \to \infty$ because of the term $t\sin\omega t$.

   (b) If there is a positive resistance R, then the ODE is $Lq'' + Rq' + q/C = E_0\cos\omega t$. The roots of the characteristic polynomial have negative real parts: $r_1 = -R/2L \pm (R^2/4L^2 - 1/LC)^{1/2}$, so there is no frequency ω term in the solution set of the undriven system. A particular solution of the driven system may be found of the form $q_p(t) = A\cos\omega t + B\sin\omega t$. There are no terms of the form $At\cos\omega t$ or $Bt\sin\omega t$ in the solution $q(t)$. In this case $q(t) = C_1 e^{r_1 t} + C_2 e^{r_2 t} + A\cos\omega t + B\sin\omega t$, for arbitrary C_1 and C_2 and certain A and B. For $t \geq 0$, we have that $|C_1 e^{r_1 t}| \leq |C_1|$ and $|C_2 e^{r_2 t}| \leq |C_2|$, since r_1 and r_2 have negative real parts. Also, $|A\cos\omega t| \leq |A|$ and $|B\sin\omega t| \leq |B|$. So, $|q(t)|$ is bounded for $t \geq 0$.

Multiloop Circuit.

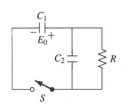

8. Two capacitors ($C_1 = 10^{-6}$ F and $C_2 = 2 \times 10^{-6}$ F) and a resistor ($R = 3 \times 10^6$ Ω) are arranged in a circuit as shown. The capacitor C_1 is initially charged to E_0 volts with polarity as shown. The switch S is closed at time $t = 0$. Determine the current that flows through the capacitor C_1 as a function of time. [*Hint*: apply Kirchhoff's Voltage Law to the left loop, then to the right loop. Note that $q_1(0) = C_1 E_0$, $q_2(0) = -C_2 E_0$, where q_1 and q_2 are the respective charges on the capacitors C_1 and C_2.]

9. Set up, but do not solve, three ODEs for the currents I_1, I_2, I_3 in the circuit below:

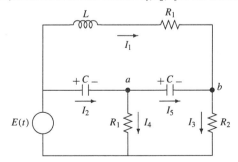

Answer 9: Kirchhoff's Current Law applied to the points a and b in the circuit implies that $I_2 = I_4 + I_5$ and $I_3 = I_1 + I_5$. So, $I_5 = I_3 - I_1$ and $I_4 = I_2 - I_3 + I_1$, and the currents through the various circuit elements may all be expressed in terms of the state variables I_1, I_2, and I_3. Apply Kirchhoff's Voltage Law to the outer loop of the circuit to obtain $E(t) = LI'_1 + R_1 I_1 + R_2 I_3$. Apply the derivative of Kirchhoff's Voltage Law to the lower left and lower right loops, respectively, to obtain $E'(t) = I_2/C + R_1 I'_4 = I_2/C + R_1 (I'_2 - I'_3 + I'_1)$ and $I_5/C + R_2 I'_3 = R_1 I'_4$ or $(I_3 - I_1)/C + R_2 I'_3 = R_1 (I'_2 - I'_3 + I'_1)$. This gives us three first-order, coupled rate equations for the currents I_1, I_2, and I_3:

$$LI'_1 + R_1 I_1 + R_2 I_3 = E(t)$$

$$R_1 I'_1 + R_1 I'_2 - R_1 I'_3 + I_2/C = E'(t)$$

$$R_1 I'_1 + R_1 I'_2 - (R_1 + R_2)I'_3 + I_1/C - I_3/C = 0$$

  10. Show how to reduce the two-loop circuit model ODEs (18) to the system (19) with specific formulas for the rate functions for f_1 and f_2. Then let the driving voltage $E(t)$ be a specific periodic and differentiable voltage of your choice and choose specific positive values for the circuit parameters R_1, R_2, L, and C. Use a solver to plot solution curves $I_1 = I_1(t)$ and $I_2 = I_2(t)$ and orbits for various initial values of $I_1(0)$ and $I_2(0)$. What is the long-term behavior of the currents, given your circuit parameters?

On-Off Voltages.

11. In this problem you will start with IVP (i) for the charge $q(t)$ on the capacitor of the simple circuit shown below:

$$Lq'' + Rq' + \frac{1}{C}q = E(t), \qquad q(0) = 10^{-2}, \quad q'(0) = 0 \tag{i}$$

where $L = 20$ H, $R = 80$ Ω, $C = 10^{-2}$ F, and $E = 50 \sin 2t$. Time is measured in seconds, charge in coulombs, and voltage E in volts.

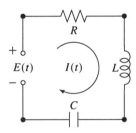

Use a numerical solver to carry out each of the following procedures:

- Graph the solution $q(t)$ of IVP (i).

- Replace E by each of the functions (refer to the inside front cover) given below and graph the solution $q = q(t)$ of IVP (i). Compare and explain what you see.

    - $E = 50 \, \text{SqWave}(t, \pi, \pi/2)$

    - $E = \text{TRWave}(t, \pi, \pi/2, \pi, 50)$

    - $E = 50 \, \text{Step}(t, 10) + 20 \, (1 - \text{Step}(t, 30))$

    - $E = 50(1 - \text{Step}(t, 10))$

- Carry out a parameter study of the effects on $q(t)$ of varying the resistance R over the range $0 \le R \le 200$. Explain what you see.

Answer 11: Group project.

Spotlight on Modeling: Tuning a Circuit

Comments:

The text shows how to home in on a particular frequency in the input to an *RLC*-circuit. Using the given values of inductance and resistance we vary the capacitance to accomplish the frequency-selecting task.

PROBLEMS

 1. *Tuning a Circuit: Sensitivity to Resistance* Using ODE (4) for $I(t)$ with $L = 1$ H, $C = 1/25$ F, $R = 1.0, 0.1, 0.01$ Ω, graphically study the effect of the resistance on the output current when tuning the circuit to the input frequency 5. What do you conclude? Explain.

Answer 1: With $L = 1$, $R = 0.1$ and $C = 1/25$ we know from the text that the tuned circuit will home in on the frequency $\omega = 5$ in the input signal since $\omega^2 = 25 = 1/LC$. The question is what will various values of the resistance do to the amplitude of the response? See Graph 1 for the graphs of the current vs. time in the three cases $R = 1.0$ (solid), $R = 0.1$ (short dashes), and $R = 0.01$ (long dashes). Apparently the lower the resistance, the larger the amplitude of the response; in the language of Section 4.3, the greater the gain. We have tuned the circuit to the input frequency $\omega = 5$ by choosing C so that $1/LC = \omega^2 = 25$, and the resulting steady-state current is $I_p = (E_0/R)\cos(\omega t + \varphi)$ where $\omega E_0 = 4$ so $E_0 = 4/\omega = 4/5$. Since the amplitude of the input for the current ODE is ωE_0 (*not* E_0, which is the amplitude of the charge ODE), the gain is $(E_0/R)/E_0\omega = 1/(R\omega)$, i.e., $1/(5R)$ since $\omega = 5$.

The amplitudes of the response currents seen in Graph 1 approach $E_0/R = 4/5$ when $R = 1$ (short dashes), $(4/5)/(0.1) = 8$ when $R = 0.1$ (long dashes), and $(4/5)/(0.01) = 80$ when $R = 0.01$ (solid). The graphs are not of I_p, but include some transients as well. As t increases the graphs approach the graph of I_p. This is particularly evident if $R = 1.0$ and 0.1, where over the interval $90 \le t \le 100$, we are essentially looking at $I_p(t)$ with respective amplitudes $4/5$ ($R = 1.0$), 8 ($R = 0.1$). But, the graph corresponding to $R = 0.01$ has a long way to go before its amplitude approaches 80. Graph 2 shows the current if $R = 0.01$; the ODE is solved over $0 \le t \le 1000$, but plotted only for $975 \le t \le 1000$, by which point the amplitude has approached 80.

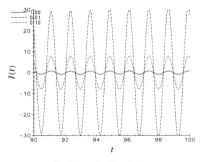

Problem 1, Graph 1.

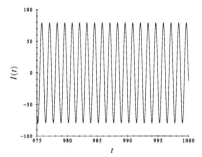

Problem 1, Graph 2.

 2. Tune in the Grateful Dead by varying the resistance, inductance, or capacitance of a circuit. Look at the case where the hard rock and classical stations broadcast at almost the same frequency.

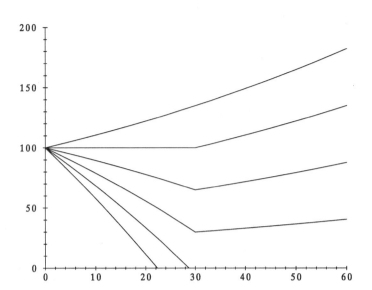

The effect of a 30 day harvest at rate H on a changing population. How much harvesting is too much? See Sections 5.1 and 5.2.

The Laplace Transform

There are differences of opinion about the place of Laplace transforms in an ODE course. Institutions with an engineering program often offer a "baby systems" course which contains a healthy dose of Laplace transforms. This reflects the central role played by the Laplace transform in the work of practicing engineers. Because of this, some feel that the time in an ODE course can be better spent in doing some state plane analysis, stability, and bifurcations, subjects that engineers may not otherwise see in the classroom. On the other hand, institutions without an engineering program may want to put Laplace transforms in their ODE course. This chapter has been included just in case it is needed, but no harm is done in skipping the chapter altogether.

The chapter is quite demanding technically, requiring, as it does, a great deal of algebraic pyrotechnics and much skill in using tables and the calculus of transforms. For convenience in calculating Laplace transforms, we put a table of transforms of commonly used functions on the inside back cover. Partial fractions play a big role in using transform tables to solve IVPs, and the basic expansions are listed on the inside back cover.

In using Laplace transforms to solve IVPs, the attempt will fail unless the ODEs are linear, and will be very difficult, even if linear, if the coefficients are not constant. The big advantage of transforms lies in their ability to handle discontinuous driving terms such as square waves or sawtooth waves and in the simple way the transform involves initial data right from the start of the solution process. The disadvantage (other than the restriction to linear ODEs) is that the formula for the inverse transform is beyond the scope of this book. And so we must be satisfied with inversion by pattern matching from tables and by the calculus of transforms. It is a remarkable fact, however, that Laplace transforms can be used to solve constant-coefficient linear operator equations other than linear ODEs. See Problem 34 in Section 5.2 for an example of the transform technique to solve difference equations and the SPOTLIGHT ON MODELING: TIME DELAYS AND COLLISIONS for an unusual car-following model that involves differential-delay equations.

5.1 The Laplace Transform: Solving IVPs

Comments:

The purpose of this section is to give a quick introduction to the Laplace transform, the ability to calculate the transform of some elementary functions, and a glimpse at applications of the transform to solving IVPs. The inverse transform is introduced (but not by formula). We note that although distinct continuous functions have distinct transforms, two different discontinuous functions may have the same transform. How do you set about finding transforms—brain only, calculus book, tables of integrals, or tables of Laplace transforms? Our preference is to work out a few transforms from scratch, but after that to use tables; we have included a table of transforms on the inside back cover. It is not a good idea to spend so much time working out integrals that you lose sight of what the transform is all about.

PROBLEMS

Laplace Transform. Find the Laplace transform of each function of t. Find the largest interval $s > s_0$ on which each transform is defined. Use the Transform Tables on the inside back cover to verify your answers.

1. $3t - 5$ **2.** t^2 **3.** $2t + 5e^{-8t}$ **4.** $e^{-3t} + t^2 - t + 7$

5. $3\sin 2t$ **6.** $-2\cos 3t$ **7.** $\cos at$ **8.** $\sin at$

9. $7\cos t + 8\sin 2t$ **10.** $\cos 2t - 6\sin 3t$ **11.** $e^{2t}\cos 3t$ **12.** $e^{2t}\sin 3t$

13. $t^2 e^{-3t} - 5t\cos 2t$ **14.** $t\sinh 2t$ **15.** $t^2\cosh \pi t$

Answer 1:

$$\mathcal{L}[3t - 5] = \int_0^\infty e^{-st}(3t - 5)dt = \lim_{\tau \to +\infty} \int_0^\tau e^{-st}(3t - 5)dt$$

$$= \lim_{\tau \to +\infty} \left[\frac{3e^{-st}}{s^2}(-st - 1) + \frac{5}{s}e^{-st} \right]_{t=0}^{t=\tau}$$

$$= \lim_{\tau \to +\infty} \left[\frac{3e^{-s\tau}}{s^2}(-s\tau - 1) + \frac{5}{s}e^{-s\tau} \right] + \frac{3}{s^2} - \frac{5}{s}$$

$$= \frac{3}{s^2} - \frac{5}{s}$$

since for $s > 0$ the terms inside the square brackets tend to 0 as $\tau \to +\infty$. The transform above checks with Table Entry 2 with $n = 1$ and 0, respectively. The transform is defined for $s > 0$.

Answer 3:

$$\mathcal{L}[2t + 5e^{-8t}] = \int_0^\infty e^{-st}(2t + 5e^{-8t})dt = \lim_{\tau \to +\infty} \int_0^\tau e^{-st}(2t + 5e^{-8t})dt$$

$$= \lim_{\tau \to +\infty} \left[\frac{2e^{-st}}{s^2}(-st - 1) + \frac{5}{s+8}e^{-t(s+8)} \right]_0^{t=\tau} + \frac{2}{s^2} + \frac{5}{s+8}$$

$$= \frac{2}{s^2} + \frac{5}{s+8}$$

since for $s > 0$ the terms inside the square brackets tend to 0 as $\tau \to +\infty$. See Table Entries 2 and 3. The transform is defined for $s > 0$.

Answer 5:

$$\mathcal{L}[3\sin 2t] = \int_0^\infty e^{-st}(3\sin 2t)dt = \lim_{\tau\to+\infty}\int_0^\tau e^{-st}(3\sin 2t)dt$$

$$= \lim_{\tau\to+\infty}\left[\frac{3e^{-st}}{s^2+4}(-s\sin 2t - 2\cos 2t)\right]_{t=0}^{t=\tau}$$

$$= \frac{6}{s^2+4}$$

since for $s > 0$ the terms inside the square brackets evaluated at $t = \tau$ tend to 0 as $\tau \to +\infty$. See Table Entry 11. The transform is defined for $s > 0$.

Answer 7:

$$\mathcal{L}[\cos at] = \int_0^\infty e^{-st}\cos at\, dt = \lim_{\tau\to+\infty}\int_0^\tau e^{-st}\cos at\, dt$$

$$= \lim_{\tau\to+\infty}\left[\frac{e^{-st}}{s^2+a^2}[-s\cos at + a\sin at]\right]_{t=0}^{t=\tau}$$

$$= 0 + \frac{1}{s^2+a^2}\cdot s = \frac{s}{s^2+a^2},$$

where we have used the fact that for $t = \tau$, $\lim_\tau[[\cdot]] \to 0$ as $\tau \to +\infty$ if $s > 0$. This agrees with Table Entry 12.

Answer 9:

$$\mathcal{L}[7\cos t + 8\sin 2t] = \int_0^\infty e^{-st}(7\cos t + 8\sin 2t)dt = \lim_{\tau\to+\infty}\int_0^\tau e^{-st}(7\cos t + 8\sin 2t)dt$$

$$= \lim_{\tau\to+\infty}\left[\frac{7e^{-st}}{s^2+1}[-s\cos t + \sin t] + \frac{8e^{-st}}{s^2+4}(-\sin 2t - 2\cos 2t)\right]_{t=0}^{t=\tau}$$

$$= \frac{7s}{s^2+1} + \frac{16}{s^2+4},$$

where we have used the fact that for $t = \tau$, $\lim_\tau[[\cdot]] \to 0$ as $\tau \to +\infty$ if $s > 0$. See Table Entries 11 and 12.

Answer 11:

$$\mathcal{L}[e^{2t}\cos 3t] = \int_0^\infty e^{-st}(e^{2t}\cos 3t)dt = \lim_{\tau\to+\infty}\int_0^\tau e^{-st}(e^{2t}\cos 3t)dt$$

$$= \lim_{\tau\to+\infty}\left[\frac{e^{-t(s-2)}}{(s-2)^2+9}(-(s-2)\cos 3t + 3\sin 3t)\right]_{t=0}^{t=\tau}$$

$$= \frac{s-2}{(s-2)^2+9}$$

since for $s > 0$ and $t = \tau$ the terms inside the square brackets tend to 0 as $\tau \to +\infty$. See Table Entry 16.

Answer 13:

$$\mathcal{L}[t^2 e^{-3t} - 5t\cos 2t] = \int_0^\infty e^{-st}(t^2 e^{-3t} - 5t\cos 2t)dt = \lim_{\tau\to+\infty}\int_0^\tau e^{-st}(t^2 e^{-3t} - 5t\cos 2t)dt$$

$$= \lim_{\tau\to+\infty}\left[\frac{e^{-t(s+3)}}{(s+3)^3}(-t^2(s+3)^2 - 2t(s+3) - 2)\right]_{t=0}^{t=\tau} - 5\int_0^\infty e^{-st}t\cos 2t\, dt$$

$$= \frac{2}{(s+3)^2} - \frac{5(s^2-4)}{(s^2+4)^2}$$

since for $s > 0$ and $t = \tau$ the terms inside the square brackets tend to 0 as $\tau \to +\infty$. Rather than carry out the lengthy process to evaluate $\mathcal{L}[-5t\cos 2t]$ directly, we just used Table Entry 14.

Answer 15: Use the formula $\cosh x = (e^x + e^{-x})/2$:

$$\mathcal{L}[t^2 \cosh \pi t] = \int_0^\infty e^{-st}(t^2 \cosh \pi t)dt = \lim_{\tau \to +\infty} \frac{1}{2}\int_0^\tau t^2 e^{-st}(e^{\pi t} + e^{-\pi t})dt$$

$$= \lim_{\tau \to +\infty} \frac{1}{2}\left[\frac{-e^{-t(s-\pi)}}{(s-\pi)^3}(t^2(s-\pi)^2 + 2t(s-\pi) + 2) \right.$$

$$\left. + \frac{-e^{-t(s+\pi)}}{(s+\pi)^3}(t^2(s+\pi)^2 + 2t(s+\pi) + 2)\right]_{t=0}^{t=\tau}$$

$$= \frac{1}{(s-\pi)^3} + \frac{1}{(s+\pi)^3}$$

since for $s > \pi$ and $t = \tau$ the terms inside the square brackets tend to 0 as $\tau \to +\infty$.

Inverse Laplace Transform. Use the Laplace Transform Table on the inside back cover to find the inverse Laplace transform of each function of s. [*Hint*: use the partial fraction formulas on the inside back cover as needed to reduce a function to a linear combination of functions in the Transform Table. For Problems 25 and 26 use the identity $s - 2as + a^2 + b^2 = (s-a)^2 + b^2$.]

16. $2/s^6$

17. $-7/(s+1)^3$

18. $\dfrac{7s}{s^2+9}$

19. $\dfrac{1}{(s+1)(s-2)}$

20. $\dfrac{-5}{(s-5)(s+3)}$

21. $\dfrac{3}{(s+1)(s-2)^2}$

22. $\dfrac{-5}{(s-3)(s^2+25)}$

23. $\dfrac{2s^2-s}{(s+1)(s-1)^2}$

24. $\dfrac{-1+s}{(s-3)(s^2+1)}$

25. $\dfrac{1}{s^2-2s+1}$

26. $\dfrac{2s}{s^2+4s+13}$

27. $\dfrac{8s}{(s^2+1)^2}$

Answer 17: $\mathcal{L}^{-1}\left[-7/(s+1)^3\right] = \dfrac{-7t^2 e^{-t}}{2}$ by Table Entry 4.

Answer 19: $\mathcal{L}^{-1}\left[\dfrac{1}{(s+1)(s-2)}\right] = -(1/3)e^{-t} + (1/3)e^{2t}$ by Table Entry 6.

Answer 21: $F = \dfrac{3}{(s+1)(s-2)^2} = \dfrac{1/3}{s+1} + \dfrac{-1/3}{s-2} + \dfrac{1}{(s-2)^2}$

$\mathcal{L}^{-1}[F](s) = (1/3)e^{-t} - (1/3)e^{2t} + te^{2t}$ by Table Entries 3 and 4.

Answer 23: $F = \dfrac{2s^2-s}{(s+1)(s-1)^2} = \dfrac{3/4}{s+1} + \dfrac{5/4}{s-1} + \dfrac{1/2}{(s-1)^2}$

$\mathcal{L}^{-1}[F](s) = (3/4)e^{-t} + (5/4)e^t + (1/2)te^t$ by Table Entry 3 and 4.

Answer 25: $F = \dfrac{1}{s^2-2s+s} = \dfrac{1}{(s-1)^2}$

$\mathcal{L}^{-1}[F](s) = te^t$ by Table Entry 4.

Answer 27: $\mathcal{L}^{-1}[\dfrac{8s}{(s^2+1)^2}] = 4t\sin t$ by Table Entry 13.

Laplace Transforms and First-Order IVPs. Use the Laplace transform to solve each IVP. [*Hint*: see Examples 5.1.7–5.1.10 and the three-step procedure for using the Laplace transform to solve an IVP. Use the Integral Table, the Transform Table and the partial fraction formulas on the inside back cover.]

28. $y' + 2y = 0,$ $y(0) = 1$

29. $y' + 2y = e^{-3t},$ $y(0) = 5$

30. $y' - 3y = 5,$ $y(0) = 2$

31. $y' - y = 5\cos t,$ $y(0) = 1$

32. $y' - 7y = 49t,$ $y(0) = 2$

33. $y' + y = 3t,$ $y(0) = 1$

34. $y' + 2y = e^{it}$, $y(0) = 0$ [*Hint*: treat i as a constant.]

35. $y' + ay = \sin \omega t$, $y(0) = y_0$, a, ω positive constants

36. $y' + ay = e^{bt}$, $y(0) = y_0$, a, b positive constants

Answer 29: Applying $\mathcal{L}$, we have that $s\mathcal{L}[y] - 5 + 2\mathcal{L}[y] = 1/(s+3)$; $\mathcal{L}[y] = 1/((s+3)(s+2)) + 5/(s+2)$. So, from the Transform Table, $y = 6e^{-2t} - e^{-3t}$

Answer 31: Applying $\mathcal{L}$, we have that $s\mathcal{L}[y] - 1 - \mathcal{L}[y] = 5s/(s^2+1)$; $\mathcal{L}[y] = 5s/((s^2+1)(s-1)) + 1/(s-1)$. First, use a partial fractions formula to split apart $s/((s^3+1)(s-1))$. Then, from the Transform Table, $y = (7/2)e^t + (5/2)(\sin t - \cos t)$

Answer 33: Applying $\mathcal{L}$, we have that $s\mathcal{L}[y] - 1 + \mathcal{L}[y] = 3/s^2$; $\mathcal{L}[y] = 3/(s^2(s+1)) + 1/(s+1)$. First, use a partial fractions formula to split apart $1/(s^2(s+1))$. Then, from the Transform Table, $y = 4e^{-t} + 3t - 3$

Answer 35: Apply $\mathcal{L}$: $s\mathcal{L}[y] - y_0 + a\mathcal{L}[y] = \omega/(s^2+\omega^2)$; $\mathcal{L}[y] = \omega/((s^2+\omega^2)(s+a)) + y_0/(s+a)$. First, use a partial fractions formula to split apart $1/(s^2+\omega^2)(s+a)$. Then, from the Transform Table, $y = (-\omega/(a^2+\omega^2))\cos \omega t + (a/(a^2+\omega^2))\sin \omega t + (\omega/(a^2+\omega^2))e^{-at} + y_0 e^{-at}$

Modeling Problems. For each of the population problems with harvesting do the following:

(a) Create a model IVP of the form $y' = ky -$ harvest rate, $y(0) = A$.

(b) Find the Laplace transform of the solution $y(t)$. (Problems 29–31 in Section 5.2 ask for $y(t)$.)

37. The rate constant k is 0.05/yr, the initial population is 10 tons and the harvest rate is 2 tons for one year, followed by two years of no harvesting. Take the time span to be 3 years. [*Hint*: see the beginning of the section and Example 5.1.11.]

38. Repeat Problem 37 but with a time span of 6 years. [*Hint*: your model IVP needs to allow for two harvesting years, each harvesting year followed by two years of no harvesting.]

39. *Harvesting/Restocking* The rate constant is 0.02/year, the initial population is 2 tons and the population is harvested in alternate years at the rate of 2 tons/year and restocked in the off-years at the rate of 2 tons/year over a span of 4 years.

Answer 37:

(a) From Example 5.1.11, we have the system:

$$
\begin{aligned}
y' &= (.05)y - 2 &\quad 0 \leq t \leq 1 \quad y(0) = 10 \\
&= (.05)y - 0 &\quad 1 < t \leq 3
\end{aligned}
$$

(b) $\mathcal{L}[y'] = s\mathcal{L}[y] - 10 = .05\mathcal{L}[y] - 2\int_0^1 e^{-st}\,dt$, which after some algebra leads to $\mathcal{L}[y] = \dfrac{10}{s-.05} + \dfrac{2}{s(s-.05)}(e^{-s} - 1)$

Answer 39: **(a)** From Example 5.1.11, we see the system:

$$
\begin{aligned}
y' &= (.02)y - 2 &\quad 0 \leq t \leq 1 \quad y(0) = 2 \\
&= (.02)y + 2 &\quad 1 < t \leq 2 \\
&= (.02)y - 2 &\quad 2 < t \leq 3 \\
&= (.02)y + 2 &\quad 3 < t \leq 4
\end{aligned}
$$

(b)

$$\mathcal{L}[y'] = s\mathcal{L}[y] - 2 = .02\mathcal{L}[y] - 2\int_0^1 2e^{-st}\,dt + 2\int_1^2 e^{-st}\,dt - 2\int_2^3 e^{-st}\,dt + 2\int_3^4 e^{-st}\,dt$$

$$\mathcal{L}[y](s - .02) = 2 + \frac{2e^{-st}}{s}\bigg|_0^1 - \frac{2e^{-st}}{s}\bigg|_1^2 + \frac{2e^{-st}}{s}\bigg|_2^3 - \frac{2e^{-st}}{s}\bigg|_3^4$$

So, $\mathcal{L}[y] = \dfrac{2}{s(s - .02)}\left[s - 1 + 2e^{-s} - 2e^{-2s} + 2e^{-3s} + e^{-4s}\right].$

5.2 Working with the Transform

Comments:

Laplace transforms are a useful tool for solving IVPs, but, as with any tool, you have to learn the operational rules first. We have put some of the most important of the rules in this section (e.g., the transform of a derivative or an on-off function). Going backward, how do you inverse transform a quotient of polynomials or a quotient with exponentials in the numerator and a polynomial in the denominator? An important part of the answer to the last question is the algebraic technique of partial fractions (see the inside back cover). The Shifting Theorem (the three formulas in Theorem 5.2.2) are essential when dealing with functions where the time scale (or the s-variable scale) has been shifted. You may want to use a CAS to avoid lengthy and exacting algebraic manipulations.

PROBLEMS

The Laplace Transform.   Find the Laplace transform of each of the following functions:

1. $\sinh at$ **2.** $\cosh at$ **3.** $t^2 e^{at}$ **4.** $(1 + 6t)e^{at}$

5. $te^{2t}f(t)$ **6.** $(D^2 + 1)f(t)$ **7.** $(t + 1)\,\text{Step}(t, 1)$ **8.** $te^{at}\,\text{Step}(t, 1)$

9. $e^{at}[\text{Step}(t, 1) - \text{Step}(t, 2)]$ **10.** $(t - 2)[\text{Step}(t, 1) - \text{Step}(t, 3)]$

Answer 1:   Since $\sinh at = (e^{at} - e^{-at})/2$, $\mathcal{L}[\sinh at] = a/(s^2 - a^2)$

Answer 3: $\mathcal{L}[t^2 e^{at}] = \mathcal{L}[t^2](s - a) = 2/(s - a)^3$, using text formula (1).

Answer 5: $\mathcal{L}[te^{2t}f(t)] = \mathcal{L}[tf(t)](s - 2) = -\phi'(s - 2)$, where we denote $\mathcal{L}[f](s)$ by $\phi(s)$, using text formula (1) and Table Entry 29.

Answer 7: $\mathcal{L}[(t + 1)\,\text{Step}(t, 1)] = \mathcal{L}[(t + 2 - 1)\,\text{Step}(t - 1)] = e^{-s}((1/s^2) + (2/s))$, using text formula (2).

Answer 9: $\mathcal{L}[e^{at}\left[\text{Step}(t, 1) - \text{Step}(t, 2)\right]] =$
$e^{-s}\mathcal{L}[e^{a(t+1)}] - e^{-2s}\mathcal{L}[e^{a(t+2)}] = \left(e^{-(s-a)} - e^{-2(s-a)}\right)/(s - a)$

Use Transform Tables.   Use the formulas and examples of this section or the Transform Tables on the inside back cover to find $\mathcal{L}[f]$, if f is given, or f if $\mathcal{L}[f]$ is given.

11. $f(t) = 3\,\text{Step}(t, 2) - 3\,\text{Step}(t, 5)$ **12.** $f(t) = 2\sin t\,(1 - \text{Step}(t, \pi))$

13. $f(t) = t^{12}e^{5t}$ **14.** $f(t) = 6t\sin 3t$

15. $\mathcal{L}[f] = \ln(s + 5) - \ln(s - 2)$ **16.** $\mathcal{L}[f] = (s + 1)/\left((s + 1)^2 - 4\right)$

Answer 11: $\mathcal{L}[f] = 3\left(e^{-2s} - e^{-5s}\right)/s$ by text formula (3).

Answer 13: $f(t) = t^{12}e^{5t}$, and so $\mathcal{L}[f] = 12!/(s - 5)^{13}$.

Answer 15: $\mathcal{L}[f] = \ln(s+5) - \ln(s-2)$, and so $f = (e^{2t} - e^{-5t})/t$ by Table Entry 8.

IVPs. Use the Laplace transform to solve each IVP. [*Hint*: use the Transform Tables and partial fraction formulas on the inside back cover, and the Shifting Theorem.]

17. $y'' - y' - 6y = 0$, $y(0) = 1, \quad y'(0) = -1$
18. $y'' + y = \sin t$, $y(0) = 0, \quad y'(0) = 1$
19. $y'' - 2y' + 2y = 0$, $y(0) = 0, \quad y'(0) = 1$
20. $y'' + 4y' + 4y = e^t$, $y(0) = 1, \quad y'(0) = 1$
21. $y'' - 2y' + y = \text{Step}(t, 1)$, $y(0) = 1, \quad y'(0) = 0$
22. $y'' + 2y' - 3y = \text{Step}(t, 1)$, $y(0) = 1, \quad y'(0) = 0$

Answer 17: Applying the transform, $s^2 \mathcal{L}[y] - s + 1 - s\mathcal{L}[y] + 1 - 6\mathcal{L}[y] = 0$. Using partial fractions

$$\mathcal{L}[y] = \frac{s-2}{s^2 - s - 6} = \frac{s-2}{(s-3)(s+2)} = \frac{1}{5(s-3)} + \frac{4}{5(s+2)}$$

So, $y(t) = (e^{3t} + 4e^{-2t})/5$ by Table Entry 3.

Answer 19: Applying the transform, $s^2 \mathcal{L}[y] - 1 - 2s\mathcal{L}[y] + 2\mathcal{L}[y] = 0$:

$$\mathcal{L}[y] = \frac{1}{s^2 - 2s + 2} = \frac{1}{(s-1)^2 + 1}$$

By Table Entry 15, $y(t) = e^t \sin t$.

Answer 21: Applying the transform, $s^2 \mathcal{L}[y] - s - 2s\mathcal{L}[y] + 2 + \mathcal{L}[y] = \mathcal{L}[\text{Step}(t, 1)] = e^{-s}/s$ by Table Entry 18. Using partial fractions, we have

$$\mathcal{L}[y] = \frac{s-2}{s^2 - 2s + 1} + \frac{e^{-s}}{s(s^2 - 2s + 1)} = \frac{1}{s-1} - \frac{1}{(s-1)^2} + e^{-s}\left(\frac{1}{s} - \frac{1}{s-1} + \frac{1}{(s-1)^2}\right)$$

Then using Table Entries 1 and 4 and text formula (2) we have

$$y = e^t(1 - t) + \text{Step}(t, 1)[1 - e^{t-1} + (t-1)e^{t-1}] = e^t(1 - t) + \text{Step}(t, 1)[1 + (t-2)e^{t-1}]$$

Laplace Transform of Integrals. Find the Laplace transform of each integral. [*Hint*: see Table Entry 30 on the inside back cover.]

23. $\displaystyle\int_0^t (x-1)e^x \, dx$ 24. $\displaystyle\int_0^t (x^2 - 2x) \, dx$

25. $\displaystyle\int_0^t \sin(x - \pi/4)e^x \, dx$ 26. $\displaystyle\int_0^t e^{(x-a)} \cos x \, dx$

Answer 23: The transform is (by Table Entries 30 and 4)

$$\mathcal{L}\left[\int_0^t (x-1)e^x \, dx\right] = \frac{1}{s}\mathcal{L}[(t-1)e^t] = \frac{1}{s}\left[\frac{1}{(s-1)^2} - \frac{1}{(s-1)}\right]$$

Answer 25: Since $\sin(t - \pi/4) = (\sin t - \cos t)/\sqrt{2}$ by a trigonometric identity, the transform of $\int_0^t \sin(x - \pi/4)e^x \, dx$ is (using Table Entries 30, 15, and 16)

$$\frac{1}{\sqrt{2}s}\mathcal{L}\left[e^t \sin t - e^t \cos t\right] = \frac{1}{\sqrt{2}s}\left[\frac{1}{(s-1)^2 + 1} - \frac{s-1}{(s-1)^2 + 1}\right]$$

 27. Prove that the sum, product, and antiderivative of functions in **E** are also in **E**.

Answer 27:

Sum: $|f(t) + g(t)| \leq |f(t)| + |g(t)| \leq Me^{at} + Ne^{bt} \leq (M + N)e^{at}$ if $a > b$ and $|f(t)| \leq Me^{at}$, $|g(t)| \leq Ne^{bt}$, a and b constants.

Product: $|f(t) \cdot g(t)| = |f(t)| \cdot |g(t)| \le Me^{at} \cdot Ne^{bt} = MNe^{t(a+b)}$

Antiderivative: $|\int f(t)\,dt| \le \int |f(t)|\,dt \le \int Me^{at} \le \dfrac{Me^{at}}{|a|}$ where a is a nonzero constant.

Modeling Problems.

28. *Radioactive Decay* A sample of a radioactive element decays at a rate proportional to the amount y present, where k_1 is the proportionality constant. Suppose that at some time $t = a$ more of the element is allowed to enter the sample from outside at a constant rate k_2, and at a later time $t = b$ this flow is stopped.

    (a) Show that this system can be modeled by the IVP

    $$y' = -k_1 y + k_2 \left(\text{Step}(t, a) - \text{Step}(t, b) \right), \qquad y(0) = y_0, \quad 0 < a < b$$

    (b) Use the Laplace transform to find $y(t)$. Plot the graph of $y(t)$ with $k_1 = 1$, $k_2 = 0.5$, $a = 1$, $b = 10$, $y_0 = 1$, $0 \le t \le 20$.

Harvested Populations. Problems 37–39 in Section 5.1 have to do with model IVPs of the form $y' = ky-$harvest/restock rate, $y(0) = A$. These problems are revisited below. In each case create the model IVP, transform it, and then use the Laplace transform to find $y(t)$. Finally, interpret your solution formula in terms of the fate of the population being harvested.

29. Problem 37 in Section 5.1.   30. Problem 38 in Section 5.1.

31. Problem 39 in Section 5.1.

    Answer 29: The Laplace transform of the IVP with $y(0) = 10$,

    $$y' = 0.5y - \begin{cases} 2 & 0 \le t \le 1 \\ 0 & 1 < t \le 3 \end{cases} = 0.5y - 2(1 - \text{Step}(t, 1)),$$

    is $\mathcal{L}[y] = \dfrac{10}{s - .05} + \dfrac{2(e^{-s} - 1)}{s(s - .05)}$. Using Table Entry 5 and Text formula (2), we see that

    $$y(t) = 10e^{.05t} - 2(20)(e^{.05t} - 1) + 2(20)\left[e^{.05(t-1)} - 1 \right] \text{Step}(t, 1)$$

    which resolves to

    $$y(t) = \begin{cases} -30e^{.05t} + 40 & 0 \le t \le 1 \\ -30e^{.05t} + 40e^{.05(t-1)} & 1 < t \le 3 \end{cases}$$

    Answer 31: The model IVP is

    $$y' - .02y = \begin{cases} -2, & 0 \le t \le 1, \quad 2 < t < 3, \\ +2, & 1 < t \le 2, \quad 3 < t \le 4 \end{cases}$$

    and $y(0) = 2$. Then

    $$\mathcal{L}[y] = \frac{2}{s - .02} + \frac{2(e^{-s} - 1)}{s(s - .02)} + \frac{2(e^{-3s} - e^{-2s})}{s(s - .02)} - \frac{2(e^{-2s} - e^{-s})}{s(s - .02)} - \frac{2(e^{-4s} - e^{-3s})}{s(s - .02)}$$

    $$= \frac{2}{s - .02} + \frac{2}{s(s - .02)} \left[-1 + 2e^{-s} - 2e^{-2s} + 2e^{-3s} - e^{-4s} \right]$$

    So $y(t) = 2e^{.02t} - 100(e^{.02t} - 1) +$ a sum of four other terms. For example, one of the other terms is $\mathcal{L}^{-1}\left[4e^{-3s}/(s(s - .02)) \right] = 200 \left[e^{.02(t-3)} - 1 \right] \text{Step}(t, 3)$, where we used Table Entry 5 and text formula (2).

32. *Maintenance of a Game Species* In the model of the 30-day harvest, find a relation among k, H, and A that will ensure exactly 330 days after the end of the harvest, the population will once more be at the initial level A.

33. Formulate and solve a harvesting problem where January is the harvest season each year for five years. [*Hint*: use the single-season model in the text as your guide.]

Answer 33: Group problem.

Other Uses for the Transform.

 34. *Difference Equations* Solve the difference equation $3x(t) - 4x(t-1) = 1$, where $x(t) = 0$ if $t \leq 0$. [*Hint*: first show that $L[x(t-1)](s) = e^{-s}L[x]$. Next show that

$$L[x] = 1/[s(3 - 4e^{-s})] = [1 + 4e^{-s}/3 + \cdots + (4e^{-s}/3)^n + \cdots]/(3s)$$

by using the series $1/(1-z) = 1 + z + \cdots + z^n + \cdots$, where $z = 4e^{-s}/3$ and s is large enough that $|z| < 1$. Use the Shifting Theorem and Transform Tables to find a series for $x(t)$.]

Laplace Transform for Linear Systems. Laplace transforms may be used to solve some linear systems. For example, if $x_1' = ax_1 + bx_2 + f_1(t)$, $x_2' = cx_1 + dx_2 + f_2(t)$, with a, b, c, and d constants, then applying the transform to the first ODE, we have $sL[x_1](s) - x_1(0) = aL[x_1](s) + bL[x_2](s) + L[f_1](s)$. There is a similar equation for $sL[x_2](s) - x_2(0)$. These two equations may be solved for $L[x_1]$ and for $L[x_2]$ in terms of a, b, c, d, $L[f_1]$, and $L[f_2]$. Then $x_1(t)$ and $x_2(t)$ are obtained by inverse transforms. Use this approach to solve the following systems:

35. $x_1' = 3x_1 - 2x_2 + t$, $\quad x_2' = 5x_1 - 3x_2 + 5$, $\quad x_1(0) = x_2(0) = 0$

Answer 35: Using the initial data and applying the Laplace transform to each term in each equation, we have that $sL[x_1] = 3L[x_1] - 2L[x_2] + s^{-2}$, $sL[x_2] = 5L[x_1] - 3L[x_2] + 5s^{-1}$. Solving these equations for the transforms $L[x_1]$ and $L[x_2]$, we have (using a partial fractions decomposition)

$$L[x_1] = \frac{3 - 9s}{s^2(s^2 + 1)} = \frac{-9}{s} + \frac{3}{s^2} + \frac{9s - 3}{s^2 + 1}$$

$$L[x_2] = \frac{5s^2 - 15s + 5}{s^2(s^2 + 1)} = \frac{-15}{s} + \frac{5}{s^2} + \frac{15s}{s^2 + 1}$$

Using the Laplace transform tables, we can invert the transforms to obtain $x_1(t) = -9 + 3t + 9\cos t - 3\sin t$, $x_2(t) = -15 + 5t + 15\cos t$.

36. $x_1' = x_1 + 3x_2 + \sin t$, $\quad x_2' = x_1 - x_2$, $\quad x_1(0) = 0$, $x_2(0) = 1$

The Transform of Higher Derivatives. Derive the indicated transform formulas.

37. Show that $L[f'''](s) = s^3L[f](s) - s^2 f(0) - sf'(0) - f''(0)$ if $f'''(t)$ is continuous for $t \geq 0$ and if $e^{-s\tau}f''(\tau)$, $e^{-s\tau}f'(\tau)$ and $e^{-s\tau}f(\tau)$ both tend to zero as $\tau \to +\infty$ (for $s > s_0$).

Answer 37: Use integratiion by parts and the formulas for $L[f'']$.

$$L[f'''](s) = \int_0^\infty e^{-st} f''' dt = \lim_{\tau \to +\infty} e^{-st} f'' \Big|_0^\infty + s \int_0^\infty e^{-st} f'' dt$$

$$= -f''(0) + s^3 L[y] - s^2 f(0) - sf'(0) \quad \text{for } t \geq 0$$

38. Prove the following formula for the transform of the nth derivative of a function:

$$L[f^{(n)}] = s^n L[f] - s^{n-1} f(0) - s^{n-2} f'(0) - \cdots - sf^{(n-2)}(0) - f^{(n-1)}(0)$$

Assume that $f^{(n)}$ is continuous for $t \geq 0$ and that the limits of functions such as $e^{-s\tau} f^{(k)}(\tau)$ for $k < n$ are 0, as $\tau \to +\infty$. [*Hint*: use mathematical induction on the order of the derivative. *Assume* the result is true for order k. Then *prove* that the result is true for order $k + 1$.]

Higher-Order IVPs. Use transforms to solve higher-order IVPs.

39. $y''' - y'' + y' - y = 0$, $\quad y(0) = 2$, $\quad y'(0) = 1$, $\quad y''(0) = -1$

40. $y''' + y'' + 4y' - 6y = 0$, $\quad y(0) = 0$, $\quad y'(0) = 1$, $\quad y''(0) = -1$

41. $y^{(4)} - y = 0$, $\quad y(0) = 0$, $\quad y'(0) = 1$, $\quad y''(0) = 0$, $\quad y'''(0) = 0$

42. $y^{(4)} + 3y'' + 2y = 0$, $\quad y(0) = 1$, $\quad y'(0) = 0$, $\quad y''(0) = 0$, $\quad y'''(0) = 0$

Answer 39:   Use the result of Problem 37 and the formula for $\mathcal{L}[y'']$ and $\mathcal{L}[y']$

$$0 = s^3 \mathcal{L}[y] - 2s^2 - s + 1 - \left(s^2 \mathcal{L}[y] - 2s - 1\right) + s\mathcal{L}[y] - 2 - \mathcal{L}[y]$$

$$= \mathcal{L}[y](s^3 - s^2 + s - 1) - 2s^2 + s$$

$$\mathcal{L}[y] = \frac{2s^2 - s}{(s^1 + 1)(s - 1)}$$

Using the partial fraction technique, we have

$$\frac{2s^2 - s}{(s^2 + 1)(s - 1)} = \frac{1/2}{s - 1} + \frac{(3/2)s}{s^2 + 1} + \frac{1/2}{s^2 + 1}$$

Therefore

$$y(t) = \frac{1}{2}e^t + \frac{3}{2}\cos t + \frac{1}{2}\sin t$$

Answer 41:   Use the formula in Problem 38 for $\mathcal{L}[y^{(4)}]$:

$$0 = s^4 \mathcal{L}[y] - s^3 y(0) - s^2 y'(0) - sy''(0) - y'''(0) - \mathcal{L}[y] = \mathcal{L}[y](s^4 - 1) - s^2$$

$$\mathcal{L}[y] = \frac{s^2}{(s^2 + 1)(s^2 - 1)} = \frac{1/2}{s^2 + 1} + \frac{1/2}{s^2 - 1}$$

$$y = \frac{1}{2}\sin t + \frac{1}{2}\sinh t$$

5.3 Transforms of Periodic Functions

Comments:

The main topic of this section is the Laplace transform of a periodic function. Of course, we have already shown how to transform sinusoids, but now we derive a formula for the transform of any periodic function. That formula is then applied to an LC circuit modeled by an ODE in the current $I(t)$ with a square wave driving function. The details are all worked out, and they are far from trivial since geometric expansions and partial fractions are required. Partial fractions formulas are introduced in a table on the inside back cover. As is usually the case when working with transforms, there is a lot of algebra. However, you will see how the work pays off in a clear, step-by-step formula for the response to an input such as a square wave. The geometric series expansion $(1 - x)^{-1} = 1 + x + x^2 + \cdots + x^n + \cdots$ is used throughout the section and its problems. It is valid for any x whose magnitude is less than 1; for example, it is valid for $x = e^{-s}$, $s > 0$.

PROBLEMS

Periodic Functions.   Use Theorem 5.3.1 to find the Laplace transform of each function. [*Hint*: see the inside front cover for the engineering functions. See Example 5.3.3 for Problems 10 and 11.]

<u>1.</u> $\cos 5t$ **2.** $3\sin 2t + 5\cos 3t$ **3.** $\sin^2 2t$ **4.** $|\sin t|$

5. $|\cos 2t|$ **<u>6.</u>** $3\,\text{SqWave}(t, 4, 2)$ **7.** $\text{SqWave}(t, 2\pi, \pi) - 2$ **8.** $\text{TRWave}(t, 4, 2, 3, 5)$

9. $\text{TRWave}(t, 4, 2, 3, -1) + 1$ **10.** $\sin \pi t\, \text{SqWave}(t, 2, 1)$ **11.** $\sin \pi t\, \text{SqWave}(t, 4, 1)$

Answer 1:   Use Theorem 5.3.1 with $T = 2\pi/5$:

$$\mathcal{L}[\cos 5t] = \frac{1}{1 - e^{-Ts}} \int_0^T e^{-su} \cos 5u \, du$$

$$= \frac{1}{1 - e^{-2\pi s/5}} \left[\frac{e^{-su}}{s^2 + 25} [-s \cos 5u + 5 \sin 5u] \right]\Big|_{u=0}^{u=2\pi/5}$$

$$= \frac{s}{s^2 + 25}$$

Answer 3:   First note that $\frac{1}{2}(1 - \cos 4t) = \sin^2 2t$. Use Theorem 5.3.1 with $T = \pi/2$:

$$\mathcal{L}[\frac{1}{2}(1 - \cos 4t)] = \mathcal{L}\left[\frac{1}{2}\right] - \mathcal{L}\left[\frac{1}{2}\cos 4t\right]$$

$$= \frac{1}{2s} - \frac{1/2}{1 - e^{-Ts}} \int_0^T e^{-su} \cos 4u \, du$$

$$= \frac{1}{2s} - \frac{1/2}{1 - e^{-\pi s/2}} \left[\frac{e^{-su}}{s^2 + 16} [-s \cos 4u + 4 \sin 4u] \right]\Big|_{u=0}^{u=\pi/2}$$

$$= \frac{1}{2s} - \frac{s}{2(s^2 + 16)}$$

Answer 5:   Use Theorem 5.3.1 with $T = \pi/2$:

$$\mathcal{L}[|\cos 2t|] = \frac{1}{1 - e^{-\pi/2s}} \int_0^{\pi/2} e^{-su} \cos 2u \, du$$

$$= \frac{1}{1 - e^{-\pi s/2}} \left[\frac{e^{-su}}{s^2 + 4} (-s \cos 2u + 2 \sin 2u) \right]\Big|_{u=0}^{u=\pi/2}$$

$$= \frac{1}{1 - e^{-\pi s/2}} \cdot \frac{s}{s^2 + 4} \cdot \left(1 + e^{-\pi s/2}\right)$$

$$= \frac{s}{s^2 + 4} + \frac{s}{s^2 + 4} \sum_{n=1}^{\infty} e^{-n\pi s}$$

where we used a geometric series expansion (1) for $\left(1 - e^{-\pi s/2}\right)^{-1}$.

Answer 7:   First note that

$$\text{SqWave}(t, 2\pi, \pi) = \begin{cases} 1, & 0 \le t < \pi \\ 0, & \pi \le t \le 2\pi \end{cases}$$

Use Theorem 5.3.1 with $T = 2\pi$:

$$\mathcal{L}[\text{SqWave}(t, 2\pi, \pi) - 2] = \frac{1}{1 - e^{-2\pi s}} \int_0^\pi e^{-su} \, du - \mathcal{L}[2]$$

$$= -\frac{1}{s} \cdot \frac{1}{1 - e^{-2\pi s}} (e^{-su})\Big|_{u=0}^{u=\pi} - \frac{2}{s}$$

$$= \frac{1}{s} \cdot \frac{1}{1 - e^{-2\pi s}} \cdot (1 - e^{-\pi s})$$

$$= \frac{1}{s} \sum_{n=0}^{\infty} (-1)^n e^{-n\pi s} - \frac{2}{s}$$

where we have used $1 - e^{-2\pi s} = (1 - e^{-\pi s})(1 + e^{-\pi s})$ and a geometric expansion (1) for $(1 + e^{-\pi s})^{-1}$.

Answer 9: Let

$$f(t) = \text{TRWave}(t, 4, 2, 3, A) = \begin{cases} At/2, & 0 \le t \le 2 \\ A(3 - t), & 2 \le t \le 3 \\ 0, & 3 \le t \le 4 \end{cases}$$

which is then repeated with period 4 (see inside front cover). Then, by Theorem 5.3.1

$$\mathcal{L}[f](s) = \frac{1}{1 - e^{-4s}} \int_0^4 e^{-st} f(t)\, dt = \frac{A}{1 - e^{-4s}} \left[\frac{1}{2} \int_0^2 e^{-st} t\, dt + \int_2^3 e^{-st}(3 - t)\, dt \right]$$

$$= \frac{A}{1 - e^{-4s}} \cdot \frac{1}{2s^2} \left[1 - 3e^{-2s} + e^{-3s} \right]$$

where we have used the Integral Tables on the inside back cover. In this problem, $A = -1$. We also have $\mathcal{L}[1] = 1/s$, so

$$\mathcal{L}[f + 1](s) = \frac{1}{s} - \frac{1}{1 - e^{-4s}} \cdot \frac{1}{2s^2} \left[1 - 3e^{-2s} + e^{-3s} \right]$$

Answer 11: Use Theorem 5.3.1 with $T = 4$:

$$\mathcal{L}[\sin \pi t \, \text{SqWave}(t, 4, 1)] = \frac{1}{1 - e^{-4s}} \int_0^1 e^{-su} \sin \pi u \, du$$

$$= \frac{1}{1 - e^{-4s}} \left[\frac{e^{-su}}{s^2 + \pi^2} (-s \sin \pi u - \pi \cos \pi u) \right]_{u=0}^{u=1}$$

$$= \frac{\pi}{1 - e^{-4s}} \cdot \frac{1}{s^2 + \pi^2} \cdot (1 + e^{-s})$$

$$= \frac{\pi}{s^2 + \pi^2} \sum_{n=0}^{\infty} \left(e^{-4ns} + e^{-(4n+1)s} \right)$$

where we have used a geometric series expansion (1) for $(1 - e^{-4s})^{-1}$.

Find the Laplace Transform. Use the material in Sections 5.1–5.3 and the Integral Tables or the Trigonometric Identities on the inside back cover to find the Laplace transform of each function. [*Hint*: use the Transform Tables on the inside back cover to check your answers.]

12. $e^t \sin t$ 13. $\sin^2 t$ 14. $\cos^2 t$ 15. $t \sin at$

16. $t^2 \sin at$ 17. $\cos^3 t$ **18.** $e^{-3t} \cos(2t + \pi/4)$ 19. $t^2 e^t \cos t$

Answer 13: $\mathcal{L}[\sin^2 t] = 2/[s(s^2 + 4)]$ since $\sin^2 t = (1 - \cos 2t)/2$, $\mathcal{L}[1] = 1/s$, and $\mathcal{L}[\cos 2t] = s/(s^2 + 4)$. Note that $\mathcal{L}[\sin^2 t] \ne (\mathcal{L}[\sin t])^2$.

Answer 15: Use integration by parts to evaluate $\int e^{-st} t \sin at \, dt$ with $u = t$, $dv = e^{-st} \sin at \, dt$: then $\mathcal{L}[t \sin at](s) = 2as/(s^2 + a^2)^2$. Use Table Entry 29 to check.

Answer 17: Use the formula for $\cos \alpha \cos \beta$ twice to show that $\cos^3 t = (3 \cos t)/4 + (\cos 3t)/4$. Then use $\mathcal{L}[\cos at] = s/(s^2 + a^2)$ to obtain $\mathcal{L}[\cos^3 t] = \left[3s(s^2 + 1)^{-1} + s(s^2 + 9)^{-1} \right]/4$.

Answer 19: Using Table Entry 29 and (1) of the Shifting Theorem 5.2.2, we see that

$$\mathcal{L}[t^2 e^t \cos t](s) = \frac{d^2}{ds^2} \{ \mathcal{L}[e^t \cos t](s) \} = \frac{d^2}{ds^2} \{ \mathcal{L}[\cos t](s - 1) \}$$

So, using Table Entry 12, we see that

$$\mathcal{L}[t^2 e^t \cos t](s) = \frac{d^2}{ds^2} \left[\frac{s - 1}{(s - 1)^2 + 1} \right] = \frac{4s(s - 1)(s - 2)}{((s - 1)^2 + 1)^3}$$

Find the Inverse Laplace Transform.   [*Hint*: use the Shifting Theorem 5.2.2.]

20. $\dfrac{1+e^{-s}}{s}$ **21.** $\dfrac{3e^{-2s}}{3s^2+1}$ **22.** $\dfrac{1}{(s-a)^n}$

23. $\dfrac{1-e^{-s}}{s^2}$ **24.** $\dfrac{(s-1)e^{-s}+1}{s^2}$ **25.** $\ln\dfrac{s+3}{s+2}$

Answer 21:   Using formula (2) of the Shifting Theorem 5.2.2, we have that

$$\mathcal{L}^{-1}\left[\frac{3e^{-2s}}{3s^2+1}\right]=\mathcal{L}^{-1}\left[e^{-2s}\frac{1}{s^2+1/3}\right]=\text{Step}(t,2)f(t-2)$$

where $\mathcal{L}[f]=1/(s^2+1/3)$, so $f(t)=\sqrt{3}\sin(t/\sqrt{3})$ by Table Entry 11. Then

$$\mathcal{L}^{-1}\left[\frac{3e^{-2s}}{3s^2+1}\right]=\sqrt{3}\,\text{Step}(t,2)\sin((t-2)/\sqrt{3})$$

Answer 23: $\mathcal{L}^{-1}[1/s^2-e^{-s}/s^2]=t-(t-1)\,\text{Step}(t,1)$ by formula (2) of the Shifting Theorem 5.2.2.

Answer 25:   By Table Entry 8

$$\mathcal{L}^{-1}\left[\ln\frac{s+3}{s+2}\right]=\frac{1}{t}(e^{-2t}-e^{-3t})$$

Modeling Problems.

LC **Circuit.** The current $I(t)$ of the *LC* circuit modeled by $LI''+I/C=4\,\text{SqWave}(t,T,T/2)$ $-\,2,\ I(0)=0,\ I'(0)=0$, is sensitive to changes in the parameters L, C, and T. In particular, set $L=1$ and plot the square wave input and the current for $0\le t\le 50$ if C and T are as given below. Does $I(t)$ appear to be periodic? If it is, what is the period? If it is not periodic, describe and explain the behavior of $I(t)$ as time passes.

26. $C=1,\ T=5\pi$ **27.** $C=0.05,\ T=5\pi$ **28.** $C=1,\ T=2\pi$

Answer 27:   ($L/C=\omega^2=20,\ T=5\pi$. The natural period is $2\pi/\sqrt{20}=\pi/\sqrt{5}$. The period of the input is 5π and the ratio if the two periods is $\sqrt{5}$, which is irrational. So, the sum of a natural oscillation of period $\pi/\sqrt{5}$ and a driven oscillation of period 5π is not periodic. Note the appearance of beats (see Section 4.2). The middle and lower figures show I and I' respectively.

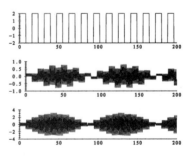

Problem 27.

29. *Square Wave Inputs*   Solve each IVP below. Plot the input and the output.

☞ See the SPOTLIGHT ON STEADY STATES: LINEAR ODEs

(a) *Low-Pass Filter* $y'+y=\text{SqWave}(t,T,T/2),\ y(0)=0$. First let $T=30,\ 0\le t\le 100$; then let $T=1,\ 0\le t\le 10$. The IVP models a low-pass communication channel that transmits low-frequency signals with much less distortion than high-frequency signals.

☞ See Section 4.2 for a discussion of beats.

(b) *Beats* $y'' + 4y = \text{SqWave}(t, T, T/2)$, $y(0) = 0$, $y'(0) = 0$, $0 \le t \le 60$. First let $T = 7$ and then $T = 3$. Plot the input, $y(t)$, and $y'(t)$. Explain why pronounced beats appear when $T = 3$. [*Hint*: the natural period is $\pi = 3.14\cdots$, while the period of the driving term is 3.]

Answer 29:

(a) The IVP is $y' + y = \text{SqWave}(t, T, T/2)$, $y(0) = 0$. $\mathcal{L}[y' + y] = s\mathcal{L}[y] + \mathcal{L}[y] = \mathcal{L}[\text{SqWave}(t, T, T/2)]$. So,

$$\mathcal{L}[y] = \frac{1}{s+1} \cdot \frac{1}{s} \sum_{n=0}^{\infty} (-1)^n e^{-nTs/2}$$

where we have used Table Entry 20 with $T = a$. Since

$$\frac{1}{(s+1)s} = \frac{1}{s} - \frac{1}{s+1}$$

we have

$$\mathcal{L}[y] = \sum_{n=0}^{\infty} (-1)^n e^{-nTs/2} \left[\frac{1}{s} - \frac{1}{s+1} \right]$$

Inverting, we have

$$y = \sum_{n=0}^{\infty} (-1)^n \mathcal{L}^{-1} \left\{ e^{-nTs/2} \left[\frac{1}{s} - \frac{1}{s+1} \right] \right\}$$

$$= 1 - e^{-t} + \sum_{n=1}^{\infty} (-1)^n \text{Step}(t - nT/2)[1 - e^{-(t-nT/2)}]$$

where we have used formula (2) in Theorem 5.2.2 and Table Entries 2 and 3. See Graph 1 where the period of the driving term is $T = 30$, and Graph 2 for $T = 1$. In each case, the response tends to a forced oscillation of period T. Graph 1 and Graph 2 show (from top down) the square wave input and the output $y(t)$.

If the ODE models a communication channel that receives square waves of various periods as inputs, then the long period (i.e., low frequency) input comes out of the channel with much less distortion and amplitude loss (Graph 1 where $T = 30$) than the short period and high-frequency input (Graph 2 where $T = 1$).

(b) The IVP is $y'' + 4y = \text{SqWave}(t, T, T/2)$, $y(0) = 0$, $y'(0) = 0$. $\mathcal{L}[y'' + 4y] = s^2\mathcal{L}[y] + 4\mathcal{L}[y] = \mathcal{L}[\text{SqWave}(t, T, T/2)]$. As in **(a)**,

$$\mathcal{L}[y] = \frac{1}{s^2 + 4} \cdot \frac{1}{s} \sum_{n=0}^{\infty} (-1)^n e^{-nTs/2}$$

Since

$$\frac{1}{s(s^2 + 4)} = \frac{1}{4} \left(\frac{1}{s} - \frac{s}{s^2 + 4} \right)$$

we have

$$\mathcal{L}[y] = \frac{1}{4} \sum_{n=0}^{\infty} (-1)^n e^{-nTs/2} \left[\frac{1}{s} - \frac{s}{s^2 + 4} \right]$$

So,

$$y = \frac{1}{4} \sum_{n=0}^{\infty} (-1)^n \mathcal{L}^{-1} \left\{ e^{-nTs/2} \left[\frac{1}{s} - \frac{s}{s^2 + 4} \right] \right\}$$

$$= \frac{1}{4} - \frac{1}{4} \cos 2t + \frac{1}{4} \sum_{n=1}^{\infty} (-1)^n \text{Step}(t - nT/2)[1 - \cos(2t - nT)]$$

where we have used formula (2) in Theorem 5.2.2. See Graph 1 (for $T = 7$) and Graph 2 (for $T = 3$) where the graphs of the input and $y(t)$ and $y'(t)$ are shown. The natural period is π, and the driving period is 7 in the first case. Since $\pi/7$ is not a rational number, one would not expect to see a periodic response, and indeed Graph 1 does not show one. However, if $T = 3$, that period is close to the natural period of π, and one might expect to see beats. Graph 2 shows one beat; solving over $0 \leq t \leq 200$ would show more beats. The three graphs in each case show (from top down) the square wave input and the component graphs of y and y'.

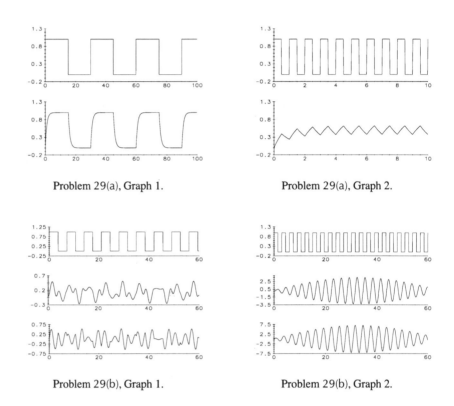

Problem 29(a), Graph 1.   Problem 29(a), Graph 2.

Problem 29(b), Graph 1.   Problem 29(b), Graph 2.

30. *Communication Channel*   Solve the IVP $y' + y = \text{TRWave}(t, T, T/2, T, 1)$, $y(0) = 0$. First let $T = 30$, $0 \leq t \leq 10$; then let $T = 1$, $0 \leq t \leq 10$. Plot the input and the output. If the IVP models a communication channel, which input wave is transmitted with less distortion? Explain. [*Hint*: see the SPOTLIGHT ON STEADY STATES: LINEAR ODES.]

5.4 Convolution

Comments:

The Convolution Theorem says that if $\mathcal{L}[f] = F$ and $\mathcal{L}[g] = G$ then $\mathcal{L}^{-1}[FG]$ is the *convolution product* $f * g$ of f and g, *not* the algebraic product $f \cdot g$. At the same time, the convolution product gives a practical way to construct the solution of $P(D)[y] = f$ (with zero initial data) for any constant coefficient operator $P(D)$ and integrable input f. The solution is $y = g * f$, where $g(t) = \mathcal{L}^{-1}[1/P(s)]$. The convolution product is a neat idea and new to many readers. The more theoretically inclined like to prove its various properties, and all should appreciate its power in representing solutions of $P(D)[y] = f$ (with zero initial data). Because the approach in this section is to focus on the idea of the convolution, we have deliberately kept calculations to a minimum in the text and in the problems. This will come as a welcome relief after all of the laborious computations of the previous sections.

PROBLEMS

Find Laplace Transforms. Use the Transform Table on the inside back cover to find the Laplace transform of each integral. [*Hint*: each integral is the convolution product of two functions.]

1. $\int_0^t (t - u)u\,du$

2. $\int_0^t \sin u\,du$ [*Hint*: $\sin u = 1 \cdot \sin u$.]

3. $\int_0^t (t^2 - 2tu + u^2)\,du$

4. $\int_0^t (\sin t \sin u \cos u - \cos t \sin^2 u)\,du$

Answer 1: Using the Convolution Theorem with $f(t) = t$, $g(t) = t$, we see that

$$L\left[\int_0^t (t - u)u\,du\right] = L[t]L[t] = 1/s^4$$

Answer 3: Since $t^2 - 2tu + u^2 = (t - u)^2$, let $f(t) = t^2$, $g(t) = 1$. The transform of the integral is (by the Convolution Theorem)

$$L\left[\int_0^t (t^2 - 2tu + u^2)\,du\right] = L[t^2]L[1] = 2/s^4$$

Inverse Laplace Transforms. Use the Convolution Theorem to find the inverse Laplace transform of each function. Write your answers as convolution products, but don't evaluate the integrals. [*Hint*: write each expression as the product of two functions whose inverse transforms you can find in the Transform Table on the inside back cover.]

5. $\dfrac{s}{(s^2 + 1)^2}$

6. $\dfrac{s}{(s^2 + 10)^2}$

7. $\dfrac{1}{s^2(s + 1)}$

8. $\dfrac{s}{(s^2 + 9)^3}$

9. $\dfrac{s}{(s + 1)(s + 2)^3}$

10. $\dfrac{s^2 + 4s + 4}{(s^2 + 4s + 13)^2}$

11. $\dfrac{L[f]}{s^2 + 1}$

12. $\dfrac{e^{-3s}L[f]}{s^3}$

Answer 5:

$$L^{-1}\left[\frac{s}{(s^2 + 1)^2}\right] = L^{-1}\left[\frac{s}{s^2 + 1}\frac{1}{s^2 + 1}\right] = \cos t * \sin t = \int_0^t \cos(t - u)\sin u\,du$$

Answer 7:

$$L^{-1}\left[\frac{1}{s^2(s + 1)}\right] = L^{-1}\left[\frac{1}{s^2}\frac{1}{s + 1}\right] = t * e^{-t} = \int_0^t (t - u)e^{-u}\,du$$

Answer 9:

$$L^{-1}\left[\frac{s}{(s + 1)(s + 2)^3}\right] = L^{-1}\left[\frac{s}{(s + 1)(s + 2)}\frac{1}{(s + 2)^2}\right]$$

$$= L^{-1}\left[\left(\frac{2}{s + 2} - \frac{1}{s + 1}\right) \cdot \frac{1}{(s + 2)^2}\right]$$

$$= (2e^{-2t} - e^{-t}) * te^{-2t} = \int_0^t \left[2e^{-2(t - u)} - e^{-(t - u)}\right]ue^{-2u}\,du$$

Answer 11:

$$\mathcal{L}^{-1}\left[\frac{\mathcal{L}[f]}{s^2+1}\right] = \mathcal{L}^{-1}\left[\mathcal{L}[f]\frac{1}{s^2+1}\right] = f * \sin t = \int_0^t f(t-u)\sin u\, du$$

Inverse Laplace Transforms. Construct $\mathcal{L}^{-1}[1/P(s)]$ for each differential operator $P(D)$.

13. $D^2 + 6D + 13$ **14.** $D^2 + (1/3)D + 1/36$ **15.** $D^2 - 1$

Answer 13: $P(D) = D^2 + 6D + 13$, and so $\mathcal{L}^{-1}[1/(s^2 + 6s + 13)] = \mathcal{L}^{-1}[1/((s+3)^2 + 2^2)] = (1/2)e^{-3t}\sin 2t$. Alternatively, use Table 5.4.1 with $r_1 = \overline{r_2} = 3 + 2i$.

Answer 15:

$$\mathcal{L}^{-1}\left[\frac{1}{s^2-1}\right] = \sinh t = \frac{e^t - e^{-t}}{2}$$

Construct a Convolution Product. Write the solution of each IVP as a convolution product (don't evaluate the integral). Use the initial conditions $y(0) = y'(0) = 0$. [*Hint*: Theorem 5.2.2, Theorem 5.4.3, and the partial fraction formulas on the inside back cover may be useful here.]

16. $y'' + y = t\,\text{Step}(t, 1)$ **17.** $y'' + y = f(t)$ **18.** $2y'' + y' - y = f(t)$

19. $y'' + 4y = f(t) = \begin{cases} 0, & 0 \le t < 1 \\ t - 1, & 1 \le t < 2 \\ 1, & t \ge 2 \end{cases}$ **20.** $y'' + 2y' + y = f(t)$

Answer 17: The IVP is $y'' + y = f(t)$, $y(0) = y'(0) = 0$.

$$\mathcal{L}[y] = \frac{1}{s^2+1} \cdot \mathcal{L}[f]; \quad y(t) = \int_0^t \sin(t-u)f(u)\,du$$

Answer 19: The IVP is $y'' + 4y = \begin{cases} 0, & 0 \le t < 1 \\ t - 1, & 1 \le t < 2 \\ 1, & t \ge 2 \end{cases}$, $\quad y(0) = y'(0) = 0$.

$$\mathcal{L}[y] = \frac{1}{s^2+4}\mathcal{L}\big[(t-1)\,\text{Step}(t, 1) - (t-2)\,\text{Step}(t, 2)\big]$$

$$y(t) = \frac{1}{2}\int_0^t \sin 2(t-u)\big[(u-1)\,\text{Step}(u, 1) - (u-2)\,\text{Step}(u, 2)\big]\,du$$

where we used the fact that $f(t)$ reduces to $(t-1)\,\text{Step}(t, 1) - (t-2)\,\text{Step}(t, 2)$.

Solve IVPs. Use Theorem 5.4.3 to solve the following IVPs. The function $f(t)$ belongs to **E**.

21. $y'' + 6y' + 13y = f(t)$, $y(0) = y_0$, $y'(0) = v_0$

22. $y'' + 6y' + 10y = 7e^{3t}$, $y(0) = 1$, $y'(0) = -1$

23. $y'' + 5y' + 6y = f(t)$, $y(0) = y_0$, $y'(0) = v_0$

24. $y'' - 7y' + 12y = 6e^{3t}$, $y(0) = -1$, $y'(0) = 2$

25. $y'' + y'/3 + y/36 = f(t)$, $y(0) = y_0$, $y'(0) = v_0$

26. $y'' - 2y' + y = te^t$, $y(0) = 2$, $y'(0) = -1$

27. $y'' - y = (t-1)\,\text{Step}(t, 1)$, $y(0) = 1$, $y'(0) = 1$

Answer 21: Here, $P(s) = (s^2 + 6s + 13)^{-1}$, $g(t) = \mathcal{L}^{-1}\left[1/P(s)\right] = \frac{1}{2}e^{-3t}\sin 2t$, and $h(t) = \mathcal{L}^{-1}[s/P(s)] = e^{-3t}\left(-\frac{3}{2}\sin 2t + \cos 2t\right)$ where we have used Table 5.4.1 with $r_1 = \overline{r_2} = -3 + 2i$. So by formula (12), the solution is $y = (6y_0 + v_0)g(t) + y_0h(t) + [(e^{-3t}\sin 2t)/2] * f(t)$.

Answer 23: The characteristic polynomial $r^2 + 5r + 6$ has the roots $r_1 = -3$ and $r_2 = -2$. Using Table 5.4.1, we see that since $r_1 \neq r_2$ and both are real that

$$g(t) = \frac{e^{-3t} - e^{-2t}}{-1} \quad \text{and} \quad h(t) = \frac{-3e^{-3t} + 2e^{-2t}}{-1}$$

By (12), $y = (5y_0 + v_0)g(t) + y_0 h(t) + (g * f)(t)$

Answer 25: The double root is $-1/6$, so by Table 5.4.1, $g = te^{-t/6}$, $h = (1 - t/6)e^{-t/6}$. By (12), the solutions is $y = (y_0/3 + v_0)g + y_0 h + te^{-t/6} * f(t)$.

Answer 27: The characteristic equation $r^2 - 1$ gives the roots $r_1 = 1$ and $r_2 = -1$. Using Table 5.4.1, we see that since $r_1 \neq r_2$ and both are real that

$$g(t) = \frac{e^t - e^{-t}}{2} \quad \text{and} \quad h(t) = \frac{e^t + e^{-t}}{2}$$

Finding $(g * f)(t)$:

$$(g * f)(t) = 0, \quad 0 \leq t \leq 1$$

$$= \int_1^t \left[\frac{e^{t-u} - e^{-(t-u)}}{2} \right] (u - 1)\, du, \quad t \geq 1$$

$$= 1 - t + \frac{1}{2}\left(e^{t-1} - e^{1-t} \right), \quad t \geq 1$$

Therefore,

$$y(t) = g(t) + h(t) + (g * f)(t)$$

$$= \begin{cases} g(t) + h(t), & 0 \leq t \leq 1 \\ g(t) + h(t) + 1 - t + \frac{1}{2}\left(e^{t-1} - e^{1-t} \right), & t \geq 1 \end{cases}$$

Properties of the Convolution Product.

28. Show that $(e^{at} f) * (e^{at} g) = e^{at}(f * g)$ if f and g are in **E**.

29. Suppose that f, g, and h are in **E**. Show that

(a) $f * g$ is in **E**. (b) $(f * g) * h = f * (g * h)$. (c) $(f + g) * h = f * h + g * h$.

Answer 29:

(a) Let's show that $f * g$ is in **E** if f and g are (the closure property). If f and g belong to **E** then there are positive constants M_1, M_2, and constants a_1, a_2 such that $|f(t)| \leq M_1 e^{a_1 t}$ and $|g(t)| \leq M_2 e^{a_2 t}$ for $t \geq 0$; in addition, f and g are assumed to be piecewise continuous on $[0, \infty)$. Then (assuming for simplicity that $a_1 \neq a_2$) we have

$$|f * g| \leq \int_0^t |f(t - u)||g(u)|\, du \leq M_1 M_2 \int_0^t e^{a_1(t-u)} e^{a_2 u}\, du$$

$$= M_1 M_2 e^{a_1 t} \int_0^t e^{(a_2 - a_1)u}\, du \leq \frac{M_1 M_2}{|a_2 - a_1|} e^{a_2 t}$$

which therefore belongs to **E**. If $a_1 = a_2$, then $|f * g| \leq M_1 M_2 t e^{a_1 t} \leq M_1 M_2 e^{(a_1 + a_2)t}$, which belongs to **E**.

(b) To show that $(f * g) * h = f * (g * h)$ (associativity), we proceed as follows:

$$f * (g * h) = \int_0^t f(t-u)(g * h)(u)\, du = \int_0^t f(t-u) \left\{ \int_0^u g(u-v)h(v)\, dv \right\} du$$

$$= \int_0^t \left\{ \int_v^t f(t-u)g(u-v)\, du \right\} h(v)\, dv$$

$$= \int_0^t \left\{ \int_0^{t-v} f(t-v-w)g(w)\, dw \right\} h(v)\, dv = (f * g) * h$$

where the order of integration has been interchanged and the variable $w = u - v$ has been introduced.

(c) To show that $(f + g) * h = f * h + g * h$ (distributivity), we have

$$(f + g) * h = \int_0^t (f + g)(t-u)h(u)\, du = \int_0^t f(t-u)h(u)\, du + \int_0^t g(t-u)h(u)\, du$$

$$= f * h + g * h$$

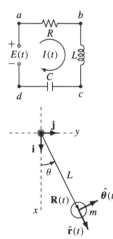

Modeling Problems. Construct the IVP that models the phenomenon. Write the solution of the IVP as a convolution product but don't evaluate the integral. What happens as $t \to +\infty$? Explain.

30. *An RLC Circuit*   In a simple *RLC* series circuit, $L = 20$ H, $R = 80\ \Omega$ and $C = 10^{-2}$ F. The source voltage is the continuously differentiable function $E(t)$. [*Hint:* write the IVP for the current $I(t)$ if $I(0) = 2$ amps and $I'(0) = 3$ amps/sec. See Section 4.4.]

31. *Linearized Pendulum*   A pendulum executes lightly damped, small-amplitude oscillations about the downward equilibrium $\theta = 0$, $\theta' = 0$. Suppose that $L = 2$, $m = 1$, $\theta(0) = 0.5$ rad, $\theta'(0) = -0.1$ rad/sec, and the external force tangent to the circle of motion is $-e^{-t} \sin t$.

Answer 31:   From Section 4.1 and assuming that the damping coefficient $c \approx 0$,

$$mL\theta'' + cL\theta' + mg\sin\theta = 2\theta'' + g\sin\theta = -e^{-t}\sin t$$

Suppose, first, that $\theta(0) = 0$ and $\theta'(0) = 0$. Since $g \approx 32$ ft/sec^2, the ODE is $\theta_d'' + 16\theta_d = -(e^{-t}\sin t)/2 = f(t)$, and the transformed IVP is

$$\mathcal{L}[\theta_d] = \mathcal{L}[f] \cdot \frac{1}{s^2 + 16}$$

By the Convolution Theorem $\theta_d = -(e^{-t}\sin t)/2 * (\sin t)/2$. This convolution product tends to 0 as $t \to +\infty$. Since $\theta_u = -0.05\sin 2\theta + 0.5\cos\theta$ solves the IVP, $\theta'' + 16\theta = 0$, $\theta(0) = 0.5$, $\theta'(0) = 0 = -0.1$, the desired solution is $\theta_u + \theta_d = \theta$.

Spotlight on the Delta Function

Comments

The elusive delta function finally makes its appearance, and the we are faced with a hard choice: define the delta function in a slightly shady way as we do, just "wave one's hands," or do enough with distributions to give the whole thing a mathematically respectable aura. At various times in various courses, we have used each of these approaches—and with varying success. The approach we take in this section is as good (or as bad) as any other we have tried. The essential properties of the delta function are contained in Theorem 1. The model of a vibrating spring that is subjected to an impulsive force helps to flesh out the idea of the delta function.

PROBLEMS

The Impulse Response. Find the solution $y_\delta(t, u)$ of the IVP, $y'' + Ay' + By = \delta(t - u)$, $y(0) = 0$, $y'(0) = 0$, where A and B are as indicated.

1. $A = 3$, $B = 2$ 2. $A = -1$, $B = 2$ 3. $A = 2$, $B = 1$
4. $A = -2$, $B = 1$ 5. $A = 1$, $B = 0$ 6. $A = 0$, $B = 0$
7. $A = 2$, $B = 2$ 8. $A = -2$, $B = 2$ 9. $A = 0$, $B = 4$

Answer 1: The characteristic polynomial is $r^2 + 3r + 2$, with roots $r_1 = -1$ and $r_2 = -2$. That makes $g(t) = e^{-t} - e^{-2t}$. Therefore by formula (14)

$$y_\delta(t, u) = \left[e^{-(t-u)} - e^{-2(t-u)}\right] \text{Step}(t, u)$$

Answer 3: The characteristic polynomial is $r^2 + 2r + 1$, with roots $r_1 = r_2 = -1$. That makes $g(t) = te^{-t}$. Therefore by formula (14)

$$y_\delta(t, u) = (t - u)e^{-(t-u)} \, \text{Step}(t, u)$$

Answer 5: The characteristic polynomial is $r^2 + r$, with roots $r_1 = 0$ and $r_2 = -1$. That makes $g(t) = 1 - e^{-t}$. Therefore by formula (14)

$$y_\delta(t, u) = \left[1 - e^{-(t-u)}\right] \text{Step}(t, u)$$

Answer 7: The characteristic polynomial is $r^2 + 2r + 2$, with roots $r_1 = \overline{r_2} = -1 + i$. That makes $g(t) = e^{-t} \sin t$. Therefore by formula (14)

$$y_\delta(t, u) = e^{-(t-u)} \sin(t - u) \, \text{Step}(t, u)$$

Answer 9: The characteristic polynomial is $r^2 + 4$, with roots $r_1 = \overline{r_2} = 2i$. That makes $g(t) = \dfrac{\sin 2t}{2}$. Therefore by formula (14)

$$y_\delta(t, u) = \frac{\sin 2(t - u)}{2} \, \text{Step}(t, u)$$

Convolution and Impulse. Use Theorem 3 to find the solution $y_d(t)$ of the IVP $y'' + Ay' + By = f(t)$, $y(0) = 0$, $y'(0) = 0$, where A, B, and f are as given. Then solve the same ODE but with the general initial data, $y(0) = y_0$, $y'(0) = v_0$. [*Hint*: if you have already solved the matching Problems 1–9, then you have saved yourself some work.]

Note: The Integral Tables and the trigonometric identities on the inside back cover are useful here.

10. $A = 3$, $B = 2$, $f = e^{-t}$ 11. $A = -1$, $B = 2$, $f = t$
12. $A = 2$, $B = 1$, $f = \sin t$ 13. $A = -2$, $B = 1$, $f = \cos t$
14. $A = 1$, $B = 0$, $f = te^t$ 15. $A = 0$, $B = 0$, $f = \cos 2t$
16. $A = 2$, $B = 2$, $f = e^t$ 17. $A = -2$, $B = 2$, $f = \sin t$
18. $A = 0$, $B = 4$, $f = \sin t$

Answer 11: There was a misprint in the text in the first printing. The coefficients are $A = -1$, $B = -2$.

$$y_d(t) = \int_0^t \left[e^{2(t-u)} - e^{-(t-u)}\right] \frac{u}{3} \, du = \frac{-t}{2} + \frac{1}{4} + \frac{e^{2t}}{12} - \frac{e^{-t}}{3}$$

$$y = y_d + (y_0 + v_0)e^{2t}/3 + (2y_0 - v_0)e^{-t}/3$$

Answer 13: The characteristic polynomial is $r^2 - 2r + 1 = (r-1)^2$. Hence $r_1 = r_2 = 1$ and $g(t) = te^t$. So

$$y_d(t) = \int_0^t (t-u)e^{t-u}\cos u \, du = \frac{1}{2}te^t - \frac{1}{2}\sin t$$
$$y = y_d + y_0 e^t + (v_0 - y_0)te^t$$

Answer 15:

$$y_d(t) = \int_0^t (t-u)\cos 2u \, du = \frac{-\cos 2t}{4} + \frac{1}{4}$$
$$y = y_d + y_0 + v_0 t$$

Answer 17:

$$y_d(t) = \int_0^t e^{t-u}\sin(t-u)\sin u \, du = \frac{2}{5}\cos t + \frac{1}{5}\sin t + \frac{1}{5}e^t\sin t - \frac{2}{5}e^t\cos t$$
$$y = y_d + e^t(y_0\cos t + (v_0 - y_0)\sin t)$$

Convolution and Impulse. Use the Laplace transform to find the solution of an IVP, where $y(0) = y'(0) = 0$.

19. $y'' + 2y' + 2y = \delta(t - \pi)$ 20. $y'' + 4y = \delta(t - \pi) - \delta(t - 2\pi)$
21. $y'' + 3y' + 2y = \sin t + \delta(t - \pi)$ 22. $y'' + y = \delta(t - \pi)\cos t$
23. $y'' + y = e^t + \delta(t - 1)$ 24. $y'' - y = 8(t - 1)e^t$

Answer 19: We have that

$$\mathcal{L}[y](s) = \frac{e^{-\pi s}}{s^2 + 2s + 2} = \frac{e^{-\pi s}}{(s+1)^2 + 1}$$

By formula (2) of Theorem 5.2.2, $y(t) = \text{Step}(t, \pi)f(t - \pi)$, where $\mathcal{L}[f](s) = 1/[(s+1)^2 + 1]$, and $f(t) = e^{-t}\sin t$. So, $y(t) = \text{Step}(t, \pi)e^{-(t-\pi)}\sin(t - \pi) = -\text{Step}(t, \pi)e^{-(t-\pi)}\sin t$.

Answer 21: The ODE is $y'' + 3y' + 2y = \sin t + \delta(t - \pi)$. Here, we have

$$\mathcal{L}[y](s) = \frac{1}{(s^2+1)(s+1)(s+2)} + \frac{e^{-\pi s}}{(s+1)(s+2)}$$

After using partial fractions, we have

$$y(t) = [-3\cos t + \sin t + 5e^{-t} - 2e^{-2t}]/10 + \text{Step}(t, \pi)[e^{-(t-\pi)} - e^{-2(t-\pi)}]$$

Answer 23: The ODE is $y'' + y = e^t + \delta(t - 1)$. In this case

$$\mathcal{L}[y] = \frac{1}{(s^2+1)(s-1)} + \frac{e^{-s}}{s^2+1}$$

After using partial fractions, we have $y = [e^t - \cos t - \sin t]/2 + \text{Step}(t, 1)\sin(t - 1)$.

Properties of the Delta Function.

25. Show that $\delta(at) = (1/|a|)\delta(t)$, $a \neq 0$.

Answer 25: Let f be any function in $\mathbf{E}_1$. We will make the substitution $v = au$ to obtain

$$\int_{-\infty}^{\infty} \delta(at - au) f(u) du = \frac{1}{a} \int_{-\infty}^{\infty} \delta(at - v) f(v/a) dv$$

$$= \frac{1}{a} f(\frac{at}{a}) = \frac{1}{a} f(t) \text{ if } a > 0$$

if $a > 0$. If $a < 0$, then we have

$$\frac{1}{a} \int_{\infty}^{-\infty} (\cdot) = -\frac{1}{a} \int_{-\infty}^{\infty} (\cdot) = \frac{1}{|a|} \int_{\infty}^{-\infty} (\cdot)$$

Since the above holds for all functions f, we have that $\delta(at) = |a|^{-1} \delta(t)$ if $a \neq 0$.

26. Suppose that $f(t)$ is in $\mathbf{E}_{\text{ext}}$. Show that $\delta(t) * f(t) = f(t)$.

Modeling Problems.

27. *Driven Spring*   A Hooke's Law spring with spring constant k supports a block of mass 1 and is subject to a force $f(t) = A \sin \omega t$, $\omega^2 \neq k$, for $t \geq 0$. Hit the block with a sharp upward blow at $t = 2$ that yields an impulse of 2 units, so the force of the blow is $2\delta(t - 2)$. Determine the motion if the object is initially at rest at its static equilibrium. Plot the solution when $k = 1$, $A = 1$, $\omega = 2$.

Answer 27: The motion is modeled by $y'' + ky = A \sin \omega t + 2\delta(t - 2)$, $y(0) = y'(0) = 0$, $\omega^2 \neq k$. Applying $\mathcal{L}$, we have that

$$\mathcal{L}[y](s) = \frac{A\omega}{s^2 + \omega^2} \frac{1}{s^2 + k} + \frac{2e^{-2s}}{s^2 + k} = \frac{A\omega}{k - \omega^2} \left[\frac{1}{s^2 + \omega^2} - \frac{1}{s^2 + k} \right] + \frac{2e^{-2s}}{s^2 + k}$$

Inverting, we have

$$y(t) = \frac{A}{k - \omega^2} \left(\sin \omega t - \frac{\omega}{\sqrt{k}} \sin \sqrt{k} t \right) + \frac{2}{\sqrt{k}} \text{Step}(t, 2) \sin \sqrt{k}(t - 2)$$

See the figure for the solution curve where $k = 1$, $A = 1$, $\omega = 2$. Note the "angle" in the curve at $t = 2$ when the blow is struck, and observe the consequences of the resultant increase in energy producing higher amplitude oscillations.

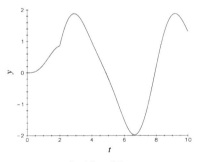

Problem 27.

28. *Driven LC Circuit*   Suppose that an *LC* circuit is subject to a constant voltage F_0. At time $t = 1$ the circuit is dealt a sharp voltage burst of amplitude $2F_0$ (i.e., $2F_0\delta(t - 1)$). Find the voltage as a function of time if the change $q(t)$ satisfies $q(0) = q'(0) = 0$. Plot the voltage drop across the capacitor as a function of time if $F_0 = 10$.

Spotlight on Modeling: Time Delays and Collisions

Comments:

Watch cars in a line at a stoplight accelerate as the light turns green. Some drivers are quick off the mark, others delay until the blast of a horn gets them going. In this Spotlight you will find that differential-delay equations provide one way to model this everyday dynamical event. We use the transform to solve the equations. The parameters of car/driver response time and sensitivity play a critical role in the model's prediction of a rear-end crash a few seconds after the light turns green.

PROBLEMS

Velocity Control: Two Cars.

1. *Velocity Control: A Rear-Ender?* Use the velocity control model in the text to decide which of the following sets of values of T and λT will result in a collision between the two cars at some time t in the time interval from T to $2T$ seconds? [*Hint*: see inequality (13).]

   (a) $T = 1$, $\lambda T = 20$ **(b)** $T = 5$, $\lambda T = 12.3$ **(c)** $T = 2$, $\lambda T = 14$

   Answer 1: The data in (a), (b) and (c) all satisfy inequality (13), so there is a collision in the time span $T \leq t < 2T$ after the light turns.

2. Show that if $T = 3$ sec, $\lambda = 2$ sec^{-1}, then a collision occurs between 6 and 9 sec after the light turns, but not before.

Velocity Control: Three Cars.

3. Suppose that there are three identical cars in line: each car is 15 ft long, there is a 5 ft separation between cars, all have sensitivity coefficient λ sec^{-1} and time delay T sec. The differential delay ODEs for the third car are $y_3' = v_3$, $v_3'(t+T) = \lambda[v_2(t) - v_3(t)]$, with initial data $y_3(0) = -40$, $v_3(0) = 0$, $v_3'(t) = 0$ for $0 \leq t \leq 2T$.

   (a) Show that $\mathcal{L}[v_3] = \dfrac{\lambda}{\lambda + se^{Ts}} \mathcal{L}[v_2] = \left(\dfrac{\lambda}{\lambda + se^{Ts}}\right)^2 \cdot \dfrac{\alpha}{s^2}$

   (b) Show that $\mathcal{L}[v_3] = \alpha\lambda^2 \left[\dfrac{e^{-2Ts}}{s^4} - \dfrac{2\lambda e^{-3Ts}}{s^5} + \dfrac{\lambda^2 e^{-4Ts}}{3s^6} - \cdots \right]$

   [*Hint*: first show that $\mathcal{L}[v_3] = \dfrac{\alpha\lambda^2}{s^4 e^{2Ts}} \left[1 + \dfrac{\lambda e^{-Ts}}{s} \right]^{-2}$ and then apply the Binomial Theorem to obtain a series expansion for $[\cdot]^{-2}$.]

   (c) Show that

   $$v_3 = \alpha\lambda^2 \left[\frac{1}{6}(t-2T)^3 \, \text{Step}(t, 2T) - \frac{\lambda}{12}(t-3T)^4 \, \text{Step}(t, 3T) \right.$$
   $$\left. + \frac{\lambda^2}{40}(t-4T)^5 \, \text{Step}(t, 4T) - \cdots \right]$$

   (d) Show that

   $$y_3(t) = \begin{cases} -40, & 0 \leq t \leq 2T \\[2mm] -40 + \dfrac{\alpha\lambda^2}{24}(t-2T)^4, & 2T \leq t < 3T \end{cases}$$

and that

$$
y_2(t) - y_3(t) = \begin{cases} 20, & 0 \le t \le T \\ 20 + \dfrac{1}{6}\alpha\lambda(t-T)^3, & T \le t \le 2T \\ 20 + \dfrac{1}{6}\alpha\lambda(t-T)^3 - \dfrac{1}{12}\alpha\lambda^2(t-2T)^4, & 2T \le t \le 3T \end{cases}
$$

(e) Suppose that $\alpha = 6$ ft/sec^2. For a time delay T, show that the third car rear-ends the second car for some value of t between $2T$ and $3T$ if the first and second cars have not collided and if $10 \le T^2(\lambda T - 16)\lambda T$.

(f) Show that if the first two cars don't collide during the time span $0 \le t \le 3T$, then $8 + 8(1 + 10/(64T^2))^{1/2} < \lambda T < 12 + 5/T^2$. Plot the following three curves in the $T, \lambda T$-plane: $\lambda T = 8 + 8(1 + 10/(64T^2))^{1/2}$, $\lambda T = 12 + 5/T^2$, and $T^2(\lambda T - 16)\lambda T = 10$ over the time span $0.4 \le T \le 0.7$. Shade the region of points $(T, \lambda T)$ where the second and third cars crash during the time span $0 \le t \le 3T$, if the first two cars haven't collided. Choose three points in the region and discuss the reality of each corresponding physical situation.

Answer 3:

(a) Taking the Laplace transform of $v_3'(t+T) = \lambda[v_2(t) - v_3(t)]$ we get

$$
\mathcal{L}[v_3'(t+T)] = e^{Ts}\mathcal{L}[v_3'(t)\,\text{Step}(t,T)] = \begin{cases} 0, & 0 \le t \le T \\ e^{Ts}\mathcal{L}[v_3'], & t \ge T \end{cases}
$$

From formula (3) in Theorem 5.2.2, $e^{Ts}\mathcal{L}[v_3'(t)] = e^{Ts}s\mathcal{L}[v_3]$, since $v_3(0) = 0$. Setting $\mathcal{L}[v_3] = \lambda[\mathcal{L}[v_2] - \mathcal{L}[v_3]]$ and solving for $\mathcal{L}[v_3]$ we get

$$
\mathcal{L}[v_3] = \frac{\lambda}{\lambda + se^{Ts}} \cdot \mathcal{L}[v_2]
$$

From this section we see that

$$
\mathcal{L}[v_2] = \frac{\lambda}{\lambda + se^{Ts}} \cdot \frac{\alpha}{s^2}
$$

Therefore,

$$
\mathcal{L}[v_3] = \left(\frac{\lambda}{\lambda + se^{Ts}}\right)^2 \cdot \frac{\alpha}{s^2}, \quad t \ge 2T
$$

Note that $\mathcal{L}[v_3]$ and v_3 are zero for $0 \le t \le 2T$.

(b) From **(a)**,

$$
\mathcal{L}[v_3] = \frac{\lambda^2\alpha}{s^4 e^{2Ts}}\left[1 + \frac{\lambda e^{-Ts}}{s}\right]^{-2}
$$

Apply the Binomial Theorem to get

$$
\mathcal{L}[v_3] = \alpha\lambda^2\left[\frac{e^{-2Ts}}{s^4} - \frac{2\lambda e^{-3Ts}}{s^5} + \frac{3\lambda^2 e^{-4Ts}}{s^6} + \cdots\right], \quad t \ge 2T
$$

(c) Using formula (2) of Theorem 5.2.2 and the table of Laplace transforms, we get

$$
\mathcal{L}^{-1}\left[e^{-2Ts}\frac{1}{s^4}\right] = \frac{(t-2T)^3}{6}\,\text{Step}(t-2T)
$$

Apply a similar formula to each term in $\mathcal{L}[v_3]$ to get

$$
v_3 = \alpha\lambda^2\left[\frac{(t-2T)^3}{6}\,\text{Step}(t-2T) - \frac{\lambda}{12}(t-2T)^4(t-3T) + \frac{\lambda^2}{40}(t-4T)^5\,\text{Step}(t-4T) - \cdots\right]
$$

Note that $v_3(t) = 0$ for $0 \le t \le 2T$.

(d) The formulas for $y_3(t)$ are obtained by integrating $v_3(t)$, as given in **(c)**. The formula for $y_2(t) - y_3(t)$ follows directly from the formula for y_3 and text formula (11) for y_2.

(e) Insert $t = 3T$ into the formula in **(d)** for $y_2 - y_3$. The desired inequality is just a rearrangement of the inequality $y_2(3T) - y_3(3T) \leq 15$. Since it is assumed that $y_2(t) - y_3(t) > 15$ for $0 \leq t \leq 2T$, there must have been a rear-ender sometime between $2T$ and $3T$.

Separation Control.

4. The distance between cars (i.e., their separation), rather than the relative velocity, may be taken as the control mechanism for cars at a stoplight. Assume for simplicity that the cars are "points" rather than the 15-ft cars used in the text.

   (a) Write out a car-following model in which the stimulus to the second driver is the separation from the lead car. Assume a response time of T sec and a sensitivity of λ sec^{-2}. [*Hint*: compare with the model in (1), and make the needed changes.]

   (b) Assume that the lead car accelerates from a stop with constant acceleration $\alpha = 6$ ft/sec^2 and that the initial separation between cars is 5 ft. Find the motion of the second car. Are there values of λ, given the delay time $T = 2$, that will result in a rear-end collision for some t^*, $2 \leq t^* \leq 4$? [*Hint*: as in the text, you will have to use a geometric series expansion before resorting to the inverse transform to get $v_2(t)$.]

 5. *Velocity/Separation Control*   Repeat Problem 4 but in a situation where the stimulus is a linear combination of the velocity control model treated in the text and the separation control model of Problem 4. Use sensitivity coefficients λ_1 and λ_2 for the respective models.

   Answer 5:   Group problem.

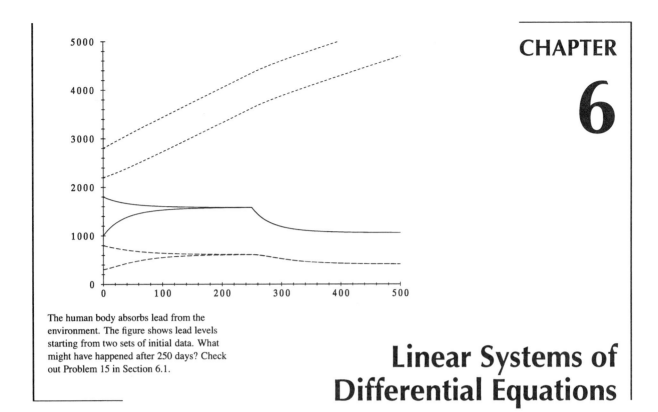

The human body absorbs lead from the environment. The figure shows lead levels starting from two sets of initial data. What might have happened after 250 days? Check out Problem 15 in Section 6.1.

Linear Systems of Differential Equations

It is hard to judge how much linear algebra readers of this text know before beginning a serious study of linear differential systems. There is a growing trend in many engineering curricula to drop linear algebra as a required course, expecting people to pick it up along the way in other courses. So some will have had a lot of exposure to linear algebra and a growing number may not. Our experience is that even students who have had a linear algebra course are often quite rusty when they have to use it. For these reasons, we have included several sections and Spotlights on matrices, systems of linear (algebraic) equations, determinants, eigenvalues, and eigenvectors. We have tried to make the matrix material as self-contained as possible, and there are examples of every kind of computation needed for the chapter. We recommend that these concepts be thoroughly reviewed.

After the section on eigenvalues, we treat undriven linear systems with constant coefficients, $x' = Ax$. In Section 6.5 we display a gallery of orbital portraits of planar systems with constant coefficients. This is a place to have some fun with solvers, exploring the behavior of the orbits of planar systems. Later, we introduce the exponential matrix as a convenient way of finding a formula for the solution of the IVP $x' = Ax$, $x(0) = x^0$. The exponential matrix comes in handy when the method of Variation of Parameters is used to solve the IVP $x' = Ax + F(t)$, $x(0) = 0$. An interesting feature of this chapter is our treatment of the steady-state solution where F is a constant or periodic driving term. The solutions of a linear system whose system matrix A has only eigenvalues with negative real parts have some appealing features. Finally, we extend everything to the general linear system $x' = A(t)x + F(t)$.

There are two models in this chapter. Lead in the human body is the subject of the entire first section. We study the low-pass filter (also used as a noise filter) in the WEB SPOTLIGHT ON MODELING: NOISE FILTER. The lead and filter models offer instructive examples of the sensitivity of linear systems to changing data.

6.1 Compartment Models: Tracking Lead

Comments:

This section is devoted to a single model as the title indicates. It provides an example of applying the Balance Law to a three-compartment model to obtain a linear differential system. Students seem to enjoy this model, perhaps because it shows the vulnerability of living in Southern California (at least in the bad old days of leaded gasoline and leaded paint). The model provides an excellent backdrop for illustrating the asymptotic approach to equilibrium, the effect of a discontinuous driving term, and the sensitivity of the system as parameters in the model change. In later sections, the text has more on the lead model. A discussion of general compartment models appears in this Resoure Manual at the end of this section. Here are two more references on the lead model: *Getting the Lead Out*, Irene Kessel and John T. O'Connor [Plenum, New York, 1997]; "Biological Modeling for Predictive Purposes," Naomi H. Harley, pp. 27–36 in *Methods for Biological Monitoring*, T.J. Kneip and J.V. Crable (editors) [American Public Health Association, Washington, 1988].

PROBLEMS

From Boxes and Arrows to ODEs. Model each linear cascade with a linear IVP based on the Balance Law and first-order rate laws for each box [*Hint*: see Example 6.1.1]. Solve and describe what happens in each compartment as $t \to +\infty$.

1. 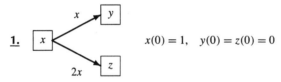 $x(0) = 1, \quad y(0) = z(0) = 0$

Answer 1: Analysis of the cascade model as in Example 6.1.1 produces the linear system $x' = -3x$, $y' = x$, $z' = 2x$. The system is solved from the top down. Using the techniques of Section 2.1, we see that the general solution to the first ODE is $x = Ce^{-3t}$. Since $x(0) = 1$, $C = 1$ and the solution is $x = e^{-3t}$. Substituting this into the other equations of the system, we have $y' = e^{-3t}$ and $z' = 2e^{-3t}$. Integrating each of these gives $y = -e^{-3t}/3 + C_1$ and $z = -2e^{-3t}/3 + C_2$. Using the initial data, $y(0) = 0 = -1/3 + C_1$, so $C_1 = 1/3$; and $z(0) = 0 = -2/3 + C_2$, so $C_2 = 2/3$. The solutions are $x = e^{-3t}$, $y = (1 - e^{-3t})/3$, and $z = 2(1 - e^{-3t})/3$. As $t \to +\infty$, $x \to 0$, $y \to 1/3$, and $z \to 2/3$.

2. $I = 1$ $x(0) = y(0) = z(0) = 0$

3. $I = 1 + \sin t$ $3x$ y $x(0) = y(0) = 0$

Answer 3: The system, now with a sinusoidal input, is $x' = 1 + \sin t - 3x$, $y' = 3x - y$. We will solve from the top down. Using the integrating factor e^{3t}, $(xe^{3t})' = (1 + \sin t)e^{3t}$. Integrating, $x = 1/3 + (3\sin t - \cos t)/10 + Ce^{-3t}$. Using $x(0) = 0$, we get $C = -7/30$ and

$$x = 1/3 + (3\sin t - \cos t)/10 - 7e^{-3t}/30$$

Since $y' = 3x - y$, another use of the integrating factor method (after replacing x by the expression in t just derived) gives $y = 1 + 3(\sin t - 2\cos t)/10 + 7e^{-3t}/20 + Ce^{-t}$. So, since $y(0) = 0$, $y =$

$1 + 3(\sin t - 2\cos t)/10 + 7e^{-3t}/20 - 3e^{-t}/4$. As $t \to +\infty$, x oscillates about $1/3$ and y oscillates about 1.

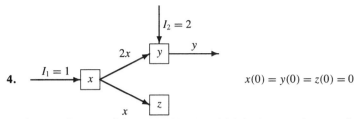

4. $x(0) = y(0) = z(0) = 0$

From ODEs to Boxes and Arrows. Sketch and label a boxes-and-arrows diagram that is described by the system of ODEs.

5. $x' = 5 - x$, $y' = x - 5y$ **6.** $x' = -1 - x + y$, $y' = -2y + x$

7. $x_1' = -k_{21}x_1 + k_{12}x_2$, $x_2' = I + k_{21}x_1 - k_{12}x_2$

8. $x' = -x/2$, $y' = 1 - y/3$, $z' = x/2 + y/3$

9. $x' = -x$, $y' = x/2 - 3y$, $z' = x/2 + 3y - 2z$

Answer 5: The system is $x' = 5 - x$, $y' = x - 5y$. The first compartment (i.e., the top of the cascade) is x, which has a constant rate input of 5 units of substance per unit of time. The substance leaves the x compartment at the rate of x units of substance per unit of time and enters the y compartment at the same rate. The substance leaves the y compartment at the rate of $5y$ units of substance per time unit. The corresponding compartment diagram is

Answer 7: The system is $x_1' = -k_{21}x_1 + k_{12}x_2$, $x_2' = I + k_{21}x_1 - k_{12}x_2$. The substance exits the x_1-compartment and enters the x_2-compartment at the rate $k_{21}x_1$. Also, the substance leaves the x_2-compartment and enters the x_1-compartment at rate $k_{12}x_2$. The substance enters the x_2-compartment at rate I (assuming $I > 0$). See the diagram.

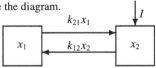

Answer 9: The system is $x' = -x$, $y' = x/2 - 3y$, $z' = x/2 + 3y - 2z$. The substance exits the top x compartment and enters the y compartment at rate $x/2$ and the z compartment also at the rate $x/2$ units of substance per unit of time. The substance also leaves the y compartment and enters the z compartment at the rate $3y$, and leaves the z compartment at the rate $2z$. The diagram is

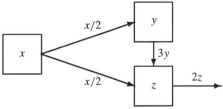

Method of Elimination. Find the general solution of each system. [*Hint*: solve the first ODE for x_2 in terms of x_1 and then eliminate x_2 from the second ODE.]

10. $x_1' = -3x_1 + x_2, \ x_2' = -4x_1 + 2x_2$ **11.** $x_1' = -3x_1 - 2x_2, \ x_2' = 2x_1 + x_2$

Answer 11: Solve the first ODE for x_2 to get $x_2 = (-x_1' - 3x_1)/2$. Use this to eliminate x_2 from the second ODE to obtain the second-order ODE $x_1'' + 2x_1' + x_1 = 0$ which has the characteristic roots $r_1 = r_2 = -1$. Theorem 3.2.1 says that the general solution of this second-order ODE is $x_1 = C_1 e^{-t} + C_2 t e^{-t}$, where C_1 and C_2 are arbitrary constants. So $x_1 = (-x_1' - 3x_1)/2 = -(C_1 + C_2/2)e^{-t} - C_2 t e^{-t}$, and we have found the general solution of the system.

Lead in the Body.

12. *Unique Equilibrium Levels* For any positive values of the input I and the rate constants k_{ji} show that system (3) has a unique equilibrium solution $x_1 = a, \ x_2 = b, \ x_3 = c$, where a, b, and c are all positive.

13. *Equilibrium Levels* Equilibrium levels for an autonomous system are constant solutions. For the lead-model system of ODEs in IVP (5), the amounts of lead in the blood and tissues quickly approach equilibrium levels. The same is true for lead in the bones, but it takes much longer. The outline below relates to how quickly the system approaches equilibrium.

(a) Find the equilibrium levels by setting the rate functions of IVP (5) equal to zero and solving for x_1, x_2, and x_3.

(b) Using IVP (5) with initial values of zero and a numerical solver, plot the component curves $x_i(t)$, $i = 1, 2, 3$ and estimate how long it takes for the lead in the blood to reach 85% of its equilibrium level. Repeat for tissues. Try to repeat for bones (it is incredibly long). Plot the component curves when the initial values are the equilibrium values?

(c) Replace the lead ingestion rate of 49.3 micrograms/day used in IVP (5) by the positive constant I_1 (unspecified). Now find the equilibrium levels in terms of I_1. Explain why the equilibrium levels are reduced by 50% if I_1 is reduced by 50%.

(d) For $I_1 = 10, 20, 30, 40, 50$ plot the $x_3(t)$-component curve, and then for each value of I_1, estimate the time it takes the amount of lead in the bones to reach 1000 micrograms.

Answer 13: We want to find the equilibrium points of the system in IVP (5):

$$x_1' = -0.0361x_1 + 0.0124x_2 + 0.000035x_3 + 49.3, \quad x_1(0) = 0$$
$$x_2' = 0.0111x_1 - 0.0286x_2, \quad x_2(0) = 0$$
$$x_3' = 0.0039x_1 - 0.000035x_3, \quad x_3(0) = 0$$

(a) Set the rate equations of IVP (5) equal to 0 and solve simultaneously for x_1, x_2, x_3, The equilibrium levels are found to be $x_1 \approx 1800$, $x_2 \approx 699$, and $x_3 \approx 200582$.

(b) Using the answers to (a), we see that 85% of the respective equilibrium levels of lead in the blood, tissues, and bones is 1530, 594, and 170496. If we plot each of the x_1, x_2, and x_3 component curves over an appropriate time span, and then zoom in on each curve over a time interval enclosing the time when the 85% level is attained, we get Graphs 1, 2, and 3. Reading from those graphs, we see that the 85% level is reached in the blood, tissues, and bones after (approximately) 147 days, 206 days, and 61500 days respectively. **Warning!** These are only approximations because the time interval is so long. More accurate answers would be obtained if the problem were first rescaled to dimensionless form. **Second warning!** Don't be fooled by the apparent leveling off of your component curves long before the equilibrium levels are reached. In exponential decay phenomena, the approach to equilibrium can be extremely slow near the equilibrium. We have chosen to use zoom graphs in order to show that the curves are actually still rising and to make it easier to find the desired times. **Third warning!** Your solver may crash trying to solve over the long time span of 75000 days (205 years) of lead accumulation. In any case, the subject of this case study would hardly be alive by then.

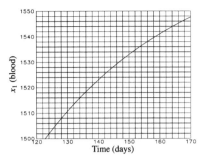

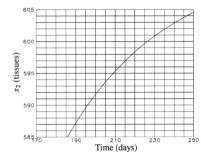

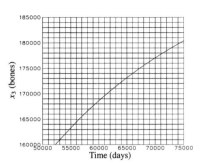

Problem 13(b), Graph 1.   Problem 13(b), Graph 2.   Problem 13(b), Graph 3.

(c) Replace 49.3 by I_1 in the above rate equations. The equilibrium levels, $x_1 = 36.5I_1$, $x_2 = 14.16I_1$, and $x_3 = 4068.6I_1$, are obtained by setting the right-hand sides of the ODEs in IVP (5) (with 49.3 replaced by I_1) equal to 0, and solving for x_1, x_2, and x_3. Since we are solving a linear algebraic system of the form $Ax = F$ with solution $x = \hat{x}$, say, then $Ax = F/2$ has solution $x = \hat{x}/2$ by linearity.

(d) See Graphs 1–5. From these graphs, we see that it takes approximately 850, 440, 305, 237, and 197 days, respectively, for the lead level in the bones to reach 1000 micrograms if $I_1 = 10, 20, 30, 40, 50$ micrograms/day.

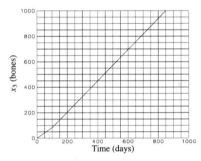

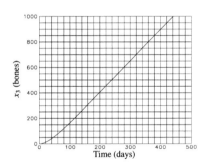

Problem 13(d), Graph 1 ($I_1 = 10$)   Problem 13(d), Graph 2 ($I_1 = 20$)

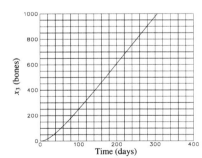

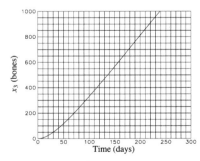

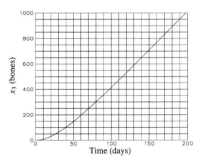

Problem 13(d), Graph 3 ($I_1 = 30$)   Problem 13(d), Graph 4 ($I_1 = 40$)   Problem 13(d), Graph 5 ($I_1 = 50$)

14. *Lead-Free Environment* Suppose that the initially lead-free subject in Example 6.1.1 [i.e., $x_1(0) = x_2(0) = x_3(0) = 0$] is exposed to lead for 400 days, and then removed to a lead-free environment

(Figure 6.1.5). Use a computer to estimate how long it takes the amount x_3 of lead in the bones to decline to 50% of $x_3(400)$; repeat for 25% and 10%.

15. *Get the Lead out of the Body*   Suppose that the lead intake level is $I_1 = 49.3$ micrograms/day, but a massive dose of lead-removal medication is given from the 400th day on.

(a) The effect of the medication is to increase the clearance coefficient k_{01} tenfold from 0.0211 to 0.211. We model this change by replacing k_{01} in the system (3) with $k_{01} = 0.0211(1 - \text{Step}(t, 400)) + 0.211\,\text{Step}(t, 400)$, while retaining the other values given in (4). Show graphically what happens.

(b) A safer dosage increases the effective value of k_{01} by 50%, from 0.0211 to 0.0316. Show graphically what happens to the lowered lead levels with this more realistic dosage.

Answer 15:

(a) Using this new value of k_{01} for $t \geq 400$, we see from the figure that massive medication from day 400 on causes a sharp drop in the lead levels in the blood and tissues.

(b) Replace k_{01} in system (3) by $0.0211(1 - \text{Step}(t, 400)) + 0.0316\,\text{Step}(t, 400)$, and let the other rate constants have the values given in data set (4). Using this new value for k_{01}, we see from the figure that a little medication from day 400 on lowers the lead levels in the blood and tissues only by a small amount.

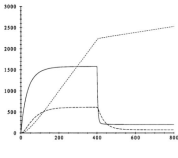

Problem 15(a).   Problem 15(b).

Other Linear Compartment Models.

16. The figure below shows a compartment system where a substance Q is injected at a rate of I_3 units of substance per unit of time into compartment 3. The substance Q moves out of compartment i into compartment j at a rate proportional to the amount $x_i(t)$ of Q in compartment i; the constant of proportionality is k_{ji}.

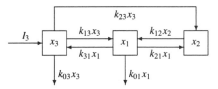

(a) Use the data to find the matrix A and driving function vector F of the linear system $x' = Ax + F$ for this compartment model using the data: $I_3 = 1$, $k_{31} = 3$, $k_{21} = 1$, $k_{01} = 16$, $k_{13} = 1$, $k_{03} = 5$, $k_{23} = 2$, $k_{12} = 4$.

(b) Find the equilibrium levels of Q in each of the three compartments.

(c) Plot the three component curves over a long enough time span that each is within 90% of its equilibrium level. Assume that $t \geq 0$ and $x_1(0) = x_2(0) = x_3(0) = 0$.

17. *The Tracer Inulin*   Water is the major component of blood and this water moves out of the bloodstream into the urinary system, but also moves back and forth between the blood (compartment 1) and intercellular areas (compartment 2). Since molecules of inulin attach to molecules of water and x-rays can detect inulin, we can track this motion of the water by injecting inulin into the blood.

(a) Draw a boxes-and-arrows diagram for the compartment model. Assume that inulin leaves compartment i and enters compartment j at the rate $k_{ji}x_i(t)$ for some constant $k_{ji} > 0$, where $x_i(t)$ is the amount of inulin in compartment i at time t. If inulin is injected into the bloodstream at rate I, find ODEs that model inulin levels in the system.

(b) Find the equilibrium levels of inulin in the blood and in the intercellular areas in terms of the rate constants and I (assume that I is a positive constant).

(c) Graph the solutions $x_1(t)$ and $x_2(t)$ of your IVP if

$$I = 1, \quad k_{21} = 0.01, \quad k_{12} = 0.02, \quad k_{01} = 0.005, \quad x_1(0) = x_2(0) = 0$$

where I is measured in grams/hour, k_{ij} in $(\text{hour})^{-1}$, and x_i in grams. Use a long enough time span that the levels of inulin in the blood and in the intercellular areas have reached at least 90% of their equilibrium values.

Answer 17:

(a) Here is the compartment diagram for the flow of inulin through the bloodstream (compartment 1) and the intercellular areas (compartment 2):

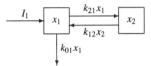

Suppose that $x_1(t)$, $x_2(t)$ are the respective amounts of inulin at time t in the blood and in the intercellular area. The Balance Law (1) implies that $x_1' = I_1 - (k_{21} + k_{01})x_1 + k_{12}x_2$, $x_2' = k_{21}x_1 - k_{12}x_2$.

(b) The equilibrium levels of inulin in the blood and intercellular areas are found by setting $x_1' = 0$ and $x_2' = 0$ and solving for x_1 and x_2. This gives $x_1 = I_1/k_{01}$ and $x_2 = (k_{21}/k_{12})(I_1/k_{01})$.

(c) The respective equilibrium levels of inulin from the formulas of **(b)** are $x_1 = 200$, $x_2 = 100$. The x_1 curve (solid) and x_2 curve (dashed) in the figure are graphed over the time span $0 \le t \le 1500$, but each reaches the 90% equilibrium level long before $t = 1500$ hours (around $t = 750$).

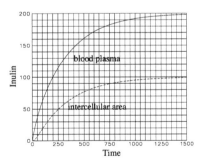

Problem 17(c).

Background Material: Compartment Models

The lead system is an example of a compartment model in which every compartment is "open to the environment." That is, the environment outside the three compartments of blood, tissues, and bones can be reached by a directed sequence of arrows from each compartment. Let's give a definition of a general compartmental matrix, and then specialize that to an open compartmental matrix, a particular example of which is the lead model coefficient matrix.

The $n \times n$ square matrix $A = [a_{ij}]$ of real numbers is a *compartmental matrix* if:

- Every diagonal entry in A is nonpositive.

- Every nondiagonal entry in A is nonnegative.

- The sum of the i-th column entries, $\sum_{j=1}^{n} a_{ji}$, is nonpositive.

Here is why our experience with compartment models in the SPOTLIGHT ON MODELING: COLD MEDICATION II, Section 6.1, and the WEB SPOTLIGHT ON MODELING: NOISE FILTER suggests that these three conditions are reasonable. Thinking of the matrix we would construct for an assemblage of boxes and arrows, we see that the first condition corresponds to the fact that the total rate out of each compartment and into other compartments and the environment must be nonpositive just because it is an *exit* rate. The second condition says that a_{ji}, $j \neq i$, which measures the entrance rate into compartment j from compartment i can't be negative (after all, it is an *entrance* rate). The third condition is less evident. But, $a_{ii}x_{ii}$ measures the rate of everything leaving compartment i. So, $a_{ii} = -\sum_{j=1, j\neq i}^{n} a_{ji} - a_{0i}$, where a_{0i} is the exit rate constant from compartment i to the environment. Since a_{0i}, a_{ji} ($j \neq i$) are all nonnegative, a_{ii} is nonpositive.

We have the following theorem about the eigenvalues of any compartmental matrix.

☞ Check that the system matrix of System (3) in Section 6.1 meets the three conditions for a compartmental matrix.

Compartmental Matrix Theorem

All eigenvalues of a compartmental matrix have nonpositive real parts, and none is pure imaginary.

We can justify this very important result by using the Gerschgorin disks of the WEB SPOTLIGHT ON LOCATING EIGENVALUES. As noted above, every diagonal entry a_{ii} is nonpositive and $a_{ii} = -\sum_{j=1, \neq i}^{n} a_{ji} - a_{0i}$. But the a_{ji} terms, $j \neq i$, are nonnegative, and so $\sum_{j=1, \neq i}^{n} a_{ji}$ is just the radius of the i-th Gerschgorin column disk whose center is at $a_{ii} \leq 0$ on the real line in the complex plane. This means that the i-th column disk lies entirely inside the left half of the complex plane, except possibly for the origin, which lies on the edge of the disk if $a_{0i} = 0$. This proves the theorem.

It is also known that if A is a compartmental matrix and $x_i(0) \geq 0$, then $x_i(t) \geq 0$ for all $t \geq 0$, $i = 1, \ldots, n$, if $x(t) = [x_1(t) \ldots x_n(t)]^T$ is a solution of $x' = Ax$.

Now let's define a compartmental system (and matrix) where every compartment is "open to the environment." If for every i there is a directed chain of arrows with positive exit rate constants which leads from compartment i to the external environment, then the system is *open to the environment*. The corresponding compartmental matrix A is *open to the environment* (often called an *open compartmental matrix*) if there is a chain of compartments $j_1, \ldots, j_r$ for which $a_{j_1 i}, a_{j_2 j_1}, \ldots$, and $a_{0 j_r}$ are positive. It isn't hard to verify that the lead matrix is open. We have the following theorem (proof omitted) for any open compartmental matrix:

Washout Theorem

If A is an open compartmental matrix, then all its eigenvalues have negative real parts, so all solutions of $x' = Ax$ tend to 0 as $t \to +\infty$.

We would expect the Washout Theorem to be true because the substance being tracked through the system leads out into the environment from each compartment either directly, or indirectly through a chain of other compartments.

All of this can be extended to compartmental systems with "traps," which are compartments or groups of compartments which "absorb," but never disgorge. In a way, the outside environment is a trap, and we could add it to the set of boxes of the model as an additional compartment (the "zero-th" compartment). The enlarged model is no longer open; in fact, it is a *closed compartmental model*. The corresponding matrix A is singular (i.e., 0 is an eigenvalue), and it can be shown that each solution $x(t)$ of $x' = Ax$ tends to some equilibrium point b, where $b_i \geq 0$ if $x_i(0) \geq 0$, $i = 1, \ldots, n$.

Well, that's all we have to say here about compartment models, but here are some sources for further study:

1. *Linear Models in Biology*, M.R. Cullen [Ellis Horwood Limited, Chichester, 1985].

2. *Compartmental Modeling and Tracer Kinetics*, David H. Anderson [v. 50 of *Lecture Notes in Biomathematics*, Springer-Verlag, Berlin].

3. *Compartmental Analysis in Biology and Medicine*, John A. Jacquez [Univ. of Michigan Press, Ann Arbor, 1985 (2nd ed.)].

6.2 Eigenvalues, Eigenvectors, and Eigenspaces of Matrices

Comments:

Look at the discussion just before Example 6.2.1 about the need to do eigen-analysis in order to generate the solution set of $x' = Ax$, where A is a constant matrix. Hence, everyone needs to understand eigenspaces and eigenvalues (and at least have a basic idea about generalized eigenvectors). Most of this section is devoted to the process of finding eigenvalues and eigenvectors. Students sometimes have trouble finding a basis for an eigenspace, probably because of having to deal with $(A - \lambda I)v = 0$, where $A - \lambda I$ is singular. Several examples will usually clear this up. See the worked-out examples in the SPOTLIGHT ON LINEAR ALGEBRAIC EQUATIONS.

PROBLEMS

Eigenvalues and Eigenspaces.   Find the eigenvalues and their multiplicities. Find a basis of eigenvectors for each eigenspace and identify any deficient eigenspace.

1. $\begin{bmatrix} 1 & 0 \\ 2 & 1 \end{bmatrix}$ 2. $\begin{bmatrix} 1 & 1 \\ 1 & 1 \end{bmatrix}$ 3. $\begin{bmatrix} 0 & 3 \\ 3 & 0 \end{bmatrix}$

4. $\begin{bmatrix} 4 & 2 \\ -1 & 1 \end{bmatrix}$ 5. $\begin{bmatrix} 6 & -7 \\ 1 & -2 \end{bmatrix}$ 6. $\begin{bmatrix} 1 & -2 \\ 1 & 4 \end{bmatrix}$

7. $\begin{bmatrix} 3 & -5 \\ 5 & 3 \end{bmatrix}$ 8. $\begin{bmatrix} -1 & 2 \\ -2 & -1 \end{bmatrix}$ 9. $\begin{bmatrix} 4 & -6 \\ 1 & -3 \end{bmatrix}$

☞ For Problem 12 see Example 6.2.6. For Problem 14 see Example 6.2.5.

10. $\begin{bmatrix} 5 & 3 \\ -6 & -4 \end{bmatrix}$ 11. $\begin{bmatrix} 1 & 0 & 1 \\ 0 & 1 & 0 \\ 1 & 0 & 1 \end{bmatrix}$ 12. $\begin{bmatrix} 2 & 2 & 1 \\ 1 & 3 & 1 \\ 1 & 2 & 2 \end{bmatrix}$

13. $\begin{bmatrix} -1 & 36 & 100 \\ 0 & -1 & 27 \\ 0 & 0 & 5 \end{bmatrix}$ 14. $\begin{bmatrix} -3 & 1 & -2 \\ 0 & -1 & -1 \\ 2 & 0 & 0 \end{bmatrix}$ 15. $\begin{bmatrix} 5 & -6 & -6 \\ -1 & 4 & 2 \\ 3 & -6 & -4 \end{bmatrix}$ 16. $\begin{bmatrix} -1 & 1 & 0 \\ -1 & 0 & 1 \\ 1 & 0 & -2 \end{bmatrix}$

Answer 1:   The characteristic polynomial is

$$p(\lambda) = \det \begin{bmatrix} 1-\lambda & 0 \\ 2 & 1-\lambda \end{bmatrix} = (1-\lambda)^2$$

Hence, $\lambda = 1$ is a double eigenvalue. The corresponding eigenspace V_1 may have dimension 1 or 2. In this case the eigenvector equation $(A - I)v = 0$ reduces to the single equation $2v_1 = 0$. V_1 is spanned by the eigenvector $[0 \;\; 1]^T$ and is one-dimensional and is, hence, deficient.

Answer 3:   The characteristic polynomial is

$$p(\lambda) = \det \begin{bmatrix} -\lambda & 3 \\ 3 & -\lambda \end{bmatrix} = \lambda^2 - 9 = (\lambda - 3)(\lambda + 3)$$

Hence, the eigenvalues 3 and -3 are simple. V_3 is spanned by $[1 \ 1]^T$, V_{-3} by $[1 \ -1]^T$.

Answer 5: The characteristic polynomial is

$$p(\lambda) = \det \begin{bmatrix} 6-\lambda & -7 \\ 1 & -2-\lambda \end{bmatrix} = (6-\lambda)(-2-\lambda)+7 = \lambda^2 - 4\lambda - 5 = (\lambda+1)(\lambda-5)$$

Hence, the eigenvalues -1 and 5 are simple. V_{-1} is spanned by $[1 \ 1]^T$ because the eigenvector equation $(A+I)v$ reduces to $7v_1 - 7v_2 = 0$. V_5 is spanned by $[7 \ 1]^T$.

Answer 7: The characteristic polynomial is

$$p(\lambda) = \det \begin{bmatrix} 3-\lambda & -5 \\ 5 & 3-\lambda \end{bmatrix} = (3-\lambda)^2 + 25 = \lambda^2 - 6\lambda + 34$$

Hence, the simple eigenvalues are the complex conjugates $3 + 5i$ and $3 - 5i$. V_{3+5i} is spanned by $[1 \ -i]^T$ and V_{3-5i} by the conjugate $[1 \ i]^T$.

Answer 9: The characteristic polynomial is

$$p(\lambda) = \det \begin{bmatrix} 4-\lambda & -6 \\ 1 & -3-\lambda \end{bmatrix} = (4-\lambda)(-3-\lambda)+6 = \lambda^2 - \lambda - 6$$

Hence, the eigenvalues are 3 and -2. We then have that V_3 is spanned by $[6 \ 1]^T$, and V_{-2} is spanned by $[1 \ 1]^T$.

Answer 11: The characteristic polynomial is

$$p(\lambda) = \det \begin{bmatrix} 1-\lambda & 0 & 1 \\ 0 & 1-\lambda & 0 \\ 1 & 0 & 1-\lambda \end{bmatrix} = (1-\lambda)^3 - (1-\lambda) = (1-\lambda)(\lambda^2 - 2\lambda)$$

Hence, the eigenvalues 0, 1, 2 are simple. V_0 has basis vector $[1 \ 0 \ -1]^T$, V_1 has basis vector $[0 \ 1 \ 0]^T$, and V_2 has the basis vector $[1 \ 0 \ 1]^T$.

Answer 13: The characteristic polynomial is

$$p(\lambda) = \det \begin{bmatrix} -1-\lambda & 36 & 100 \\ 0 & -1-\lambda & 27 \\ 0 & 0 & 5-\lambda \end{bmatrix} = (-1-\lambda)^2(5-\lambda)$$

Hence, -1 is a double eigenvalue and 5 is a simple eigenvalue. V_{-1} is spanned by $[1 \ 0 \ 0]^T$ and V_5 is spanned by $[262 \ 27 \ 6]^T$ since the eigenvector equation $(A - 5I)v = 0$ reduces to $-6v_1 + 36v_2 + 100v_3 = 0$, $-6v_2 + 27v_3 = 0$.

Answer 15: Use cofactors to expand $\det(A - \lambda I)$:

$$\det(A - \lambda I) = \det \begin{bmatrix} 5-\lambda & -6 & -6 \\ -1 & 4-\lambda & 2 \\ 3 & -6 & -4-\lambda \end{bmatrix}$$

(see the SPOTLIGHT ON VECTORS, MATRICES, INDEPENDENCE) to obtain

$$p(\lambda) = -\lambda^3 + 5\lambda^2 - 8\lambda + 4 = -(\lambda-1)(\lambda-2)^2$$

So 1 is a simple eigenvalue and 2 is a double eigenvalue. V_1 is spanned by $[3 \ -1 \ 3]^T$ and V_2 has basis $\{[2 \ 1 \ 0]^T, [2 \ 0 \ 1]^T\}$. Any two independent vectors in V_2 form a basis, so there is nothing unique about the given basis.

Working with Eigenvalues.

17. *Eigenvalues of cA*   Characterize the eigenvalues of the matrix cA, where c is a nonzero constant, in terms of the eigenvalues of A.

Answer 17: We know that $Av = \lambda v$, where v is any eigenvector of A corresponding to eigenvalue λ. Hence, $(cA)v = (c\lambda)v$ and so $c\lambda$ is an eigenvalue of cA corresponding to the same eigenvector *v.* Conversely, if μ is an eigenvalue of cA corresponding to the eigenvector w, then $(cA)w = \mu w$. Dividing by c, $Aw = (\mu/c)w$ and so μ/c is an eigenvector of A corresponding to the eigenvector w. Hence, the eigenvalues of cA are just c times the eigenvalues of A.

18. *Eigenvalues of a Triangular Matrix* Show that the eigenvalues of a triangular matrix A are the diagonal entries, the multiplicity of each eigenvalue equaling the number of times the eigenvalue appears on the diagonal. [*Hint*: in this case, det A is the product of the diagonal entries in $A - \lambda I$.]

☞ A^k for a positive integer k is the product of k factors of A.

19. *Eigenvalues of A^k* Show that if λ is an eigenvalue of A, then λ^k is an eigenvalue of A^k. If μ is an eigenvalue of A^k, is $\mu^{1/k}$ always an eigenvalue of A?

Answer 19: If $Av = \lambda v$ for some vector $v \neq 0$, then $A^k v = A^{k-1}(Av) = A^{k-1}(\lambda v) = \lambda A^{k-1}(v) = \cdots = \lambda^k v$. So λ^k is an eigenvalue of A^k if λ is an eigenvalue of A. The other way around, however, is not always true. For example, $A = \begin{bmatrix} 2 & 0 \\ 0 & -3 \end{bmatrix}$ has eigenvalues 2 and -3, while $A^2 = \begin{bmatrix} 4 & 0 \\ 0 & 9 \end{bmatrix}$ has eigenvalues 4 and 9, but $+3 = 9^{1/2}$ is *not* an eigenvalue of A.

20. *Eigenvalues of A^T* Show that a square matrix A and its transpose A^T have the same characteristic polynomial, and so the same eigenvalues with the same multiplicities. [*Hint*: expand $\det(A - \lambda I)$ on some row and $\det(A^T - \lambda I)$ on the corresponding column.]

21. *Use Eigenvalues to Find* det A Suppose that $\lambda_1, \ldots, \lambda_n$ are the eigenvalues of a matrix A, each repeated according to its multiplicity. Show that $\det A = \lambda_1 \cdots \lambda_n$. [*Hint*: write $p(\lambda)$ as $(-1)^n(\lambda - \lambda_1) \cdots (\lambda - \lambda_n)$, expand, and compare with the form of $p(\lambda)$ in formula (12).]

Answer 21: We have $\det[A - \lambda I] = p(\lambda) = (-1)^n(\lambda - \lambda_1) \cdots (\lambda - \lambda_n)$ if $\lambda_1, \ldots, \lambda_n$ are the eigenvalues of A, each repeated according to its multiplicity (see the formula in the table on the inside back cover). So $p(0) = (-1)^n(-\lambda_1) \cdots (-\lambda_n) = \lambda_1 \cdots \lambda_n = \det(A - 0I) = \det A$. On the other hand, the constant term of $p(\lambda) = \det(A - \lambda I)$ is $p(0) = \det A$, so $\det A$ is the product of its eigenvalues.

22. *Trace* The *trace* of an $n \times n$ matrix A is the sum of the diagonal terms.

$A = \begin{bmatrix} 1 & 5 & 3 \\ 6 & -7 & 10 \\ 37 & 56 & 2 \end{bmatrix}$

(a) Show that trace A is the sum of the eigenvalues of A. [*Hint*: see formula (12).]

(b) Use **(a)** to prove that the matrix A in the margin has an eigenvalue with a negative real part without actually finding the characteristic polynomial and its roots.

23. If A is an $n \times n$ matrix, show that A is nonsingular if and only if $\lambda = 0$ is not an eigenvalue.

Answer 23: By Problem 21, $\det A = \lambda_1 \cdots \lambda_n$, where $\lambda_1, \ldots, \lambda_n$ are the eigenvalues of A. Since A is nonsingular if and only if its determinant is nonzero, we see that A is nonsingular if and only if none of its eigenvalues are zero.

Generalized Eigenbases. Find a generalized eigenbasis for $\mathbb{R}^2$ in Problems 24–27.

24. $\begin{bmatrix} 1 & -1 \\ 1 & 3 \end{bmatrix}$ **25.** $\begin{bmatrix} 4 & -2 \\ 2 & 0 \end{bmatrix}$ **26.** $\begin{bmatrix} 2 & -1 \\ 1 & 4 \end{bmatrix}$ **27.** $\begin{bmatrix} 1 & -2 \\ 0 & 1 \end{bmatrix}$

Answer 25: The characteristic polynomial $p(\lambda) = \lambda^2 - 4\lambda + 4 = (\lambda - 2)^2$ has a double root corresponding to the multiplicity two eigenvalue $\lambda_1 = 2$. Solving $(A - \lambda_1 I)v = 0$ gives only a one dimensional eigenspace with eigenvectors of the form $[k \ k]^T$ for some scalar k. We will use $v = [1 \ 1]^T$ to form our generalized eigenbasis for $\mathbb{R}^2$. Solving $(A - \lambda_1 I)u = v$, to find our generalized eigenvector u, gives generalized eigenvectors of the form $[j \ (2j-1)/2]^T$ for some arbitrary scalar j. Our generalized eigenbasis, will then be $[1 \ 1]^T, [1 \ 1/2]^T$.

Answer 27: The characteristic polynomial is

$$p(\lambda) = \det \begin{bmatrix} 1 - \lambda & -2 \\ 0 & 1 - \lambda \end{bmatrix} = (1 - \lambda)^2$$

Hence, $\lambda_1 = 1$ has multiplicity 2. We then have that V_1 is spanned by $[1 \ 0]^T$, implying that V_1 is deficient. We use v to find a generalized eigenvector u as a solution of the system $(A - I)u = v$; that is,

$$\begin{bmatrix} 0 & -2 \\ 0 & 0 \end{bmatrix} u = \begin{bmatrix} 1 \\ 0 \end{bmatrix}$$

Hence, $u = [0 \ -1/2]^T$ is a generalized eigenvector and a generalized eigenbasis is $\{u, v\}$.

28. Let the 2×2 matrix A have a double eigenvalue λ_1 with $\dim V_{\lambda_1} = 1$. If u^1 is an eigenvector for A, show that there is a vector u^2 such that $(A - \lambda_1 I)u^2 = u^1$. [*Hint:* if w is any vector such that $(A - \lambda_1 I)w \neq 0$, then $\{u^1, w\}$ is an independent set.]

6.3 Undriven Linear Differential Systems: Real Eigenvalues

Comments:

With the matrix and eigen machinery of Section 6.2 and the SPOTLIGHT ON VECTORS, MATRICES, INDEPENDENCE in hand, we can solve undriven, constant coefficient linear systems of ODEs, and that is what we do in this section. We show how the IVP $x' = Ax$, $x(0) = x^0$, can be solved if A is a matrix of constants which has all real eigenvalues and the eigenvalues of A (and their multiplicities) as well as the corresponding eigenspaces are known. The only real algebraic problem in all this occurs when the dimension of the eigenspace corresponding to a multiple eigenvalue is less than the multiplicity of the eigenvalue. The generalized eigenvectors introduced in Section 6.2 come to the rescue in this case. We work out a few examples using a chain of generalized eigenvectors. We don't dwell on this awkward case, but focus on the more common situation of full eigenspaces (we call these eigenspaces nondeficient in the text).

PROBLEMS

General Solutions. Find the general solution for each of the systems below.

1. $x_1' = 5x_1 + 3x_2$; $\ x_2' = -x_1 + x_2$
2. $x_1' = 5x_1 - 2x_2$; $\ x_2' = -2x_1 + 8x_2$
3. $x_1' = x_1 - x_2$; $\ x_2' = x_1 + 3x_2$
4. $x_1' = x_1 + 2x_2$; $\ x_2' = 2x_1 + 4x_2$
5. $x_1' = 3x_1 + 4x_2$; $\ x_2' = 2x_2 + x_1$
6. $x_1' = x_1/2 + 9x_2$; $\ x_2' = x_1/2 + 2x_2$

Answer 1: The system matrix is $A = \begin{bmatrix} 5 & 3 \\ -1 & 1 \end{bmatrix}$. Since $\det(A - \lambda I) = p(\lambda) = \lambda^2 - 6\lambda + 8 = (\lambda - 2)(\lambda - 4)$, the eigenvalues are $\lambda_1 = 2$ and $\lambda_2 = 4$. To find the eigenspace for λ_1, note that $(A - 2I)v = 0$ reduces to $3v_1 + 3v_2 = 0$, so V_2 is spanned by $[1 \ -1]^T$. For λ_2, note that $(A - 4I)v = 0$ reduces to $v_1 + 3v_2 = 0$ so V_4 is spanned by $[3 \ -1]^T$. The general real solution of the given system is

$$x = c_1 \begin{bmatrix} 1 \\ -1 \end{bmatrix} e^{2t} + c_2 \begin{bmatrix} 3 \\ -1 \end{bmatrix} e^{4t} \qquad c_1, c_2 \text{ arbitrary reals}$$

Answer 3: The system matrix is $A = \begin{bmatrix} 1 & -1 \\ 1 & 3 \end{bmatrix}$ and since $\det(A - \lambda I) = p(\lambda) = (\lambda - 2)^2$, $\lambda_1 = 2$ is the only eigenvalue and it has multiplicity $m_1 = 2$. To find the eigenspace for λ_1, note that $(A - 2I)v = 0$ becomes $v_1 + v_2 = 0$. So the eigenspace for λ_1 is spanned by $[1 \ -1]^T$. So $x^1 = e^{2t}[1 \ -1]^T$ is one solution of the given system. Since the eigenspace is deficient, we need to find a generalized eigenvector in order to construct a fundamental set. Any vector w which solves $(A - 2I)w = [1 \ -1]^T$ is a generalized eigenvector. Solving, we have $w = [0 \ -1]^T$ as one such vector. So the general solution is

$$x = c_1 \begin{bmatrix} 1 \\ -1 \end{bmatrix} e^{2t} + c_2 \left(\begin{bmatrix} 0 \\ 1 \end{bmatrix} + t \begin{bmatrix} 1 \\ -1 \end{bmatrix} \right) e^{2t}$$

where c_1 and c_2 are arbitrary reals.

Answer 5: Since

$$\det \begin{bmatrix} 3-\lambda & 4 \\ 2 & 1-\lambda \end{bmatrix} = \lambda^2 - 4\lambda - 5 = (\lambda - 5)(\lambda + 1)$$

it follows that $\lambda_1 = 5$, $\lambda_2 = -1$, are the eigenvalues of the system matrix. For $\lambda_1 = 5$ we get an eigenvector $u = [2 \ 1]^T$, and for $\lambda_2 = -1$ we get an eigenvector $v = [1 \ -1]^T$. Hence,

$$x = c_1 \begin{bmatrix} 2 \\ 1 \end{bmatrix} + c_2 \begin{bmatrix} 1 \\ -1 \end{bmatrix} e^{-t}$$

where c_1 and c_2 are arbitrary constants, is the general solution of the system.

Solving IVPs. Solve the IVP with $x_1(0) = 1$, $x_2(0) = 1$. What happens as $t \to +\infty$?

<u>**7.**</u> Problem 1   **8.** Problem 2   **9.** Problem 3

10. Problem 4   **11.** Problem 5   **12.** Problem 6

Answer 7: Imposing the initial conditions $x_1(0) = 1$, $x_2(0) = 1$ on the general solution given in Problem 1, we have that c_1 and c_2 must satisfy the condition

$$\begin{bmatrix} 1 \\ 1 \end{bmatrix} = c_1 \begin{bmatrix} 1 \\ -1 \end{bmatrix} + c_2 \begin{bmatrix} 3 \\ -1 \end{bmatrix}$$

or in component form, $1 = c_1 + 3c_2$, $1 = -c_1 - c_2$. Solving this system, we obtain $c_1 = -2$ and $c_2 = 1$, and so the unique solution of the given IVP is given by $x = -2[1 \ -1]^T e^{2t} + [3 \ -1]^T e^{4t}$. As $t \to +\infty$, $x(t)$ becomes unbounded.

Answer 9: Imposing the initial conditions on the general solution derived in Problem 3, we obtain

$$\begin{bmatrix} 1 \\ 1 \end{bmatrix} = c_1 \begin{bmatrix} 1 \\ -1 \end{bmatrix} + c_2 \begin{bmatrix} 0 \\ -1 \end{bmatrix}$$

or in component form $1 = c_1$, $1 = -c_1 - c_2$. Solving, we have $c_1 = 1$ and $c_2 = -2$, and so the unique solution of the given IVP is

$$x = e^{2t}[1 \ -1]^T - 2e^{2t}([0 \ -1]^T + t[1 \ -1]^T)$$

As $t \to +\infty$, $x(t)$ becomes unbounded.

Answer 11: Imposing the initial conditions on the general solution derived in Problem 5,

$$\begin{bmatrix} 1 \\ 1 \end{bmatrix} = c_1 \begin{bmatrix} 2 \\ 1 \end{bmatrix} + c_2 \begin{bmatrix} 1 \\ -1 \end{bmatrix}$$

Write the vector equation as the system $1 = 2c_1 + c_2$, $1 = c_1 - c_2$, which has the unique solution $c_1 = 2/3$, $c_2 = -1/3$. Hence, the solution of the IVP is $x = (2/3)[2 \ 1]^T e^{5t} - (1/3)[1 \ -1]^T e^{-t}$. As $t \to +\infty$, $x(t)$ becomes unbounded.

Nondeficient Eigenspaces. Find all solutions of $x' = Ax$, where the matrix A is as given. Write your answers in the form given in the general solution formula (4).

13. $\begin{bmatrix} 0 & 1 \\ 8 & -2 \end{bmatrix}$ <u>**14.**</u> $\begin{bmatrix} 3 & -2 \\ 2 & -2 \end{bmatrix}$ **15.** $\begin{bmatrix} 2 & 1 \\ -3 & 6 \end{bmatrix}$ **16.** $\begin{bmatrix} -8 & 4 \\ -10 & 4 \end{bmatrix}$

17. $\begin{bmatrix} -1 & 4 \\ 2 & 1 \end{bmatrix}$ **18.** $\begin{bmatrix} 3 & 2 & 2 \\ 1 & 4 & 1 \\ -2 & -4 & -1 \end{bmatrix}$ **19.** $\begin{bmatrix} 2 & 2 & 1 \\ 1 & 3 & 1 \\ 1 & 2 & 2 \end{bmatrix}$ **20.** $\begin{bmatrix} -1 & 1 & 0 \\ 1 & 2 & 1 \\ 0 & 3 & -1 \end{bmatrix}$

Answer 13: The eigenvalues of the system matrix $\begin{bmatrix} 0 & 1 \\ 8 & -2 \end{bmatrix}$ are $\lambda_1 = -4$, $\lambda_2 = 2$. The eigenspace for λ_1 is spanned by $[1 \ -4]^T$, and the eigenspace for λ_2 is spanned by $[1 \ 2]^T$. The general real-valued solution of the system is $x = c_1 e^{-4t}[1 \ -4]^T + c_2 e^{2t}[1 \ 2]^T$, where c_1 and c_2 are arbitrary reals.

Answer 15:   The eigenvalues of the system matrix $\begin{bmatrix} 2 & 1 \\ -3 & 6 \end{bmatrix}$ are $\lambda_1 = 5$ and $\lambda_2 = 3$, and the corresponding eigenspaces are spanned, respectively, by $[1\ 3]^T$ and $[1\ 1]^T$. So the general real-valued solution is $x = c_1 e^{5t}[1\ 3]^T + c_2 e^{3t}[1\ 1]^T$, where c_1 and c_2 are arbitrary reals.

Answer 17:   The eigenvalues of the system matrix $\begin{bmatrix} -1 & 4 \\ 2 & 1 \end{bmatrix}$ are $\lambda_1 = 3$ and $\lambda_2 = -3$, and the corresponding eigenspaces are spanned, respectively, by $[1\ 1]^T$ and $[2\ -1]^T$. The general real-valued solution is $x = c_1 e^{3t}[1\ 1]^T + c_2 e^{-3t}[2\ -1]^T$, where c_1 and c_2 are arbitrary reals.

Answer 19:   The characteristic polynomial of the system matrix $\begin{bmatrix} 2 & 2 & 1 \\ 1 & 3 & 1 \\ 1 & 2 & 2 \end{bmatrix}$ is $p(\lambda) = -(\lambda -$ $1)^2(\lambda - 5)$, and so the eigenvalue $\lambda_1 = 1$ has multiplicity 2, and $\lambda_2 = 5$ is a simple eigenvalue. The eigenspace corresponding to $\lambda_1 = 1$ has dimension 2 (and is nondeficient) and has the basis $\{[1\ 0\ -1]^T, [2\ -1\ 0]^T\}$. The eigenspace corresponding to $\lambda_2 = 5$ is spanned by $[1\ 1\ 1]^T$. The general real-valued solution is

$$x = c_1[1\ 0\ -1]^T e^t + c_2[2\ -1\ 0]^T e^t + c_3[1\ 1\ 1]^T e^{5t}$$

where c_1, c_2, c_3 are arbitrary reals.

IVPs.   Solve the IVP with $x(0) = [-1\ 1]^T$ or $[1\ 1\ 1]^T$ as appropriate. Plot component curves.

21. Problem 13	**22.** Problem 14	**23.** Problem 15	**24.** Problem 16
25. Problem 17	**26.** Problem 18	**27.** Problem 19	**28.** Problem 20

Answer 21:   Imposing the initial conditions on the general solution in Problem 13, we have $[-1\ 1]^T = c_1[1\ -4]^T + c_2[1\ 2]^T$, or in component form $-1 = c_1 + c_2,\ 1 = -4c_1 + 2c_2$. Solving, we have $c_1 = c_2 = -1/2$, and so the unique solution of the IVP is $x = (-1/2)(e^{-4t}[1\ -4]^T + e^{2t}[1\ 2]^T)$. See the figure.

Answer 23:   Imposing the initial conditions on the general solution given in Problem 15, we obtain $-1 = c_1 + c_2,\ 1 = 3c_1 + c_2$. Solving, we have $c_1 = 1,\ c_2 = -2$. The unique solution of the given IVP is $x = e^{5t}[1\ 3]^T - 2e^{3t}[1\ 1]^T$. See the figure.

Answer 25:   Imposing the initial conditions on the general solution given in Problem 17, we obtain $-1 = c_1 + 2c_2,\ 1 = c_1 - c_2$. Solving, we have $c_1 = 1/3,\ c_2 = -2/3$. The unique solution of the given IVP is $x = (1/3)e^{3t}[1\ 1]^T + (-2/3)e^{-3t}[2\ -1]^T$. See the figure.

Answer 27:   Imposing the initial conditions on the general solution given in Problem 19, we obtain $1 = c_1 + 2c_2 + c_3,\ 1 = -c_2 + c_3$, and $1 = -c_1 + c_3$. Solving, we have $c_1 = 0,\ c_2 = 0$, and $c_3 = 1$.
The unique solution of the given IVP is $x = e^{5t}[1\ 1\ 1]^T$. See the figure.

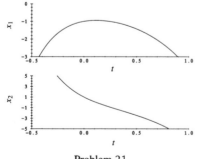

Problem 21.

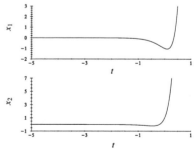

Problem 23.

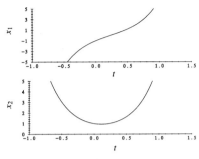

Problem 25.   Problem 27.

Complex Eigenvalues. Find all real-valued solutions for each given system. Plot several component curves. [*Hint*: solve the first ODE for x_2 in terms of x_1 and x_1' and insert into the second ODE. Use the techniques of Chapter 3 to solve the resulting second-order ODE for x_1.]

29. $x_1' = x_2$, $\quad x_2' = -64x_1$ $\qquad\qquad$ **30.** $x_1' = 2x_1 - x_2$, $\quad x_2' = 8x_1 - 2x_2$

Answer 29: Solving the first ODE for x_2 gives us $x_2 = x_1'$. Inserting into this into the second ODE we get $(x_1')' = -64x_1$ which is equivalent to $x_1'' + 64x_1 = 0$, a linear constant-coefficient, undriven second order ODE. Using techniques in Chapter 3, we plug in Ce^{rt} as a solution and we find that $Ce^{rt}(r^2 + 64) = 0$ must be satisfied. r, then, must equal $\pm 8i$. From Chapter 3, we know the general solution is $x_1 = c_1 \sin 8t + c_2 \cos 8t$. We can differentiate x_1 to find that $x_2 = 8c_1 \cos 8t - 8c_2 \sin 8t$. See the figure.

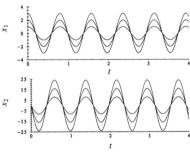

Problem 29.

Deficient Eigenspaces. Find the general solution of $x' = Ax$ for the given matrix A. [*Hint*: see Theorem 6.3.3.]

31. $\begin{bmatrix} 1 & 1 \\ -1 & 3 \end{bmatrix}$ $\qquad$ **32.** $\begin{bmatrix} 3 & -2 \\ 2 & -1 \end{bmatrix}$ $\qquad$ **33.** $\begin{bmatrix} 2 & -1 \\ 1 & 4 \end{bmatrix}$ $\qquad$ **34.** $\begin{bmatrix} -8 & -1 \\ 16 & 0 \end{bmatrix}$

35. $\begin{bmatrix} -1 & 1 & 0 \\ 0 & -1 & 0 \\ 0 & 0 & -2 \end{bmatrix}$ $\qquad$ **36.** $\begin{bmatrix} 5 & 3 & -3 \\ 2 & 4 & -5 \\ -4 & 2 & -3 \end{bmatrix}$ $\qquad$ **37.** $\begin{bmatrix} 0 & 1 & 0 \\ 2 & 3 & 4 \\ 1 & 1 & 2 \end{bmatrix}$

Answer 31: The characteristic polynomial $p(\lambda) = \det(A - \lambda I) = \lambda^2 - 4\lambda + 4 = (\lambda - 2)^2$ has one root and A has one eigenvalue $\lambda = 2$ with multiplicity two. Solving the eigenvalue equation $(A - 2I)v = 0$ we find eigenvectors of the form $v = [k \ \ k]^T$ where k is an arbitrary constant. For simplicity, we will choose $k = 1$. We need a generalized eigenvector to find an eigenbasis for $\mathbb{R}^2$. So, we solve $(A - 2I)u = v$ for the generalized eigenvector u. We find that u can be $[1 \ \ 2]^T$. From Example 6.3.6, then, we see that the general solution will be $x = C_1 \begin{bmatrix} 1 \\ 1 \end{bmatrix} e^{2t} + C_2 e^{2t} \left(\begin{bmatrix} 1 \\ 2 \end{bmatrix} + \begin{bmatrix} 1 \\ 1 \end{bmatrix} t \right)$, where C_1 and C_2 are arbitrary constants.

Answer 33: The eigenvalue(s) of $\begin{bmatrix} 2 & -1 \\ 1 & 4 \end{bmatrix}$ is $\lambda = 3$, again with multiplicity 2. The eigenspace is deficient with vectors $v = [k \;\; -k]^T$ for arbitrary k. As usual, for simplicity we will choose $k = 1$; remember that eigenvectors are nonzero. Because the eigenspace is deficient, we need to find a generalized eigenvector to solve the system of ODE's. The generalized eigenvector $u = [a \;\; b]^T$ must satisfy $(A - \lambda I)u = v$ or in this case $\left(\begin{bmatrix} 2 & -1 \\ 1 & 4 \end{bmatrix} - 3I \right) u = \begin{bmatrix} 1 \\ -1 \end{bmatrix}$. This gives us two equations which are equivalent to $a + b = -1$. To satisfy this condition, we chose $u = [0 \;\; -1]^T$. The solution, then, is $x = c_1 \begin{bmatrix} 1 \\ -1 \end{bmatrix} e^{3t} + c_2 e^{3t} \left(\begin{bmatrix} 0 \\ -1 \end{bmatrix} + \begin{bmatrix} 1 \\ -1 \end{bmatrix} t \right)$

Answer 35: The characteristic polynomial is $(-1 - \lambda)(-1 - \lambda)(-2 - \lambda)$. So $\lambda_1 = \lambda_2 = -1$ and $\lambda_3 = -2$ are the eigenvalues. For $\lambda_3 = -2$, the eigenspace V_{-2} is the span of $u = [0 \; 0 \; 1]^T$. For $\lambda_1 = \lambda_2 = -1$ the eigenspace V_{-1} is the span of $v = [1 \; 0 \; 0]^T$, so V_{-1} is deficient. We find the generalized eigenvector $w = [0 \; 1 \; 0]^T$. Hence, the general solution is

$$x = C_1 e^{-t} \begin{bmatrix} 1 \\ 0 \\ 0 \end{bmatrix} + C_2 \left(\begin{bmatrix} 0 \\ 1 \\ 0 \end{bmatrix} + t \begin{bmatrix} 1 \\ 0 \\ 0 \end{bmatrix} \right) + C_3 e^{-2t} \begin{bmatrix} 0 \\ 0 \\ 1 \end{bmatrix}$$

where C_1, C_2, and C_3 are arbitrary constants.

Answer 37: The characteristic polynomial is

$$p(\lambda) = \det \begin{bmatrix} 0 - \lambda & 1 & 0 \\ 2 & 3 - \lambda & 4 \\ 1 & 1 & 2 - \lambda \end{bmatrix} = -(\lambda - 5)\lambda^2$$

The eigenvalues are $\lambda_1 = 5$, $\lambda_2 = \lambda_3 = 0$. Respective eigenvectors are $[1 \; 5 \; 2]^T$ and $[2 \; 0 \; -1]^T$; so V_0 is deficient. One solution of $(A - 0I)v = [2 \; 0 \; -1]^T$ is $v = [-1 \; 2 \; -1]^T$. The general solution is

$$x = C_1[1 \; 5 \; 2]^T e^{5t} + C_2[2 \; 0 \; -1]^T + C_3 \left([-1 \; 2 \; -1]^T + t[2 \; 0 \; -1]^T \right)$$

where C_1, C_2, and C_3 are arbitrary real constants.

A Scalar Linear ODE and Its Equivalent Linear System.

38. Follow the outline below to find the "characteristic polynomial" of the nth-order constant coefficient ODE $y^{(n)} + a_{n-1} y^{(n-1)} + \cdots + a_1 y' + a_0 y = 0$ by converting it to a system.

(a) Show that if $x_1 = y$, $x_2 = y'$, ..., $x_n = y^{(n-1)}$, the ODE is equivalent to the first-order linear system $x' = Ax$, where A is the *companion matrix*

$$A = \begin{bmatrix} 0 & 1 & 0 & \cdots & \cdots & 0 \\ 0 & 0 & 1 & & & \\ \vdots & & \ddots & & & \vdots \\ \vdots & & & \ddots & & \vdots \\ & & & & 0 & 1 \\ -a_0 & -a_1 & -a_2 & \cdots & -a_{n-2} & -a_{n-1} \end{bmatrix}$$

(b) Show that the characteristic polynomial of A is $(-1)^n (\lambda^n + a_{n-1}\lambda^{n-1} + \cdots + a_1\lambda + a_0)$. [*Hint:* use column operations to transform $A - \lambda I$ into a matrix, where the submatrix of the top $(n - 1)$ rows starts with a column of zeros followed by an $(n - 1) \times (n - 1)$ triangular matrix with ones on the diagonal. Then calculate the determinant by cofactors.]

(c) Show that the term "characteristic polynomial" used in Chapter 3 for the ODE $y'' + a_1 y' + a_0 y = 0$, where a_1 and a_0 are real constants, is consistent with that given in (b) for A.

 3×3 **System Matrices with a Deficient Eigenspace.** Carry out the tasks below to find a general solution formula for each of the systems.

•Find the eigenvalues and their multiplicities

- Find the eigenspaces and their dimensions
- Find a chain of generalized eigenvectors
- Find a basis of generalized eigenvectors for $\mathbb{R}^3$
- Use the matrix $U = [u^1 \ u^2 \ u^3]$ to make the change of state variables $x = Uy$ and obtain the equivalent system $Uy' = AUy$.
- Find a general solution formula for $Uy' = AUy$. Then construct a general solution formula for $x' = Ax$ from $x = Uy$.

39. $x' = \begin{bmatrix} -1 & 1 & 0 \\ -1 & 0 & 1 \\ 1 & 0 & -2 \end{bmatrix} x$ **40.** $x' = \begin{bmatrix} 1 & 1 & 1 \\ 2 & 1 & -1 \\ -3 & 2 & 4 \end{bmatrix} x$

Answer 39: The characteristic polynomial of the system matrix A is $-(\lambda + 1)^3$, and hence A has just the triple eigenvalue $\lambda_1 = -1$. The eigenspace $V_{\lambda_1} = \text{Span}\left([1 \ 0 \ 1]^T\right)$ and thus is deficient. Denote the eigenvector $[1 \ 0 \ 1]^T$ by u^1, and find a chain of generalized eigenvectors based on u^2. Denote the chain by u^2, u^3 by solving the algebraic systems $(A + I)u^2 = u^1$ and $(A + I)u^3 = u^2$. Doing this we find that

$$u^1 = [1 \ 0 \ 1]^T, \quad u^2 = [1 \ 1 \ 0]^T, \quad u^3 = [1 \ 1 \ 1]^T$$

and note that

$$Au^2 = u^1 - u^2, \quad Au^3 = u^2 - u^3$$

Note that $\{u^1, u^2, u^3\}$ is a basis for $\mathbb{R}^3$. For the system $x' = Ax$ make the change of variables $x = [u^1 \ u^2 \ u^3]y = Uy$ to obtain the equivalent system

$$Uy' = AUy$$

Using the Matrix Product Identities, this system becomes

$$
\begin{aligned}
y_1'u^1 + y_2'u^2 + y_3'u^3 &= \begin{bmatrix} Au^1 & Au^2 & Au^3 \end{bmatrix} y \\
&= Au^1 y_1 + Au^2 y_2 + Au^3 y_3 \\
&= -u^1 y_1 + (u^1 - u^2)y_2 + (u^2 - u^3)y_3
\end{aligned}
$$

Equating the coefficients of u^1, u^2, and u^3 we obtain the linear cascade

$$
\begin{aligned}
y_1' &= y_2 - y_1 \\
y_2' &= y_3 - y_2 \\
y_3' &= -y_3
\end{aligned}
$$

The general solution of this cascade is

$$
\begin{aligned}
y_1 &= (C_3 t^2/2 + C_2 t + C_1)e^{-t} \\
y_2 &= (C_3 t + C_2)e^{-t} \\
y_3 &= C_3 e^{-t}
\end{aligned}
$$

where C_1, C_2, and C_3 are arbitrary real numbers. Using the Matrix Product Identities again we get the general solution of $x' = Ax$:

$$
\begin{aligned}
x = Uy = [u^1 \ u^2 \ u^3]y &= u^1 y_1 + u^2 y_2 + u^3 y_3 \\
&= u^1(C_3 t^2/2 + C_2 t + C_1)e^{-t} + u^2(C_3 t + C_2)e^{-t} + u^1 C_3 e^{-t} \\
&= C_1 e^{-t} u^1 + C_2(u^2 + tu^1)e^{-t} + C_3(u^3 + tu^2 + (t^2/2)u^1)e^{-t}
\end{aligned}
$$

where C_1, C_2, and C_3 are arbitrary reals.

6.4 Undriven Linear Systems: Complex Eigenvalues

Comments:

The reason why we put this material in a separate section is because students are often uncomfortable working with matrices and vectors whose entries are complex numbers or complex-valued functions. The matter-of-fact approach in Section 6.3 also works in the complex eigenvector case if the algebra of complex numbers is used. Our "proof" of a solution formula is by construction, once a basis of eigenvectors and generalized eigenvectors is known. We did not prove the existence of such a basis for the general case (because that would be beyond the scope of this text—or even a first course in Linear Algebra). This is not a problem because our constructive approach works just fine when $n \leq 4$, and these are the examples that occur in this text. Basic solution sets are important, so is Theorem 6.4.4 because it tells us the form of every term in every component of every solution vector of $x' = Ax$.

PROBLEMS

General Real-Valued Solutions. Find the general real-valued solution of $x' = Ax$ for the given matrix A. For Problems 1–10, solve the IVP $x' = Ax$, $x(0) = [-1 \ \ 1]^T$. Describe the behavior of each component of the solution as time increases. For Problems 13–19, find the general real-valued solution.

1. $\begin{bmatrix} 0 & 1 \\ -64 & 0 \end{bmatrix}$ 2. $\begin{bmatrix} 2 & -1 \\ 8 & -2 \end{bmatrix}$ 3. $\begin{bmatrix} 2 & -1 \\ 1 & 2 \end{bmatrix}$ 4. $\begin{bmatrix} 1 & 1 \\ -5 & 3 \end{bmatrix}$

5. $\begin{bmatrix} 1 & 2 \\ -2 & 1 \end{bmatrix}$ 6. $\begin{bmatrix} 5 & 1 \\ -2 & 3 \end{bmatrix}$ 7. $\begin{bmatrix} 6 & -1 \\ 5 & 2 \end{bmatrix}$ 8. $\begin{bmatrix} 1 & 1 \\ -2 & 1 \end{bmatrix}$

9. $\begin{bmatrix} 0 & 1 \\ -13 & 4 \end{bmatrix}$ 10. $\begin{bmatrix} -2 & \pi \\ -\pi & -2 \end{bmatrix}$ 11. $\begin{bmatrix} 3 & -2\pi \\ 2\pi & 3 \end{bmatrix}$ 12. $\begin{bmatrix} 0 & 1 \\ -2 & -2 \end{bmatrix}$

13. $\begin{bmatrix} 1 & 2 & 3 \\ 0 & 1 & 0 \\ 2 & 1 & 2 \end{bmatrix}$ 14. $\begin{bmatrix} 1 & 0 & 0 \\ 2 & 1 & -2 \\ 3 & 2 & 1 \end{bmatrix}$ 15. $\begin{bmatrix} 1 & 2 & 3 \\ 0 & 1 & 2 \\ 0 & -2 & 1 \end{bmatrix}$ 16. $\begin{bmatrix} 0 & 1 & 0 \\ 2 & 0 & 4 \\ 1 & -1 & 2 \end{bmatrix}$

17. $\begin{bmatrix} -3 & 1 & -2 \\ 0 & -1 & -1 \\ 2 & 0 & 0 \end{bmatrix}$ 18. $\begin{bmatrix} 5 & 5 & 2 \\ -6 & -6 & -5 \\ 6 & 6 & 5 \end{bmatrix}$ 19. $\begin{bmatrix} 0 & 1 & 1 \\ -1 & 0 & 1 \\ -1 & -1 & 0 \end{bmatrix}$

Answer 1:   The eigenvalues of the system matrix $A = \begin{bmatrix} 0 & 1 \\ -64 & 0 \end{bmatrix}$ are $\lambda_1 = 8i$, $\lambda_2 = -8i$. The eigenspaces are spanned, respectively, by $[1 \ \ 8i]^T$ and $[1 \ \ -8i]^T$. So the general (complex-valued) solution of $x' = Ax$ is $x = C_1 e^{8it}[1 \ \ 8i]^T + C_2 e^{-8it}[1 \ \ -8i]^T$ where C_1 and C_2 are arbitrary complex numbers. To find all real-valued solutions, put

$$x^1 = \text{Re}\left[e^{8it}[1 \ \ 8i]^T\right] = [\cos 8t \ \ -8\sin 8t]^T$$
$$x^2 = \text{Im}\left[e^{8it}[1 \ \ 8i]^T\right] = [\sin 8t \ \ 8\cos 8t]^T$$

Then the general real-valued solution is $x = C_1 x^1 + C_2 x^2$, where C_1 and C_2 are arbitrary reals. Now,

$$x(0) = \begin{bmatrix} -1 \\ 1 \end{bmatrix} = C_1 \begin{bmatrix} 1 \\ 0 \end{bmatrix} + C_2 \begin{bmatrix} 0 \\ 8 \end{bmatrix}$$

which implies that $C_1 = -1$, $C_2 = 1/8$. Hence the solution of the IVP is

$$x = -[\cos 8t \ \ -8\sin 8t]^T + \frac{1}{8}[\sin 8t \ \ 8\cos 8t]^T$$

Each component is periodic with period $\pi/4$.

Answer 3: The eigenvalues of the system matrix are $\lambda_1 = 2 + i$ and $\lambda_2 = 2 - i$ with eigenvectors $[1 \ -i]^T$ and $[1 \ i]^T$, respectively. Put

$$x^1 = \mathrm{Re}\left[e^{(2+i)t}[1 \ -i]^T\right], \quad x^2 = \mathrm{Im}\left[e^{(2+i)t}[1 \ -i]^T\right]$$

Hence, $x^1 = e^{2t}[\cos t \ \sin t]^T$, and $x^2 = e^{2t}[\sin t \ -\cos t]^T$, and the general real-valued solution is $x = C_1x^1 + C_2x^2$, for arbitrary real numbers C_1 and C_2. Now, $x(0) = [-1 \ 1]^T = C_1[1 \ 0]^T + C_2[0 \ -1]^T$. Hence, $C_1 = -1$ and $C_2 = -1$ and the solution of the IVP is $x = -x^1 - x^2$. Each component becomes unbounded as $t \to +\infty$ and oscillates with oscillation time 2π.

Answer 5: To get the eigenvalues, we need the characteristic polynomial,

$$p(\lambda) = \det\begin{bmatrix} 1-\lambda & 2 \\ -2 & 1-\lambda \end{bmatrix} = (1-\lambda)^2 + 4$$

and so $\lambda_1 = \overline{\lambda_2} = 1 + 2i$. The eigenspace for λ_1 is given by all vectors $[a \ b]^T$ such that $-2ia + 2b = 0$, and so $[1 \ i]^T$ is an eigenvector corresponding to λ_1.

Consider the complex-valued solution

$$\begin{bmatrix} 1 \\ i \end{bmatrix} e^{(1+2i)t} = \begin{bmatrix} e^t(\cos 2t + i\sin 2t) \\ e^t i(\cos 2t + i\sin 2t) \end{bmatrix}$$

Extracting real and imaginary parts we have that

$$x^1 = \begin{bmatrix} e^t \cos 2t \\ -e^t \sin 2t \end{bmatrix} \quad \text{and } x^2 = \begin{bmatrix} e^t \sin 2t \\ e^t \cos 2t \end{bmatrix}$$

form a fundamental set and so $x = C_1x^1 + C_2x^2$, for arbitrary reals C_1, C_2, is the general real-valued solution. Now

$$x(0) = \begin{bmatrix} -1 \\ 1 \end{bmatrix} = C_1\begin{bmatrix} 1 \\ 0 \end{bmatrix} + C_2\begin{bmatrix} 0 \\ 1 \end{bmatrix}$$

which implies that $C_1 = -1$, $C_2 = 1$. Hence the solution of the IVP is $x = -x^1 + x^2$. Each component becomes unbounded as $t \to +\infty$ and oscillates with oscillation time π.

Answer 7: The characteristic polynomial is $p(\lambda) = \lambda^2 - 8\lambda + 17$ and so the eigenvalues are $\lambda_1 = \overline{\lambda_2} = 4 + i$ with eigenvectors $[1 \ 2 - i]^T$ and $[1 \ 2 + i]^T$ respectively. Let

$$x^1 = \mathrm{Re}\left[e^{(4+i)t}[1 \ 2 - i]^T\right], \quad x^2 = \mathrm{Im}\left[e^{(4+i)t}[1 \ 2 - i]^T\right]$$

The general real-valued solution is $x = C_1x^1 + C_2x^2$, for arbitrary real numbers C_1 and C_2. Now $x(0) = [-1 \ 1]^T = C_1[1 \ 2]^T + C_2[0 \ -1]^T$. Hence, $C_1 = -1$ and $C_2 = -3$ and the solution of the IVP is $x = -x^1 - 3x^2$. Each component becomes unbounded as $t \to +\infty$ and oscillates with oscillation time 2π.

Answer 9: The characteristic polynomial is $p(\lambda) = \lambda^2 - 4\lambda + 13$ and so $\lambda_1 = \lambda_2 = 2 + 3i$ with eigenvectors $[1 \ 2 + 3i]^T$ and $[1 \ 2 - 3i]^T$ respectively. Let $x^1 = \mathrm{Re}\left[e^{(2+3i)t}[1 \ 2 + 3i]^T\right]$, and $x^2 = \mathrm{Im}\left[e^{(2+3i)t}[1 \ 2 + 3i]^T\right]$, and the general real-valued solution is $x = C_1x^1 + C_2x^2$, for arbitrary real numbers C_1 and C_2. Now $x(0) = [-1 \ 1]^T = C_1[1 \ 2]^T + C_2[0 \ 3]^T$. Hence, $C_1 = -1$ and $C_2 = 1$ and the solution of the IVP is $x = -x^1 + x^2$. Each component becomes unbounded as $t \to +\infty$ and oscillates with oscillation time $2\pi/3$.

Answer 11: The characteristic polynomial is $p(\lambda) = \lambda^2 - 6\lambda + 9 + 4\pi^2$ and so $\lambda_1 = \overline{\lambda_2} = 3 + 2\pi i$ with eigenvectors $[i \ 1]^T$ and $[-i \ 1]^T$, respectively. Let $x^1 = \mathrm{Re}\left[e^{(3+2\pi i)t}[i \ 1]^T\right]$, and $x^2 = \mathrm{Im}\left[e^{(3+2\pi i)t}[i \ 1]\right]$ Hence,

$$x^1 = e^{3t}[-\sin 2\pi t \ \cos 2\pi t]^T, \quad x^2 = e^{3t}[\cos 2\pi t \ \sin 2\pi t]^T$$

and the general real-valued solution is $x = C_1x^1 + C_2x^2$, for arbitrary real numbers C_1 and C_2. Now $x(0) = [-1 \ 1]^T = C_1[0 \ 1]^T + C_2[1 \ 0]^T$. Hence, $C_1 = 1$ and $C_2 = -1$ and the solution of the IVP is $x = x^1 - x^2$. Each component becomes unbounded as $t \to +\infty$ and oscillates with oscillation time $2\pi/3$.

Answer 13: The characteristic polynomial of the matrix is $p(\lambda) = -(\lambda - 1)(\lambda + 1)(\lambda - 4)$, and the eigenvalues are $\lambda_1 = 1$, $\lambda_2 = -1$ and $\lambda_3 = 4$. Corresponding eigenvectors are $[1/2 \ \ -3 \ \ 2]^T$, $[-3 \ \ 0 \ \ 2]^T$ and $[1 \ \ 0 \ \ 1]^T$, respectively. Therefore, all real-valued solutions of the system are

$$x = C_1 \begin{bmatrix} 1/2 \\ -3 \\ 2 \end{bmatrix} e^t + C_2 \begin{bmatrix} -3 \\ 0 \\ 2 \end{bmatrix} e^{-t} + C_3 \begin{bmatrix} 1 \\ 0 \\ 1 \end{bmatrix} e^{4t}$$

where C_1, C_2, and C_3 are arbitrary real numbers.

Answer 15: The characteristic polynomial is $p(\lambda) = (1 - \lambda)(\lambda^2 - 2\lambda + 5)$ and so $\lambda_1 = 1$ and $\lambda_2 = \overline{\lambda_3} = 1 + 2i$ with eigenvectors $[1 \ \ 0 \ \ 0]^T$, $[-i + 3/2 \ \ 1 \ \ i]^T$, and $[i + 3/2 \ \ 1 \ \ -i]^T$, respectively. Let $x^1 = e^t[1 \ \ 0 \ \ 0]^T$, $x^2 = \text{Re}\left[e^{(1+2i)}[-i + 3/2 \ \ 1 \ \ i]^T\right]$ and $x^3 = \text{Im}\left[e^{(1+2i)}[-i + 3/2 \ \ 1 \ \ i]^T\right]$. Hence,

$$x^2 = e^t \left[\frac{3}{2}\cos 2t + \sin 2t \ \ \cos 2t \ \ -\sin 2t\right]^T, \quad x^3 = e^t \left[-\cos 2t + \frac{3}{2}\sin 2t \ \ \sin 2t \ \ \cos 2t\right]^T$$

The general real-valued solution is $x = C_1 x^1 + C_2 x^2 + C_3 x^3$, for arbitrary real numbers C_1, C_2 and C_3.

Answer 17: The characteristic polynomial is $p(\lambda) = -(\lambda^3 + 4\lambda^2 + 7\lambda + 6) = -(\lambda + 2)(\lambda^2 + 2\lambda + 3)$, and hence the eigenvalues are $\lambda_1 = -2$, $\lambda_2 = -1 + \sqrt{2}i$ and $\lambda_3 = -1 - \sqrt{2}i$. The corresponding eigenvectors are $[-1 \ \ 1 \ \ 1]^T$, $[1 - \sqrt{2}i \ \ -\sqrt{2}i \ \ -2]^T$, and $[1 + \sqrt{2}i, \sqrt{2}i, -2]^T$, respectively. Put

$$x^1 = e^{-2t}[-1 \ \ 1 \ \ 1]^T$$

$$x^2 = \text{Re}\left[e^{(-1+\sqrt{2}i)t}[1 - \sqrt{2}i \ \ -\sqrt{2}i \ \ -2]^T\right]$$

$$x^3 = \text{Im}\left[e^{(-1+\sqrt{2}i)t}[1 - \sqrt{2}i \ \ -\sqrt{2}i \ \ -2]^T\right]$$

Notice that

$$x^2 = e^{-t}\left(\cos\sqrt{2}t[1 \ \ 0 \ \ -2]^T + \sin\sqrt{2}t[\sqrt{2} \ \ \sqrt{2} \ \ 0]^T\right)$$

$$x^3 = e^{-t}\left(\cos\sqrt{2}t[-\sqrt{2} \ \ -\sqrt{2} \ \ 0]^T + \sin\sqrt{2}t[1 \ \ 0 \ \ -2]^T\right)$$

The general real-valued solution is $x = C_1 x^1 + C_2 x^2 + C_3 x^3$, where C_1, C_2 and C_3 are arbitrary real numbers.

Answer 19: The characteristic polynomial is $p(\lambda) = \lambda(\lambda^2 + 3)$ and so the eigenvalues are $\lambda_1 = 0$, $\lambda_2 = \overline{\lambda_3} = \sqrt{3}i$ with eigenvectors $[1 \ \ -1 \ \ 1]^T$, $[1 - \sqrt{3}i \ \ 2 \ \ 1 + \sqrt{3}i]^T$, and $[1 + \sqrt{3}i \ \ 2 \ \ 1 - \sqrt{3}i]^T$, respectively. Put

$$x^1 = [1 \ \ -1 \ \ 1]^T$$

$$x^2 = \text{Re}\left[e^{(\sqrt{3}i)t}[1 - \sqrt{3}i \ \ 2 \ \ 1 + \sqrt{3}i]^T\right]$$

$$x^3 = \text{Im}\left[e^{(\sqrt{3}i)t}[1 - \sqrt{3}i \ \ 2 \ \ 1 + \sqrt{3}i]^T\right]$$

Notice that

$$x^2 = e^{2t}\left[\cos\sqrt{3}t + \sqrt{3}\sin\sqrt{3}t \ \ 2\cos\sqrt{3}t \ \ \cos\sqrt{3}t - \sqrt{3}\sin\sqrt{3}t\right]^T$$

$$x^3 = e^{2t}\left[\sin\sqrt{3}t - \sqrt{3}\cos\sqrt{3}t \ \ 2\sin\sqrt{3}t \ \ \sin\sqrt{3}t + \sqrt{3}\cos\sqrt{3}t\right]^T$$

and the general real-valued solution is $x = C_1 x^1 + C_2 x^2 + C_3 x^3$, where C_1, C_2 and C_3 are arbitrary real numbers.

☐ **Solutions of a Second-Order Linear Undriven ODE.**   Use the Method of Eigenvectors to find all real-valued
☞ Compare your
results with the
solutions given in
Theorem 3.2.1.

solutions of the ODEs below. Plot some component curves. [*Hint*: first convert the second-order ODE to an
equivalent first-order system by using the state variables $x_1 = y, \quad x_2 = y'$.]

20. $y'' + y' + 2y = 0$ **21.** $y'' + 2y' + 5y = 0$ **22.** $y'' + 2y' + 2y = 0$

Answer 21:   Solving this problem is very similar to solving Problem 20. As such, this solution will
be much less expansive. We let $x_1 = y$, $x_2 = y'$, and $x = [x_1 \ x_2]^T$, giving us the system $x' = Ax$ with
$A = \begin{bmatrix} 0 & 1 \\ -5 & -2 \end{bmatrix}$. The eigenvalues of A are $\lambda = -1 \pm 2i$. Taking the positive imaginary case, we get

eigenvectors that are a constant complex multiple of $\begin{bmatrix} (-1-2i)/5 \\ 1 \end{bmatrix}$. The corresponding solution is

$$x^1(t) = \begin{bmatrix} \dfrac{-1-2i}{5} \\ 1 \end{bmatrix} e^{(-1+2i)t} = e^{-t} \left(\begin{bmatrix} -1/5 \\ 1 \end{bmatrix} \cos 2t + \begin{bmatrix} 2/5 \\ 0 \end{bmatrix} \sin 2t \right)$$

$$+ ie^{-t} \left(\begin{bmatrix} -1/5 \\ 1 \end{bmatrix} \sin 2t + \begin{bmatrix} -2/5 \\ 0 \end{bmatrix} \cos 2t \right)$$

To find real solutions to $x' = Ax$ we take a linear combination of the real and imaginary parts of x^1
with arbitrary constant coefficients C_1, C_2. However, as above, we only care about what y is and not
what y' is, so we only use the top row of the linear combination. Replacing x_1 with y and simplifying,
we get $(-C_1 - 2C_2)/5 e^{-t} \cos 2t + (2C_1 - C_2)/5 e^{-t} \sin 2t$ that, as noted, is in the form predicted by
Theorem 3.2.1.

General Real-Valued Solutions.   Find the general real-valued solution for the system below using the indicated values for the constants a and b.

$$x' = Ax, \quad \text{where} \quad A = \begin{bmatrix} 0 & 1 & 0 & 0 \\ -a & 0 & b & 0 \\ 0 & 0 & 0 & 1 \\ a & 0 & -a & 0 \end{bmatrix} \tag{i}$$

23. $a = 1/2, \quad b = 1/8$ **24.** $a = 2, \quad b = 1/2$ **25.** $a = 3, \quad b = 1/3$

Answer 23:   The characteristic polynomial of the given system matrix A is $\lambda^4 + 2a\lambda^2 + a^2 - ab$.
Using the quadratic formula, find that λ^2 is either $-a + \sqrt{ab}$ or $-a - \sqrt{ab}$. Here $a = 1/2, b = 1/8$
and the characteristic polynomial $p(\lambda)$ is $\lambda^4 + \lambda^2 + 3/16$ To find the roots of $p(\lambda)$, we can factor. This
gives us $p(\lambda) = (\lambda^2 + 1/4)(\lambda^2 + 3/4)$, which gives us eigenvalues , $\pm i/2, \pm(\sqrt{3}i)/2$. Since we want
real valued solutions we will only find the eigenvector for the positive imaginary cases (the eigenvector
for the negative case is just the complex conjugate, anyway). We do this in the usual way: solving $(A -
\lambda I)x = 0$. For $\lambda = i/2$, we get eigenvectors of the form $[-ik \ k/2 \ -2ik \ k]^T$ where k is an arbitrary
constant. For $\lambda = (\sqrt{3}i)/2$ we get eigenvectors of the form $[(j\sqrt{3}i)/3 \ -j/2 \ -(2j\sqrt{3}i)/3 \ j]^T$
where j is a second arbitrary constant. For convenience, we will choose $j = k = 1$. Now we need to
find the solutions corresponding to each eigenvector to build a fundamental set. Let

$$x^1 = \left(\begin{bmatrix} 0 \\ 1/2 \\ 0 \\ 1 \end{bmatrix} + i \begin{bmatrix} -1 \\ 0 \\ -2 \\ 0 \end{bmatrix} \right) (\cos t/2 + i \sin t/2),$$

$$x^2 = \left(\begin{bmatrix} 0 \\ -1/2 \\ 0 \\ 1 \end{bmatrix} + i \begin{bmatrix} \dfrac{\sqrt{3}}{3} \\ 0 \\ -\dfrac{2\sqrt{3}}{3} \\ 0 \end{bmatrix} \right) \left(\cos \dfrac{\sqrt{3}}{2} t + i \sin \dfrac{\sqrt{3}}{2} t \right)$$

The fundamental set for the general real valued solution, then, is the real and imaginary parts of x^1 and x^2. This makes the general real-valued solution

$$x = C_1 \operatorname{Re}[x^1] + C_2 \operatorname{Im}[x^1] + C_3 \operatorname{Re}[x^2] + C_4 \operatorname{Im}[x^2]$$

$$\operatorname{Re}[x^1] = \begin{bmatrix} 0 \\ 1/2 \\ 0 \\ 1 \end{bmatrix} \cos\frac{t}{2} + \begin{bmatrix} 1 \\ 0 \\ 2 \\ 0 \end{bmatrix} \sin\frac{t}{2}$$

$$\operatorname{Im}[x^1] = \begin{bmatrix} -1 \\ 0 \\ -2 \\ 0 \end{bmatrix} \cos\frac{t}{2} + \begin{bmatrix} 0 \\ 1/2 \\ 0 \\ 1 \end{bmatrix} \sin\frac{t}{2}$$

$$\operatorname{Re}[x^2] = \begin{bmatrix} 0 \\ -1/2 \\ 0 \\ 1 \end{bmatrix} \cos\frac{\sqrt{3}t}{2} + \begin{bmatrix} -\sqrt{3}/3 \\ 0 \\ 2\sqrt{3}/3 \\ 0 \end{bmatrix} \sin\frac{\sqrt{3}t}{2}$$

$$\operatorname{Im}[x^2] = \begin{bmatrix} \sqrt{3}/3 \\ 0 \\ -2\sqrt{3}/3 \\ 0 \end{bmatrix} \cos\frac{\sqrt{3}t}{2} + \begin{bmatrix} 0 \\ -1/2 \\ 0 \\ 1 \end{bmatrix} \sin\frac{\sqrt{3}t}{2}$$

Answer 25: From Problem 23, the characteristic polynomial for the system matrix is $(\lambda^2 + 4)(\lambda^2 + 2)$ and the eigenvalues are $\lambda_1 = 2i$, $\lambda_2 = -2i$, $\lambda_3 = \sqrt{2}i$, $\lambda_4 = -\sqrt{2}i$. The eigenspace V_{λ_1} is spanned by the eigenvector $[1 \ \ 2i \ \ -3 \ \ -6i]^T$ and the eigenspace V_{λ_3} is spanned by the eigenvector $[1 \ \ \sqrt{2}i \ \ 3 \ \ 3\sqrt{2}i]^T$. Hence we have the two complex-valued solutions

$$z^1 = e^{2it} \begin{bmatrix} 1 \\ 2i \\ -3 \\ -6i \end{bmatrix} = (\cos 2t + i \sin 2t) \left(\begin{bmatrix} 1 \\ 0 \\ -3 \\ 0 \end{bmatrix} + i \begin{bmatrix} 0 \\ 2 \\ 0 \\ -6 \end{bmatrix} \right)$$

$$z^2 = e^{\sqrt{2}it} \begin{bmatrix} 1 \\ \sqrt{2}i \\ 3 \\ 3\sqrt{2}i \end{bmatrix} = (\cos \sqrt{2}t + i \sin \sqrt{2}t) \left(\begin{bmatrix} 1 \\ 0 \\ 3 \\ 0 \end{bmatrix} + i \begin{bmatrix} 0 \\ \sqrt{2} \\ 0 \\ 3\sqrt{2} \end{bmatrix} \right)$$

and a real-valed basic solution set for $x' = Ax$,

$$x^1 = \operatorname{Re}[z^1], \quad x^2 = \operatorname{Im}[z^1], \quad x^3 = \operatorname{Re}[z^2], \quad x^4 = \operatorname{Im}[z^2]$$

Explicitly,

$$x^1 = [\cos 2t \quad -2\sin 2t \quad -3\cos 2t \quad 6\sin 2t]^T$$

$$x^2 = [\sin 2t \quad 2\cos 2t \quad -3\sin 2t \quad -6\cos 2t]^T$$

$$x^3 = [\cos \sqrt{2}t \quad -\sqrt{2}\sin \sqrt{2}t \quad 3\cos \sqrt{2}t \quad -3\sqrt{2}\sin \sqrt{2}t]^T$$

$$x^4 = [\sin \sqrt{2}t \quad \sqrt{2}\cos \sqrt{2}t \quad 3\sin \sqrt{2}t \quad 3\sqrt{2}\cos \sqrt{2}t]^T$$

Modeling Problems.

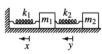

26. *Coupled Springs* Two blocks of masses m_1 and m_2 are coupled with two Hooke's Law springs with spring constants k_1 and k_2 and slide frictionlessly in a straight line on a horizontal table. Measure the position of the blocks from the neutral position by x and y with the indicated positive direction for these variables (see the margin figure).

☞ We discussed springs in Section 3.1.

(a) Show that the following ODEs model the motion of the blocks:

$$m_1 x'' = -k_1 x + k_2(y - x), \qquad m_2 y'' = -k_2(y - x) \tag{i}$$

(b) Write system (i) as a normal first-order system [*Hint*: set $x_1 = x$, $x_2 = x'$, $x_3 = y$, $x_4 = y'$.]

(c) Find the general solution of system (i) if $k_1/m_2 = 2$, $k_2/m_1 = 2$, and $k_2/m_2 = 1$.

 These numbers are unrealistic but they make the math easy.

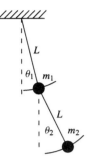

27. *The Double Pendulum*   Suppose that one pendulum is suspended from another (see the margin figure). The following linear system models the small-amplitude oscillations about the equilibrium position $x_1 = x_2 = x_3 = x_4 = 0$: $x_1' = x_2$, $x_2' = -x_1 + \alpha x_3$, $x_3' = x_4$, $x_4' = x_1 - x_3$ where $\alpha = (m_2/m_1)(1 + m_2/m_1)^{-1}$ is the dimensionless *reduced mass*, x_1 and x_3 are the angles θ_1 and θ_2, and x_2 and x_4 are the angular velocities, θ_1' and θ_2'.

(a) Build the general real-valued solution of this system using the methods of this section.

(b) Set $\alpha = 0.3$. Does the system have any periodic solutions? If so, what are they?

(c) For $\alpha = 0.3$, find the unique solution of the system for each set of initial conditions:
(1) $x_1(0) = (0.15)\sqrt{0.3}$, $\quad x_2(0) = 0$, $\quad x_3(0) = -0.15$, $\quad x_4(0) = 0$
(2) $x_1(0) = (0.15)\sqrt{0.3}$, $\quad x_2(0) = 0$, $\quad x_3(0) = 0.15$, $\quad x_4(0) = 0$
(3) $x_1(0) = (0.3)\sqrt{0.3}$, $\quad x_2(0) = 0$, $\quad x_3(0) = 0$, $\quad x_4(0) = 0$
[*Hint*: the initial data in **(3)** are the sums of the initial data in **(1)** and **(2)**.]

(d) For each initial data set given in **(c)**, overlay several plots of x_1 against t on the screen $0 \le t \le 25$, $|x_1| \le 0.2$. Repeat for plots of x_3 against t on the screen $0 \le t \le 25$, $|x_3| \le 0.3$.

(e) For each IVP for the system with initial data given in **(c)**, overlay plots of x_1 against x_3 on the screen $|x_1| \le 0.2$, $|x_3| \le 0.3$. Explain what you see.

Answer 27:

(a) The system matrix A and characteristic polynomial $p(\lambda)$ for the pendulum system in the text are

$$A = \begin{bmatrix} 0 & 1 & 0 & 0 \\ -1 & 0 & \alpha & 0 \\ 0 & 0 & 0 & 1 \\ 1 & 0 & -1 & 0 \end{bmatrix}, \quad p(\lambda) = \lambda^4 + 2\lambda^2 + 1 - \alpha$$

Solving $p(\lambda) = 0$, $\lambda^2 = -1 \pm \sqrt{\alpha}$. Since $0 < \alpha < 1$, the four eigenvalues are pure imaginary:

$$\lambda_1 = i\sqrt{1 + \sqrt{\alpha}} = \bar{\lambda}_2 = \beta i, \quad \text{where } \beta = \sqrt{1 + \sqrt{\alpha}}$$

$$\lambda_3 = i\sqrt{1 - \sqrt{\alpha}} = \bar{\lambda}_4 = \gamma i, \quad \text{where } \gamma = \sqrt{1 - \sqrt{\alpha}}$$

Corresponding eigenvectors, v^1, $v^2 = \bar{v}^1$, v^3, and $v^4 = \bar{v}^3$, have complex components, but we can follow the technique of this section to generate the real solution space by taking the real and the imaginary parts of $e^{\lambda_1 t} v^1$ and $e^{\lambda_3 t} v^3$. Note that the double pendulum has the two *natural frequencies*, β and γ.

After carrying out the lengthy calculations, we find that the general real solution of the system is given by

$$\begin{bmatrix} x_1 \\ x_2 \\ x_3 \\ x_4 \end{bmatrix} = c_1 \begin{bmatrix} (1 + \lambda_1^2)\cos\beta t \\ -\beta(1 + \lambda_1^2)\sin\beta t \\ \cos\beta t \\ -\beta\sin\beta t \end{bmatrix} + c_2 \begin{bmatrix} (1 + \lambda_1^2)\sin\beta t \\ \beta(1 + \lambda_1^2)\cos\beta t \\ \sin\beta t \\ \beta\cos\beta t \end{bmatrix}$$

$$+ c_3 \begin{bmatrix} (1 + \lambda_3^2)\cos\gamma t \\ -\gamma(1 + \lambda_3^2)\sin\gamma t \\ \cos\gamma t \\ -\gamma\sin\gamma t \end{bmatrix} + c_4 \begin{bmatrix} (1 + \lambda_3^2)\sin\gamma t \\ \gamma(1 + \lambda_3^2)\cos\gamma t \\ \sin\gamma t \\ \gamma\cos\gamma t \end{bmatrix}$$

$$= c_1 y^1 + c_2 y^2 + c_3 y^3 + c_4 y^4$$

where c_1, c_2, c_3, and c_4 are arbitrary real constants.

(b) The solutions y^1 and y^2 are periodic of period $2\pi/\beta$, while y^3 and y^4 have period $2\pi/\gamma$. So if $\alpha = 0.3$ and $\beta = \sqrt{1 + \sqrt{\alpha}}$, we have $\beta = \sqrt{1 + \sqrt{0.3}} \approx 1.24$ and $\gamma = \sqrt{1 - \sqrt{0.3}} \approx 0.67$ So y^1 and y^2 have period ≈ 5.06 and y^3 and y^4 have approximate period 9.37.

(c) As in **(b)**, if $\alpha = 0.3$, then $\beta = \sqrt{1 + \sqrt{0.3}}$, $\gamma = \sqrt{1 - \sqrt{0.3}}$, and solving we find the solution $x(t) = -0.15y^1(t)$, and the solution $x(t) = 0.15y^3(t)$, and finally the solution $x(t) = -0.15y^1(t) + 0.15y^3(t)$.

(d) In Graph 1 we have plotted the x_1 component curves of the three solutions $-0.15y^1$ (long dashed curve), $0.15y^3$ (short dashed curve), and the sum $-0.15(y^1 - y^3)$ (solid curve). Observe that the first curve has period $2\pi/\beta$, the second has period $2\pi/\gamma$, while the third is not periodic at all since it is a linear combination of solutions of period β and of period γ where β/γ is irrational. In Graph 2 we have plotted the x_3 component curves for the same three solutions.

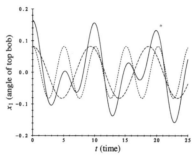

Problem 27(d), Graph 1.

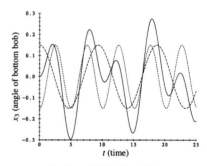

Problem 27(d), Graph 2.

(e) Now let's try to visualize the orbits, but since orbits live in a four-dimensional state space we will have to be content with orbital projections into lower dimensional "state spaces."

We can visualize the three solutions given in **(d)** by projecting the three orbits from four-dimensional state space into the three-dimensional $x_1x_2x_3$-space (Graph 1) and into the two-dimensional x_1x_3-space of the two angles (Graph 2).

The periodic orbits of the solutions $-0.15y^1$ and $0.15y^3$ project into the two oval-shaped simple closed curves shown in Graph 1 and onto the crossed line segments through the origin in Graph 2. The orbit of the sum $-0.15(y^1 - y^3)$ projects onto what appears to be a curve winding around a torus (Graph 1) and onto a curve wandering through a parallelogram determined by the crossed line segments just mentioned (Graph 2). The latter curve is called a *Lissajous figure*. These projections of the third orbit look periodic, but they are not because the third solution itself is not periodic.

The orbits of the solutions $C_1y^1 + C_2y^2$ (called the *β-normal modes*) project as line segments along $-0.15y^1$, but the segment length depends on C_1 and C_2. Similarly the orbits of the solutions $C_3y^3 + C_4y^4$ (the *γ-normal modes*) project onto the other line, a segment of which is shown in Graph 2. See the WEB SPOTLIGHT ON COUPLED SPRINGS: NORMAL MODES for another interpretation of normal modes.

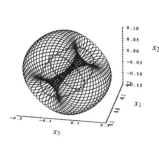

Problem 27(e), Graph 1

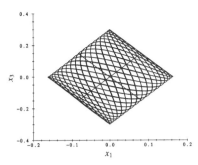

Problem 27(e), Graph 2

Basic Real-Valued Solution Set.

28. Show that the solutions $\{y^1, y^2\}$ defined in (8) of Example 6.4.2 form a basic real-valued solution set for system (5) by following the outline below, where $y^1 = \text{Re}[ue^{\lambda t}]$, $y^2 = \text{Im}[ue^{\lambda t}]$, λ is a complex eigenvalue of the real system matrix, and u is an eigenvector.

(a) Show that $y^1 = (ue^{\lambda t} + \bar{u}e^{\bar{\lambda}t})/2$ and $y^2 = (ue^{\lambda t} - \bar{u}e^{\bar{\lambda}t})/2i$.

(b) Let the scalars c_1 and c_2 be such that $c_1 y^1(t) + c_2 y^2(t) = 0$, for all t. Show this implies that $(c_1 - ic_2)u + (c_1 + ic_2)\bar{u} = 0$.

(c) Use the fact that $\{u, \bar{u}\}$ is an independent set to show that $c_1 - ic_2 = 0$ and $c_1 + ic_2 = 0$, and hence that $c_1 = c_2 = 0$.

29. Verify that the solutions y_1, y_2, y_3, y_4 of Example 6.4.4 form a basic real-valued solution set for the system given in that example. [*Hint*: follow the outline given in Problem 28.]

Answer 29: Direct extensions of the technique used in the solution to Problem 28.

30. *Verification of the Procedure for a* 2×2 *Matrix*   Suppose that A is a 2×2 matrix with real entries and a pair of complex conjugate eigenvalues $\lambda = \alpha + i\beta$, $\bar{\lambda} = \alpha - i\beta$ (with $\beta > 0$) and corresponding eigenvectors v and $\bar{v}$. The outline below shows that the procedure does indeed produce all real-valued solutions of the system $x' = Ax$.

(a) Find a solution formula for all complex-valued solutions of $x' = Ax$.

(b) Extract all real-valued solutions of $x' = Ax$ from the general complex-valued solution formula obtained in (a). [*Hint*: a solution $x(t)$ is real-valued if and only if $x(t) = \bar{x}(t)$, all t.]

(c) For any complex number z, $2\,\text{Re}\,[z] = z + \bar{z}$. Use this fact and (b) to find the general real-valued solution formula for $x' = Ax$.

6.5 Orbital Portraits for Planar Systems

Comments:

The focus of this section is on the gallery of the portraits of the orbits of linear planar autonomous systems. There are six qualitatively different portraits here (more if we include system matrices with one or more zero eigenvalues). The saddle, spiral point, center, and three kinds of nodal portraits are by now visually distinct icons for the systems they portray. Then we show how a planar orbital portrait evolves as one of the coefficients in the rate function changes.

The emphasis throughout the section is on relating the orbital portraits to the nature of the eigensets of the system matrix.

PROBLEMS

General Real-Valued Solutions. Find the general real-valued solution of each system. Classify the origin as a saddle, center, spiral, or one of the nodal types (identify the type). Identify as neutrally stable, unstable, or asymptotically stable. [*Hint*: use Tables 6.5.1 and 6.5.2.]

1. $x' = x + y$, $\quad y' = 4x - 2y$

2. $x' = 6x - 8y$, $\quad y' = 2x - 2y$

3. $x' = 7x + 6y$, $\quad y' = 2x + 6y$

4. $x' = 4y$, $\quad y' = -x$

5. $x' = -x - 4y$, $\quad y' = x - y$

6. $x' = -2x$, $\quad y' = -2y$

7. $x' = 2x + y$, $\quad y' = -x + 4y$

8. $x' = x - y$, $\quad y' = x + 3y$

9. $x' = 2x - y$, $\quad y' = x + 4y$

10. $x' = x - 2y$, $\quad y' = x + 4y$

11. $x' = 3x - 5y$, $\quad y' = 5x + 3y$

12. $x' = x - 2y$, $\quad y' = 2x + y$

Answer 1: The characteristic polynomial of the system matrix $\begin{bmatrix} 1 & 1 \\ 4 & -2 \end{bmatrix}$ is $p(\lambda) = \lambda^2 + \lambda - 6$, and the eigenvalues are $\lambda_1 = 2$ and $\lambda_2 = -3$; corresponding eigenvectors are $v^1 = [1 \quad 1]^T$ and $v^2 = [1 \quad -4]^T$. The general real-valued solution is $[x \quad y]^T = c_1 v^1 e^{2t} + c_2 v^2 e^{-3t}$, for arbitrary reals c_1, c_2, and there is a saddle point (unstable) at the origin.

Answer 3: The characteristic polynomial of the system matrix $\begin{bmatrix} 7 & 6 \\ 2 & 6 \end{bmatrix}$ is $p(\lambda) = \lambda^2 - 13\lambda + 30$, and the eigenvalues are $\lambda_1 = 10$ and $\lambda_2 = 3$; corresponding eigenvectors are $v^1 = [2 \quad 1]^T$ and $v^2 = [-3 \quad 2]^T$. The general real-valued solution is $[x \quad y]^T = c_1 v^1 e^{10t} + c_2 v^2 e^{3t}$ for arbitrary reals c_1, c_2: we have an improper node (unstable).

Answer 5: The characteristic polynomial of the system matrix $\begin{bmatrix} -1 & -4 \\ 1 & -1 \end{bmatrix}$ is $p(\lambda) = \lambda^2 + 2\lambda + 5$, and the eigenvalues are $\lambda_1 = -1 + 2i$ and $\lambda_2 = -1 - 2i$; corresponding eigenvectors are $v^1 = [2 \quad -i]^T$ and $v^2 = [2 \quad i]^T$. The general real-valued solution is

$$\begin{bmatrix} x \\ y \end{bmatrix} = c_1 e^{-t} \begin{bmatrix} 2\cos 2t \\ \sin 2t \end{bmatrix} + c_2 e^{-t} \begin{bmatrix} 2\sin 2t \\ -\cos 2t \end{bmatrix}.$$

for arbitrary reals c_1, c_2: this is a spiral point (asymptotically stable).

Answer 7: The characteristic polynomial of the system matrix is

$$p(\lambda) = \det \begin{bmatrix} 2-\lambda & 1 \\ -1 & 4-\lambda \end{bmatrix} = (\lambda - 3)(\lambda - 3)$$

So the eigenvalues are $\lambda_1 = \lambda_2 = 3$. Corresponding eigenspace is spanned by the eigenvector $v^1 = [1 \quad 1]^T$. Hence the eigenspace is deficient. Now solve for v^2 where $(A - 3I)v^2 = v^1$ to get $v^2 = [0 \quad 1]^T$. The general real-valued solution is $[x \quad y]^T = c_1 v^1 e^{3t} + c_2 e^{3t}(v^1 t + v^2)$, where c_1 and c_2 are arbitrary reals. The origin is a deficient node, and it is unstable.

Answer 9: The characteristic polynomial of the system matrix is

$$p(\lambda) = \det \begin{bmatrix} 2-\lambda & -1 \\ 1 & 4-\lambda \end{bmatrix} = (\lambda - 3)(\lambda - 3)$$

and the eigenvalues are $\lambda_1 = \lambda_2 = 3$. Corresponding eigenspace is spanned by eigenvector $u = [1 \quad -1]^T$. Hence, the eigenspace is deficient, we solve for v where $(A - 3I)v = u$ to get $v = [-2 \quad 1]^T$. The general real-valued solution is $[x \quad y]^T = c_1 u e^{3t} + c_2 e^{3t}(ut + v)$, where c_1 and c_2 are arbitrary reals. The origin is a deficient node, and it is unstable.

Answer 11: The characteristic polynomial of the system matrix is

$$p(\lambda) = \det \begin{bmatrix} 3-\lambda & -5 \\ 5 & 3-\lambda \end{bmatrix} = \lambda^2 - 6\lambda + 34$$

and the eigenvalues are $\lambda_1 = \overline{\lambda}_2 = 3 + 5i$. Corresponding eigenvectors are $v^1 = [i \ \ 1]^T$ and $v^2 = [1 \ \ i]^T$. The general real-valued solution is

$$[x \ \ y]^T = c_1 \begin{bmatrix} -\sin 5t \\ \cos 5t \end{bmatrix} e^{3t} + c_2 \begin{bmatrix} \cos 5t \\ \sin 5t \end{bmatrix} e^{3t}$$

where c_1 and c_2 are arbitrary reals. The origin is a spiral point, and it is unstable.

Plotting Orbital Portraits. Plot orbital portraits in the frame $|x| \leq 3, |y| \leq 4$ for each system from the indicated problems. Plot the eigenlines (if any) and representative orbits in sectors bounded by eigenlines. Plot component curves and time-state curves as well. [*Hint*: look at Figures 6.5.1–6.5.8.]

13. Problem 1	**14.** Problem 2	**15.** Problem 3	**16.** Problem 4
17. Problem 5	**18.** Problem 6	**19.** Problem 7	**20.** Problem 8
21. Problem 9	**22.** Problem 10	**23.** Problem 11	**24.** Problem 12

Answer 13: The system is $x' = x + y$, $y' = 4x - 2y$; see the figure.

Answer 15: The system is $x' = 7x + 6y$, $y' = 2x + 6y$; see the figure.

Answer 17: The system is $x' = -x - 4y$, $y' = x - y$; see the figure.

Answer 19: The system is $x' = 2x + y \ \ y' = -x + 4y$; see the figures.

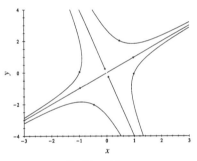

Problem 13.

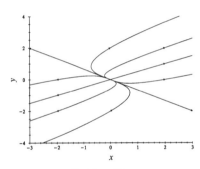

Problem 15.

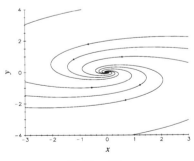

Problem 17.

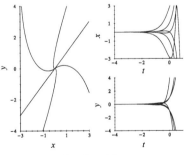

Problem 19.

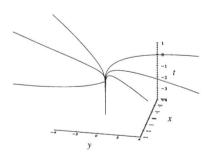

Problem 19.

Answer 21: The system is $x' = 2x - y \ \ y' = x + 4y$; see the figures.

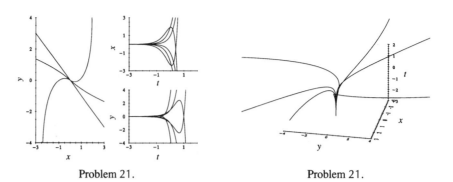

Problem 21.   Problem 21.

Answer 23:   The system is $x' = 3x - 5y$ $y' = 5x + 3y$; see the figures.

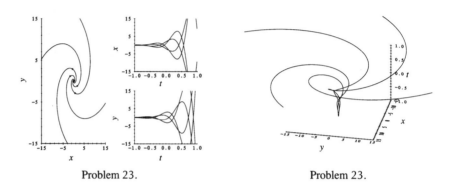

Problem 23.   Problem 23.

Orbital Portraits of Affine Planar Systems.

25. *Affine Planar Systems*   The driven system $x' = ax + by + r$, $y' = cx + dy + s$, where a, b, r, c, d, s are real constants, is an *affine system*. Suppose that $ad - bc \neq 0$.

(a) Show that the affine system has a unique equilibrium point.

(b) Show that the change of variables from x and y to u and v given by $x = u + x_0$, $y = v + y_0$, where (x_0, y_0) is an equilibrium point, yields the undriven system $u' = au + bv$, $v' = cu + dv$.

(c) Show that if $u = u(t)$, $v = v(t)$ is the general solution of the u', v' system in **(b)**, then the general solution of the affine system is $x(t) = u(t) + x_0$, $y(t) = v(t) + y_0$.

(d) Find the general solution of the system $x' = x + y + 1$, $y' = 4x - 2y - 1$. Classify the system (is it nodal, spiral, ... ?). Plot an orbital portrait in $|x - x_0| \leq 3$, $|y - y_0| \leq 4$, where (x_0, y_0) is the equilibrium point. Plot all eigenlines through the equilibrium point.

Answer 25:

(a) We wish to find (x_0, y_0) such that $x' = 0$ and $y' = 0$. That is, we want to solve the system $Ax = -[r \ s]^T$:

$$A \begin{bmatrix} x_0 \\ y_0 \end{bmatrix} = \begin{bmatrix} a & b \\ c & d \end{bmatrix} \begin{bmatrix} x_0 \\ y_0 \end{bmatrix} = \begin{bmatrix} -r \\ -s \end{bmatrix}$$

Since $\det \begin{bmatrix} a & b \\ c & d \end{bmatrix} \neq 0$ the system has a unique solution,

$$[x_0 \quad y_0]^T = \begin{bmatrix} a & b \\ c & d \end{bmatrix}^{-1} \begin{bmatrix} -r \\ -s \end{bmatrix}$$

(b) We have $ax_0 + by_0 + r = 0$ and $cx_0 + dy_0 + s = 0$ if (x_0, y_0) is the equilibrium point of the planar affine system. Substituting $x = u + x_0$ and $y = v + y_0$ into the affine system, we obtain the homogeneous system $u' = au + bv, \ v' = cu + dv$.

(c) This follows directly from **(b)**.

(d) Calculations show that the equilibrium point is $(-1/6, -5/6)$. The general solution of the corresponding homogeneous system is $[u \quad v]^T = c_1 w^1 e^{2t} + c_2 w^2 e^{-3t}$, where $w^1 = [1 \quad 1]^T$ and $w^2 = [1 \quad -4]^T$ are eigenvectors of the homogeneous system corresponding to $\lambda_1 = 2, \ \lambda_2 = -3$, respectively. The general real-valued solution of the system $x' = x + y + 1, \ y' = 4x - 2y - 1$ is $[x \quad y]^T = c_1 w^1 e^{2t} + c_2 w^2 e^{-3t} + [-1/6 \quad -5/6]^T$, for arbitrary reals c_1, c_2. The equilibrium point is a saddle. See the figure.

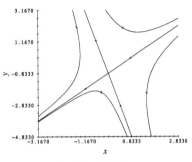

Problem 25(d).

Planar Autonomous Systems.

26. *Zero Eigenvalues* Consider a planar autonomous system where the system matrix A of real constants has at least one zero eigenvalue. Explain why either there is one eigenline consisting solely of equilibrium points or else all points are equilibrium points. Plot orbital portraits of the various distinct cases where at least one of the eigenvalues is zero. Explain why your gallery of portraits represents all the possible cases. Classify each equilibrium point. as neutrally stable, unstable, or asymptotically stable.

27. *Second-Order ODE* The scalar ODE $y'' + 2ay' + by = 0$, where a and b are real constants, is equivalent to the system $x_1' = x_2, \ x_2' = -bx_1 - 2ax_2$.

(a) Explain why the system has a saddle point if $a = 1$ and $b < 0$ and an improper node if $a = 1$ and $0 < b < 1$.

(b) Show that the system has a deficient node if $a = 1$, $b = 1$ and a spiral point if $a = 1, b > 1$.

(c) Find all values for a and b for which the system has a center.

(d) Show that no values of a and b lead to a system that has a star node.

Answer 27: The system matrix is $\begin{bmatrix} 0 & 1 \\ -b & -2a \end{bmatrix}$. The characteristic polynomial is $p(\lambda) = \lambda^2 + 2a\lambda + b$, and the eigenvalues are $\lambda_{1,2} = -a \pm \sqrt{a^2 - b}$.

(a) If $a = 1$ and $b < 0$, then $\lambda_{1,2} = -1 \pm \sqrt{1 - b}$, and $\lambda_1 > 0$ while $\lambda_2 < 0$, and so the origin is a saddle point. If $0 < b < 1$, then $\lambda_1, \lambda_2 < 0$ and $\lambda_1 \neq \lambda_2$ so the origin is an improper node.

(b) If $a = 1, \ \lambda_{1,2} = -1 \pm \sqrt{1 - b}$. So if $b = 1$, then $\lambda_1 = \lambda_2 = -1$ is a double eigenvalue with a one-dimensional eigenspace (spanned by $[1 \quad -1]^T$); the origin is a deficient node. If $b > 1$, then $1 - b < 0$, and the eigenvalues are complex with nonzero real part, making the origin a spiral point.

(c) For the system to have a center, the eigenvalues must be pure imaginary. This only happens for $a = 0, \ b > 0$.

(d) For the system to have a star node, there must be a double eigenvalue, so $a^2 = b > 0$ and $\lambda = -a$. However, the eigenspace corresponding to $\lambda = -a$ is one-dimensional (spanned by $[1 \quad -a]^T$). So the system cannot have a star node.

28. *An Inverse Problem*   Suppose it is known that $x_1 = 2\sin(2t - \pi)$, $x_2 = \cos(2t - \pi)$ is one solution of a 2×2 linear system $x' = Ax$ for a constant matrix A.

   (a) Find the general, real-valued solution of the system.

   (b) Plot some representative orbits of the system near the origin.

29. Show that the orbits in Example 6.5.5 are the tilted ellipses $x^2 + (2y - 3x)^2 = C$.

   Answer 29:   As indicated in Example 6.5.5, the general solution of the system is

   $$x = 2(c_1 \cos t + c_2 \sin t)$$

   $$y = c_1(3\cos t + \sin t) + c_2(3\sin t - \cos t)$$

   where c_1 and c_2 are arbitrary reals. First, observe that $2y - 3x = 2(c_1 \sin t - c_2 \cos t)$. It follows that $x^2 + (2y - 3x)^2 = 4(c_1^2 + c_2^2)$ which is a nest of ellipses with common axes and centered at the origin.

30. *Solution by Elimination*   Follow the outline below to derive the general solution formula for the planar autonomous system

   $$x' = ax + by$$

   $$y' = cx + dy$$

   - If $b = 0$ or $c = 0$, then the system is a linear cascade. In this case, solve the system one ODE at a time.

   - If $b \neq 0$, then solve the first ODE for y to obtain $y = (x' - ax)/b$. Use this formula to eliminate y in the second ODE and obtain the second-order ODE $x'' - (a + d)x' + (ad - bc)x = 0$.

   - Let $p = a + d$, $q = ad - bc$. Show that the characteristic roots of the second-order ODE are $r_1 = (p + \sqrt{p^2 - 4q})/2$ and $r_2 = (p - \sqrt{p^2 - 4q})/2$. Use Theorem 3.2.1 to find $x(t)$, and back substitution to find $y(t)$.

   Use this technique to verify the solutions to the systems in Examples 6.5.1–6.5.6.

Orbital Portraits in Three Dimensions.

31. Consider the system $x' = Ax$, where $x = [x_1 \ x_2 \ x_3]^T$ and A is a 3×3 matrix of real constants with nonzero eigenvalues. Construct a gallery of representative orbital portraits in $x_1x_2x_3$ state space that covers all the cases. For example, if $\lambda_1, \lambda_2,$ and λ_3 are the eigenvalues of A, you will need to consider the case where the three eigenvalues are real, distinct, and have a common sign. Plot orbits in three dimensions of a system such as $x_1' = -x_1$, $x_2' = -2x_2$, $x_3' = -3x_3$. Another one of the cases you should consider is $\lambda_1 = \alpha + i\beta$, $\lambda_2 = \alpha - i\beta$, $\alpha, \beta,$ and λ_3 real and nonzero. The representative system in this case might be $x_1' = -x_1 + 5x_2$, $x_2' = -5x_1 - x_2$, $x_3' = -x_3$. Since any system with eigenvalues $\lambda_1, \lambda_2,$ and λ_3 behaves as $t \to +\infty$ exactly the same as the system with eigenvalues $-\lambda_1, -\lambda_2,$ and $-\lambda_3$ as $t \to -\infty$, their orbital portraits are identical, so don't treat them as distinct cases. Your gallery should have more than 10 distinct types of orbital portraits. [*Hint:* to construct your example system, start with the six planar cases of Examples 6.5.1–6.5.6 and add a third ODE such as $x_3' = kx_3$.]

   Answer 31:   Group project.

6.6 Driven Systems: The Matrix Exponential

Comments:

Operator notation is used first to simplify matters, and to help explain why the process of finding all solutions of $x' = Ax + F(t)$ splits two subproblems: find all solutions of $L[x] = 0$, where L is the linear operator $L = d/dt - A$ (i.e., first find the general solution of $x' = Ax$), and then find a particular solution of $L[x] = f$.

In this section we characterize the solution set of the linear system $x' = Ax + F(t)$, where A is a matrix of constants, using a variation of parameters approach. Basic matrices and transition matrices make their appearance—all quite abstract, but leading to the elegant formula (15) for the general solution and the even nicer formula (16) for the unique solution of the corresponding IVP. To find all solutions of the undriven system, $x' = Ax$, one can use any basic solution matrix $X(t)$, and write the general solution as $x(t) = X(t)c$, where c is any constant vector. On the other hand, in order to characterize the solution of the driven IVP as the sum of the responses to the initial data and the driving force as we do in formula (16), you need to use the transition matrix. Given any basic solution matrix $X(t)$, one can construct the transition matrix as $X(t)X^{-1}(t_0)$ [see formula (8)].

PROBLEMS

The Matrix Exponential. Find e^{tA} for the given matrix A. Then solve the IVP $x' = Ax$, $x(0) = [1\ \ 2]^T$. [*Hint*: find the eigenvalues of A and a basis for each eigenspace. Then construct a basic solution matrix $X(t)$ as in formula (8); $e^{tA} = X(t)X^{-1}(0)$.]

1. $\begin{bmatrix} 1 & 3 \\ 1 & -1 \end{bmatrix}$ **2.** $\begin{bmatrix} 0 & 1 \\ 9 & 0 \end{bmatrix}$ **3.** $\begin{bmatrix} -1 & 1 \\ 0 & -1 \end{bmatrix}$ **4.** $\begin{bmatrix} 1 & 1 \\ -1 & 1 \end{bmatrix}$

5. $\begin{bmatrix} 2 & 1 \\ -1 & 4 \end{bmatrix}$ **6.** $\begin{bmatrix} 1 & -1 \\ 0 & 3 \end{bmatrix}$ **7.** $\begin{bmatrix} 2 & -1 \\ 1 & 4 \end{bmatrix}$ **8.** $\begin{bmatrix} 1 & -2 \\ 2 & 1 \end{bmatrix}$

Answer 1: The eigenvalues of $A = \begin{bmatrix} 1 & 3 \\ 1 & -1 \end{bmatrix}$ are $\lambda_1 = -2$, $\lambda_2 = 2$; eigenspaces are $V_{-2} = \text{Span}\{[1\ \ -1]^T\}$, $V_2 = \text{Span}\{[3\ \ 1]^T\}$. So

$$X(t) = \begin{bmatrix} e^{-2t} & 3e^{2t} \\ -e^{-2t} & e^{2t} \end{bmatrix}, \quad X^{-1}(0) = \frac{1}{4}\begin{bmatrix} 1 & -3 \\ 1 & 1 \end{bmatrix}$$

$$e^{tA} = X(t)X^{-1}(0) = \frac{1}{4}\begin{bmatrix} e^{-2t} + 3e^{2t} & -3e^{-2t} + 3e^{2t} \\ -e^{-2t} + e^{2t} & 3e^{-2t} + e^{2t} \end{bmatrix}$$

The solution of the IVP is

$$x(t) = e^{tA}\begin{bmatrix} 1 \\ 2 \end{bmatrix} = \frac{1}{4}[-5e^{-2t} + 9e^{2t}, \ \ 5e^{-2t} + 3e^{2t}]^T$$

Answer 3: The eigenvalue of $A = \begin{bmatrix} -1 & 1 \\ 0 & -1 \end{bmatrix}$ is -1 and it is a double eigenvalue, but V_{-1} is only one-dimensional. Since the system is upper triangular, solve the system directly rather than construction $X(t)$ by first finding a generalized eigenbasis of $\mathbb{R}^2$. Start with $x_2' = -x_2$, inserting its solution $c_2 e^{-t}$ into the ODE, $x_1' = -x_1 + x_2$, and solving that ODE by integrating factors. We have $x_2 = c_2 e^{-t}$, $x_1 = (c_1 + c_2 t)e^{-t}$. A fundamental matrix is

$$X(t) = \begin{bmatrix} e^{-t} & te^{-t} \\ 0 & e^{-t} \end{bmatrix}$$

obtained by first setting $c_1 = 1$, $c_2 = 0$ then $c_1 = 0$, $c_2 = 1$. Since $X(0) = I = X^{-1}(0)$, we have $X(t) = e^{tA}$, and $x(t) = e^{-t}[1 + 2t\ \ 2]^T$.

Answer 5: The number $\lambda = 3$ is the double eigenvalue of A, but V_3 is one-dimensional and is given by $V_3 = \text{Span}\{[1\ \ 1]^T\}$. Hence V_3 is deficient. Here is one way to find e^{tA} in this case of a deficient eigenspace, a way that does not depend on finding a generalized eigenbasis of $\mathbb{R}^2$. By definition, the first column of e^{tA} is the unique solution of the IVP

$$x' = 2x + y \qquad x(0) = 1$$
$$y' = -x + 4y \qquad y(0) = 0$$

Solve for x in the second ODE to get $x = 4y - y'$ and then substitute back into the first ODE to obtain $y'' - 6y' + 9y = 0$. The general real-valued solution is $y(t) = c_1 e^{3t} + c_2 t e^{3t}$ and hence $x(t) = c_1 e^{3t} + c_2 t e^{3t} - c_2 e^{3t}$ where c_1 and c_2 are arbitrary constants. Applying the initial conditions, we get $c_1 = 0$ and $c_2 = -1$.

The second column in e^{tA} is the unique solution of the IVP

$$x' = 2x + y \qquad x(0) = 0$$
$$y' = -x + 4y \qquad y(0) = 1$$

Applying the initial conditions, we finally get the solution $x(t) = t e^{3t}$ and $y(t) = t e^{3t} + e^{3t}$. Then

$$e^{tA} = \begin{bmatrix} e^{3t} - t e^{3t} & t e^{3t} \\ -t e^{3t} & t e^{3t} + e^{3t} \end{bmatrix}$$

and the solution of the system with the initial conditions $[x(0) \ y(0)]^T = [1 \ 2]^T$ is

$$\begin{bmatrix} x(t) \\ y(t) \end{bmatrix} = e^{tA} \begin{bmatrix} 1 \\ 2 \end{bmatrix} = \begin{bmatrix} e^{3t} + t e^{3t} \\ 2e^{3t} + t e^{3t} \end{bmatrix}$$

Answer 7: The characteristic polynomial is $p(\lambda) = (\lambda - 3)(\lambda - 3)$, so the eigenvalues are $\lambda_1 = \lambda_2 = 3$, and $V_{\lambda_1} = \text{Span}\{[1 \ -1]^T\}$, and hence this eigenspace is deficient. Note that $(A - 3I)[-1 \ 0]^T = [1 \ -1]^T$. Thus, using the approach in Example 6.3.6, the general solution of the system is

$$x = C_1 \begin{bmatrix} 1 \\ -1 \end{bmatrix} e^{3t} + C_2 \left(\begin{bmatrix} -1 \\ 0 \end{bmatrix} + t \begin{bmatrix} 1 \\ -1 \end{bmatrix} \right) e^{3t}$$

where C_1 and C_2 are arbitrary real numbers. Now find a solution $x^1(t)$ with $x^1(0) = [1 \ 0]^T$:

$$\begin{bmatrix} 1 \\ 0 \end{bmatrix} = C_1 \begin{bmatrix} 1 \\ -1 \end{bmatrix} + C_2 \begin{bmatrix} -1 \\ 0 \end{bmatrix}$$

Hence, $C_1 = 0$ and $C_2 = -1$, and so

$$x^1(t) = -\left(\begin{bmatrix} -1 \\ 0 \end{bmatrix} + t \begin{bmatrix} 1 \\ -1 \end{bmatrix} \right) e^{3t} = \begin{bmatrix} e^{3t} - t e^{3t} \\ t e^{3t} \end{bmatrix}$$

Now find a solution $x^2(t)$ with $x^2(0) = [0 \ 1]^T$:

$$\begin{bmatrix} 0 \\ 1 \end{bmatrix} = C_1 \begin{bmatrix} 1 \\ -1 \end{bmatrix} + C_2 \begin{bmatrix} -1 \\ 0 \end{bmatrix}$$

Hence, $C_1 = -1$ and $C_2 = -1$, and so

$$x^2(t) = -\begin{bmatrix} 1 \\ -1 \end{bmatrix} e^{3t} - \left(\begin{bmatrix} -1 \\ 0 \end{bmatrix} + t \begin{bmatrix} 1 \\ -1 \end{bmatrix} \right) e^{3t} = \begin{bmatrix} -t e^{3t} \\ e^{3t} + t e^{3t} \end{bmatrix}$$

Hence,

$$e^{tA} = \begin{bmatrix} e^{3t} - t e^{3t} & -t e^{3t} \\ t e^{3t} & e^{3t} + t e^{3t} \end{bmatrix}$$

The solution of the system for the initial data $[1 \ 2]^T$ is

$$\begin{bmatrix} x(t) \\ y(t) \end{bmatrix} = e^{tA} \begin{bmatrix} 1 \\ 2 \end{bmatrix} = \begin{bmatrix} e^{3t} - 3t e^{3t} \\ 3t e^{3t} + 2e^{3t} \end{bmatrix}$$

Solving an IVP. Solve the IVP $x' = Ax$, $x(0) = [1\ \ 2\ \ 3]^T$ for the given system matrix A. [*Hint:* see Examples 6.6.7 and 6.6.8.]

9. $\begin{bmatrix} 0 & 1 & 1 \\ 0 & 0 & 1 \\ 0 & 0 & 0 \end{bmatrix}$ **10.** $\begin{bmatrix} 2 & 0 & 0 \\ 0 & -3 & 0 \\ 0 & 0 & 7 \end{bmatrix}$ **11.** $\begin{bmatrix} 0 & 0 & 0 \\ 2 & 0 & 0 \\ 3 & 4 & 0 \end{bmatrix}$

Answer 9: In this case A is nilpotent, and the series expansion for e^{tA} terminates after three terms since

$$A = \begin{bmatrix} 0 & 1 & 1 \\ 0 & 0 & 1 \\ 0 & 0 & 0 \end{bmatrix}, \quad A^2 = \begin{bmatrix} 0 & 0 & 1 \\ 0 & 0 & 0 \\ 0 & 0 & 0 \end{bmatrix}, \quad A^3 = \begin{bmatrix} 0 & 0 & 0 \\ 0 & 0 & 0 \\ 0 & 0 & 0 \end{bmatrix}$$

So

$$e^{tA} = I + tA + t^2 A^2/2 = \begin{bmatrix} 1 & t & t + t^2/2 \\ 0 & 1 & t \\ 0 & 0 & 1 \end{bmatrix}$$

$$x(t) = \begin{bmatrix} 1 + 5t + 3t^2/2 & 2 + 3t & 3 \end{bmatrix}^T$$

Answer 11: The series expansion for e^{tA} terminates after three terms since

$$A = \begin{bmatrix} 0 & 0 & 0 \\ 2 & 0 & 0 \\ 3 & 4 & 0 \end{bmatrix}, \quad A^2 = \begin{bmatrix} 0 & 0 & 0 \\ 0 & 0 & 0 \\ 8 & 0 & 0 \end{bmatrix}, \quad A^k = \begin{bmatrix} 0 & 0 & 0 \\ 0 & 0 & 0 \\ 0 & 0 & 0 \end{bmatrix} \text{ for } k \geq 3$$

So

$$e^{tA} = I + tA + t^2 A^2/2 = \begin{bmatrix} 1 & 0 & 0 \\ 2t & 1 & 0 \\ 3t + 4t^2 & 4t & 1 \end{bmatrix}$$

$$x(t) = \begin{bmatrix} 1 & 2t + 2 & 3 + 11t + 4t^2 \end{bmatrix}^T$$

The Matrix Exponential for Block Matrices.

12. Show that if A is the block matrix $\begin{bmatrix} B & 0 \\ 0 & C \end{bmatrix}$, then e^{tA} is the block matrix $\begin{bmatrix} e^{tB} & 0 \\ 0 & e^{tC} \end{bmatrix}$. [*Hint:* the system $x' = Ax$ decouples into two subsystems.]

13. Use the results of Problem 12 to find e^{tA} if

(a) $A = \begin{bmatrix} 1 & 3 & 0 \\ 1 & -1 & 0 \\ 0 & 0 & 2 \end{bmatrix}$ (b) $A = \begin{bmatrix} 1 & 3 & 0 & 0 \\ 1 & -1 & 0 & 0 \\ 0 & 0 & 0 & 1 \\ 0 & 0 & 0 & 0 \end{bmatrix}$

Answer 13:

(a) Using the result from Problem 12 and part of the answer to Problem 1, we have

$$e^{tA} = \frac{1}{4} \begin{bmatrix} e^{-2t} + 3e^{2t} & -3e^{-2t} + 3e^{2t} & 0 \\ -e^{-2t} + e^{2t} & 3e^{-2t} + e^{2t} & 0 \\ 0 & 0 & 4e^{2t} \end{bmatrix} \quad \text{if } A = \begin{bmatrix} 1 & 3 & 0 \\ 1 & -1 & 0 \\ 0 & 0 & 2 \end{bmatrix}$$

(b) For the second matrix observe that $C = \begin{bmatrix} 0 & 1 \\ 0 & 0 \end{bmatrix}$ is nilpotent and $e^{tC} = \begin{bmatrix} 1 & t \\ 0 & 1 \end{bmatrix}$:

$$e^{tA} = \frac{1}{4}\begin{bmatrix} e^{-2t}+3e^{2t} & -3e^{-2t}+3e^{2t} & 0 & 0 \\ -e^{-2t}+e^{2t} & 3e^{-2t}+e^{2t} & 0 & 0 \\ 0 & 0 & 4 & 4t \\ 0 & 0 & 0 & 4 \end{bmatrix} \quad \text{if } A = \begin{bmatrix} 1 & 3 & 0 & 0 \\ 1 & -1 & 0 & 0 \\ 0 & 0 & 0 & 1 \\ 0 & 0 & 0 & 0 \end{bmatrix}$$

Solve Driven Systems. For each IVP $x' = Ax + F$, $x(0) = x^0$, calculate e^{tA}, $e^{tA}x^0$, $e^{-sA}F(s)$. Write the solution in form (16), but do not evaluate the integrals.

14. $\quad x' = \begin{bmatrix} 0 & 2 \\ -2 & 0 \end{bmatrix} x + \begin{bmatrix} 1 \\ 0 \end{bmatrix}, \quad x(0) = \begin{bmatrix} a \\ b \end{bmatrix}$

15. $\quad x' = \begin{bmatrix} 2 & -1 \\ 3 & -2 \end{bmatrix} x + \begin{bmatrix} 3e^t \\ t \end{bmatrix}, \quad x(0) = \begin{bmatrix} 1 \\ 2 \end{bmatrix}.$ Plot the component curves.

16. $\quad x' = \begin{bmatrix} 2 & -5 \\ 1 & -2 \end{bmatrix} x + \begin{bmatrix} \cos t \\ 0 \end{bmatrix}, \quad x(0) = \begin{bmatrix} a \\ b \end{bmatrix}$

17. $\quad x' = \begin{bmatrix} -1 & -4 \\ 1 & -1 \end{bmatrix} x + \begin{bmatrix} e^{-3t} \\ 1 \end{bmatrix}, \quad x(0) = \begin{bmatrix} 0 \\ 0 \end{bmatrix}.$ Plot the component curves.

18. $\quad x' = \begin{bmatrix} 3 & -1 & 1 \\ 2 & 0 & 1 \\ 1 & -1 & 2 \end{bmatrix} x + \begin{bmatrix} f_1(t) \\ f_2(t) \\ f_3(t) \end{bmatrix}, \quad x(0) = \begin{bmatrix} a \\ b \\ c \end{bmatrix}$

Answer 15: $\lambda_1 = -1$, $\lambda_2 = 1$, $V_{-1} = \text{Span}\{[1 \quad 3]^T\}$, $V_1 = \text{Span}\{[1 \quad 1]^T\}$. So

$$X(t) = \begin{bmatrix} e^{-t} & e^t \\ 3e^{-t} & e^t \end{bmatrix}, \qquad e^{tA} = \frac{1}{2}\begin{bmatrix} -e^{-t}+3e^t & e^{-t}-e^t \\ -3e^{-t}+3e^t & 3e^{-t}-e^t \end{bmatrix}$$

$$e^{tA}x^0 = e^{tA}\begin{bmatrix} 1 \\ 2 \end{bmatrix} = \frac{1}{2}\begin{bmatrix} e^{-t}+e^t \\ 3e^{-t}+e^t \end{bmatrix}$$

$$e^{-sA}F(s) = \frac{1}{2}\begin{bmatrix} -e^s+3e^{-s} & e^s-e^{-s} \\ -3e^s+3e^{-s} & 3e^s-e^{-s} \end{bmatrix}\begin{bmatrix} 3e^s \\ s \end{bmatrix} = \frac{1}{2}\begin{bmatrix} -3e^{2s}+9+s(e^s-e^{-s}) \\ -9e^{2s}+9+s(3e^s-e^{-s}) \end{bmatrix}$$

From formula (16) the solution of the IVP is

$$x(t) = e^{tA}\begin{bmatrix} 1 \\ 2 \end{bmatrix} + \int_0^t e^{(t-s)A}\begin{bmatrix} 3e^s \\ s \end{bmatrix} ds$$

$$= \frac{1}{2}\begin{bmatrix} e^{-t}+e^t \\ 3e^{-t}+e^t \end{bmatrix} + \frac{1}{2}\int_0^t \begin{bmatrix} 3e^s\left(-e^{-(t-s)}+3e^{t-s}\right)+s(e^{-(t-s)}-e^{t-s}) \\ 3e^s\left(-3e^{-(t-s)}+3e^{t-s}\right)+s(3e^{-(t-s)}-e^{t-s}) \end{bmatrix} ds$$

See the figure for the component curves.

Answer 17: $\lambda_1 = -1 + 2i = \bar{\lambda}_2$, $V_{-1+2i} = \text{Span}\{[2i \quad 1]^T\} = \bar{V}_{-1-2i}$. So

$$X(t) = \begin{bmatrix} 2ie^{(-1+2i)t} & -2ie^{(-1-2i)t} \\ e^{(-1+2i)t} & e^{(-1-2i)t} \end{bmatrix}, \qquad e^{tA} = X(t)X^{-1}(0) = e^{-t}\begin{bmatrix} \cos 2t & -2\sin 2t \\ (\sin 2t)/2 & \cos 2t \end{bmatrix}$$

$$e^{tA}x^0 = e^{tA}\begin{bmatrix} 0 \\ 0 \end{bmatrix} = \begin{bmatrix} 0 \\ 0 \end{bmatrix}$$

$$e^{-sA}F(s) = e^s\begin{bmatrix} \cos 2s & 2\sin 2s \\ -(\sin 2s)/2 & \cos 2s \end{bmatrix}\begin{bmatrix} e^{-3s} \\ 1 \end{bmatrix} = \begin{bmatrix} e^{-2s}\cos 2s + 2e^s\sin 2s \\ -e^{-2s}(\sin 2s)/2 + e^s\cos 2s \end{bmatrix}$$

From formula (16), the solution of the IVP is

$$x(t) = e^{-t} \int_0^t \begin{bmatrix} e^{-3s}\cos 2(t-s) - 2\sin 2(t-s) \\ e^{-3s}(\sin 2(t-s))/2 + \cos 2(t-s) \end{bmatrix} ds$$

See the figure for the component graphs.

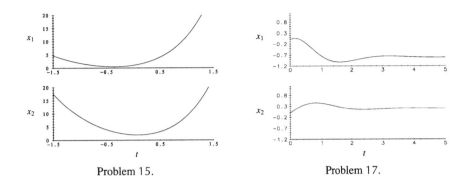

Problem 15.   Problem 17.

Solve IVPs.   Find the solution of each IVP (evaluate the integrals in the solution formulas).

19. $x' = \begin{bmatrix} 5 & 3 \\ -6 & -4 \end{bmatrix} x + \begin{bmatrix} 1 \\ e^t \end{bmatrix}, \quad x(0) = \begin{bmatrix} 1 \\ 0 \end{bmatrix}$

20. $x' = \begin{bmatrix} 6 & -4 \\ 4 & 2 \end{bmatrix} x + \begin{bmatrix} e^{-2t} \\ 0 \end{bmatrix}, \quad x(0) = \begin{bmatrix} 1 \\ 1 \end{bmatrix}$

21. $x' = \begin{bmatrix} 1 & 2 \\ -2 & 1 \end{bmatrix} x + \begin{bmatrix} e^{-t} \\ 1 \end{bmatrix}, \quad x(0) = \begin{bmatrix} 0 \\ 1 \end{bmatrix}$

22. $x' = \begin{bmatrix} 1 & -2 & -2 \\ -2 & 1 & 2 \\ 2 & -2 & -3 \end{bmatrix} x + \begin{bmatrix} e^{-t} \\ 0 \\ e^t \end{bmatrix}, \quad x(0) = \begin{bmatrix} 1 \\ 0 \\ 0 \end{bmatrix}$

Answer 19:   The characteristic polynomial of the system matrix A is $\lambda^2 - \lambda - 2$ with the real roots $\lambda_1 = 2$ and $\lambda_2 = -1$. The eigenspace V_2 is spanned by $[1 \quad -1]^T$ and the eigenspace V_{-1} is spanned by $[1 \quad -2]^T$. So we have the basic solution matrix

$$X(t) = \begin{bmatrix} e^{2t} & e^{-t} \\ -e^{2t} & -2e^{-t} \end{bmatrix}$$

Hence

$$e^{tA} = X(t)X^{-1}(0) = \begin{bmatrix} e^{2t} & e^{-t} \\ -e^{2t} & -2e^{-t} \end{bmatrix} \begin{bmatrix} 2 & 1 \\ -1 & -1 \end{bmatrix}$$

$$= \begin{bmatrix} 2e^{2t} - e^{-t} & e^{2t} - e^{-t} \\ -2e^{2t} + 2e^{-t} & -e^{2t} + 2e^{-t} \end{bmatrix}$$

and so a particular solution of the driven ODE is

$$x^d(t) = \int_0^t e^{(t-s)A} \begin{bmatrix} 1 \\ e^s \end{bmatrix} ds$$

$$= \int_0^t \begin{bmatrix} 2e^{2t}e^{-2s} - e^{-t}e^s + e^{2t}e^{-s} - e^{-t}e^{2s} \\ -2e^{2t}e^{-2s} + 2e^{-t}e^s - e^{2t}e^{-s} + 2e^{-t}e^{2s} \end{bmatrix} ds$$

$$= \begin{bmatrix} -2 + \dfrac{3}{2}e^{-t} - \dfrac{3}{2}e^t + 2e^{2t} \\ 3 - 3e^{-t} + 2e^t - 2e^{2t} \end{bmatrix}$$

Therefore, the solution to the IVP is

$$x(t) = e^{tA} \begin{bmatrix} 1 \\ 0 \end{bmatrix} + x^d(t)$$

$$= \begin{bmatrix} 2e^{2t} - e^{-t} \\ -2e^{2t} + 2e^{-t} \end{bmatrix} + x^d(t)$$

Answer 21:   The characteristic polynomial is $p(\lambda) = \lambda^2 - 2\lambda + 5$, therefore the eigenvalues are $\lambda_1 = 1 + 2i$ and $\lambda_3 = 1 - 2i$. The eigenspaces are $V_{1+2i} = \text{Span}\{[1 \ \ i]^T\} = \overline{V_{1-2i}}$. Put

$$x^1 = \text{Re}\left[e^{(1+2i)t}[1 \ \ i]^T\right] = e^t \left[\cos 2t \ \ -\sin 2t\right]^T$$

$$x^2 = \text{Im}\left[e^{(1+2i)t}[1 \ \ i]^T\right] = e^t \left[\sin 2t \ \ \cos 2t\right]^T$$

Since $[x^1(0) \ \ x^2(0)] = $ the 2×2 identity matrix, it follows that

$$e^{tA} = \left[x^1(t) \ \ x^2(t)\right]$$

Hence,

$$x^d(t) = \int_0^t e^{(t-s)A} \begin{bmatrix} e^{-s} \\ 1 \end{bmatrix} ds$$

$$= \int_0^t \left(e^{-s}x^1(t-s) + x^2(t-s)\right) ds$$

Make the change of variables $r = t - s$

$$\int_0^t e^{-s}x^1(t-s)\, ds = \int_0^t e^{-s}e^{t-s} \begin{bmatrix} \cos 2(t-s) \\ -\sin 2(t-s) \end{bmatrix} ds$$

$$= \int_0^t e^{r-t}e^r \begin{bmatrix} \cos 2r \\ -\sin 2r \end{bmatrix} dr$$

$$= e^{-t} \int_0^t \begin{bmatrix} e^{2r}\cos 2r \\ -e^{2r}\sin 2r \end{bmatrix} dr$$

$$= e^{-t} \begin{bmatrix} \int_0^t e^{2r}\cos 2r\, dr \\ -\int_0^t e^{2r}\sin 2r\, dr \end{bmatrix}$$

Using the integral table on the inside back cover, we get

$$\int_0^t e^{-s} x^1(t-s)\,ds = e^{-t} \left[\begin{array}{c} \left(\dfrac{1}{4}e^{2r}\cos 2r + \dfrac{1}{4}e^{2r}\sin 2r\right)\Big|_{r=0}^{r=t} \\[4mm] -\left(\dfrac{1}{4}e^{2r}\sin 2r - \dfrac{1}{4}e^{2r}\cos 2r\right)\Big|_{r=0}^{r=t} \end{array} \right]$$

$$= \left[\begin{array}{c} \dfrac{1}{4}e^{t}\cos 2t + \dfrac{1}{4}e^{t}\sin 2t - \dfrac{1}{4}e^{-t} \\[4mm] -\dfrac{1}{4}e^{t}\sin 2t + \dfrac{1}{4}e^{t}\cos 2t - \dfrac{1}{4}e^{-t} \end{array} \right]$$

$$\int_0^t x^2(t-s)\,ds = \int_0^t x^2(r)\,dr$$

$$= \left[\begin{array}{c} \displaystyle\int_0^t e^{r}\sin 2r\,dr \\[4mm] \displaystyle\int_0^t e^{r}\cos 2r\,dr \end{array} \right] = \left[\begin{array}{c} \left(\dfrac{1}{5}e^{r}\sin 2r - \dfrac{2}{5}e^{r}\cos 2r\right)\Big|_{r=0}^{r=t} \\[4mm] \left(\dfrac{1}{5}e^{r}\cos 2r + \dfrac{2}{5}e^{r}\sin 2r\right)\Big|_{r=0}^{r=t} \end{array} \right]$$

$$= \left[\begin{array}{c} \dfrac{1}{5}e^{t}\sin 2t - \dfrac{2}{5}e^{t}\cos 2t + \dfrac{2}{5} \\[4mm] \dfrac{1}{5}e^{t}\cos 2t + \dfrac{2}{5}e^{t}\sin 2t - \dfrac{1}{5} \end{array} \right]$$

Hence

$$x^d(t) = \left[\begin{array}{c} \dfrac{9}{20}e^{t}\sin 2t - \dfrac{3}{20}e^{t}\cos 2t - \dfrac{1}{4}e^{-t} + \dfrac{2}{5} \\[4mm] \dfrac{3}{20}e^{t}\sin 2t + \dfrac{9}{20}e^{t}\cos 2t - \dfrac{1}{4}e^{-t} - \dfrac{1}{5} \end{array} \right]$$

The solution to the IVP is

$$x(t) = e^{tA} \left[\begin{array}{c} 0 \\ 1 \end{array} \right] + x^d(t) = x^2(t) + x^d(t)$$

where x^1, x^2, and x^d are as above.

Undetermined Coefficients. Sometimes a particular solution $x^d(t)$ of $x' = Ax + F$ can be found by a method of undetermined coefficients. Use that method to find all solutions for each system.

23. $x' = \left[\begin{array}{cc} 2 & -1 \\ 5 & -2 \end{array} \right] x + \left[\begin{array}{c} e^t \\ 1 \end{array} \right]$. [*Hint*: assume x^d has the form $x_1 = ae^t + b$, $x_2 = ce^t + k$ for constants a, b, c, k. Find a, b, c, k by inserting x_1 and x_2 into the ODEs and matching coefficients of corresponding terms. Add x^d to the general solution of the undriven system.]

24. $x' = \left[\begin{array}{cc} 1 & 3 \\ 1 & -1 \end{array} \right] x + \left[\begin{array}{c} \cos t \\ 2t \end{array} \right]$. [*Hint*: assume x^d has the form $x_1 = a_1\cos t + b_1\sin t + c_1 + k_1 t$, $x_2 = a_2\cos t + b_2\sin t + c_2 + k_2 t$.]

Answer 23: The eigenvalues of $A = \left[\begin{array}{cc} 2 & -1 \\ 5 & -2 \end{array} \right]$ are $\lambda_1 = i = \overline{\lambda}_2$; $V_i = \text{Span}\{[1 \quad 2-i]^T\} = \overline{V}_{-i}$.

So,

$$X(t) = \left[\begin{array}{cc} e^{it} & e^{-it} \\ (2-i)e^{it} & (2+i)e^{-it} \end{array} \right]$$

To find a particular solution $[x_1(t) \quad x_2(t)]^T$, let $x_1 = ae^t + b$, $x_2 = ce^t + k$. Then we must have that $x_1' = ae^t = 2(ae^t + b) - (ce^t + k) + e^t$, $x_2' = ce^t = 5(ae^t + b) - 2(ce^t + k) + 1$. Matching coefficients of corresponding terms, we have that $a = 2a - c + 1$, $c = 5a - 2c$, and $0 = 2b - k$, $0 = 5b - 2k + 1$. Solving, we have $a = 3/2$, $c = 5/2$, $b = -1$, and $k = -2$. So $x_1 = 3e^t/2 - 1$, $x_2 = 5e^t/2 - 2$ gives the components of a particular solution $x^d(t)$. The general solution is $x = X(t)c + x^d(t)$, where c is any constant vector $[c_1 \ c_2]^T$.

Matrix Exponentials. Here are more properties of e^{tA}; A and B are the matrices in the margin.

$$A = \begin{bmatrix} 0 & 1 \\ 0 & 0 \end{bmatrix}$$

$$B = \begin{bmatrix} 1 & 0 \\ 0 & 0 \end{bmatrix}$$

25. Show that $e^{tA}e^{tB}$ doesn't have to be either $e^{t(A+B)}$ or $e^{tB}e^{tA}$ by calculating all three.

Answer 25: With A and B as given, we find

$$e^{tA} = \begin{bmatrix} 1 & t \\ 0 & 1 \end{bmatrix}, \quad e^{tB} = \begin{bmatrix} e^t & 0 \\ 0 & 1 \end{bmatrix}, \quad e^{tA}e^{tB} = \begin{bmatrix} e^t & t \\ 0 & 1 \end{bmatrix}, \quad e^{tB}e^{tA} = \begin{bmatrix} e^t & te^t \\ 0 & 1 \end{bmatrix}$$

Since $A + B = \begin{bmatrix} 1 & 1 \\ 0 & 0 \end{bmatrix}$, and $\begin{bmatrix} 1 & 1 \\ 0 & 0 \end{bmatrix}^k = \begin{bmatrix} 1 & 1 \\ 0 & 0 \end{bmatrix}$ for all k, we have

$$e^{t(A+B)} = I + \begin{bmatrix} 1 & 1 \\ 0 & 0 \end{bmatrix} t + \begin{bmatrix} 1 & 1 \\ 0 & 0 \end{bmatrix} t^2/2! + \dots$$

$$= I + \begin{bmatrix} 1 & 1 \\ 0 & 0 \end{bmatrix}(e^t - 1) = \begin{bmatrix} e^t & e^t - 1 \\ 0 & 1 \end{bmatrix}$$

Since none of the three matrices $e^{tA}e^{tB}$, $e^{tB}e^{tA}$, $e^{t(A+B)}$ match, we see that the usual multiplication rules of products of exponentials need not hold for matrix exponentials.

26. Suppose that $AB = BA$. Show that $e^{t(A+B)} = e^{tA}e^{tB} = e^{tB}e^{tA}$ for all t. [*Hint:* show that if $P(t) = e^{t(A+B)}e^{-tA}e^{-tB}$, then $P'(t) = 0$ for all t. Since $P(0) = I$, we must have $P(t) = I$.]

27. *Another Way to Find e^{tA}*   Let A be an $n \times n$ constant matrix and suppose that M is a non-singular $n \times n$ constant matrix. Show that if $B = M^{-1}AM$, then $e^{tA} = Me^{tB}M^{-1}$. [*Hint:* recall that e^{tA} is a special solution matrix for the system $x' = Ax$. Find an equivalent system by using the change of state variables $x = My$.]

Answer 27: By definition, we know that e^{tA} is the unique solution of the matrix system

$$X' = AX, \quad X(0) = I$$

where I is the identity matrix. Now if M is a nonsingular constant matrix, and we put $X = MY$, then Y satisfies the IVP

$$Y' = M^{-1}AMY, \quad Y(0) = M^{-1} \tag{i}$$

Put $B = M^{-1}AM$. The matrix $Y = e^{tB}M^{-1}$ is also a solution of IVP (i), and since IVP (i) has a unique solution, we must have that $M^{-1}e^{tA} = e^{tB}M^{-1}$, or $e^{tA} = Me^{tB}M^{-1}$.

Laplace Transforms.

28. *Laplace Transforms for Systems*   The following steps outline the method of Laplace transforms for solving the IVP $x' = Ax + F$, $x(0) = x^0$, where A is an $n \times n$ matrix of constants. The Laplace transform $\mathcal{L}[x(t)]$ of a vector function is the vector of the Laplace transform of the individual components. [*Hint:* review Laplace Transforms in Chapter 5.]

(a) Apply the operator $\mathcal{L}$ to the above system and obtain $(sI - A)\mathcal{L}[x] = x^0 + \mathcal{L}[F]$.

(b) Show that the matrix $sI - A$ is invertible for all real s that are large enough. [*Hint:* $sI - A$ is singular only when s is an eigenvalue of A, and A has a finite number of eigenvalues.]

(c) Show that the solution of the IVP above is

$$x = \mathcal{L}^{-1}[(sI - A)^{-1}x^0] + \mathcal{L}^{-1}\left[(sI - A)^{-1}\mathcal{L}[F]\right]$$

Compare with (16) and show that

$$\mathcal{L}^{-1}[(sI - A)^{-1}] = e^{tA}, \quad \mathcal{L}^{-1}\left[(sI - A)^{-1}\mathcal{L}[F]\right] = e^{tA} * F(t)$$

☞ $e^{tA} * F(t)$ is the *convolution* of e^{tA} and $F(t)$.

(d) Use the Laplace transform to solve the IVP $x' = \begin{bmatrix} 1 & 3 \\ 1 & -1 \end{bmatrix} x$, $x(0) = [1 \ 2]^T$.

6.7 Steady States

Comments:

This section has a distinctly applications-oriented slant. We focus on linear systems where all solutions tend to a constant or to a periodic steady-state and on conditions that assure us that this will happen. So we require that the input vector F for the system $x' = Ax + F$ be either constant or periodic, and that all eigenvalues of the constant system matrix A have negative real parts. Under these conditions, we get a constant or a periodic steady-state that attracts all solutions as $t \to +\infty$. We go even further and state a bounded input-bounded output theorem (BIBO) that assures that if the system is driven by a bounded input $F(t)$, then the output can never be unbounded. This is a crude kind of stability is often termed *engineering stability*.

As we mention briefly in this section, there is an ambiguity in the definition of steady-state. We have taken the hard-line and defined steady-state only when the input vector is a constant or else is periodic. On the other hand, if all eigenvalues of A have negative real parts (as we require), and if $F(t)$ is any bounded column vector for $t \geq 0$, then all solutions of $x' = Ax + F(t)$ tend to one another as $t \to +\infty$. So if $x^1(t)$ and $x^2(t)$ are any two solutions, then $\|x^1(t) - x^2(t)\| \to 0$ as $t \to +\infty$, which suggests that all solutions are steady-state. This is just too inclusive for us, and that is why we have taken the hard-line. Some people define a steady-state as any bounded solution of $x' = Ax + F$, whether or not all other solutions approach it. This also is too inclusive, and certainly is contrary to common-sense applications, where steady-state means a regular and predictable state approached by all other solutions.

PROBLEMS

Constant Steady States. Is the constant solution $x = 0$ a steady state for $x' = Ax$ if A is as listed? Give reasons. [*Hint*: look at the eigenvalues of A; tr $A = \lambda_1 + \lambda_2 + \lambda_3$, det $A = \lambda_1\lambda_2\lambda_3$.]

1. $\begin{bmatrix} -10 & 1 & 2 \\ -3 & 4 & 5 \\ -7 & -6 & 7 \end{bmatrix}$
2. $\begin{bmatrix} -10 & 1 & 2 \\ -3 & 4 & -2 \\ -7 & -6 & -40 \end{bmatrix}$
3. $\begin{bmatrix} -10 & 1 & 2 \\ 0 & -1 & 100 \\ 0 & 0 & -1 \end{bmatrix}$

Answer 1:   The trace of A is $-10 + 4 + 7 = 1$, which is positive; hence, at least one eigenvalue has positive real part. So, $x = 0$ is a constant solution, but (by Theorem 6.7.1), it is not a steady-state.

Answer 3:   Since the matrix is triangular, its eigenvalues are the diagonal entries $-10, -1, -1$. So, $x = 0$ is the unique steady-state by Theorem 6.7.1.

Constant Steady States. Find all equilibrium solutions for each autonomous system. Identify those equilibrium solutions that are constant steady states.

4. $x' = \begin{bmatrix} 5 & 3 \\ -6 & -4 \end{bmatrix} x + \begin{bmatrix} -2 \\ 2 \end{bmatrix}$

5. $x' = \begin{bmatrix} 4 & 8 \\ 1 & 2 \end{bmatrix} x + \begin{bmatrix} -4 \\ -1 \end{bmatrix}$

6. $x' = \begin{bmatrix} 2 & 3 \\ -3 & -4 \end{bmatrix} x + \begin{bmatrix} -1 \\ 1 \end{bmatrix}$

7. $x' = \begin{bmatrix} -6 & 4 \\ -4 & 2 \end{bmatrix} x + \begin{bmatrix} -2 \\ 2 \end{bmatrix}$

Answer 5:   To find the equilibrium solution we solve for x where

$$\begin{bmatrix} 4 & 8 \\ 1 & 2 \end{bmatrix} x + \begin{bmatrix} -4 \\ -1 \end{bmatrix} = \begin{bmatrix} 0 \\ 0 \end{bmatrix}$$

We get $x = [1 \ \ 0]^T$. This equilibrium solution is not a steady state since the characteristic polynomial $p(\lambda) = \lambda(\lambda - 6)$ has a positive root.

Answer 7:   To find the equilibrium solution we solve for x where

$$\begin{bmatrix} -6 & 4 \\ -4 & 2 \end{bmatrix} x + \begin{bmatrix} -2 \\ 2 \end{bmatrix} = \begin{bmatrix} 0 \\ 0 \end{bmatrix}$$

to find $x = [3 \ 5]^T$. This equilibrium solution is a steady state since the characteristic polynomial $p(\lambda) = (\lambda + 2)(\lambda + 2)$ has a double root, and it is negative.

Tune a System to Create a Steady State.   Find the values of the real constant α for which $x = 0$ is the steady state of $x' = Ax$. [*Hint*: look at the eigenvalues of A. In Problems 9 and 10, adapt the analysis used in the text for the lead model.]

8. $\begin{bmatrix} \alpha & 2 \\ 3 & -4 \end{bmatrix}$ **9.** $\begin{bmatrix} 0 & 1 & 0 \\ 0 & 0 & 1 \\ \alpha & -1 & -1 \end{bmatrix}$ **10.** $\begin{bmatrix} 0 & 1 & 0 \\ 0 & 0 & 1 \\ -2 & -3 & \alpha \end{bmatrix}$

Answer 9:   Adapting the argument in this section for the lead model, we see that the roots of the polynomial $\lambda^3 + a\lambda^2 + b\lambda + c$ all have negative real parts if and only if $a > 0, b > 0, c > 0$, and $ab > c$. The characteristic polynomial of the given matrix is (after dividing through by -1) $\lambda^3 + \lambda^2 + \lambda - \alpha$. Therefore, $a = 1, b = 1$, and $c = -\alpha$. So all the characteristic roots have negative real parts if and only if $-\alpha > 0$ and $1 > -\alpha$, i.e., if and only if $-1 < \alpha < 0$.

Bounded Input, Bounded Output.

11.   *BIBO*   The system $x' = y$, $y' = -bx - ay + f(t)$ is equivalent to the single ODE $x'' + ax' + bx = f(t)$. Show that every solution of the system is bounded for $t \geq 0$ if $a > 0$, $b > 0$, and $f(t)$ is continuous and bounded for $t \geq 0$.

Answer 11:   The problem statement is somewhat unclear in the first printing of this edition. It should have read Show that all the solutions of the constant-coefficient linear ODE $x'' + ax' + bx = f(t)$ are bounded on $t \geq 0$ if $a > 0, b > 0$, and $f(t)$ is bounded on $t \geq 0$. Let $x(t)$ be a solution of the ODE and put $y(t) = x'(t)$. Then the function vector $[x(t) \ y(t)]^T$ solves the system

$$\begin{bmatrix} x \\ y \end{bmatrix}' = \begin{bmatrix} 0 & 1 \\ -b & -a \end{bmatrix} \begin{bmatrix} x \\ y \end{bmatrix} + \begin{bmatrix} 0 \\ f(t) \end{bmatrix}$$

The characteristic polynomial of the undriven system is $\lambda^2 + a\lambda + b$. Since $a > 0$, $b > 0$, both roots of the polynomial have negative real parts (see, for example, Problem 16). Hence all solutions of the undriven system are transients. Also, the driving term of the system $[0 \ f(t)]^T$ is bounded. Hence, by BIBO (Theorem 6.7.4) all solutions of the system are bounded on $t \geq 0$. This implies in particular that $x(t)$ is bounded on $t \geq 0$, and we are done.

Unbounded Solutions.

12.   Show that the system $x' = y$, $y' = -x + \cos t$ has unbounded solutions as $t \to +\infty$, but that this does not contradict Theorem 6.7.4.

13.   Show that the system $x' = -x + e^{3t}y$, $y' = -y$ has unbounded solutions as $t \to +\infty$ even though the eigenvalues of the nonconstant system matrix are negative constants. Why is this not a contradiction of Theorem 6.7.4?

14.   Suppose that the matrix A has an eigenvalue with nonnegative real part. Show that there always exists a bounded driving function $F(t)$ such that $x' = Ax + F$ has unbounded solutions. In other words, show that a bounded input need *not* result in a bounded output in this setting. [*Hint*: follow the outline below.]

   (a) Show that $x' = Ax$ has unbounded solutions if A has an eigenvalue with a positive real part.

   (b) Suppose that the matrix A has an eigenvalue λ with a real part of zero. Show that if v is an eigenvector corresponding to λ, the system $x' = Ax + e^{\lambda t}v$ has the unbounded solution $x = te^{\lambda t}v$, $t \geq 0$.

15.   Let all eigenvalues of A have negative real parts. Show that if one solution of $x' = Ax + F(t)$ is unbounded as $t \to +\infty$, then so are all its solutions.

Answer 13:   The general solution of the system $x' = -x + e^{3t}y$, $y' = -y$ is $y = c_2 e^{-t}$, $x = c_1 e^{-t} +$

$(c_2/3)e^{2t}$, and if $c_2 \neq 0$, $|x(t)| \to \infty$ as $t \to +\infty$. The eigenvalues of the system matrix $\begin{bmatrix} -1 & e^{3t} \\ 0 & -1 \end{bmatrix}$ are $-1, -1$, but the Bounded Input-Bounded Output Theorem (Theorem 6.7.4) does not apply since the system matrix is *not* a matrix of constants.

Answer 15: Suppose all eigenvalues of A have negative real parts and that $x^1(t)$ is a solution of $x' = Ax + F(t)$ which is unbounded as $t \to +\infty$. Let $x^2(t)$ be any other solution. Then $x = x^2 - x^1$ is a solution of the undriven system $x' = Ax$. Since the eigenvalues of A have negative real parts, $\|x(t)\| \to 0$ as $t \to +\infty$. So, $x^2(t) = x^1(t) + x(t)$ is the sum of a vector function which is unbounded and another which decays to 0 as $t \to +\infty$, and $x^2(t)$ is also unbounded as $t \to \infty$.

Locate the Eigenvalues.

16. Show that both roots of the polynomial $\lambda^2 + a\lambda + b = 0$ have negative real parts if and only if $a > 0$, $b > 0$.

17. Show that all the roots of the polynomial with real coefficients $\lambda^3 + a\lambda^2 + b\lambda + c = 0$ have negative real parts if and only if $a > 0$, $b > 0$, $c > 0$ and $ab - c > 0$. [*Hint*: see the subsection "The Lead Model of Section 6.1 Revisited."]

Answer 17: Since a, b, and c are real numbers the cubic equation $\lambda^3 + a\lambda^2 + b\lambda + c = 0$ either has three real roots or one real root and a conjugate pair of roots. Denote the roots by λ_1, λ_2 and λ_3 and factorize the polynomial as $(\lambda - \lambda_1)(\lambda - \lambda_2)(\lambda - \lambda_3)$.

Expanding and comparing coefficients we have that $-a = \lambda_1 + \lambda_2 + \lambda_3$, $b = \lambda_1\lambda_2 + \lambda_1\lambda_3 + \lambda_2\lambda_3$, and $-c = \lambda_1\lambda_2\lambda_3$. If λ_1, λ_2, λ_3 all have negative real parts, then $a > 0$, $b > 0$ and $c > 0$. Now assume that $a > 0$, $b > 0$, $c > 0$ and $c - ab > 0$. We will show that the roots λ_1, λ_2, λ_3 all have negative real parts. At least one of the roots is real, call it λ_1. Since $\lambda_1\lambda_2\lambda_3 < 0$, we may as well assume that $\lambda_1 < 0$ and that λ_2 and λ_3 have the same sign if they are real. Now put $\mu = \lambda_2 + \lambda_3$. If we can show that $\mu < 0$, then the real parts of λ_2 and λ_3 are both negative. Since $\lambda_1 = -a - \mu$ and $\lambda_1 b = \lambda_1^2\mu - c$, it follows that $(-a - \mu)b = \lambda_1^2\mu - c$ or $c - ab = (\lambda_1^2 + b)/\mu$. Because $c - ab > 0$ and $b > 0$, it follows that $\mu > 0$ and we are done.

Locating Eigenvalues. All the roots of the polynomial $\lambda^4 + a\lambda^3 + b\lambda^2 + c\lambda + d$ have negative real parts if and only if all the coefficients are positive, $ab > c$, and $abc > a^2 d + c^2$. Which of the polynomials below have the property that all their roots have negative real parts? Justify your answer.

18. $\lambda^4 + 7\lambda^3 + 21\lambda^2 + 37\lambda + 30$ **19.** $\lambda^4 + 5\lambda^3 + 7\lambda^2 + 5\lambda + 6$

Answer 19: In this case we have

$$a = 5, \quad b = 7, \quad c = 5, \quad d = 6$$

So certainly the condition that $a > 0$, $b > 0$, $c > 0$, and $d > 0$ has been met. Next note that

$$ab - c = 5(7) - 5 = 30 > 0$$

$$abc - a^2 d - c^2 = 5(7)(5) - 5^2(6) - 5^2 = 0$$

Hence, all the stated conditions are not met and so the real parts of the roots are not all negative.

Steady State. Explain why each system has a steady state. Then find a formula for it. Plot the component graphs of the steady state and of several other solutions.

20. $x' = y + 5$, $\quad y' = -2x - 3y + 10$

21. $x' = y$, $\quad y' = -x - 2y + \cos t$. [*Hint*: write as an equivalent second-order ODE.]

22. $x' = -x + y + \sin t$, $\quad y' = -x - y$. [*Hint*: assume a particular solution of the form $x = A\cos t + B\sin t$, $y = C\cos t + D\sin t$ and match coefficients.]

Answer 21: The system is equivalent to the scalar equation $x'' + 2x' + x = \cos t$, which has the particular periodic solution $x = 0.5\sin t$. The steady-state solution of the system is the periodic forced solution, $x = 0.5\sin t$, $y = x' = 0.5\cos t$. Since the eigenvalues of the system matrix are -1 and -1, all solutions tend to the periodic forced oscillation as $t \to \infty$. See the figure.

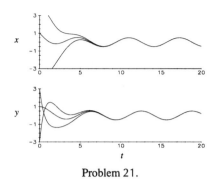

Problem 21.

 Periodic Steady State. Explain why each system has a periodic steady state. Then plot the components of several solutions over a long enough time span that the periodic steady state is clearly visible. Highlight the steady state on your graph.

23. $x' = -x + y/2 + \cos^2 t, \quad y' = -x - 2y + \sin t$

24. $x' = -2x + 5y + \text{SqWave}(t, 2, 1), \quad y' = -5x - 2y + \cos \pi t$

25. $x' = y + \text{TRWave}(t, 1, 1/4, 1/2, 1), \quad y' = -x - y + \text{SqWave}(t, 2, 1)$

26. $x' = y + (\sin t) \text{SqWave}(t, 2\pi, \pi), \quad y' = -4x - y + 1$

Answer 23: The eigenvalues of the system matrix $\begin{bmatrix} -1 & 1/2 \\ -1 & -2 \end{bmatrix}$ are $-3/2 \pm i/2$, and $F(t)$ is periodic with period 2π. So by Theorem 6.7.3 the system has a unique periodic steady state solution of the same period as $F(t)$. See the figure.

Answer 25: The eigenvalues The eigenvalues of the system matrix $\begin{bmatrix} 0 & 1 \\ -1 & -1 \end{bmatrix}$ are $-1/2 \pm \sqrt{3}i/2$, and $F(t)$ is periodic with period 2. So by Theorem 6.7.3 the system has a unique periodic steady state solution of the same period as $F(t)$. See the figure.

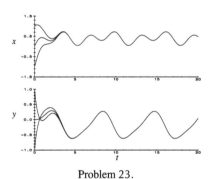

Problem 23.

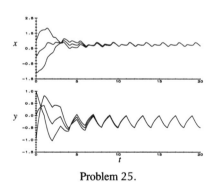

Problem 25.

Periodic Forced Oscillations.

27. A periodic forced oscillation of the system $x' = Ax + F(t)$ is a nonconstant solution $x(t)$ of period T, where A is an $n \times n$ matrix of real constants and $F(t)$ has period T. Show that the system has a unique periodic forced oscillation of period T if $2k\pi i/T$ is *not* an eigenvalue of A for any positive integer k. [*Hint:* $x(t) = e^{tA}x^0 + e^{tA} \int_0^t e^{-sA} F(s) \, ds$ is periodic with period T if and only if the initial value x^0 satisfies the equation $x^0 = x(T) = e^{TA}x^0 + e^{TA} \int_0^T e^{-sA} F(s) \, ds$, that is, if and only if $(I - e^{TA})x^0 = e^{TA} \int_0^T e^{-sA} F(s) \, ds$. Show that if $2\pi i/T$ is *not* an eigenvalue of A, then $I - e^{TA}$ is nonsingular, so a unique x^0 with the desired property exists.]

Answer 27:   Group project. We know $x' = Ax + F(t)$ where $F(t + T) = F(t)$ has a forced oscillation $x(t)$ of period T if $x(0) = x(T) = e^{TA}\left[x(0) + \int_0^T e^{-sA}F(s)ds\right]$, i.e., if $(I - e^{TA})x(0) = e^{TA}\int_0^T e^{-sA}F(s)ds$. Such an equation is uniquely solvable for $x(0)$ if and only if the matrix $I - e^{TA}$ is nonsingular. So we need to show that $I - e^{TA}$ is nonsingular if $2\pi i/T$ is not an eigenvalue of A. Suppose, to the contrary, that $I - e^{TA}$ is singular. Then for some nonzero vector v, $[I - e^{TA}]v = 0$, i.e., $e^{TA}v = v$. Then $y = e^{tA}v$ is a periodic solution of $y' = Ay$ of period T since the solution returns to its initial state v when $t = T$. But by Theorem 6.4.4 all components of all solutions of $y' = Ay$ are polynomial-exponentials, where the exponential factors have the form $e^{\lambda_j t}$, λ_j an eigenvalue of A. The only way any of these functions could be periodic of period T would be if some $\lambda_j = 2\pi i/T$ (leading to terms such as $\cos(2\pi t/T)$ or $\sin(2\pi t/T)$), but by assumption $2\pi i/T$ is *not* an eigenvalue of A. So the supposition that $I - e^{TA}$ is singular is wrong, and $I - e^{TA}$ is nonsingular. So there is a unique initial state $x(0) = [I - e^{TA}]^{-1}e^{TA}\int_0^T e^{-sA}F(s)ds$ which gives a periodic forced oscillation $x(t)$.

6.8 The Theory of General Linear Systems

Comments:

We avoided general linear differential systems until this final section because almost all the models we consider in this text lead to linear systems with constant coefficients. The reason for this is that there is a lot of theory for linear systems with nonconstant coefficients, but not many practical methods for finding solution formulas. Similarly, we didn't do much in Chapter 3 either with second-order linear ODEs with nonconstant coefficients (except at the end of the chapter). The one really useful solution formula technique for general second-order linear ODEs is the power series approach described in Chapter 11. If you are going to do that chapter, we recommend that you first read Section 3.7.

PROBLEMS

Basic Solution Set, Basic Solution Matrix, Transition Matrix.

1. Show that there is a basic solution set for the linear system of ODEs $x' = A(t)x$, where the entries in A are continuous on a t-interval I. [*Hint*: suppose that u^j is the column n-vector with 1 in the jth place and zeros elsewhere. Use Theorem 6.8.1 to show that each of the n IVPs $x' = Ax$, $x(t_0) = u^j$, $j = 1, 2, \ldots, n$, has a unique solution. Call these solutions $x^1, \ldots, x^n$ and consider $W[x^1, \ldots, x^n](t_0)$.]

    Answer 1:   The Fundamental Theorem 6.8.1 says that the IVP $x' = A(t)x$, $x(t_0) = u^j$, has a unique solution $x^j(t)$, for each $j = 1, 2, \ldots, n$. Note that the matrix $[x^1(t) \; \ldots \; x^n(t)]$ evaluated at $t = t_0$ is just the identity matrix I_n. At $t = t_0$ the Wronskian

    $$W(t) = \det[x^1(t) \; \ldots \; x^n(t)]$$

    has the value $W(t_0) = 1 \neq 0$ and so $\{x^1(t), \ldots, x^n(t)\}$ is a basic solution set.

2. Look at the system

    $$x' = \begin{bmatrix} 0 & 1 \\ -2t^{-2} & 2t^{-1} \end{bmatrix} x, \qquad t > 0$$

    (a) Show that $M(t) = \begin{bmatrix} t^2 & t \\ 2t & 1 \end{bmatrix}$ is a basic solution matrix for the system. [*Hint*: verify directly that each column vector of $M(t)$ is a solution of the system.]

    (b) Find the transition matrix $\Phi(t, t_0)$, $t > 0$, $t_0 > 0$. [*Hint*: use $M(t)$ and formula (4).] Verify that $\Phi(t_0, t_0) = I_2$ and that $\Phi^{-1}(t, t_0) = \Phi(t_0, t)$.

☞ More on Euler ODEs in Problem 6, Section 3.7, page 128.

(c) Show that the system in **(a)** is equivalent to the *Euler ODE*: $t^2 y'' - 2ty' + 2y = 0$.

3. Look at the system

$$x' = \begin{bmatrix} t^{-1} & 0 & -1 \\ 0 & 2t^{-1} & -t^{-1} \\ 0 & 0 & t^{-1} \end{bmatrix} x, \qquad t > 0$$

(a) Show that $M(t) = \begin{bmatrix} t & t & -t^2 \\ 0 & t^2 & t \\ 0 & 0 & t \end{bmatrix}$ is a basic solution matrix for the system.

(b) Find the transition matrix $\Phi(t, t_0)$, t, $t_0 > 0$. [*Hint:* use $M(t)$ and formula (4).]

Answer 3: The system is $x' = \begin{bmatrix} t^{-1} & 0 & -1 \\ 0 & 2t^{-1} & -t^{-1} \\ 0 & 0 & t^{-1} \end{bmatrix} x, \; t > 0$.

(a) Direct substitution shows that the columns of $M(t)$ are solutions of the given system. Since $\det M(t)$ is the Wronskian of these three solutions and since $\det M(1) = 1$, it follows that $M(t)$ is a basic solution matrix.

(b) Using equation (4) we have that $\Phi(t, t_0) = M(t)M^{-1}(t_0)$, let's find $M^{-1}(t)$:

$$\begin{bmatrix} t & t & -t^2 \;\| & 1 & 0 & 0 \\ 0 & t^2 & t \;\| & 0 & 1 & 0 \\ 0 & 0 & t \;\| & 0 & 0 & 1 \end{bmatrix} \to \begin{bmatrix} t & t & 0 \;\| & 1 & 0 & t \\ 0 & t^2 & 0 \;\| & 0 & 1 & -1 \\ 0 & 0 & t \;\| & 0 & 0 & 1 \end{bmatrix}$$

$$\to \begin{bmatrix} 1 & 1 & 0 \;\| & t^{-1} & 0 & 1 \\ 0 & 1 & 0 \;\| & 0 & t^{-2} & -t^{-2} \\ 0 & 0 & 1 \;\| & 0 & 0 & t^{-1} \end{bmatrix}$$

$$\to \begin{bmatrix} 1 & 0 & 0 \;\| & t^{-1} & -t^{-2} & 1+1/t^2 \\ 0 & 1 & 0 \;\| & 0 & t^{-2} & -1/t^2 \\ 0 & 0 & 1 \;\| & 0 & 0 & t^{-1} \end{bmatrix}$$

So

$$M^{-1}(t) = \begin{bmatrix} t^{-1} & -t^{-2} & 1+1/t^2 \\ 0 & t^{-2} & -1/t^2 \\ 0 & 0 & t^{-1} \end{bmatrix}$$

and $\Phi(t, t_0) = M(t)M^{-1}(t_0)$, t, $t_0 > 0$.

IVPs. Solve the following initial value problems:

4. $x' = \begin{bmatrix} 0 & 2 \\ -2 & 0 \end{bmatrix} x + \begin{bmatrix} \cos \omega t \\ \sin \omega t \end{bmatrix}$, $x(0) = \begin{bmatrix} a \\ b \end{bmatrix}$, $\omega \neq 0$. Are there any values of ω for which the response of the system is unbounded?

5. $x' = \begin{bmatrix} 0 & 1 \\ -2t^{-2} & 2t^{-1} \end{bmatrix} x + \begin{bmatrix} t^3 \\ t^4 \end{bmatrix}$, $x(1) = \begin{bmatrix} 0 \\ 0 \end{bmatrix}$, $t > 0$. [*Hint:* $[t^2 \; 2t]^T$ and $[t \; 1]^T$ are solutions of the undriven system.]

Answer 5: The IVP is $x' = \begin{bmatrix} 0 & 1 \\ -2t^{-2} & 2t^{-1} \end{bmatrix} x + \begin{bmatrix} t^3 \\ t^4 \end{bmatrix}$, $x(1) = \begin{bmatrix} 0 \\ 0 \end{bmatrix}$. Using the given solutions $[t^2 \; 2t]^T$ and $[t \; 1]^T$ of the undriven system, form the solution matrix

$$M(t) = \begin{bmatrix} t^2 & t \\ 2t & 1 \end{bmatrix}$$

Now det $M(t)$ is the Wronskian of the undriven system and since det $M(1) = -1 \neq 0$, it follows that $M(t)$ is a basic solution matrix. Using (4), we have that for t, $t_0 > 0$:

$$\Phi(t, t_0) = M(t)M^{-1}(t_0)$$

$$= \begin{bmatrix} t^2 & t \\ 2t & 1 \end{bmatrix} \begin{bmatrix} -t_0^{-2} & t_0^{-1} \\ 2t_0^{-1} & -1 \end{bmatrix}$$

$$= \begin{bmatrix} -t^2 t_0^{-2} + 2tt_0^{-1} & t^2 t_0^{-1} - t \\ -2tt_0^{-2} + 2t_0^{-1} & 2tt_0^{-1} - 1 \end{bmatrix}$$

Using solution formula (9) with $t_0 = 1$, we have the solution of the IVP:

$$x(t) = \begin{bmatrix} -t^2 + 2t & t^2 - t \\ -2t + 2 & 2t - 1 \end{bmatrix} \begin{bmatrix} 0 \\ 0 \end{bmatrix} + \int_1^t \begin{bmatrix} -t^2 s^{-2} + 2s^{-1}t & t^2 s^{-1} - t \\ -2ts^{-2} + 2s^{-1} & 2ts^{-1} - 1 \end{bmatrix} \begin{bmatrix} s^3 \\ s^4 \end{bmatrix} ds$$

$$= \int_1^t \begin{bmatrix} -t^2 s + 2s^2 t + t^2 s^3 - ts^4 \\ -2ts + 2s^2 + 2ts^3 - s^4 \end{bmatrix} ds$$

$$= \begin{bmatrix} \left(-\dfrac{t^2 s^2}{2} + \dfrac{2}{3}s^3 t + \dfrac{t^2 s^4}{4} - \dfrac{ts^5}{5} \right)\Big|_1^t \\ \left(-ts^2 + \dfrac{2s^3}{3} + \dfrac{ts^4}{2} - \dfrac{s^5}{5} \right)\Big|_1^t \end{bmatrix} = \begin{bmatrix} \dfrac{t^6}{20} + \dfrac{t^4}{6} + \dfrac{t^2}{4} - \dfrac{7t}{15} \\ -\dfrac{t^3}{3} + \dfrac{3t^5}{10} + \dfrac{t}{2} - \dfrac{7}{15} \end{bmatrix}$$

6. $x' = \begin{bmatrix} t^{-1} & 0 & 0 \\ -t^{-2} & t^{-1} & 0 \\ t^{-3} & -t^{-2} & t^{-1} \end{bmatrix} x + \begin{bmatrix} t^3 \\ t^2 \\ t^3 \end{bmatrix}$, $\quad x(1) = \begin{bmatrix} 1 \\ 0 \\ 1 \end{bmatrix}$, $\quad t > 0$. [*Hint*: $[t \ 1 \ 0]^T$, $[0 \ t \ 1]^T$,

and $[0 \ 0 \ t]^T$ are solutions of the undriven system.]

7. Consider the system $x' = y$, $y' = -2x/t^2 + 2y/t$, $\quad t > 0$.

(a) Show that $X(t) = \begin{bmatrix} t^2 & t \\ 2t & 1 \end{bmatrix}$ is a basic solution matrix.

(b) Find $\Phi(t, t_0)$, and verify that $\Phi(t_0, t_0) = I$, $\Phi^{-1}(t, t_0) = \Phi(t_0, t)$, $\quad t, t_0 > 0$.

(c) Plot the component curves and the orbits for the system in **(a)**, given the initial data $x(1) = 1$, $y(1) = -5, -4, \ldots, 4, 5$.

Answer 7:

(a) That $X(t) = \begin{bmatrix} t^2 & t \\ 2t & 1 \end{bmatrix}$ is a solution matrix of $x' = y$, $y' = -2x/t^2 + 2y/t$, $t > 0$, follows from a straightforward calculation, i.e., show that $x = t^2$, $y = 2t$ is a solution, and then that $x = t$, $y = 1$ is also a solution. Since det $X(1) = -1 \neq 0$, it follows that $X(t)$ is a basic solution matrix.

(b) The transition matrix is

$$\Phi(t, t_0) = X(t)X^{-1}(t_0) = \begin{bmatrix} t^2 & t \\ 2t & 1 \end{bmatrix} \frac{1}{t_0^2} \begin{bmatrix} -1 & t_0 \\ 2t_0 & -t_0^2 \end{bmatrix} = \frac{1}{t_0^2} \begin{bmatrix} -t^2 + 2tt_0 & (t^2 - tt_0)t_0 \\ -2t + 2t_0 & (2t - t_0)t_0 \end{bmatrix}$$

Clearly, $\Phi(t_0, t_0) = I$. A straightforward calculation (just interchange t and t_0 above) shows that

$$\Phi(t_0, t) = \frac{1}{t^2} \begin{bmatrix} -t_0^2 + 2tt_0 & (t_0^2 - tt_0)t \\ -2t_0 + 2t & (2t_0 - t)t \end{bmatrix}$$

Calculating $\Phi(t, t_0)\Phi(t_0, t)$ directly, we get I, and so $\Phi(t_0, t) = \Phi^{-1}(t, t_0)$.

(c) See Graph 1 for the orbits and Graph 2 for the component curves, given the initial data $x(1) = 1$, $y(1) = -5, -4, \ldots, 4, 5$.

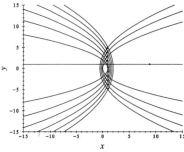

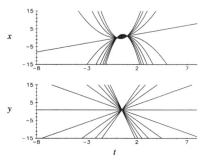

Problem 7(c), Graph 1. Problem 7(c), Graph 2.

Properties of Linear Systems. Suppose that $A(t)$ is an $n \times n$ matrix with real-valued entries.

8. If $x'(t) = A(t)x(t)$ and $(z^T(t))' = -z^T(t)A(t)$, show that $\sum_{i=1}^{n} z_i(t)x_i(t) = z^T(t)x(t)$ is constant for all t. [*Hint*: $(z^T x)' = (z^T)'x + z^T x' = 0$.]

9. Suppose that A is *skew symmetric* (i.e., $A(t) = -A^T(t)$). Show that $\|x(t)\|$ is constant for all t if $x'(t) = A(t)x(t)$; that is, show that each orbit of this differential equation lies on a sphere of radius $\|x(0)\|$ centered at the origin in state space. [*Hint*: see Problem 8.]

Answer 9: Since $x(t)$ solves the system $x' = Ax$, we see by taking the transpose of both sides that $x^T(t)$ solves the system

$$(x^T)' = x^T A^T = -x^T A$$

Therefore by Problem 8, $x^T x = \|x\|^2$ is a constant for all t, but $\|x\|^2 =$ constant describes a "sphere" centered at the origin.

Variation of Parameters.

 10. Show that the Variation of Parameter Method outlined in Section 3.7 is a special case of Theorem 6.8.5. [*Hint*: transform the second-order scalar ODE $y'' + a(t)y' + b(t)y = f$ to the system $u' = v$, $v' = -bu - av + f$, where $u = y$, $v = y'$.]

11. *Another Approach to Variation of Parameters* Prove Theorem 6.8.5 as follows. Assume that $x' = A(t)x + F$, $x(t_0) = 0$, has a solution of the form $x = \Phi(t, t_0)c(t)$, where Φ is the transition matrix for $x' = A(t)x$ and $c(t)$ is the unknown varied-parameter function vector. [*Hint*: insert $\Phi(t, t_0)c(t)$ for x in the ODE, use properties of Φ, and show that $c' = \Phi(t_0, t)F(t)$, from which $c(t) = \int_{t_0}^{t} \Phi(t_0, s)F(s)\,ds$ if it is desired that $c(t_0) = 0$.]

Answer 11: Let $x(t) = \Phi(t, t_0)c(t)$ be a solution of $x' = A(t)x + F$, where the "varied parameter" vector $c(t)$ is to be determined. We have $x' = \Phi'c + \Phi c' = A\Phi c + \Phi c' = A\Phi c + F$. So $\Phi c' = F$ and $c'(t) = \Phi^{-1}(t, t_0)F(t)$. We have $c(t) - c(t_0) = \int_{t_0}^{t} \Phi(t_0, s)F(s)\,ds$, where we have used the property that $\Phi^{-1}(t, t_0) = \Phi(t_0, t)$. Setting $c(t_0) = 0$ [which amounts to solving the IVP where $x(t_0) = 0$], we have that $c(t) = \int_{t_0}^{t} \Phi(t_0, s)F(s)\,ds$, and we have constructed a solution of the desired form $x = \Phi(t, t_0)c(t)$.

Periodic System Matrix.

12. Suppose that entries in the matrix A have common period T. Show that if $\Phi(t, t_0)$ is the transition matrix for $x' = A(t)x$, then $\Phi(t + T, t_0) = \Phi(t, t_0)C$ for some constant matrix C.

Spotlight on Vectors, Matrices, Independence

Comments:

We define a vector space as the collection of all column vectors with n entries along with matrix addition and multiplication by a scalar. It is important to distinguish the case where all entries in the vectors (and the scalars) are reals (i.e., the case $\mathbb{R}^n$) from the case where the entries (and the scalars) are complex numbers (i.e., the case $\mathbb{C}^n$). We use the term function vector space to mean the collection of all n-entry column vectors with function entries (the notion $\mathbf{C}^n(I)$ is used again, leaving the number of entries to the context).

It is important to take a look at matrices with complex entries and the corresponding linear systems with complex coefficients. Ditto for function vector spaces with entries which are complex-valued functions of a real variable.

PROBLEMS

Properties of Matrices. Verify all the *Properties of Matrix Addition, Multiplication, and Transposition* for the given matrices.

1. $A = \begin{bmatrix} 2 & -1 \\ 3 & 5 \end{bmatrix}$, $B = \begin{bmatrix} -2 & 2 \\ 4 & -1 \end{bmatrix}$, $C = \begin{bmatrix} 0 & 1 \\ -1 & 0 \end{bmatrix}$

2. $A = \begin{bmatrix} 2 & 0 & 1 \\ -1 & 3 & -1 \\ 0 & -2 & 0 \end{bmatrix}$, $B = \begin{bmatrix} 1 & -1 & 2 \\ 0 & -3 & 1 \\ 2 & 0 & -1 \end{bmatrix}$, $C = \begin{bmatrix} 1 & 2 & 0 \\ 0 & -1 & 1 \\ -2 & 0 & 3 \end{bmatrix}$

Answer 1: Verify by direct computation. In each case calculate $A + B$, $B + A$, $A + (B + C)$, $(A + B) + C$, $A(BC)$, $(AB)C$, $A(B + C)$, $AB + AC$, $(A^T)^T$, $(A + B)^T$, $(AB)^T$, $B^T A^T$.

Problems on Vectors and Matrices.

3. Find a 2×2 matrix A (if there is one) that satisfies the equation

$$2A + \begin{bmatrix} 0 & 1 \\ 1 & -1 \end{bmatrix} A = \begin{bmatrix} 1 & -2 \\ 0 & 1 \end{bmatrix}$$

Answer 3: Here's the matrix equation we want to solve for A:

$$2A + \begin{bmatrix} 0 & 1 \\ 1 & -1 \end{bmatrix} A = \begin{bmatrix} 1 & -2 \\ 0 & 1 \end{bmatrix}$$

Put $A = \begin{bmatrix} a & b \\ c & d \end{bmatrix}$ into the given equation to obtain

$$\begin{bmatrix} 2a & 2b \\ 2c & 2d \end{bmatrix} + \begin{bmatrix} c & d \\ a - c & b - d \end{bmatrix} = \begin{bmatrix} 1 & -2 \\ 0 & 1 \end{bmatrix}$$

Comparing entries, we conclude that

$$2a + c = 1, \quad a + c = 0, \quad 2b + d = -2, \quad b + d = 1$$

Solving this system, we obtain the unique solution $a = 1, b = -3, c = -1, d = 4$.

4. *Matrix Product Identities* Denote the columns of the $m \times n$ matrix A by $a^1, a^2, \ldots, a^n$. Show that if $x = [x_1 \ x_2 \ \cdots \ x_n]^T$, then $Ax = x_1 a^1 + x_2 a^2 + \cdots + x_n a^n$. Give examples.

5. *Matrix Product Identities* Look at an $m \times r$ matrix A and an $r \times n$ matrix B. If b^j is the jth column of B, show that $AB = [Ab^1 \ Ab^2 \ \cdots \ Ab^n]$. Give some examples.

Answer 5: The proof is by direct computation from the definition of matrix product. From the definition of matrix products note that the j-th column of AB is exactly A applied to the j-th column of B. We leave the examples to the reader.

Linear Independence, Linear Dependence, Span. Determine whether each collection X of vectors in V is linearly independent. If linearly dependent, find a linear combination of the vectors (with at least one of the coefficients not zero) that adds up to the zero vector. In either case describe $\text{Span}(X)$, and, if X is linearly dependent, find a linearly independent subset Y of X such that $\text{Span}(Y) = \text{Span}(X)$.

6. $X = \{[1 \ 1 \ 3]^T\}, \quad V = \mathbb{R}^3$

7. $X = \{[2 \ 2 \ 1]^T, [-1 \ 5 \ 2]^T, [0 \ 12 \ 5]^T\}, \quad V = \mathbb{R}^3$

Answer 7:   Suppose that there are constants C_1, C_2, and C_3 such that

$$C_1[2 \ 2 \ 1]^T + C_2[-1 \ 5 \ 2]^T + C_3[0 \ 12 \ 5]^T = [0 \ 0 \ 0]^T$$

Then we have the algebraic system

$$2C_1 - C_2 = 0$$
$$2C_1 + 5C_2 + 12C_3 = 0$$
$$C_1 + 2C_2 + 5C_3 = 0$$

Subtract two times the third equation from the first and second equations to get the equivalent system

$$-3C_2 - 10C_3 = 0$$
$$C_2 - 2C_3 = 0$$
$$C_1 + 2C_2 + 5C_3 = 0$$

Add three times the second equation to the first equation to get the equivalent system

$$-16C_3 = 0$$
$$C_2 - 2C_3 = 0$$
$$C_1 + 2C_2 + 5C_3 = 0$$

which is a cascade. We get $C_1 = C_2 = C_3 = 0$. Hence X is independent and $\text{Span}(X) = \mathbb{R}^3$.

8. $X = \{[1 \ 0 \ 0 \ 0]^T, [1 \ 1 \ 0 \ 0]^T, [1 \ 1 \ 1 \ 0]^T, [1 \ 1 \ 1 \ 1]^T\}, \quad V = \mathbb{R}^4$

9. $X = \{[1 \ i]^T, [1 \ -i]^T\}, \quad V = \mathbb{C}^2$

Answer 9:   Suppose there are constants C_1 and C_2 such that

$$C_1[1 \ i]^T + C_2[1 \ -i]^T = [0 \ 0]^T$$

Then we have the equivalent algebraic system

$$C_1 + C_2 = 0$$
$$iC_1 - iC_2 = 0$$

The only solution of this system is $C_1 = C_2 = 0$, so X is an independent set and $\text{Span}(X) = \mathbb{C}^2$.

An Equivalent Definition of Linear Independence.   Let X be a finite subset of a vector space V.

10. Suppose that $X = \{x^1, \ldots, x^m\}$ is linearly independent. Show that for every vector x in $\text{Span}(X)$ there are unique scalars $c_1, \ldots, c_m$ such that $x = c_1 x^1 + \cdots + c_m x^m$. The other way around, show that if $X = \{x^1, \ldots, x^m\}$ has the property that every vector x in $\text{Span}(X)$ can be represented as a unique linear combination over X, then X is linearly independent.

Subspaces.

11. Suppose that U and W are subspaces of a vector space V (either $\mathbb{R}^n$ or $\mathbb{C}^n$).

(a) Show that the set of all vectors common to U and W is also a subspace of V.

(b) If $V = \mathbb{R}^3$, $U = \text{Span}\{[2 \ 1 \ 0]^T, [1 \ 0 \ -1]^T\}$, and $W = \text{Span}\{[-1 \ 1 \ 1]^T, [0 \ -1 \ 1]^T\}$, describe the subspace of all vectors common to U and W.

Answer 11:

(a) Let v^1 and v^2 be any two vectors common to the subspaces U and W of the space V, and let α, β be any two scalars. Then $\alpha v^1 + \beta v^2$ must belong to U and also to W because they are subspaces so each contains all finite linear combinations of its own elements. Therefore, $\alpha v^1 + \beta v^2$ is common to U and W, and the vectors common to U and W form a subspace of V (by the definition of subspace).

(b) A vector $x = [x_1 \quad x_2 \quad x_3]^T$ in $\mathbb{R}^3$ lies in $U = \text{Span}\{[2 \quad 1 \quad 0]^T, [1 \quad 0 \quad -1]^T\}$ if and only if there are scalars α, β such that $x_1 = 2\alpha + \beta$, $x_2 = \alpha$, $x_3 = -\beta$, so if and only if $x_1 = 2x_2 - x_3$. Similarly, $[x_1 \quad x_2 \quad x_3]^T$ lies in $W = \text{Span}\{[-1 \quad 1 \quad 1]^T, [0 \quad -1 \quad 1]^T\}$ if and only if there exist scalars γ, δ such that $x_1 = -\gamma$, $x_2 = \gamma - \delta$, $x_3 = \gamma + \delta$, so if and only if $2x_1 + x_2 + x_3 = 0$. The subspace of all vectors in $\mathbb{R}^3$ common to U and W must consist of all vectors $x = [x_1 \quad x_2 \quad x_3]^T$ which satisfy the two conditions: $x_1 - 2x_2 + x_3 = 0$, $2x_1 + x_2 + x_3 = 0$. Using elementary row operations we find that the solution set of this pair of linear equations is given by $x = \alpha[-3 \quad 1 \quad 5]^T$, where α is an arbitrary constant. So the subspace of vectors common to U and W has dimension $= 1$.

12. Which of the following are subspaces of the function vector space V of all real-value continuous functions on the real line?. If not a subspace, why not?

(a) All polynomials of degree 2   **(b)** All polynomials of degree greater than 3

(c) All odd functions $[f(-x) = -f(x)]$ **(d)** All nonnegative functions

Linear Independence. Are the sets of functions below linearly dependent in the function vector V of all real-valued continuous functions on the real line?

13. $\{e^{-t}, -3e^t, \cosh t\}$ **14.** $\{e^t, te^t, -t^2 e^t\}$

15. $\{e^{-t}\cos t, e^{-t}\sin t\}$ **16.** $\{1, t-1, 3t^2 + t + 1, 1 - t^2\}$

Answer 13: The set $\{e^{-t}, 3e^t, \cosh t\}$ is dependent, since $\cosh t = \frac{1}{2}(e^{-t}) - \frac{1}{6}(-3e^t) = \frac{1}{2}(e^t + e^{-t})$.

Answer 15: The set $\{e^{-t}\cos t, e^{-t}\sin t\}$ is independent, for let scalars a, b be such that $ae^{-t}\cos t + be^{-t}\sin t = 0$ for all t. Then it would follow (after dividing by e^{-t}) that $a\cos t + b\sin t = 0$ for all t. Substituting $t = 0$ and $t = \pi/2$ into this statement we obtain that $a = b = 0$, so the set is independent.

Subspace of a Function Vector Space. Find each of the subsets (described below) of the function vector space W of all real-valued and twice continuously differentiable functions defined on the real line. If the subset is a subspace, find its dimension.

17. $\{$All functions $y(t)$ where $: y'' = 0, \ 2y(0) + y'(0) = 0, \ 2y(1) - y'(1) = 0\}$

Answer 17: By antidifferentiation the general solution of the ODE $y'' = 0$ is $y = C_1 t + C_2$, where C_1 and C_2 are arbitrary constants. The conditions $2y(0) + y'(0) = 0$ and $2y(1) - y'(1) = 0$ imply the following conditions on the constants C_1 and C_2:

$$2C_2 + C_1 = 0, \quad 2(C_1 + C_2) - C_1 = 0$$

All solutions of these equations have the form $C_1 = 2s$, $C_2 = s$, for all $-\infty < s < \infty$. Thus, the set of functions which satisfy the given conditions has the form

$$y = 2st + s = s(2t + 1), \quad -\infty < s < \infty$$

Thus this set is a one-dimensional subspace of W.

18. $\{$All functions $y(t)$ where $: y'' = 2, \ y(0) = 0, \ y(1) = 0\}$

19. $\{$All functions $y(t)$ where $: y'' = 2, \ 2y(0) + y'(0) = 0, \ 2y(1) - y'(1) = 0\}$

Answer 19: The general solution of the ODE $y'' = 2$ is $y = t^2 + C_1 t + C_2$, where C_1 and C_2 are arbitrary constants. The given endpoint conditions imply the following conditions on C_1 and C_2:

$$2C_2 + C_1 = 0, \quad 2(1 + C_1 + C_2) - (2 + C_1) = 0$$

All solutions of these conditions look like $C_1 = 2s$, $C_2 = s$, for $-\infty < s < \infty$. Thus our given set of functions is

$$y = t^2 + s(2t + 1), \quad -\infty < s < \infty$$

This set is not a subspace of W because it does not contain the zero function vector.

Spotlight on Linear Algebraic Equations

Comments:

The solution of linear algebraic systems lies at the heart of many techniques for solving and analysing linear differential systems. That is why learning how to solve linear algebraic systems is so important here!

PROBLEMS

Solve a Linear System.   Find all solutions of each system. Characterize them geometrically. [*Hint*: see Example 1.]

1. $x_1 - 2x_2 = 0, \quad 4x_2 - 2x_1 = 0$

Answer 1:   Write the system in standard matrix form $\begin{bmatrix} 1 & -2 \\ -2 & 4 \end{bmatrix} \begin{bmatrix} x_1 \\ x_2 \end{bmatrix} = \begin{bmatrix} 0 \\ 0 \end{bmatrix}$ and use elementary row operations on the augmented matrix $\left[\begin{array}{cc|c} 1 & -2 & 0 \\ -2 & 4 & 0 \end{array} \right]$ to obtain $\left[\begin{array}{cc|c} 1 & -2 & 0 \\ 0 & 0 & 0 \end{array} \right]$. So the solution set of our system is given by the single equation $x_1 - 2x_2 = 0$. This equation determines a line through the origin in the $x_1 x_2$-plane. To describe this line assign the arbitrary value s to x_2. Then $x_1 = 2s$, and the solution set is given by

$$x = \begin{bmatrix} x_1 \\ x_2 \end{bmatrix} = \begin{bmatrix} 2s \\ s \end{bmatrix} = s \begin{bmatrix} 2 \\ 1 \end{bmatrix}$$

2. $x_1 - 2x_2 - x_3 = 0, \quad x_2 + x_3 = 0, \quad x_1 + x_2 + 2x_3 = 0$

3. $x_1 + x_2 + 2x_3 = 1, \quad 3x_1 + 4x_2 - x_3 = -1$

Answer 3:   Writing the system in standard augmented matrix form we have

$$\left[\begin{array}{ccc|c} 1 & 1 & 2 & 1 \\ 3 & 4 & -1 & -1 \end{array} \right]$$

and using elementary row operations we have

$$\left[\begin{array}{ccc|c} 1 & 1 & 2 & 1 \\ 0 & 1 & -7 & -4 \end{array} \right]$$

which yields the two equations $x_1 + x_2 + 2x_3 = 1$, and $x_2 - 7x_3 = -4$. Assigning the arbitrary value s to x_3, the second equation reduces to $x_2 = -4 + 7s$, and the first equation becomes $x_1 = 5 - 9s$. So the solution set of our system is

$$x = \begin{bmatrix} x_1 \\ x_2 \\ x_3 \end{bmatrix} = \begin{bmatrix} 5 - 9s \\ -4 + 7s \\ s \end{bmatrix} = \begin{bmatrix} 5 \\ -4 \\ 0 \end{bmatrix} + s \begin{bmatrix} -9 \\ 7 \\ 1 \end{bmatrix}$$

which is a line in $x_1 x_2 x_3$-space passing through the point $(5, -4, 0)$ and parallel to the vector $-9\mathbf{i} + 7\mathbf{j} + \mathbf{k}$.

Solve a Linear System.   Find all solutions (if there are any) of the following systems:

4.
$$\begin{aligned} 3x_1 - x_2 + x_3 + 2x_4 - 2x_5 &= 1 \\ x_1 + 2x_2 - x_4 + x_5 &= -1 \\ x_2 + 2x_3 - 5x_4 + 2x_5 &= 2 \end{aligned}$$

5.
$$\begin{aligned} x_1 - 2x_2 + 4x_3 - x_4 &= -1 \\ 2x_1 - x_2 - x_3 + x_4 &= 1 \\ 5x_1 + 2x_3 + x_4 &= -1 \end{aligned}$$

Answer 5:   Applying elementary row operations (which we omit) to the augmented matrix of the system, we obtain

$$
\left[\begin{array}{cccc|c}
1 & -2 & 4 & -1 & -1 \\
0 & 1 & -3 & 1 & 1 \\
0 & 0 & 1 & -1/3 & -1/2
\end{array}\right]
$$

Set $x_4 = a$. then $x_3 = -1/2 + a/3$, $x_2 = -1/2$, and $x_1 = -a/3$: Then

$$
x = \left[\begin{array}{c}
0 \\
-1/2 \\
-1/2 \\
0
\end{array}\right] + a \left[\begin{array}{c}
-1/3 \\
0 \\
1/3 \\
1
\end{array}\right], \quad a \text{ arbitrary}
$$

Properties of Solutions of a Linear System.

$$
A = \left[\begin{array}{ccc}
1 & -1 & 0 \\
-1 & 3 & -2 \\
1 & 3 & -4
\end{array}\right]
$$

6. Suppose that A is the matrix in the margin.

(a) Find a solution of the system $Ax = [1 \ -1 \ 1]^T$.

(b) Find all solutions of the system $Ax = 0$.

(c) For what vectors $y = [a \ b \ c]^T$ does the system $Ax = y$ have a solution?

7. Find all solutions of the system $Ax = y$ if $A = \left[\begin{array}{cc} 1 & 2 \\ -i & i \end{array}\right]$ and $y = [-i \ 1]^T$.

Answer 7: The augmented 2×3 matrix for the system is

$$
[A|y] = \left[\begin{array}{cc|c}
1 & 2 & -i \\
-i & i & 1
\end{array}\right]
$$

Adding i times the first row to the second row, we obtain

$$
\left[\begin{array}{cc|c}
1 & 2 & -i \\
0 & 3i & 2
\end{array}\right]
$$

which translates to the system

$$
x_1 + 2x_2 = -i
$$
$$
3ix_2 = 2
$$

Solving this reduced system, we have the unique solution $x_2 = 2/(3i) = -(2/3)i$, and $x_1 = -i - 2x_2 = -i + (4/3)i = (1/3)i$.

8. Show that a 2×2 upper triangular matrix is invertible if and only if the elements on its diagonal are nonzero. Show that the inverse (if it exists) is also upper triangular.

Invertible Matrices.
Suppose that A and B are invertible $n \times n$ matrices.

9. Show that $(A^{-1})^T = (A^T)^{-1}$.

Answer 9: The matrix A is invertible if and only if there is a matrix M such that $MA = AM = I$. Taking the transpose of these relations, we obtain that $A^T M^T = M^T A^T = I$. So, A^T is also invertible and $(A^T)^{-1} = M^T = (A^{-1})^T$.

10. Show that A has a unique inverse.

11. Show that AB is invertible and that $(AB)^{-1} = B^{-1}A^{-1}$.

Answer 11: Let A^{-1} be the inverse of A, and B^{-1} the inverse of B. Then since (using the associativity of the matrix product)

$$
AB(B^{-1}A^{-1}) = A(BB^{-1})A^{-1} = AIA^{-1} = AA^{-1} = I
$$

it follows that $(AB)^{-1} = B^{-1}A^{-1}$.

Determinants.
Evaluate the following determinants:

$$\textbf{12.} \quad \det \begin{bmatrix} 3 & 5 & 7 & 2 \\ 2 & 4 & 1 & 1 \\ -2 & 0 & 0 & 0 \\ 1 & 1 & 3 & 4 \end{bmatrix} \qquad \textbf{13.} \quad \det \begin{bmatrix} 1 & -2 & 3 & -2 & -2 \\ 2 & -1 & 1 & 3 & 2 \\ 1 & 1 & 2 & 1 & 1 \\ 1 & -4 & -3 & -2 & -5 \\ 3 & -2 & 2 & 2 & -2 \end{bmatrix}$$

Answer 13: Before using cofactors, first introduce as many zeros as possible on some row or column by using some of the techniques listed in Properties of det A. We just give the answer here. The final value of this determinant turns out to be 118.

 Rank-Nullity Theorem. An *operator* L acts on any element v in a set V to produce an element $L[v]$ in another set W. Suppose that A is an $n \times m$ matrix and define the operator whose action is given by $L[x] = Ax$ for any vector x in $\mathbb{R}^n$, where Ax is in $\mathbb{R}^m$.

See Section 3.6 for more on operators.

14. Show that L is a linear operator.

15. The *null space* of L is the set $N(L)$ of all vectors v in $\mathbb{R}^n$ such that $L[v] = 0$. The *range* of L is the set $R(L)$ of all vectors w in $\mathbb{R}^m$ for which there is a vector v in $\mathbb{R}^n$ with $L[v] = w$. Show that $R(L)$ is a subspace of $\mathbb{R}^m$ and that $N(L)$ is a subspace of $\mathbb{R}^n$.

Answer 15: Group project.

16. The Rank-Nullity Theorem states that $\dim R(L) + \dim N(L) = n$. Verify this assertion for the matrix A given in Problem 6.

Spotlight on Bifurcations: Sensitivity

PROBLEMS

1. The system for the following problems is $x' = \alpha x + y$, $y' = -x - y$ (see Example 1).

(a) Explain why the equilibrium point at the origin is a saddle if $\alpha > 1$.

(b) Explain why the origin is a spiral point if $-3 < \alpha < 1$.

(c) Why is the origin an improper node if $\alpha < -3$?

(d) Compare orbital portraits for $|x| \le 3$, $|y| \le 4$ for pairs of α-values: -6,-5; -0.5,0; 3,4.

Answer 1: The characteristic polynomial of the system matrix $\begin{bmatrix} \alpha & 1 \\ -1 & -1 \end{bmatrix}$ is $p(\lambda) = \lambda^2 + (1 - \alpha)\lambda + 1 - \alpha$, and the eigenvalues are

$$\lambda_{1,2} = \frac{-(1 - \alpha) \pm \sqrt{(1 - \alpha)^2 - 4(1 - \alpha)}}{2} = \frac{-(1 - \alpha) \pm \sqrt{(1 - \alpha)(-\alpha - 3)}}{2}$$

(a) If $\alpha > 1$, we know $\lambda_1 = [-(1 - \alpha) + \sqrt{(1 - \alpha)(-\alpha - 3)}]/2 > 0$ and $\lambda_2 = [-(1 - \alpha) - \sqrt{(1 - \alpha)(-\alpha - 3)}]/2 < 0$, so the equilibrium point at the origin is a saddle because the eigenvalues have opposite signs.

(b) If $-3 < \alpha < 1$, we have $(1 - \alpha)(-\alpha - 3) < 0$. The eigenvalues are complex conjugates with negative real parts. So the origin is a spiral point if $(1 - \alpha)(-\alpha - 3) < 0$.

(c) If $\alpha < -3$, $\lambda_{1,2} = -\frac{1}{2}(1 - \alpha) \pm \frac{1}{2}(1 - \alpha)\sqrt{1 - 4(1 - \alpha)^{-1}} < 0$, and $\lambda_1 \ne \lambda_2$. So the origin is an improper node if $\alpha < -3$. Since the eigenvalues are real and distinct, a pair of eigenvectors (one for each eigenvalue) is independent.

(d) See Graphs 1, 2, and 3 for orbits in the respective cases $\alpha = 4, 0, -4$.

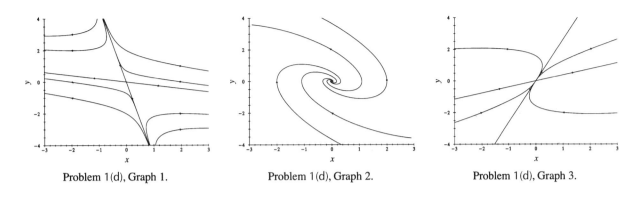

Problem 1(d), Graph 1.   Problem 1(d), Graph 2.   Problem 1(d), Graph 3.

2. *Sensitivity to Changes in System Parameters*   Describe how the orbital portrait of the system $x' = -x + \alpha y$, $y' = -x - y$ changes as the system parameter α ranges from $-\infty$ to ∞. Find the bifurcation values of α where the portrait suddenly changes its nature. Plot the orbital portraits in the frame $|x| \leq 3$, $|y| \leq 4$ for $\alpha = -2, -1, 0, 1$ and identify the type of each portrait (e.g., a spiral point if $\alpha = 1$). [*Hint*: read Example 1 for the general approach.]

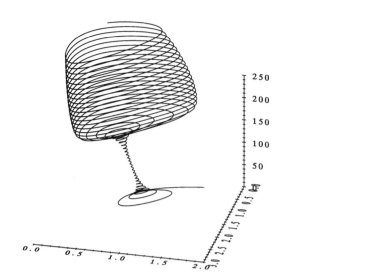

This time-state curve envelops a wineglass.
Take a look at Example 7.1.1.

Nonlinear Differential Systems

Since solution formulas for systems of ODEs (particularly when they are not linear) are even rarer than they are for first- or second-order ODEs, we start out with computer simulations. That way, you can see what component curves, time-state curves, orbits, and so on look like. Many models are presented in this chapter: coupled springs and masses, chemical reactions, models of competing or cooperating species, and models of the spread of a disease. The aim of this profusion of models is to show the naturalness of systems of ODEs as models of natural processes. Our experience is that people take to systems right away if there are models around to give some physical reality to the mathematical constructs. The chapter cover figure shows a time-state orbit of a model system for an autocatalytic reaction (see Section 7.1).

We introduce a version of the Fundamental Theorem (Theorem 7.2.1) for the existence, uniqueness, extension, and continuity/sensitivity properties of the solutions of initial value problems for systems of ODEs. The sensitivity of solutions to changes in the data again comes up in the text and in the problems. Computer simulations are critical in conveying the effects of parameter changes.

Population models (like the Lotka-Volterra models of Section 2.6) are controversial because they are often viewed as reductionist; that is, they vastly oversimplify the reality of population dynamics. Although true in many cases, these models do provide a way to think about interacting species, and they are a first step in gaining an understanding of how populations interact. The possum disease model of the WEB SPOTLIGHT ON MODELING: THE POSSUM PLAGUE is due to Professor Graeme Wake and his colleagues in New Zealand. We have included it here because it is still in the development stage and models an important problem for the New Zealand economy. We think that every ODE course should have at least one open question, model, or theorem—something where the final "answer" isn't yet known. The possum model is one of these.

7.1 Chemical Kinetics: The Fundamental Theorem

Comments:

Here we are with another Fundamental Theorem (Theorem 7.1.1). This time the theorem is so inclusive it contains the Fundamental Theorem of Section 3.8. Since we work mostly with systems of first-order ODEs in Chapters 6, 7, 8, and 9, the theorem is formulated for these systems. One very important point may get buried in the details: parameters in the rate functions can be treated as additional state variables whose rates of change are zero. Applying the Fundamental Theorem to the extended system, we conclude that solutions are continuous in all the parameters as long as the hypotheses of Theorem 7.2.1 are satisfied by the rate functions.

Systems of first-order ODEs have appeared several times in models in earlier chapters. We introduce a new kind of model in the autocatalator system—a model of chemical kinetics. The basic principle of Chemical Mass Action is given and discussed; from the principle we can write out the rate equations for each of the species involved in the chemical reaction. The rate equations for autocatalytic reactions are nonlinear, and their solutions behave in an interesting way.

The examples in this section are discussed entirely in the light of their computer simulations. The aim of the examples is to show the naturalness of systems in modeling phenomena and to give the reader some experience with computer simulation of systems. We also introduce vector notation for modeling systems, but we do not use the boldface notation of Section 4.1. Component curves, orbits, and time-state curves are again defined and graphed.

Experiments

You can illustrate the autocatalytic chemical process for yourself if you can line up some chemical equipment. If you want to do this experiment, take a look at *Chemical Demonstrations* by Lee R. Summerlin and James L. Ealy; American Chemical Society, Washington, 1985. Summerlin and Ealy give you all the information you need to do three autocatalytic experiments. In each experiment the autocatalytic oscillations are seen as color changes in a chemical solution. The actual reactions taking place are quite complicated, so the model given in this section can be viewed as just a rough caricature of what is going on in an actual reaction.

PROBLEMS

The Fundamental Theorem. For Problems 1–7 verify that each IVP satisfies the hypotheses of the Fundamental Theorem. Then solve the IVP.

1. $x_1' = x_2,$ $\quad x_2' = -x_1 - 2x_2;$ $\quad x_1(0) = 1,$ $\quad x_2(0) = 1$ [*Hint:* note $x_1'' + 2x_1' + x_1 = 0$.]
2. $x_1' = x_2,$ $\quad x_2' = -x_1 - 2x_2;$ $\quad x_1(0) = a,$ $\quad x_2(0) = b$
3. $x_1' = 2x_1 + e^{2t},$ $\quad x_2' = -4x_2;$ $\quad x_1(0) = a,$ $\quad x_2(0) = b$
4. $x_1' = x_2,$ $\quad x_2' = -9x_1;$ $\quad x_1(0) = a,$ $\quad x_2(0) = b$
5. $x_1' = x_2^3,$ $\quad x_2' = -x_1^3;$ $\quad x_1(0) = a,$ $\quad x_2(0) = b$
6. $x_1' = -x_1^3,$ $\quad x_2' = -x_2 \sin t;$ $\quad x_1(0) = 1,$ $\quad x_2(0) = 1$
7. $x_1' = x_2,$ $\quad x_2' = -26x_1 - 2x_2,$ $\quad x_3' = x_3/2;$ $\quad x_1(0) = x_2(0) = x_3(0) = 1$

Answer 1: Here, $f_1 = x_2,$ $\partial f_1/\partial x_1 = 0,$ $\partial f_1/\partial x_2 = 1,$ $f_2 = -x_1 - 2x_2,$ $\partial f_2/\partial x_1 = -1,$ and $\partial f_2/\partial x_2 = -2$ are continuous for all x_1, x_2, t. So the IVP has a unique solution. The solution of the corresponding scalar IVP, $x_1'' + 2x_1' + x_1 = 0, x_1(0) = 1, x_1'(0) = 1,$ is $x_1 = e^{-t}(1 + 2t)$; then $x_2 = e^{-t}(1 - 2t)$.

Answer 3:   The conditions of the Fundamental Theorem are satisfied since $f_1 = 2x_1 + e^{2t}$, $\partial f_1/\partial x_1 = 2$, $\partial f_1/\partial x_2 = 0$, $f_2 = 4x_2$, $\partial f_2/\partial x_1 = 0$, $\partial f_2/\partial x_2 = 4$ are continuous for all x_1, x_2, t. The general solution is $x_1 = e^{2t}(t + C_1)$, $x_2 = C_2 e^{-4t}$. Imposing the initial conditions, we get, $x_1 = e^{2t}(t + a)$, $x_2 = be^{-4t}$.

Answer 5:   The conditions of the Fundamental Theorem are satisfied since $f_1 = x_2^3$, $\partial f_1/\partial x_1 = 0$, $\partial f_1/\partial x_2 = 3x_2^2$, $f_2 = -x_1^3$, $\partial f_2/\partial x_1 = -3x_1^2$, $\partial f_2/\partial x_2 = 0$ are continuous for all x_1, x_2, t. We see that $dx_1/dx_2 = -x_2^3/x_1^3$, so $x_1^3\,dx_1 + x_2^3\,dx_2 = 0$. Integrating, we see that $x_1^4 + x_2^4 = C$. Imposing the intiial conditions, we see that $x_1^4 + x_2^4 = a^4 + b^4$ is the equation of the orbit through (a, b). The solutions $x_1 = x_1(t)$ and $x_2 = x_2(t)$ cannot be expressed in terms of simple functions.

Answer 7:   The conditions of the Fundamental Theorem are satisfied since $f_1 = x_2$, $\partial f_1/\partial x_1 = 0$, $\partial f_1/\partial x_2 = 1$, $\partial f_1/x_3 = 0$, $f_2 = -26x_1 - 2x_2$, $\partial f_2/\partial x_1 = -26$, $\partial f_2/\partial x_2 = -2$, $\partial f_2/\partial x_3 = 0$, $f_3 = x_3/2$, $\partial f_3/\partial x_1 = 0$, $\partial f_3/\partial x_2 = 0$, $\partial f_3/\partial x_3 = 1/2$, are continuous for all x_1, x_2, x_3, t. The first two ODEs are equivalent to the ODE $x_1'' + 2x_1' + 26x_1 = 0$, so $x_1 = e^{-t}(C_1 \cos 5t + C_2 \sin 5t)$, $x_2 = e^{-t}((-C_1 + 5C_2) \cos 5t - (C_2 + 5C_1) \sin 5t)$. The general solution of the third ODE is $x_3 = C_3 e^{t/2}$. The initial conditions imply that $x_1 = e^{-t}(\cos 5t + (2/5) \sin 5t)$, $x_2 = e^{-t}(\cos 5t - (27/5) \sin 5t)$, and $x_3 = e^{t/2}$.

8.   *Here Is the Solution; What Is the System?*   Alex Trebek hands you two function vectors

$$u = [\sin t \quad t]^T, \qquad v = [te^t \quad t^3]^T$$

and asks you for continuously differentiable functions $f(x, y, t)$ and $g(x, y, t)$ for which u and v solve the system of ODEs $x' = f$, $y' = g$. Can you do it? Explain why, or why not.

Orbits.   For each IVP in the indicated problem plot the nine orbits corresponding to all possible combinations of $a, b = -1, 0, 2$. Identify any orbits that are equilibrium points or cycles.

9. Problem 1   **10.** Problem 2   **11.** Problem 3

12. Problem 4   **13.** Problem 5   **14.** Problem 6 ($a = 1$, $b = 1$ only)

Answer 9:   No cycles, $(0, 0)$ is an equilibrium point. See figure for the graph with initial data $x_1(0) = x_2(0) = 1$.

Answer 11:   No cycles, no equilibrium points. See figure.

Answer 13:   All nonconstant orbits are cycles, $(0, 0)$ is an equilibrium point. See figure.

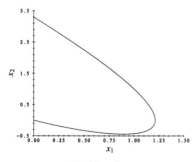

Problem 9.

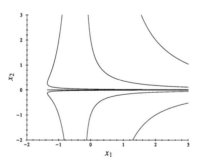

Problem 11.

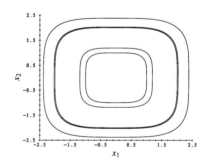

Problem 13.

Chemical Rate Equations.   Given the diagrams below for the chemical reaction steps, write out the ODEs for the concentrations. The quantities k_1, k_2, k are positive rate constants. [*Hint*: the rate equation for the concentration x of reactant X in Problem 15 is $x' = -k_1 xy$.]

15. $X + Y \xrightarrow{k_1} Z \xrightarrow{k_2} W$ **16.** $X + 2Y \xrightarrow{k} Z$ **17.** $X + 2Y \xrightarrow{k} 6Y + W$

Answer 15:   The speed of the first step of the reaction schematic, $X + Y \xrightarrow{k_1} Z \xrightarrow{k_2} W$, is $k_1 xy$, while the speed of the second step is $k_2 z$. So the ODEs are $x' = -k_1 xy$, $y' = -k_1 xy$, $z' = k_1 xy - k_2 z$, $w' = k_2 z$.

Answer 17:   The speed of the reaction with the schematic, $X + 2Y \xrightarrow{k} 6Y + W$, is kxy^2. Since 2 units of Y are consumed to generate 6 units of Y, the net gain is 4 units of Y. So the ODEs are $x' = -kxy^2$, $y' = 4kxy^2$, $w' = kxy^2$.

The Autocatalator.   The rate constant α in the autocatalator system (2) "turns on" the autocatalytic reaction if its value is positive. The reaction "turns off" if $\alpha = 0$ (see system (1)).

18.   Solve the cascade in system (1) with the initial conditions, $w(0) = w_0 > 0$, $x(0) = y(0) = z(0) = 0$.

19.   *Comparison of Linear Model ODE with Autocatalator*   Compare the w, x, y, and z-component curves of the nonlinear autocatalytic system (2) where $\alpha = 0.002$ with those of the linear system (1) as follows: Use the data in (6) and plot each component curve over the time span $0 \le t \le 1000$. Overlay the $\alpha = 0$ and $\alpha = 0.002$ curves for each of the four reactants. How much does the autocatalytic reaction affect the long-term behavior of the various concentrations? The short-term behavior? Explain what you see.

Answer 19:   As Graphs 1 and 4 suggest, the autocatalytic reaction has no effect at all on $w(t)$ and very little effect on $z(t)$ (the top curve in the latter graph corresponds to the autocatalytic sequence). The oscillations in Graphs 2 and 3 show that there is a profound short term oscillatory effect of autocatalysis (although the actual concentrations of X and Y remain low), but the effect wears off as $t \to 1000$. This occurs for $w(0) = 500$, $x(0) = y(0) = z(0) = 0$. [Note: the value of α is 1 for autocatalysis, *not* 0.002 as in the first printing. The graphs use $\alpha = 1$. The correction will be made in the second printing.]

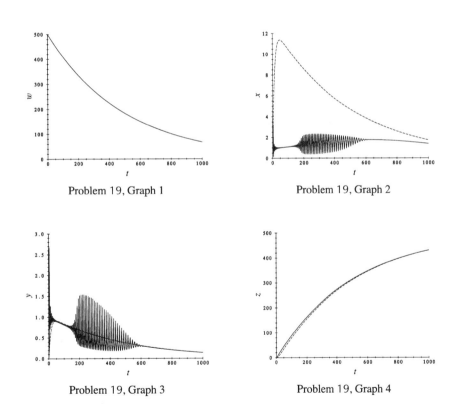

Problem 19, Graph 1   Problem 19, Graph 2

Problem 19, Graph 3   Problem 19, Graph 4

20.   *Sensitivity to Changes in Data: Chemical Intermediates*   Replace the coefficient 0.002 in IVP (7) by α and let α have the value 0.002, then 0.02, and finally 0.2. Plot the tx- and ty-component curves for $0 \le t \le 10$, then for $0 \le t \le 1000$. Are the curves sensitive to the changes in α over the short time span? Over the long time span? Explain.

Changing the Data. The two-dimensional autocatalator orbit of Figure 7.1.2 is the projection of the time-state curve of IVP (7) onto the xy-plane. Problems 21–25 extend IVP (7) by varying the initial concentration $w(0) = w_0$ of the reactant w. [*Hint*: see Multimedia Module 8 in ODE Architect; also see the Library entry, The Autocatalator Reaction, under the heading Chemical Models.]

21. Solve system (2), using the data in (6), but with $w_0 = 50$, then 100. What unusual features do you see in the x- and y-component graphs in comparison to what is visible in Figure 7.1.1?

22. Find the value of w_0 for which you first see sustained oscillations in the component graphs.

23. Plot time-state curves like the chapter opening figure, for $w_0 = 50, 250, 500, 800$.

Answer 21: Graphs 1 and 2 ($W_0 = 50$ corresponds to the lower curve in each graph, $w_0 = 100$ to the upper) show none of the oscillations visible in Figure 7.1.2. Apparently, the oscillations only appear with a high initial concentration of W (at least for the rate constants listed in (6)).

Answer 23: Graph 1 shows no oscillations in the X and Y concentrations for $0 \leq t \leq 350$ if $w_0 = 50$ and sustained oscillations if $w_0 = 500$. Graph 2 shows oscillations that decay if $w_0 = 250, 0 \leq t \leq 700$, and late-onset oscillations over the same time span if $w_0 = 800$.

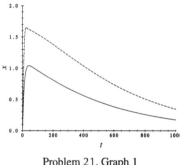

Problem 21, Graph 1   Problem 21, Graph 2

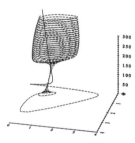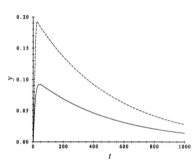

Problem 23, Graph 1   Problem 23, Graph 2

24. Describe the behavior of solutions for various values of w_0 over the range $100 \leq w_0 \leq 500$.

25. *Two-Dimensional Autocatalator, Turning Oscillations On and Off* It may be possible to turn the x and y oscillations on or off by changing any one of the five parameters w_0, a, b, c, and α. That is the aim of the project: change a parameter up or down from the value given until the oscillations of the concentrations of the intermediates X and Y disappear. Then explain why you think that happens. Some suggestions:

   • Duplicate Figures 7.1.1 and 7.1.2.

- Duplicate the wineglass in the opening figure of the chapter. What happens if you plot the solution curve for $0 \leq t \leq 2000$ rather than the truncated curve of the wineglass, $7 \leq t \leq 377$?

- Vary the coefficient b from 0.08 up to 0.14, keeping all other parameters fixed. What happens? Any explanation?

- Vary the coefficients a from 0.002 and c from 1, but keep the other parameters fixed at the values given. Can you turn off the oscillations?

Answer 25: Group problem. [*Suggestion*: see the answer to Problem 20.]

7.2 Properties of Autonomous Systems, Direction Fields

PROBLEMS

Linear Systems. Find the general solution of each system. [*Hint*: use techniques in Chapter 3 or Chapter 6.]

1. $x' = y$, $y' = -\omega^2 x$, ω a nonzero constant [*Hint*: the equivalent ODE is $x'' + \omega^2 x = 0$.]

2. $x' = x - 10y$, $y' = 10x + y$ 3. $x' = x + y - 4$, $y' = x - 2y - 1$

Answer 1: $x = C_1 \cos \omega t + C_2 \sin \omega t$, where C_1 and C_2 are arbitrary constants; $y = x' = -\omega C_1 \sin \omega t + \omega C_2 \cos \omega t$.

Answer 3:

$$\begin{bmatrix} x \\ y \end{bmatrix} = \begin{bmatrix} 3 \\ 1 \end{bmatrix} + C_1 e^{\lambda_1 t} \begin{bmatrix} 1 \\ \lambda_1 - 1 \end{bmatrix} + C_2 e^{\lambda_2 t} \begin{bmatrix} 1 \\ \lambda_2 - 1 \end{bmatrix}$$

where $\lambda_1 = (-1 + \sqrt{13})/2$, $\lambda_2 = (-1 - \sqrt{13})/2$, C_1 and C_2 are arbitrary constants.

Nullclines, Orbital Portraits, Equilibrium Points, Cycles. Find formulas for the orbits of each system. For Problems 4–8 plot (in the rectangle $|x| \leq 3$, $|y| \leq 3$) the nullclines and the nine orbits corresponding to all possible combinations of $(x(0), y(0))$, where $x(0)$ and $y(0)$ independently take on the values $-1, 0, 2$. Identify any orbits that are equilibrium points or cycles.

4. $x' = 2x$, $y' = -4y$ 5. $x' = y$, $y' = -9x$

6. $x' = y$, $y' = -x - 2y$ [*Hint*: note that $x'' + 2x' + x = 0$.]

7. $x' = y^3$, $y' = -x^3$ [*Hint*: write as $dy/dx = -x^3/y^3$.]

8. $x' = -x^3$, $y' = -y$

Answer 5: Here $dy/dx = -9x/y$, so $9xdx + ydy = 0$. Integrating, we see that $9x^2 + y^2 = $ constant. If $x(0) = a$ and $y(0) = b$, then the orbital equations are $9x^2 + y^2 = 9a^2 + b^2$. See the figure; the x- and y-nullclines are, respectively, $y = 0$ and $x = 0$. All orbits aside from the equilibrium point $(0, 0)$ are elliptical cycles.

Answer 7: From the system $x' = y^3$, $y' = -x^3$ we have that $dy/dx = -x^3/y^3$, $x^3 dx + y^3 dy = 0$, or upon solving, $x^4 + y^4 = C$. Since $x(0) = a$ and $y(0) = b$, the equation of the orbit through (a, b) is $x^4 + y^4 = a^4 + b^4$. Note that we have not found $x(t)$ and $y(t)$, but solutions are defined for all t since the nonconstant orbits are cycles. See the figure; the x- and y-nullclines are, respectively, $y = 0$ and $x = 0$. The equilibrium point is $(0, 0)$.

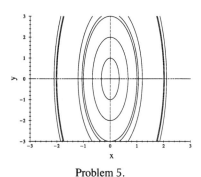

Problem 5.

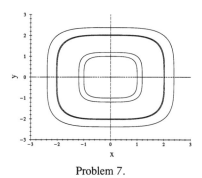

Problem 7.

9. *An Orbit in Three-Space* Find the solution $x = x(t)$, $y = y(t)$, $z = z(t)$ of the IVP $x' = y$, $y' = -26x - 2y$, $z' = -z/2$; $x(0) = y(0) = z(0) = 1$. Plot the orbit in the box $|x| \leq 1$, $-4 \leq y \leq 3$, $0 \leq z \leq 1$, using the time interval $0 \leq t \leq 10$. Then plot the projection of the orbit in each of the coordinate planes xy, xz, yz.

Answer 9: The third IVP, $z' = -z/2$, $z(0) = 1$, is uncoupled from the others and has solution $z = e^{-t/2}$. The scalar ODE equivalent to the system $x' = y$, $y' = -26x - 2y$ is $x'' + 2x' + 26x = 0$, where $x(0) = x'(0) = y(0) = 1$. The characteristic polynomial $r^2 + 2r + 26$ has roots $-1 \pm 5i$. The general solution has the form $x = C_1 e^{-t} \cos 5t + C_2 e^{-t} \sin 5t$. Applying the initial conditions, we obtain $x = e^{-t}[\cos 5t + (2/5) \sin 5t]$; then $y = e^{-t}[\cos 5t - (27/5) \sin 5t]$. So, all state variables tend to 0 as $t \to \infty$, x and y in an oscillatory manner, z monotonically decreasing. See Graphs 1–4 for the respective xy, xz, yz, and xyz graphs; $(0,0,0)$ is an equilibrium point.

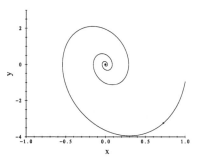

Problem 9, Graph 1.

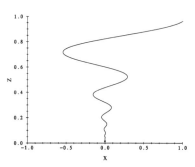

Problem 9, Graph 2.

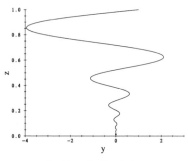

Problem 9, Graph 3.

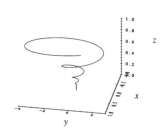

Problem 9, Graph 4.

State Plane Geometry and Systems.   Direction fields, nullclines (dashed curves), or orbits (solid curves) are shown in Problems 10–13. Match up each with a planar system from **(a)–(h)**. Explain your answers.

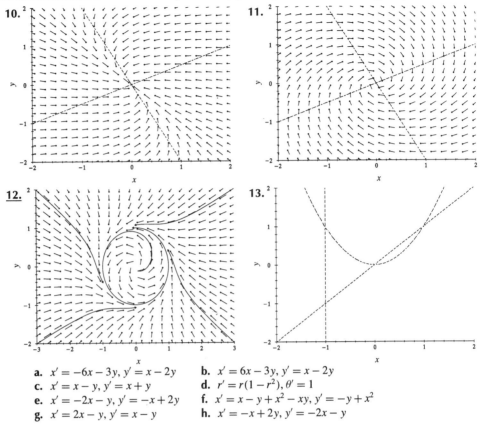

a. $x' = -6x - 3y, y' = x - 2y$ **b.** $x' = 6x - 3y, y' = x - 2y$
c. $x' = x - y, y' = x + y$ **d.** $r' = r(1 - r^2), \theta' = 1$
e. $x' = -2x - y, y' = -x + 2y$ **f.** $x' = x - y + x^2 - xy, y' = -y + x^2$
g. $x' = 2x - y, y' = x - y$ **h.** $x' = -x + 2y, y' = -2x - y$

Answer 11:   **(h)** which has the right x- and y- nullclines ($y = x/2$ and $y = -2x$, respectively) and the right character of an attracting spiral point (eigenvalues of system matrix are $-1 \pm 2i$.

Answer 13:   **(f)** The x-nullclines are $y = x$ and $x = -1$. The y-nullcline is $y = x^2$. See Figures 7.2.1 and 7.2.2.

Direction Fields, Nullclines, Orbital Portraits.   Find all equilibrium points. Plot nullclines, direction fields, and several orbits in rectangles that contain several equilibrium points.

14. $x' = x - y^2, \quad y' = x - y$ **15.** $x' = y \sin x, \quad y' = xy$ **16.** $x' = y + 1, \quad y' = \sin^2 3x$

17. $x' = 2 + \sin(x + y), \quad y' = x - y^3 + 27$ **18.** $x' = 3(x - y), \quad y' = y - x$

Answer 15:   Since either $x = 0$ or $y = 0$ makes $x' = 0$ and $y' = 0$, the two coordinate axes are lines of equilibrium points: $(a, 0)$, $(0, b)$, where a and b are arbitrary constants. See the figure where the x-nullclines are the lines given by $y = 0$ and by $x = n\pi$, n any integer, and the y-nullclines are the x- and y- axes. Orbits seem to emerge from the positive x- and y-axes as t increases from $-\infty$. They seem to approach the negative x- or y-axes as $t \to +\infty$. The axes are lines of equilibrium points.

Answer 17:   There are no equilibrium points since $2 + \sin(x + y)$ is never smaller than $+1$; x' is always positive, so no cycles. See the figure. The y-nullcline is given by $x = y^3 - 27$. Note that segments of orbits seem to approach this curve asymptotically, hiding it from view.

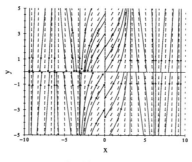

Problem 15.   Problem 17.

Maximally Extended Solutions. Find all solutions. Verify that if $x = x(t)$, $y = y(t)$ is a solution, then so is $x = x(t - T)$, $y = y(t - T)$, where T is any constant. Specify the interval on which each maximally extended solution is defined.

19. $x' = 3x$, $y' = -y$ 20. $x' = 1/x$, $y' = -y$

21. $x' = -x^3$, $y' = 1$ 22. $x' = x^2(1 + y)$, $y' = -y$

Answer 19: Since the ODEs, $x' = 3x$, $y' = -y$, are uncoupled, the equations are easily solved, $x(t) = C_1 e^{3t}$, $y(t) = C_2 e^{-t}$, $-\infty < t < \infty$. Since $x(t - T) = C_1 e^{3(t-T)}$, we have that $dx(t - T)/dt = 3C_1 e^{3(t-T)} = 3x(t - T)$, and so $x(t - T)$ is a solution of $x' = 3x$ if $x(t)$ is a solution. Similarly, $dy(t - T)/dt = -C_2 e^{-(t-T)} = -y(t - T)$, so $y(t - T)$ is a solution. The solution $x = x(t - T)$, $y = y(t - T)$ is defined for all t.

Answer 21: Separating variables and solving the first ODE, $x' = -x^3$, for x, we have that $x = C_1(1 + 2C_1^2 t)^{-1/2}$, $t > -1/2C_1^2$ while the second ODE, $y' = 1$, has the solutions $y(t) = t + C_2$, $t > -1/2C_1^2$. Since $dx(t - T)/dt = -C_1^3(1 + 2C_1^2(t - T))^{-3/2} = -x^3(t - T)$, $x(t - T) = C_1(1 + 2C_1^2(t - T))^{-1/2}$, $t > -1/2C_1^2 + T$ is a solution. Similarly, $dy(t - T)/dt = 1$, so $x(t - T)$, $y(t - T)$, $t > -1/2C_1^2 + T$ is a solution of the system.

Investigating Systems of ODEs: Behavior of Orbits. Problems 23–28 have to do with the planar autonomous system $x' = 1 - y^2$, $y' = 1 - x^2$.

23. *Equilibrium Points*   Locate all the equilibrium points of the system.

Answer 23: The four equilibrium points of $x' = 1 - y^2$, $y' = 1 - x^2$ are $(1, 1)$, $(1, -1)$, $(-1, -1)$, and $(-1, 1)$.

24. *Orbital Portrait*   Create a portrait of orbits in the rectangle $|x| \le 3$, $|y| \le 3$. [*Hint*: if (a, b), $b \ne a$ is on one of your orbits, also plot the orbit through $(-a, -b)$.] Use arrowheads on orbits to indicate the direction of increasing time. Are there any cycles?

25. *Orbital Symmetry about a Line*   Find a formula for the orbits. Use the formula (or the ODEs) to explain why the reflection through the line $y = x$ of an orbital arc is also an orbital arc.

Answer 25: $dy/dx = (1 - x^2)/(1 - y^2)$ separates to $(1 - y^2)dy = (1 - x^2)dx$. Thus $y - y^3/3 = x - x^3/3 + C$. Since $y = x$ satisfies this orbital equation for $C = 0$, it is a line of orbits. Moreover, the orbital equation is unchanged (except C is replaced by $-C$) if x and y are interchanged. This implies that orbital arcs have the stated reflection property.

26. Use geometric reasoning with the direction field to show that the line $y = x$ is composed of five orbits. Identify these orbits and use arrowheads on the orbits to show the direction of increasing time.

27. For each nonconstant orbit on the line $y = x$, find a function $f(t)$ such that $x = f(t)$, $y = f(t)$ describes the orbit. Show that the orbit that originates at $x(0) = a$, $y(0) = a$, $a < -1$, escapes to infinity in finite time.

Answer 27: Substitute $x = f(t)$ and $y = f(t)$ into the system ODEs to obtain a single first-order

ODE $f' = 1 - f^2$. Separating the variables and integrating, we obtain

$$\ln \left| \frac{1+f}{1-f} \right|^{1/2} = t + C, \quad \text{where } C \text{ is an arbitrary constant}$$

Exponentiating, squaring, and dropping the absolute value signs we have

$$\frac{1+f}{1-f} = ce^{2t}, \quad \text{where } c \text{ is an arbitrary constant.}$$

Imposing the initial conditions $x(0) = y(0) = a$, we find that $c = (1+a)/(1-a)$, and solving for $f(t)$ we have

$$f(t) = \frac{c - e^{-2t}}{e^{-2t} + c}$$

If $a < -1$, then $-1 < c < 0$, and we see that $f(t)$ reaches $-\infty$ as t increases to the finite value t_0 where $e^{-2t_0} + c = 0$, so $t_0 = (-1/2)\ln(-c)$.

28. Nests of cycles enclose each equilibrium point $(-1, 1)$ and $(1, -1)$. What happens to the periods as the amplitudes of the cycles increase? Decrease? [*Hint*: plot component curves.]

Attraction and Repulsion: Polar Coordinates. The system below (where a is a constant) illustrates the attracting and repelling properties of cycles and equilibrium points:

$$\begin{aligned} x' &= x - ay - x(x^2 + y^2) \\ y' &= ax + y - y(x^2 + y^2) \end{aligned} \tag{i}$$

29. *Cycle* Explain why the unit circle $x^2 + y^2 = 1$ is a cycle of the system and the origin is the only equilibrium point. [*Hint*: use (12) to write (i) in polarty coordinates.]

Answer 29: In polar coordinates the system is $r' = r(1 - r^2)$, $\theta' = a$. So if $a \neq 0$, the origin is the only equilibrium point and $r = 1$ solves the first ODE and is a cycle. If $a = 0$, then all points on the unit circle are also equilibrium points.

30. *Solution in Polar Coordinates* Using the polar coordinate form of system (i), solve for $r(t)$ and $\theta(t)$. Explain why all nonconstant orbits approach the cycle as time $t \to +\infty$. Why could you call the origin a repeller and the unit circle an attractor?

31. *Direction Field* Draw a direction field for system (i) with $a = 1$, plot x- and y-nullclines, and plot orbits inside, on, and outside the unit circle; use $|x| \le 3$, $|y| \le 2$. Repeat with $a = 0$ and with $a = -1$. Describe the dramatic differences in orbital behavior as a changes from a positive constant, to zero, then to a negative constant.

Answer 31: Graph 1 shows the behavior if $a = 1$: counterclockwise rotation since $\theta' = a$; all orbits (other than $(0, 0)$) spiral toward the unit circle as $t \to +\infty$. All orbits with $0 < r(0) < 1$ tend to the origin as $t \to -\infty$. The cycle $r = 1$ is traced out counterclockwise. If $a = 0$, then $\theta = C$, C an arbitrary constant; all motion is along rays from the origin, outward from inside the unit circle, inward from outside the circle (Graph 2). If $a = -1$, the direction of rotation is reversed compared with the situation when $a = 1$ (so now clockwise rotation). Otherwise, the behavior is the same (Graph 3) as in Graph 1. The dashed lines in Graphs 1 and 3 are nullclines. The nullclines in Graph 2 are the axes and the unit circle.

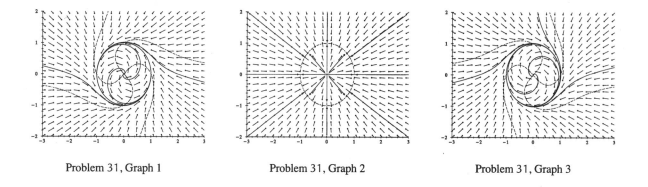

Problem 31, Graph 1   Problem 31, Graph 2   Problem 31, Graph 3

32. *Component Graphs* Plot *x*- and *y*-component graphs for the orbits plotted in Problem 31. What is the period of the cycle if *a* is a nonzero constant?

Equilibrium Points, Cycles. Suppose the autonomous system $x' = f(x)$ satisfies the conditions of the Fundamental Theorem 7.1.1 in a box S in state space. Show the following properties of solution curves, equilibrium points, and cycles in S.

33. Suppose that $x(t)$ is a nonconstant solution for which $x(t) \to P$ as $t \to T$. Show that if P is an equilibrium point, then $|T| = \infty$. [*Hint*: since $x(t)$ is a continuous function of t, $x(T)$ is P if T is finite. But this violates the uniqueness part of the Fundamental Theorem 7.1.1.]

Answer 33: Suppose that P is an equilibrium point of $x' = f(x)$, and that P is inside a region S in which the conditions of the Fundamental Theorem hold. Then $x = P$ is a solution. Suppose $x = x^1(t)$, $a < t < b$, is another maximally extended solution of the system and that $x^1(t) \to P$ as $t \to T$. By the maximal extension property for the orbits of an autonomous system, and the fact that $x^1(t)$ is a continuous vector function of t, we have that either $a < T < b$ and $x^1(T) = P$, or else $T = a = -\infty$, or else $T = b = +\infty$, or else $T = a$ (or b) and $x^1(t)$ tends to the boundary of S as $t \to a^+$ (or b^-). The first alternative is forbidden by the uniqueness property since otherwise there would be two different maximally extended orbits through P, $x = P$ for all t and $x = x^1(t)$. The last alternative would imply that P lies on the edge of S, but P is assumed to lie inside S. So one of the two middle alternatives must hold, $T = \infty$ or $T = -\infty$, and a solution reaches a constant equilibrium state inside S only as $t \to +\infty$, or $t \to -\infty$.

34. Show that if $x = x(t)$ is a nonconstant periodic solution, then the corresponding orbit is a simple closed curve (i.e., a cycle). [*Hint*: suppose that the period of $x(t)$ is T and that $x(t_1) = x(0)$ for some t_1, $0 < t_1 \leq T$. Then $y(t) = x(t + t_1)$ is also a solution (why?), and $y(t) = x(t)$ since $y(0) = x(0)$. So $x(t)$ has period t_1. Show that $t_1 = T$ and so the orbit is simple (i.e., the curve has no self-intersections).]

7.3 Interacting Species: Cooperation, Competition

Comments:

Models of the population of interacting species have been popular for some time. These models (particularly the oversimplified models introduced here) don't have much predictive value, but they are guides to thinking about how species interact. Some would call these models caricatures, rather than models. A caricature is a distortion of reality, but the original is still recognizable—and that is just what these models are like in comparison to the real thing. We restrict attention to two interacting species whose model rate equations are autonomous. That means that time variations in the rate functions have been averaged out—a dubious proposition at best, but the modeling process usually begins with oversimplifications. We give a catalogue of possible interactions between two populations, ranging from the now classical Volterra model for predator-prey interactions (see Section 2.6) to assorted models of cooperation, competition, harvesting, and satiation.

PROBLEMS

Biological Dynamics. Explain the biological dynamics: identify predators, prey, competitors, cooperators; identify harvesting and restocking terms and their nature. The constants in each ODE are assumed to be positive.

1. $x' = (\alpha - by)x,$ $y' = (\beta - cx)y + H$

2. $x' = (\alpha + by)x,$ $y' = (-\beta + cx - dy)y$

3. $x' = (\alpha - ax - by)x,$ $y' = (\beta - cx - dy)y$

4. $x' = (\alpha - ax - by)x,$ $y' = (-\beta + cx - dy)y + (2 + \cos t)$

5. $x' = (\alpha + by)x - H\,\text{SqWave}(t, 1, 0.25),$ $y' = (\beta + cx - dy)y$

6. $x' = (\alpha - ax)x,$ $y' = \left(-\beta + \dfrac{cx}{m + kx}\right)y + H$

7. $x' = (\alpha - bz)x - H\cos t,$ $y' = (\beta - my - kz)y,$ $z' = (-\gamma + ax + cy)z - Hz$

Answer 1: Each species modeled by the system $x' = (\alpha - by)x$, $y' = (\beta - cx)y + H$ would grow exponentially in isolation (the rate terms αx, βy), but when occupying the same ecological region (or *niche*, to use the technical term) they compete for common resources (the terms $-byx$ and $-cxy$). The y-species is restocked at rate H.

Answer 3: Each species modeled by the system $x' = (\alpha - ax - by)x$, $y' = (\beta - cx - dy)y$ would grow exponentially in isolation (rate terms αx and βy) except that overcrowding implies logistic change (rate terms $-ax^2$ and $-dy^2$). When occupying the same ecological niche, the two species compete for common resources (the terms $-byx$ and $-cxy$).

Answer 5: The system is $x' = (\alpha + by)x - H\,\text{SqWave}(t, 1, 0.25)$, $y' = (\beta + cx - dy)y$. Each species would grow exponentially in isolation (rate terms αx and βy), except that overcrowding implies logistic change for the y-species (the rate term $-dy^2$). When occupying the same ecological niche the two species cooperate, the presence of each being favorable for the other (the terms byx and cxy). The x-species is seasonally harvested at the rate H, but only for the first quarter of the period 1. This is a model of cooperation (the terms byx and cxy).

Answer 7: The system is $x' = (\alpha - bz)x - H\cos t$, $y' = (\beta - my - kz)y$, $z' = (-\gamma + az + cy)z - Hz$. The y-species would change logistically if isolated (the term $\beta y - my^2$), but the x-species would increase exponentially if isolated (the term αx) and not harvested/restocked (the term $-H\cos t$), and the z-species would decline exponentially if isolated and harvested (the term $(-\gamma - H)z$). The z-species is a predator on the prey species x and y (the terms $-bzx$ and axz, $-kzy$ and cyz). The z-species is harvested with constant effort harvesting (the term $-Hz$).

Nullclines and Orbits. Set the constants $a, b, c, d, k, m,$ and α in Problems 1–7 equal to 1, and the constants β, γ equal to 2. Use a computer solver to sketch orbits and direction fields in state space. Interpret the orbits in the portrait in terms of the long-term behavior of the species involved.

Note: Note: Set $H = 1$ in Problems 8, 12, 13, 14.

8. Problem 1	**9.** Problem 2	**10.** Problem 3	**11.** Problem 4
12. Problem 5	**13.** Problem 6	**14.** Problem 7	

Answer 9: See the figure for orbits of $x' = (1 + y)x$, $y' = (-2 + x - y)y$. This is a cooperation model (see the $+xy$ terms in both ODEs). The natural decay coefficient -2 for the y-species and the overcrowding term $-y^2$ suggest that the positive effect of cooperation saves the y-species from extinction. Both species eventually

grow in numbers, but only the x-species has an ever-increasing positive growth rate. The x-nullclines are the lines $y = -1$ and the y-axis. The y-nullclines are the lines $y = x - 2$ and the x-axis.

Answer 11: See the figure for some orbits of $x' = (1 - x - y)x$, $y' = (-2 + x - y)y + (2 + \cos t)$. Since the system is nonautonomous, nullclines and a direction field are not relevant. The orbits head toward a region on the y-axis, and the x-species heads toward extinctino. The y-species has a marked advantage over x because of the term $+xy$ in the y-rate equation (compared to $-xy$ in the x-rate equation). In addition, y is continually restocked at a rate that oscillates between 1 and 3. Note: the y-species does not approach an equilibrium value on the y-axis (which is a union of orbits), but satisfies the ODE $y' = (2 - y)y + (2 + \cos t)$ on the y-axis; so y appears to oscillate in the absence of x.

Answer 13: Set $H = 1$. The system $x' = (1 - x)x$, $y' = (-2 + x/(1 + x))y + 1$ has x-nullclines $x = 1$ (which consists of an equilibrium point $(1, 2/3)$ and two vertical rays that tend to the point as $t \to +\infty$) and the y-axis. The y-nullcline is the curve $y = (1 + x)/(2 + x)$. It appears that the population orbits tend to $(1, 2/3)$ as $t \to +\infty$ inside the population quadrant. See the figure.

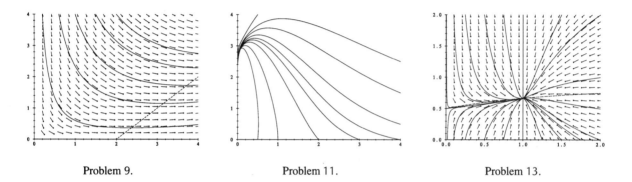

Problem 9.   Problem 11.   Problem 13.

Nullclines. Find the equations of the x-nullclines and the y-nullclines for the system and sketch their graphs by hand in the population quadrant. Find the equilibrium points. Determine the direction of orbital motion across the nullclines as time advances. Identify the nature of the x-species and of the y-species (e.g., predator, prey, cooperator, competitor).

15. $x' = (5 - x + y)x$, $y' = (10 + x - 5y)y$

16. $x' = (10 - x + 5y)x$, $y' = (5 + x - y)y$

17. $x' = (5 - x - y)x$, $y' = (10 - x - 2y)y$

18. $x' = (10 - x - 5y)x$, $y' = (5 - x - y)y$ [*Hint*: see Example 7.3.4.]

Answer 15: The x-nullclines are defined by $(5 - x + y)x = 0$. That is, the x-nullclines are the lines $x = 0$ and $y = x - 5$. The y-nullclines are defined by $(10 + x - 5y)y = 0$. That is, the y-nullclines are the lines $y = 0$ and $y = x/5 + 2$. The equilibrium points are the intersection points of the x-nullclines with the y-nullclines. For example, if $x = 0$, then we require also that $y = 0$ or $y = x/5 + 2$. This gives us the two equilibrium points $(0, 0)$ and $(0, 2)$. If $y = x - 5$, then we require (as before) that $y = 0$ (in which case $x = 5$) or $y = x/5 + 2$ (in which case $x/5 + 2 = y = x - 5$, so $x = 35/4$, and $y = 15/4$). Now we get two new equilibrium points $(5, 0)$ and $(35/4, 15/4)$. The terms $+xy$ in each rate equation show that the two species cooperate. From the behavior of the orbits (see the figure for Problem 19) the populations seem to approach the equilibrium point $(35/4, 15/4)$ as $t \to +\infty$.

Answer 17: The x-nullclines are the lines $x = 0$ and $y = -x + 5$. The y-nullclines are the lines $y = 0$ and $y = -x/2 + 5$. The intersection points of the nullclines are found as follows: If $x = 0$, then $y = 0$ or $y = -x/2 + 5 = 5$. If $y = -x + 5$, then $-x + 5 = y = 0$ (so $x = 5$, $y = 0$) or $-x + 5 = y = -x/2 + 5$ (so $x = 0$, $y = 5$). The three equilibrium points are $(0, 0)$, $(0, 5)$, and $(5, 0)$. The terms $-xy$ in the rate equations indicate that the species are competitors. From the behavior of the orbits (see the figure for Problem 21), it seems that $x \to 0$ and $y \to 5$ as time increases. So the y-species wins the competition.

Visualizing Nullclines. Use a computer solver to graph orbits, direction fields, and nullclines in the population quadrant. What does each orbit suggest about the long-term behavior of the species?

19. Problem 15   **20.** Problem 16   **21.** Problem 17   **22.** Problem 18

Answer 19:   See the figure for orbits of $x' = (5 - x + y)x$, $y' = (10 + x - 5y)y$. Each population tends to a positive equilibrium value as $t \to +\infty$.

Answer 21:   See the figure. The y-species wins the competition as $x \to 0$ and $y \to 5$ as $t \to +\infty$.

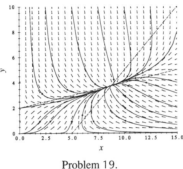

Problem 19.

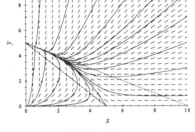

Problem 21.

Modeling Problems.

Qualitative Behavior.   The models of cooperation and competition in Examples 7.3.1–7.3.4 have specific numerical coefficients, but the general behavior of the orbits stays much the same if the coefficients are changed a little. Explain why this is so in Problems 23-26. [*Hint*: plot a direction field and the nullclines for the original example. Then change the coefficients in the example a little and plot the new direction field and nullclines. Compare the two.]

23. *Cooperation and Explosive Growth*   Suppose that the x- and y-nullclines have the same relative positions as the nullclines in Example 7.3.1, that the signs of x' and of y' are the same in the regions between the nullclines as in that example, and that the equilibrium points (other than the origin) remain on the positive axes or inside the third quadrant. Explain why the altered system still models the explosive growth of both species.

24. *Stable Cooperation*   Modify the statement of Problem 23 so that it applies to Example 7.3.2. Explain why the altered system still models stable cooperation.

25. *Stable Competition*   Modify the statement of Problem 23 so that it applies to Example 7.3.3. Explain why the altered system still models stable competition.

26. *Competitive Exclusion*   Modify the statement of Problem 23 so that it applies to Example 7.3.4. Explain why the altered system still models competitive exclusion.

Answer 23:   The respective systems are those of Example 7.3.1, $x' = (2 - x + 2y)x$, $y' = (2 + 2x - y)x$ (Graph 1) and an altered system, $x' = (1 - x + y)x$, $y' = (1 + 3x - y)y$ (Graph 2). Each axis is a nullcline and a union of orbits for each system. The nullclines have the same relative positions in the two systems and are crossed by orbits in the same way (left to right for the y-nullcline, upward for the x-nullcline). The nullclines and the direction fields of the two systems have the same general appearance. So the long-term explosive behavior of the orbits of the second system is much like that of the first. Both show cooperation that leads to explosive growth.

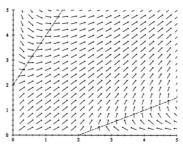

Problem 23, Graph 1   Problem 23, Graph 2

Answer 25:   The coefficients of the system, $x' = (2 - 2x - y)x$, $y' = (2 - x - 2y)y$ (Graph 1) of Example 7.3.3 are altered to those of the system $x' = (2 - 4x - y)x$, $y' = (1 - x - y)y$ (Graph 2). The axes are nullclines and unions of orbits for both systems. The nullclines have the same relative positions in both systems and divide the quadrants into four regions where the orbital behavior is similar; for example, downward and to the right at the upper left. All orbits inside the quadrant tend to an internal attracting equilibrium point for both systems; so these are models of stable competition.

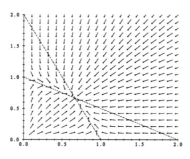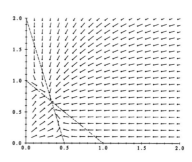

Problem 25, Graph 1   Problem 25, Graph 2

 27. *Another Model of Competitive Exclusion*   In the competition model, $x' = (\alpha - ax - by)x$, $y' = (\beta - cx)y$, only the x-species has a self-limitation term. Regardless of the values of the positive coefficients, α, β, a, b, c, the Principle of Competitive Exclusion applies (see Example 7.3.4). Explain why. Plot orbits for your choices of values for the coefficients. Explain what you see. [*Hint*: consider the two cases $\alpha/a > \beta/c$ and $\alpha/a \le \beta/c$.]

Answer 27:   Group project.

28. *Models of Cooperation: How to Control the Population Levels*   In the cooperation model with self-limiting terms $x' = (2 - x + 2y)x$, $y' = (2 + 2ax - y)y$, we can adjust the positive parameter a to "steer" the system to the explosive growth of both populations or toward a stable equilibrium.

(a) Discuss the meaning of the coefficient $2a$ in population terms.

(b) Find the critical value a_0 of a that divides the explosive growth model of Example 7.3.1 from a stable model in which both populations tend to an equilibrium inside the population quadrant as time increases. Plot orbits and a direction field in the population quadrant for a value of $a > a_0$. Repeat with $a = a_0$, and then with a value of $a < a_0$. [*Hint*: find the maximal value a_0 of a such that if $0 < a < a_0$, then there is an equilibrium point inside the population quadrant.]

29. *Knocking Out the Competition*   The system $x' = (2 - x - 2y)x + Hx$, $y' = (2 - 2x - y)y$ models competition where the x-species can be restocked ($H > 0$) at a rate proportional to the population. Using a numerical solver to plot solution curves, find the restocking coefficient H of minimal magnitude H_0 so that the y-species becomes extinct regardless of its initial population. Plot orbits for values of H below, at, and above H_0 and explain what you see. [*Hint*: find the value H_0 of the restocking coefficient that has the property that as H increases through H_0, the equilibrium point inside the population quadrant exits across an axis.]

Answer 29: The system is $x' = (2 + H - x - 2y)x$, $y' = (2 - 2x - y)y$. The equilibrium point determined by the intersection of the x-nullcline $2 + H - x - 2y = 0$ and the y-nullcline $2 - 2x - y = 0$ has coordinates $x = (2 - H)/3$, $y = 2(1 + H)/3$. This point exits the first quadrant as H increases through the value $H_0 = 2$. The other equilibrium points are $(0,0)$, $(0,2)$, and $(2 + H, 0)$. For $0 < H < 2$, it seems that we have a model of competitive exclusion. Graph 1 with $H = 1$ shows y winning if (x_0, y_0) lies in the region at the upper left, while y wins if (x_0, y_0) lies to the right. At $H = 2$, the x-species seems to win the competition (see Graph 2 where $H = 2$). For $H > 2$, the x-species wins (see Graph 3 where $H = 3$). Note that $x(t) \to H + 2 = 5$ and $y(t) \to 0$ if $x(0) > 0$ and $H = 3$.

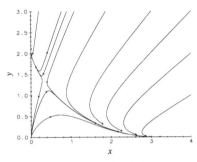

Problem 29, Graph 1.

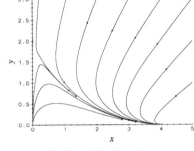

Problem 29, Graph 2.

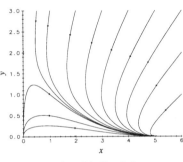

Problem 29, Graph 3.

Rescaling.

 30. *Rescaling a Competition Model* Suppose that you belong to the x-species and are able to change the rate constants in the x' equation of system (8). Analyze the effects of changing each of your rate constants on your competitor. First, however, scale the system by letting $x = ku$, $y = mv$, and $t = n\tau$ and then choosing the positive scaling constants k, m, and n so that the new rate equation for your competitor is $dv/d\tau = (1 - u - v)v$. Your scaled rate equation is $du/d\tau = (\alpha^* - a^*u - b^*v)u$. Explain what happens as you tune the system by changing α^*, a^*, and b^* in turn. What values of these parameters might be both realistic and optimal for your species? Plot and interpret the orbits.

☞ Scaling cuts the number of system parameters from 6 to 3.

 31. *Rescaling Time on Orbits* Let the functions $f(x, y)$, $g(x, y)$, and $r(t, x, y)$ be continuously differentiable on the entire txy-space, and suppose that $r(t, x, y)$ is always positive.

☞ We also discuss scaling in the WEB SPOTLIGHT ON SCALING AND UNITS and in the SPOTLIGHT ON CONTINUITY IN THE DATA after Chapter 2.

- Explain why you would expect the orbits of the autonomous planar system

$$x' = f(x, y), \qquad y' = g(x, y)$$

to be the same as the orbits for the planar system

$$x' = f(x, y)r(t, x, y), \qquad y' = g(x, y)r(t, x, y)$$

- Use a solver to verify that orbits of the two systems are identical if they have common initial points, if $t_0 = 0$, $0 \le t \le 15$, $r = [2 + \sin t + x^2]^{-1}$, $f = x - 10y - x(x^2 + y^2)$, and $g = 10x + y - y(x^2 + y^2)$. Are the corresponding component curves of the two systems identical?
- Verify that the unit circle is an orbit of the two systems given above.
- Explain why multiplying f and g by r has the effect of rescaling time. [*Hint*: set $d\tau = rdt$ in the scaled system.]

Answer 31: Group project.

Spotlight on Modeling: Destructive Competition

Comments:

These models vastly oversimplify combat, but they do provide a way to begin a quantitative and qualitative study of hypothetical and of real battles. The model ODEs used here are all linear. More sophisticated models use nonlinearities and time-dependent rate coefficients. The interpretation in terms of combat will put some people off.

PROBLEMS

Conventional Combat Without Reinforcements. Problems 1 and 2 have to do with the conventional combat models of Examples 1 and 2.

1. Reproduce Figure 1 as closely as possible. Plot the orbit with initial condition $x_0 = 10$, $y_0 = 7$. Does y win or lose? How much time does it take for the conflict to be resolved? [*Hint*: first check the y-win condition (5).]

    Answer 1:   See the figure for the orbit and component curves of the IVP $x' = -0.1y$, $y' = -0.064$, $x_0 = 10$, $y_0 = 7$. Here $a = 0.064$, $b = 0.1$ and y loses because $y_0 = 7 < (a/b)^{1/2}x_0 = (0.64)^{1/2} \cdot 10 = 8$. The combat is over and $y = 0$ by $t \approx 17$. **Warning!** We let time run on to $t = 20$, but the model is no longer valid after the y-side reaches 0. In fact, negative values of y lead to an upturn in the x-component curve; we should have stopped the x-component curve as soon as $y(t)$ became 0.

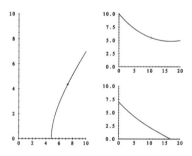

Problem 1.

2. *Who Wins?*   Answer this question for each of the model ODEs below. Then if $y_0 = 10$, estimate how long it takes the battle to end by inspecting the loser's component curve.

    (a) $x' = -0.1y$, $y' = -0.01x$, $x_0 = y_0/3$

    (b) $x' = -0.2y$, $y' = -0.05x$, $x_0 = y_0/2$

    (c) $x' = -0.02y$, $y' = -0.01x$, $x_0 = y_0$

    (d) $x' = -0.01y$, $y' = -0.04x$, $y_0 = 2x_0$

3. Which is the best precombat strategy for the y-force: to increase its average combat effectiveness coefficient b by a certain percentage α, or to increase the initial number y_0 by the same percentage?

    Answer 3:   The win condition (5) for y can be written as $y_0\sqrt{b} > x_0\sqrt{a}$. So, it would be best for y to increase the initial strength y_0 by α percent: $y_0\sqrt{b}$ becomes $y_0(1 + \alpha/100)\sqrt{b}$. Since the win condition for the y-force is that $y_0\sqrt{b} > x_0\sqrt{a}$, increasing b by α percent only increases the product $y_0\sqrt{b}$ by the factor $\sqrt{1 + \alpha/100}$, instead of by a factor of $1 + \alpha/100$.

Conventional Combat with Reinforcements.

4. Consider the conventional combat model (1) where R_1 and R_2 are positive constants.

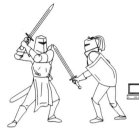

(a) Follow the procedure of Example 1 to obtain a separable first-order ODE in the variables x and y. Solve the ODE and show that $x(t)$ and $y(t)$ satisfy the equation

$$b\left(y(t) - \frac{R_1}{b}\right)^2 - a\left(x(t) - \frac{R_2}{a}\right)^2 = C$$

where C is a constant determined by the initial data.

(b) Find a y-win condition (analogous to (5)) for this kind of combat.

 (c) Take $a = b = 1$, $R_1 = 2$, and $R_2 = 3$ and plot the orbits when (i) $x(0) = 3$, $y(0) = 1$; (ii) $x(0) = 2$, $y(0) = 2$; (iii) $x(0) = 4$, $y(0) = 2$. Assign arrowheads to the orbits indicating the direction of increasing time.

(d) Who wins in each of the cases described in **(c)**?

(e) What happens when $x(0) = R_2/a$ and $y(0) = R_1/b$? Explain in terms of the development of the combat.

Spotlight on Modeling: Bifurcation and Sensitivity

Comments:

The questions of controlling a model of some physical system to some desired state were first considered in Section 2.7 in an analytical way. The methods used here are primarily based on graphical considerations, although analysis does play a role.

PROBLEMS

1. *Stabilize the Cost of Cooperation*   The rate equations $x' = (1 - x + y)x$, $y' = (1 + \alpha x - y)y$ model the interactions of a pair of cooperating species (see Examples 7.3.1 and 7.3.2). The coefficients of the x-species are hardwired and cannot be changed, but the y-species can control the positive cost/benefit parameter α, which measures how much cooperation affects the y-species. Address the following points:

- Find an interval of values of α for which the xy-orbits range from unbounded growth (unacceptable to the y-species) to low stable equilibrium levels for both species.
- Find the equations of the slanted x- and y-nullclines and the value α^* where the two nullclines are parallel.
- Explain why it makes sense to say that there is a bifurcation as α transits the value α^*.
- The y-species aims for a stable equilibrium value in the interval $3 \le y \le 6$. Find the range of values of α that will achieve this goal.
- Explain why the long-term behavior of the orbits is relatively insensitive to changes in the value of α inside the interval $\alpha^* < \alpha < \infty$. What about the interval $0 < \alpha < \alpha^*$?
- Plot representative orbital portraits and the nullclines for two values of α, where $\alpha^* < \alpha < \infty$. Repeat for the interval $0 < \alpha < \alpha^*$. Finally, plot the nullclines and an orbital portrait for the bifurcation value $\alpha = \alpha^*$. Comment on what you see in terms of the long-term behavior of the populations of each species in each of these settings.

Answer 1: Group problem.

 2. *From Stable Competition to Competitive Exclusion*   Create a one-parameter model of competition with the property that the competition changes from a stable situation to one of competitive exclusion as the parameter changes (see Examples 7.3.3 and 7.3.4). Follow the structure of Problem 1 in your analysis of the model.

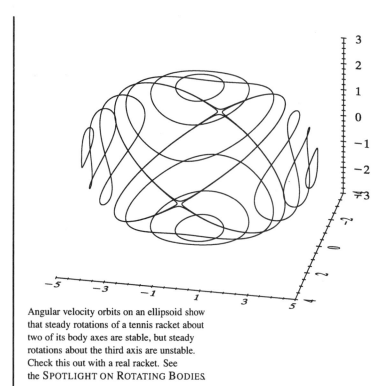

Angular velocity orbits on an ellipsoid show
that steady rotations of a tennis racket about
two of its body axes are stable, but steady
rotations about the third axis are unstable.
Check this out with a real racket. See
the SPOTLIGHT ON ROTATING BODIES.

CHAPTER

8

Stability

In Section 8.1 we define stability (both neutral and asymptotic) of a system of linear autonomous ODEs at an equilibrium point and illustrate the definition with many examples. We also give the criterion for stability for an autonomous linear system $x' = Ax$. In Section 8.2 we give the result that an autonomous linear system whose system matrix has only eigenvalues with negative real parts is not only asymptotically stable at the origin, but cannot be destabilized by any higher order perturbation. On the other hand, if at least one of the eigenvalues has positive real part, then the unstable linear system cannot be stabilized by any higher order perturbation. Theorem 8.2.1 is actually phrased as follows: If the Jacobian matrix of $f(x)$ at an equilibrium point p has all eigenvalues with negative real parts, then the system $x' = f(x)$ is asymptotically stable at p. If there is at least one eigenvalue with positive real part, then $x' = f(x)$ is unstable at p. These results form the basis of the long-standing technique of linearizing a nonlinear system at an equilibrium point in order to understand the long-term behavior of the solutions of the nonlinear system near the point. In Section 8.3 we generalize the existence of an integral given in Section 2.5 and its problem set to the existence of an integral for certain systems of ODEs, which are said to be conservative. Typically, physical systems for which no energy is lost or gained have model systems of ODEs which are conservative, and this analogy is explored. The chapter cover figure displays an integral surface and several orbits of the conservative system that models the angular velocities of a whirling tennis racket. In the SPOTLIGHT ON LYAPUNOV FUNCTIONS we present Lyapunov's tests and the so-called Lyapunov Method for determining stability. In the process we characterize the definiteness properties of quadratic forms.

Since the material of this chapter is likely to be new and rather strange for many readers, we give lots of examples.

8.1 Stability and Linear Autonomous Systems

Comments:

In this section we lay out the definitions of stability and its refinements, neutral stability and asymptotic stability, for an autonomous linear system at an equilibrium point. Theorem 8.1.3 allows us to test the stability properties simply by inspecting the eigenvalues of the system matrix and, if an eigenvalue λ with zero real part is multiple, determining the dimension of its eigenspace.

"Stability" is an everyday word that has many meanings. In science and engineering it is often used to describe something other than what we define here. For example, in engineering a system is sometimes said to be stable if it responds to bounded inputs with bounded outputs (the "BIBO" of Sections 2.7 and 6.7), but that is quite different from our definition. We give many examples here in order to help solidify understanding of what stability means.

PROBLEMS

Neutral Stability. Explain why each system is neutrally stable at the origin.

1. $x' = 2y, \quad y' = -8x$ **2.** $x' = 2x - 2y, \quad y' = 4x - 2y$

3. $x' = -x/10, \quad y' = z, \quad z' = -y$ **4.** $x' = 2x - 3y, \quad y' = 2x - 2y$

Answer 1: The system $x' = 2y, \; y' = -8x$ is linear. The system matrix has eigenvalues $\lambda_1 = 4i = \overline{\lambda}_2$, so the origin is neutrally stable by Theorem 8.1.3.

Answer 3: The system matrix has eigenvalues $-1/10, \pm i$, so the origin is neutrally stable by Theorem 8.1.3.

Asymptotic Stability. Explain why each system is asymptotically stable at the origin.

5. $x' = -4x, \quad y' = -3y$ **6.** $x' = x - 3y, \quad y' = 4x - 6y$

7. $x' = -x + 4y, \quad y' = -3x - 2y$ **8.** $x' = -x + y + z, \; y' = -2y, \; z' = -3z$

Answer 5: The system $x' = -4x, \; y' = -3y$ is linear; the system matrix has eigenvalues $\lambda_1 = -4, \; \lambda_2 = -3$. Both eigenvalues are negative, so the system is asymptotically stable at the origin by Theorem 8.1.3.

Answer 7: The system $x' = -x + 4y, \; y' = -3x - 2y$ is linear; the system matrix has eigenvalues $\lambda_1 = -(3 + i\sqrt{47})/2 = \overline{\lambda}_2$ with negative real parts; we have asymptotic stability at the origin by Theorem 8.1.3.

Instability. The following systems are unstable at the origin. Explain why.

9. $x' = 3x - 2y, \quad y' = 4x - y$ **10.** $x' = 3x - 2y, \quad y' = 2x - 2y$

11. $x' = x + y, \quad y' = -x - y$ **12.** $x' = y, \quad y' = x, \quad z' = -2x + y + z$

Answer 9: The system $x' = 3x - 2y, \; y' = 4x - y$ is linear; the system matrix has eigenvalues $\lambda, \overline{\lambda}$, where $\lambda = 1 + 2i$. Therefore, by Case 3 of Theorem 8.1.3 this system is unstable at the origin.

Answer 11: The system $x' = x + y, \; y' = -x - y$ is linear; the system matrix has 0 as a double eigenvalue and only a one-dimensional eigenspace, which is spanned by $[1, -1]^T$. The origin is unstable by Case 3 of Theorem 8.1.3.

Orbits of Stable and Unstable Planar Linear Systems. Plot several orbits of the indicated system near the origin and insert arrowheads to show the direction of increasing time. Label each system as U (unstable), AS (asymptotically stable), or NS (neutrally stable).

13. Problem 1	**14.** Problem 2	**15.** Problem 3	**16.** Problem 4
17. Problem 5	**18.** Problem 6	**19.** Problem 7	**20.** Problem 8
21. Problem 9	**22.** Problem 10	**23.** Problem 11	**24.** Problem 12

Answer 13:   See figure (neutral stability).

Answer 15:   See figure (neutral stability).

Answer 17:   Asymptotically stable at the origin; see the figure for some orbits.

Answer 19:   Asymptotically stable at the origin; see the figure for some orbits.

Answer 21:   See the figure for orbits near the unstable origin (Case 3 of Theorem 8.1.3).

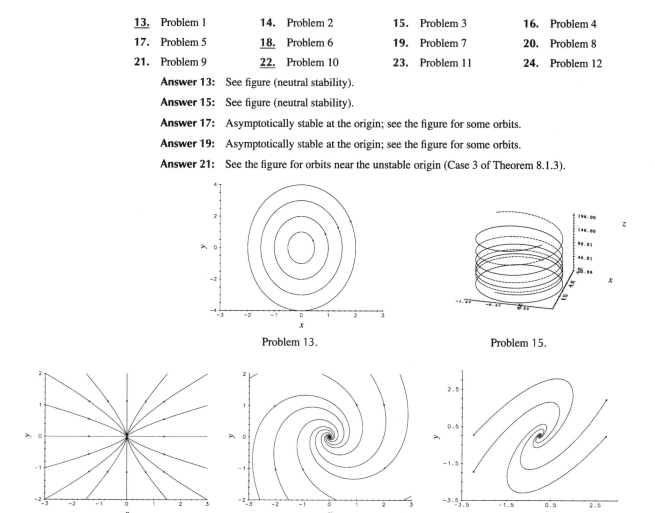

Problem 13.   Problem 15.

Problem 17.   Problem 19.   Problem 21.

Answer 23:   The origin is unstable by Case 2 of Theorem 8.1.3. See the figure for orbits; nonconstant orbits are parallel to the line $x + y = 0$ of equilibrium points. Above the line the orbits move downward and to the right since $x' > 0$ and $y' < 0$ in the region. Below the line the direction of motion is reversed.

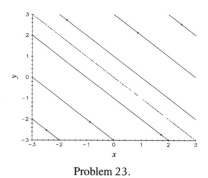

Problem 23.

Stability Properties. Determine the stability properties of $x' = Ax$, where A is the given matrix.

25. $\begin{bmatrix} -1 & 2 \\ -2 & -1 \end{bmatrix}$ **26.** $\begin{bmatrix} -3 & 1 \\ 2 & -4 \end{bmatrix}$ **27.** $\begin{bmatrix} 1 & -1 \\ 2 & -2 \end{bmatrix}$

28. $\begin{bmatrix} 0 & 1 \\ 0 & 0 \end{bmatrix}$ **29.** $\begin{bmatrix} 5 & 1 \\ 1 & -2 \end{bmatrix}$ **30.** $\begin{bmatrix} 3 & -2 \\ 5 & -3 \end{bmatrix}$

☞ For Problem 31 use the fact that all roots of $\lambda^3 - a_2\lambda^2 - a_1\lambda - a_0$ have negative real parts if all $a_j < 0$ and $a_0 + a_1 a_2 > 0$ (see Section 6.7).

31. $\begin{bmatrix} -10 & 3 & -4 \\ 1 & -12 & 9 \\ -1 & -2 & -5 \end{bmatrix}$ **32.** $\begin{bmatrix} -5 & 0 & 0 \\ 0 & 0 & -3 \\ 0 & 3 & 0 \end{bmatrix}$ **33.** $\begin{bmatrix} 0 & 1 & 1 \\ 0 & 0 & 1 \\ 0 & 0 & 0 \end{bmatrix}$

34. $\begin{bmatrix} 0 & 0 & 1 & 0 \\ 0 & 0 & 0 & 0 \\ 0 & 0 & 0 & 0 \\ 0 & 0 & 0 & 0 \end{bmatrix}$ **35.** $\begin{bmatrix} 3 & -2 & 0 & 0 \\ 5 & -3 & 0 & 0 \\ 0 & 0 & 3 & -2 \\ 0 & 0 & 5 & -3 \end{bmatrix}$ **36.** $\begin{bmatrix} 3 & -2 & 1 & 0 \\ 5 & -3 & 0 & 1 \\ 0 & 0 & 3 & -2 \\ 0 & 0 & 5 & -3 \end{bmatrix}$

Answer 25:   The eigenvalues of the matrix $\begin{bmatrix} -1 & 2 \\ -2 & -1 \end{bmatrix}$ are $\lambda_1 = \bar{\lambda}_2 = -1 + 2i$. Since each eigenvalue has negative real part, the system is asymptotically stable by Theorem 8.1.3.

Answer 27:   The eigenvalues of the matrix $\begin{bmatrix} 1 & -1 \\ 2 & -2 \end{bmatrix}$ are $\lambda_1 = 0$, $\lambda_2 = -1$. Since each eigenvalue is simple and nonpositive and λ_1 is zero, the system is neutrally stable by Theorem 8.1.3.

Answer 29:   The eigenvalues of the matrix $\begin{bmatrix} 5 & 1 \\ 1 & -2 \end{bmatrix}$ are $\lambda_1 = (3 + \sqrt{53})/2$ and $\lambda_2 = (3 - \sqrt{53})/2$. Since one of the eigenvalues is positive, the system is unstable by Theorem 8.1.3.

Answer 31:   The eigenvalues of the matrix $\begin{bmatrix} -10 & 3 & -4 \\ 1 & -12 & 9 \\ -1 & -2 & -5 \end{bmatrix}$ all have negative real parts. To see this, note that each diagonal entry is negative and that on each row the sum of the magnitudes of the off-diagonal elements has magnitude less than the magnitude of the diagonal entry on that row. So using Gerschgorin row disks in the WEB SPOTLIGHT ON LOCATING EIGENVALUES, the eigenvalues lie entirely inside the left half of the complex plane, and the system is globally asymptotically stable. Alternatively, see Problem 17 of Section 6.7 for another way to show that the eigenvalues have negative real parts.

Answer 33:   The matrix $\begin{bmatrix} 0 & 1 & 1 \\ 0 & 0 & 1 \\ 0 & 0 & 0 \end{bmatrix}$ has $\lambda = 0$ as a triple eigenvalue. However, the eigenspace is spanned by the vector $[1 \ \ 0 \ \ 0]^T$ and so is of dimension one. So by Theorem 8.1.3, the system is unstable since the dimension of the eigenspace is less than the multiplicity of the eigenvalue.

Answer 35:   For the matrix $A = \begin{bmatrix} 3 & -2 & 0 & 0 \\ 5 & -3 & 0 & 0 \\ 0 & 0 & 3 & -2 \\ 0 & 0 & 5 & -3 \end{bmatrix}$, the first two equations in the system $x' = Ax$ decouple from the last two. We find that the double eigenvalues $\pm i$ have zero real part, V_i has a basis $\{[2 \ \ 3 - i \ \ 0 \ \ 0]^T, [0 \ \ 0 \ \ 2 \ \ 3 - i]^T\}$ and V_{-i} has basis $\{[2 \ \ 3 + i \ \ 0 \ \ 0]^T, [0 \ \ 0 \ \ 2 \ \ 3 + i]^T\}$ so $m_i = d_i = m_{-i} = d_{-i} = 2$. Therefore the system here is also neutrally stable by Theorem 8.1.3.

Inequalities.

 37. Prove Theorem 8.1.1 by considering the following points:

- Cauchy–Schwarz Inequality: first show its validity if $v = 0$. Then for $v \neq 0$ and any real number α, complete the square in α, where $\|u + \alpha v\|^2 = \|u\|^2 + 2\alpha u \cdot v + \alpha^2 \|v\|^2$. Finally, set $\alpha = u \cdot v / \|v\|^2$.

- Triangle Inequality: use the Cauchy–Schwarz Inequality.

- Matrix–Norm Inequalities: first show that $\|v\|^2 \leq (\Sigma|v_i|)^2$. Then use the Cauchy–Schwarz Inequality to show that $\|AB\| \le \Sigma |A_i||B_j|$ where A_i and B_j are the i and the j column vectors of A and B, respectively.

Answer 37:   Recall that $|v| = \sum |v_i|$ for any vector $v = [v_1 \ \cdots \ v_n]^T$. First we prove the Cauchy–Schwarz Inequality for vectors in $\mathbb{R}^n$. From the definition we directly verify that $u \cdot v$ has the following properties:

1. $(\alpha u + \beta w) \cdot v = \alpha u \cdot v + \beta w \cdot v$, for all u, v, w in $\mathbb{R}^n$ and α, β in $\mathbb{R}$

2. $u \cdot v = v \cdot u$, for all u, v in $\mathbb{R}^n$

3. $u \cdot u \ge 0$ for all u in $\mathbb{R}^n$, and $= 0$ if and only if $u = 0$

If $v = 0$ then $\|v\|^2 = 0$ and $u \cdot v = 0$, so the Cauchy–Schwarz Inequality holds. Now assume that $v \ne 0$. For any real number α, use the scalar product properties **1** and **2** above to show that

$$0 \le \|u + \alpha v\|^2 = (u + \alpha v) \cdot (u + \alpha v) = \|u\|^2 + 2\alpha u \cdot v + \alpha^2 \|v\|^2$$

Complete the square in the above quadratic polynomial for α

$$0 \le \|v\|^2 \left(\alpha + \frac{u \cdot v}{\|v\|^2} \right)^2 - \frac{|u \cdot v|^2}{\|v\|^2} + \|u\|^2$$

Choose $\alpha = -u \cdot v / \|v\|^2$ to obtain the Cauchy–Schwarz Inequality.

The Cauchy–Schwarz Inequality and the scalar product properties **1** and **2** above show that

$$\|u + v\|^2 = \|u\|^2 + 2u \cdot v + \|v\|^2 \le \|u\|^2 + 2|u \cdot v| + \|v\|^2$$

$$\le \|u\|^2 + 2\|u\| \cdot \|v\| + \|v\|^2 = (\|u\| + \|v\|)^2$$

which produces one side of the Triangle Inequality. The other side follows by applying this result to $\|u\| = \|(u - v) + v\|$ and $\|v\| = \|(v - u) + u\|$.

To prove the Matrix-Norm Inequalities, first let $|v| = |v_1| + \cdots + |v_n|$, for any vector $v = [v_1 \ \cdots \ v_n]^T$. Since for any vector v,

$$\|v\|^2 = v_1^2 + \cdots + v_n^2 \le (|v_1| + \cdots + |v_n|)^2 = |v|^2$$

it follows that $\|v\| \le |v|$. Now partition the $n \times n$ matrices A and B into column vectors $A_1, \ldots, A_n$ and $B_1, \ldots, B_n$. Then

$$AB = \begin{bmatrix} A_1^T B_1 & \cdots & A_1^T B_n \\ \vdots & & \\ A_n^T B_1 & \cdots & A_n^T B_n \end{bmatrix}$$

and hence the Cauchy–Schwarz Inequality implies that

$$\|AB\| = |A_1^T B_1| + \cdots + |A_n^T B_n| \le \|A_1\| \cdot \|B_1\| + \cdots + \|A_n\| \cdot \|B_n\|$$

$$\le |A_1| \cdot |B_1| + \cdots + |A_n| \cdot |B_n| \le (|A_1| + \cdots + |A_n|)(|B_1| + \cdots + |B_n|)$$

$$= \|A\| \cdot \|B\|$$

Next, observe that $Ax = [A_1^T x \ \cdots \ A_n^T x]^T$, for any column n-vector x, so

$$\|Ax\|^2 = |A_1^T x|^2 + \cdots + |A_n^T x|^2 \le \|A_1\|^2 \cdot \|x\|^2 + \cdots + \|A_n\|^2 \cdot \|x\|^2$$

$$\le \left(|A_1|^2 + \cdots + |A_n|^2 \right) \|x\|^2 = \|A\|^2 \|x\|^2$$

which implies that $\|Ax\| \le \|A\| \cdot \|x\|$.

Modeling Problems.

Springs and Interacting Species. Plot orbits in the indicated region. Locate the equilibrium points and plot direction fields and orbits in small neighborhoods of each point. Zoom in on each neighborhood and state whether the equilibrium point has the characteristics of an asymptotically stable, neutrally stable, or unstable equilibrium point. According to the behavior of nearby orbits, label each equilibrium point as a nonlinear center, saddle, node, or spiral point.

38. *Damped Hard Spring* $x' = v, v' = -10x - 0.01x^3 - 2v; |x| \le 1, |v| \le 2.$

39. *Damped Soft Spring* $x' = v, v' = -10x + 0.2x^3 - 0.2v - 9.8; |x| \le 15, |v| \le 25.$ [*Hint*: $(-1, 0)$ is an equilibrium point. To find the other two equilibrium points, set $v = 0$ and factor the expression for v' into the form $(x + 1)g(x)$, where $g(x)$ is a quadratic whose roots you can find by using the quadratic formula.]

40. *Competitive Exclusion* $x' = 0.6x(1 - x/30 - 1.5y/30), y' = 0.1x(1 - y/30 - 1.5x/30); 0 \le x \le 50, 0 \le y \le 50.$

Answer 39: There are three equilibrium points $(-1, 0)$, $(x_2, 0)$, and $(x_3, 0)$. The values x_2 and x_3 result from the factorization $-10x + .02x^3 - 9.8 = 0.2(x + 1)(x^2 - x - 49)$; the roots of the quadratic factor are $(1 \pm \sqrt{197})/2$ and so $x_2 \approx -6.52$ and $x_3 \approx 7.52$. From the orbital portrait in Graph 1 we can visually classify the equilibrium points. Graphs 2–4 are zoomed-in portions of the orbital portrait in Graph 1.

$(-1, 0)$: nonlinear, asymptotically stable spiral point (Graph 2)

$(x_2, 0)$: nonlinear, unstable saddle (Graph 3)

$(x_3, 0)$: nonlinear, unstable saddle (Graph 4)

See also Example 3.8.1 and *Modeling a Soft Spring* under Physical Models in the ODEA Library.

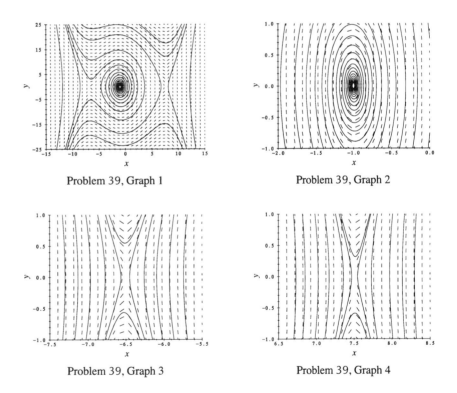

Problem 39, Graph 1   Problem 39, Graph 2

Problem 39, Graph 3   Problem 39, Graph 4

8.2 Stability and Nonlinear Autonomous Systems

Comments:

Theorem 8.2.1 partly justifies the often heard assertion in science and engineering that the linear terms of a system near an equilibrium point determine the stability properties without having to look at the nonlinearities. It isn't quite that simple, as Problem 51 shows, but the assertion comes close enough to the truth to be widely believed. We outline a part of the proof of Theorem 8.2.1 at the end of the SPOTLIGHT ON LYAPUNOV FUNCTIONS. We can't do that proof here because it uses some of the ideas of that Spotlight. The Jacobian matrix is introduced because it often is needed to determine the linearization at an equilibrium point. The table and sketches at the end of the section suggest that the terms node, saddle, and spiral point introduced in Section 6.5 for an equilibrium point of a planr autonomous linear system, still make sense for a nonlinear system. So from now on when (for example) we refer to a saddle point of a system, the system need not be linear, but the Jacobian matrix of the system at the point must have real eigenvalues of opposite sign. In order to make all of this mathematically correct, we introduce the idea of the order of a function at a point. **Warning!** Keep your eye out for situations where the Jacobian matrix at an equilibrium point has eigenvalues that are either 0 or pure imaginary. The system may be right on the edge of a stability region and the nonlinear terms can then tip the full system either way—into stability or into instability. Problems 51 and 53 show what can happen.

Asymptotic stability of a system at an equilibrium point requires that the system be an attractor (i.e., all orbits passing near the equilibrium point must approach it as $t \to +\infty$) *and also be stable.* At first glance, one might think that an attracting equilibrium point is automatically stable—but that need not be so (see Problem 47 in Section 8.2).

PROBLEMS

Jacobian Matrices and Stability at the Origin.   Determine the stability properties of each system at the origin. [*Hint*: use Theorem 8.2.1.]

1. $x' = -x + y^2,$ $y' = -8y + x^2$ **2.** $x' = 2x + y + x^4,$ $y' = x - 2y + x^3 y$

3. $x' = 2x - y + x^2 - y^2,$ $y' = x - y$ **4.** $x' = e^{x+y} - \cos(x - y),$ $y' = -\sin x$

Answer 1:   The system $x' = -x + y^2,$ $y' = -8y + x^2$ is asymptotically stable at the origin since the eigenvalues of the Jacobian matrix of the rate functions at the origin $\begin{bmatrix} -1 & 0 \\ 0 & -8 \end{bmatrix}$ are $-1, -8$ (both are negative).

Answer 3:   The system $x' = 2x - y + x^2 - y^2,$ $y' = x - y$ is unstable at the origin since the eigenvalues of the Jacobian matrix $\begin{bmatrix} 2 & -1 \\ 1 & -1 \end{bmatrix}$ at the origin are $(1 \pm \sqrt{5})/2$ (one is positive).

Jacobian Matrices and Stability.   Discuss the stability properties of each system at each of its equilibrium points. [*Hint*: use polar coordinates for Problem 8.]

5. $x' = y,$ $y' = -6x - y - 3x^2$ **6.** $x' = y^2 - x,$ $y' = x^2 - y$

7. $x' = -x - x^3,$ $y' = y + x^2 + y^2$ **8.** $x' = -y - x(x^2 + y^2),$ $y' = x - y(x^2 + y^2)$

9. $x' = x + xy^2,$ $y' = x$ **10.** $x' = -x + y^2,$ $y' = -x + y$ **11.**   System (8)

Answer 5:   The system is $x' = y,$ $y' = -6x - y - 3x^2$. $P_1(0,0)$ and $P_2(-2,0)$ are equilibrium points. The Jacobian matrix has entries $a_{11} = 0, a_{12} = 1, a_{21} = -6 - 6x, a_{22} = -1$. At the point $P_1, x = 0$ and the matrix is $\begin{bmatrix} 0 & 1 \\ -6 & -1 \end{bmatrix}$, whose eigenvalues both have negative real parts since the characteristic polynomial is $\lambda^2 + \lambda + 6$, which has roots $-1/2 \pm i\sqrt{23}/2$. At $P_2, x = -2$, and the matrix is $\begin{bmatrix} 0 & 1 \\ 6 & -1 \end{bmatrix}$. The characteristic polynomial is $\lambda^2 + \lambda - 6$, whose roots are -3 and 2. The system is asymptotically stable at P_1 and unstable at P_2.

Answer 7:   The system is $x' = -x - x^3,$ $y' = y + x^2 + y^2$. There are equilibrium points at the origin and at $(0, -1)$. Since the Jacobian matrix is $\begin{bmatrix} -1 - 3x^2 & 0 \\ 2x & 1 + 2y \end{bmatrix}$, the matrix is $\begin{bmatrix} -1 & 0 \\ 0 & 1 \end{bmatrix}$ at the origin; there is a positive eigenvalue and the system is unstable at the origin. But the eigenvalues of the

Jacobian matrix at the point $(0, -1)$ are -1 and -1; the system is asymptotically stable at the point $(0, -1)$.

Answer 9: The system is $x' = x + xy^2$, $y' = x$. The y-axis (i.e., $x = 0$) is a line of equilibrium points. The entries in the Jacobian matrix are $a_{11} = 1 + y^2$, $a_{12} = 2xy$, $a_{21} = 1$, $a_{22} = 0$. The characteristic polynomial when $x = 0$ is $\lambda^2 - (1 + y^2)\lambda$ with 0 and $1 + y^2$ as roots. The system is unstable at every equilibrium point $(0, y_0)$ because one of the eigenvalues is $1 + y_0^2$, which is positive.

Answer 11: System (8) is $x' = y$, $y' = -101x - 2y + x^2$. The Jacobian matrix is

$$J(x, y) = \begin{bmatrix} 0 & 1 \\ -101 + 2x & -2 \end{bmatrix}$$

J has eigenvalues $-1 \pm 10i$ at the equilibrium point $P_1(0, 0)$ and eigenvalues $-1 \pm \sqrt{102}$ at the equilibrium point $P_2(101, 0)$. So the system is asymptotically stable at P_1 and unstable at P_2. P_1 is a nonlinear spiral point and P_2 is a nonlinear saddle point.

Plot the Orbits. Plot several orbits each system in the neighborhood of each equilibrium point. Relying solely on the visual appearance of these orbits, what are your conclusions about the nature of the eigenvalues of the Jacobian matrix at each equilibrium point?

12. Problem 5   **13.** Problem 6   **14.** Problem 7   **15.** Problem 8

16. Problem 9   **17.** Problem 10   **18.** Problem 11

Answer 13: See the figure for some orbits of $x' = y^2 - x$, $y' = x^2 - y$. One eigenvalue of the Jacobian matrix at the apparent nonlinear saddle point must be positive, and the other must be negative. The eigenvalues of the Jacobian matrix at the apparent nonlinear node point are negative since orbits tend to the point as t increases.

Answer 15: See the figure for some orbits of $x' = -y - x(x^2 + y^2)$, $y' = x - y(x^2 + y^2)$. The eigenvalues of the Jacobian matrix at the apparent nonlinear spiral point appear to be complex conjugates with negative real parts because the direction is inward. However, the eigenvalues of the Jacobian matrix are actually $\pm i$; it is the higher order terms that pull orbits into the equilibrium point. This is one of those cases where the visual evidence is *not* sufficient to decide the nature of the eigenvalues.

Answer 17: See the figure for some orbits of $x' = -x + y^2$, $y' = -x + y$. The eigenvalues of the Jacobian matrix at the apparent nonlinear saddle point seem to be of opposite signs. The eigenvalues of the Jacobian matrix at the apparent nonlinear center seem to be pure imaginary.

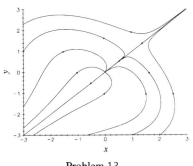

Problem 13.

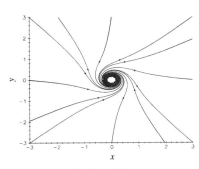

Problem 15.

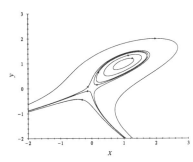

Problem 17.

Asymptotic Stability. Explain why each system is asymptotically stable at the origin. Is the asymptotic stability global?

19. $x' = -x$ **20.** $x' = -x^3$

21. $x' = -\sin x$ **22.** $x' = -2x^3$, $y' = -5y$

23. $x' = -x^3$, $y' = -y^3$ **24.** $x' = -x - y$, $y' = x - y$, $z' = -z^3$

Answer 19: The ODE $x' = -x$ is linear; the system "matrix" $[-1]$ has eigenvalue $\lambda = -1$; we have global asymptotic stability at $x = 0$.

Answer 21: Separate variables in the ODE $x' = -\sin x$ and integrate to obtain $\ln|\tan(x/2)| = -t + C_1$. So, $x = 2\arctan(Ce^{-t})$, where C is an arbitrary constant. The values of the arctangent are limited to the range $(-\pi/2, \pi/2)$ since we are only interested in the ODE near $x = 0$; $|x|$ is a decreasing function of t and tends to 0 as t increases to $+\infty$. The ODE is asymptotically stable at $x = 0$ since 0 is an attractor and we can set $\delta = \varepsilon$ in the definition of stability. We have only local asymptotic stability since, for example, equilibrium points such as $x = \pm\pi$ are *not* attracted to $x = 0$.

Answer 23: Solving $x' = -x^3$ and $y' = -y^3$ by separating variables, we have that $x = c_1(1 + 2c_1^2 t)^{-1/2}$ and $y = c_2(1 + 2c_2^2 t)^{-1/2}$, where if $c_1, c_2 \neq 0$, then t must be greater than $-1/(2c_1^2)$ and $-1/(2c_2^2)$. We see that $(0,0)$ is an attractor since $x(t) \to 0$ and $y(t) \to 0$ as $t \to +\infty$, while $(x(t), y(t))$ decreases steadily to 0 as $t \to +\infty$ [we may set $\delta = \varepsilon$ in the definition of stability]. The system is globally asymptotically stable at the origin.

Plot the Orbits and Time-State Curves. Plot several orbits near the origin. [*Hint*: the orbits in Problems 25–27 lie on state lines.]

 25. Problem 19   **26.** Problem 20   **27.** Problem 21

 28. Problem 22   **29.** Problem 23

Answer 25: See the figure for some time-state curves, and the margin for the state line.

Answer 27: See the figure for some time-state curves, and the margin for the state line.

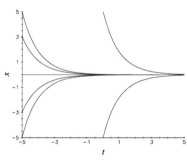

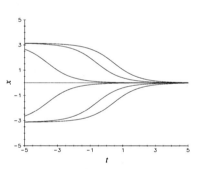

Problem 25.   Problem 27.

Answer 29: See the figure for some orbits.

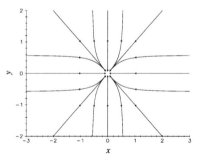

Problem 29.

Neutral Stability. Show that each of the following systems is neutrally stable at the origin. [*Hint*: find solution formulas.]

30. $x' = 0$, $y' = -y^3$

31. $x' = y$, $y' = -x$, $z' = -z^3$

Answer 31: The general solution is $x = C_1 \cos t + C_2 \sin t$, $y = -C_1 \sin t + C_2 \cos t$, and the nontrible solutions for z are $z = \pm(2t + C_3)^{-1/2}$ for $t > -C_3/2$, where C_1, C_2, and C_3 are arbitrary constants. As $t \to +\infty$, $z(t) \to 0$; the nonconstant orbits $(x(t), y(t))$ are circles. So $(0,0,0)$ is neutrally stable.

Plot the Orbits. Plot several orbits of each system near the origin.

32. Problem 30

33. Problem 31

Answer 33: See the figure.

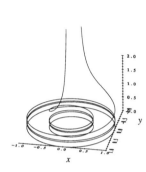

Problem 33.

Instability. The following systems are unstable at the origin. Explain why.

34. $x' = x^2$

35. $x' = \sin x$

36. $x' = |x|$

37. $x' = x^3$, $y' = -3y$

38. $x' = -2x$, $y' = x - 10z$, $z' = 10y + z$

Answer 35: The solution of $x' = \sin x$ is found to be $x = 2\arctan(Ce^t)$, where we assume $-\pi < x < \pi$. Suppose we set $\varepsilon = 1$. Then for every $C > 0$, $x(t) \to +\pi$ as $t \to +\infty$. For no positive δ is it true that $|x(t)| < \varepsilon = 1$ for all $t \geq 0$ if $|x(0)| = |2\arctan C| < \delta$. The ODE is unstable at 0.

Answer 37: The system is $x' = x^3$, $y' = -3y$. We have $y = c_2 e^{-3t}$. Separating variables in $x' = x^3$, and solving for x in terms of t, we have that $x = c_1(1 - 2c_1^2 t)^{-1/2}$, $t < 1/(2c_1^2)$ if $c_1 \neq 0$, and $-\infty < t < \infty$ if $c_1 = 0$. Even though $y(t) \to 0$ as $t \to +\infty$, since $|x(t)| \to \infty$ as t increases from 0 to $1/(2c_1^2)$ (if $c_1 \neq 0$), the system is unstable at the origin.

Plot the Orbits. Plot several orbits near the origin. Insert arrowheads on orbits to show the direction of motion as time increases. [*Hint*: the graphs in Problems 39–41 are state lines.]

39. Problem 34

40. Problem 35

41. Problem 36

42. Problem 37

43. Problem 38

Answer 39: See the state line below and the figure for solution curves.

Answer 41: See the state line below and the figure for solution curves.

Answer 43: See the figure for orbits and the corresponding tx, ty, tz component curves.

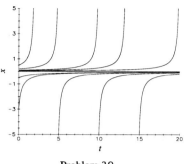

Problem 39.

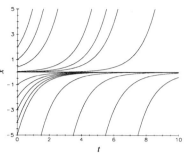

Problem 41.

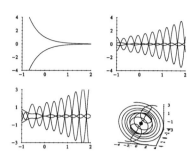

Problem 43.

Orbital Portraits with Several Equilibrium Points.

☞ We introduced this system in Example 7.2.2.

44. *Planar Portrait* Look at the nonlinear system $x' = x - y + x^2 - xy$, $y' = -y + x^2$.

(a) Find the three equilibrium points.

(b) Use Jacobian matrices and their eigenvalues to determine the stability properties of the system at each equilibrium point.

💻 (c) Create the first orbital portrait below for the given system. Plot orbits (solid) of the nonlinear system in the rectangle $|x| \leq 0.5$, $|y| \leq 0.5$. Then find the linearized system at the origin and plot some of its orbits (dashed) in the same rectangle.

💻 (d) Create the second orbital portrait below. Locate and mark the three equilibrium points. Use other orbits of the nonlinear system to create the third orbital portrait below.

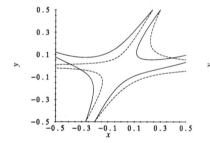

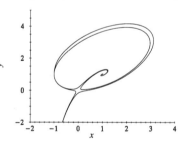

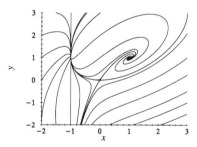

45. *Basin of Attraction* Consider the system $x' = y$, $y' = -2x - y - 3x^2$.

(a) Show that the system is asymptotically stable at the origin, but not globally.

💻 (b) Use computer graphics to find the approximate part of the basin of attraction of the origin that lies in the region $-3 \leq x \leq 2$, $-3 \leq y \leq 5$.

Answer 45:

(a) The Jacobian matrix of $x' = y$, $y' = -2x - y - 3x^2$ is $\begin{bmatrix} 0 & 1 \\ -2 - 6x & -1 \end{bmatrix}$, and its eigenvalues at the origin are $(-1 \pm \sqrt{7}i)/2$; the system is asymptotically stable at the origin. However, the system is not globally asymptotically stable at the origin since $(-2/3, 0)$ is another equilibrium point, and so the origin is not a global attractor.

(b) See the figure. To find the boundaries of the region of attraction of the origin, start near the saddle point at $(-2/3, 0)$ and compute orbits backward in time.

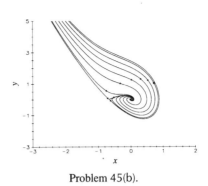

Problem 45(b).

Attractors. All *linear* attractors are asymptotically stable. Most (but not all) *nonlinear* attractors are also asymptotically stable.

46. *A Stable Attractor* Show that the nonlinear system $x' = -x - 10y - x(x^2 + y^2)$, $y' = 10x - y - y(x^2 + y^2)$ is globally asymptotically stable at the origin. [*Hint*: use polar coordinates.]

47. *An Unstable Attractor* Parts **(a)**–**(b)** outline the steps needed to show that the nonlinear system $x' = x - y - x^3 - xy^2 + xy(x^2 + y^2)^{-1/2}$, $y' = x + y - x^2y - y^3 - x^2(x^2 + y^2)^{-1/2}$ has an unstable attractor at the equilibrium point $x = 1$, $y = 0$.

(a) Show that this system is equivalent to the system $r' = r(1 - r^2)$, $\theta' = 1 - \cos\theta$, where r, θ are polar coordinates in the xy-plane. Solve the rate equations for r and θ.

(b) Show that the point $x = 1$, $y = 0$ is an unstable attractor. [*Hint*: use sign analysis to show that the unit circle consists of the equilibrium point $x = 1$, $y = 0$ and an orbit that leaves the point as t increases from $-\infty$ but returns to the point as $t \to +\infty$. So, $(1, 0)$ is unstable. Then show that $(1, 0)$ attracts all nonconstant orbits as $t \to +\infty$.]

🖥 **(c)** Plot a portrait of orbits of the xy-system in the rectangle $|x| \le 2$, $|y| \le 1.5$.

Answer 47:

(a) Using formulas (11) and (12) in Section 7.2, we see that the system $x' = x - y - x^3 - xy^2 + xy(x^2 + y^2)^{-1/2}$, $y' = x + y - x^2y - y^3 - x^2(x^2 + y^2)^{-1/2}$ can be written in polar coordinates as

$$r' = \cos\theta[r\cos\theta - r\sin\theta - r^3\cos^3\theta - r^3\cos\theta\sin^2\theta + r\cos\theta\sin\theta]$$

$$+ \sin\theta[r\cos\theta + r\sin\theta - r^3\cos^2\theta\sin\theta - r^3\sin^3\theta - r\cos^2\theta]$$

$$= r(1 - r^2)$$

where we have used the identity $\cos^2\theta + \sin^2\theta = 1$ several times. Similarly,

$$\theta' = r^{-1}\cos\theta[r\cos\theta + r\sin\theta - r^3\cos^2\theta\sin\theta - r^3\sin^3\theta - r\cos^2\theta]$$

$$- r^{-1}\sin\theta[r\cos\theta - r\sin\theta - r^3\cos^3\theta - r^3\cos\theta\sin^2\theta + r\cos\theta\sin\theta]$$

$$= 1 - \cos\theta$$

From the given ODEs in r and θ, we see that there are two equilibrium points, $r = 0$, and $r = 1$, $\theta = 2k\pi$, [i.e., the points $(0, 0)$, $(1, 0)$ in xy-rectangular coordinates]. Since the r, θ ODEs are decoupled, they are easier to solve than the x, y ODEs. We have that $-\cot(\theta/2) = t + C$, or $\theta = 2\arctan(-t - C)^{-1} \to 2k\pi$ as $t \to \pm\infty$, and that $r(t) = r_0[r_0^2 + (1 - r_0^2)e^{-2t}]^{-1/2} \to 1$ as $t \to +\infty$ if $r_0 > 0$.

(b) We conclude that if $r_0 > 0$, then as $t \to +\infty$, we have $r \to 1$ and $\theta \to 2k\pi$; i.e., aside from the equilibrium point at the origin, all orbits tend to the point $(1, 0)$ [in rectangular coordinates], which is an attractor. This point is, however, unstable since the solution $r(t) = 1$, $\theta = -2\arctan(1/t)$ has the property that it exits the point $x = 1$, $y = 0$, (the point $r = 1$, $\theta = 0$) as t increases from $-\infty$ and turns counterclockwise around the unit circle approaching the same point (but now with $r = 1$, $\theta = 2\pi$) as

$t \rightarrow +\infty$. If $\varepsilon = 1$, say, this orbit does not remain in the circular region $(x - 1)^2 + y^2 < 1$ of radius $\varepsilon = 1$ about the equilibrium point, even though it begins and ends inside that region.

(c) See the figure for some orbits. Note in particular the orbit on the unit circle that exits from the equilibrium point $x = 1$, $y = 0$ as t increases from $-\infty$, and then returns to the point as $t \rightarrow +\infty$. Note also how nonconstant orbits outside and inside the circle approach the point $x = 1$, $y = 0$ as $t \rightarrow +\infty$.

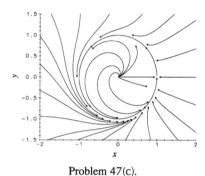

Problem 47(c).

Stability Properties.   For each system, find the Jacobian matrix and its eigenvalues at each equilibrium point. Determine the stability properties of the system at each equilibrium point.

48. $x' = -x - x^3$, $y' = -y$ **49.** $x' = \alpha x - y + x^2$, $y' = x + \alpha y + x^2$, constant $\alpha > -1$

Answer 49:   The system $x' = \alpha x - y + x^2$, $y' = x + \alpha y + x^2$, has the equilibrium points $(0, 0)$ and $(x_0, \alpha x_0 + x_0^2)$ where $x_0 = -(1 + \alpha^2)/(1 + \alpha)$, $\alpha \geq -1$. The Jacobian matrix at the origin is $\begin{bmatrix} \alpha & -1 \\ 1 & \alpha \end{bmatrix}$ with eigenvalues $\alpha \pm i$. The origin is unstable if $\alpha > 0$ and asymptotically stable if $\alpha < 0$ (Theorem 8.2.1), but Theorem 8.2.1 is *not* applicable if $\alpha = 0$.

The Jacobian matrix at $(x_0, \alpha x_0 + x_0^2)$ is $\begin{bmatrix} \alpha + 2x_0 & -1 \\ 1 + 2x_0 & \alpha \end{bmatrix}$. The eigenvalues λ are the roots of quadratic $\lambda^2 - 2(\alpha + x_0)\lambda + (1 + \alpha^2) + 2x_0(1 + \alpha)$. With the value of x_0 given above and $\alpha > -1$, we see that the constant term of the quadratic is negative, and so the eigenvalues are real and of opposite signs. The system has nonlinear saddle behavior near the second equilibrium point and is unstable there.

More Stability Problems.

50. *Stabilizing a System by Tuning a Parameter*   Find all values of the constant α such that the system $x' = z + x^2 y$, $y' = x - 4y + z + xz^2$, $z' = \alpha x + 2y - z + x^2$ is asymptotically stable at the origin.

51. *The Case of a Zero Eigenvalue*   For any constant a show that the system matrix of the linearization at the origin of the system $x' = -x$, $y' = ay^3$ has simple eigenvalues -1 and 0 and so the linearized system is neutrally stable at the origin. Show that the system is asymptotically stable at the origin if $a < 0$, unstable if $a > 0$. What happens if $a = 0$? Why does Theorem 8.2.1 *not* apply? What do you conclude about basing a stability analysis of a nonlinear system entirely on the eigenvalues of the Jacobian matrix?

Answer 51:   The matrix $\begin{bmatrix} -1 & 0 \\ 0 & 0 \end{bmatrix}$ of coefficients of the linear terms of $x' = -x$, $y' = ay^3$ has eigenvalues -1 and 0, respectively. Theorem 8.2.1 does not apply, and we must look at the nonlinear terms to determine the stability properties of the system. Let's find all solutions. First we see that $x = c_1 e^{-t}$, here c_1 is an arbitrary constant. Next, if $a \neq 0$, we can separate the variables in $y' = ay^3$ and solve to obtain $y = y_0(1 - 2y_0^2 at)^{-1/2}$. If $a > 0$, then $y \rightarrow \infty$ as $t \rightarrow (y_0^2 a/2)^-$, and we have instability. If $a < 0$, then $y \rightarrow 0$ as $t \rightarrow +\infty$. Since $x \rightarrow 0$ steadily as $t \rightarrow +\infty$, we have stability. If $a = 0$ then $y(t) = y_0$ for all t and we have neutral stability since $x(t) \rightarrow 0$ as $t \rightarrow +\infty$.

52. Plot orbital portraits of the system in Problem 51 for $a = -1$, $a = 1$, $a = 0$. Use the frame $|x| \leq 1$, $|y| \leq 1$.

53. *The Case of Pure Imaginary Eigenvalues*   The linear system $x'_1 = -x_2$, $x'_2 = x_1$ is neutrally stable (the eigenvalues of the system matrix are $\pm i$). Find functions $P_i(x_1, x_2)$, $i = 1, 2$, each of the form $(x_1^2 + x_2^2)z_i$, where z_i is 0, $\pm x_1$ or $\pm x_2$, such that the nonlinear system $x'_1 = -x_2 + P_1$, $x'_2 = x_1 + P_2$ has the indicated stability property. [*Hint*: use polar coordinates.]

(a) Globally asymptotically stable   (b) Unstable   (c) Neutrally stable

Answer 53:

(a) Let $P_1 = -x_1(x_1^2 + x_2^2)$, $P_2 = -x_2(x_1^2 + x_2^2)$. Since $r^2 = x_1^2 + x_2^2$, $rr' = x_1x'_1 + x_2x'_2 = -r^4$, so $r' = -r^3$ and the origin is globally asymptotically stable in the x_1x_2-plane.

(b) Let $P_1 = x_1(x_1^2 + x_2^2)$, $P_2 = x_2(x_1^2 + x_2^2)$. Since $r' = r^3$, the origin in the x_1x_2-plane is unstable.

(c) Let $P_1 = P_2 = 0$. Then the orbits are the circles $x_1 = C_1 \cos t + C_2 \sin t$, $x_2 = -C_1 \sin t + C_2 \cos t$, and the origin is neutrally stable.

Pendulum Model.

54. *Upended Pendulum*   Give a complete analysis of the stability properties of the damped simple pendulum system of Example 8.2.1 at the equilibrium point $(\pi, 0)$. Use your numerical solver and plot orbits of the system and its linearization near $(\pi, 0)$.

 Interacting Species Models.   Interpret the interacting species model $x' = (2 + ax + by)x$, $y' = (2 + cx + dy)y$ for the indicated cases. Give a complete stability analysis at each equilibrium point, using the categories node, spiral, saddle and also asymptotically stable, neutrally stable, unstable. Plot orbits in the population quadrant. Use arrowheads to show increasing time.

55. *Cooperation leading to explosive growth* $a = d = -1$ and $b = c = 2$. See Example 7.3.1.

56. *Stable competition* $a = d = -2$ and $b = c = -1$. See Example 7.3.3.

57. *Competitive exclusion* $a = d = -1$ and $b = c = -2$. See Example 7.3.4.

58. Fix the parameters $b = c = -1$ and take nonzero values of your choice for a and d.

Answer 55:   Group Problem.

Answer 57:   Group Problem.

59. *Predator-prey* $a = d = 0$, $b = 1$, $c = -10$.

Answer 59:   Group Problem.

Circuit Model.

60. *A Nonlinear RLC Circuit*   The ODE $Lx'' + Rx' + x/C + g(x, x') = 0$ governs a simple RLC electric circuit, where x represents the charge on the capacitor, x' is the current in the loop, and L, R, and C are the positive circuit parameters inductance, resistance, and capacitance. The twice continuously differentiable function $g(x, x')$ represents circuit nonlinearities. Suppose that $g(x, x')$ has an order of at least 2 at $(0, 0)$ in the xx'-plane. Show that the equivalent system

$$x' = y, \quad y' = -x/LC - Ry/L - g(x, y)/L$$

is asymptotically stable at the origin.

8.3   Conservative Systems

Comments:

A system of ODEs is conservative if it has an integral. The level sets of the integral in state space are unions of orbits, and, so, knowledge of an integral tells us a lot about orbital behavior. What kinds of ODEs have integrals? First-order exact ODEs do (see the WEB SPOTLIGHT ON EXACT ODES), and that is why a conservative system may be considered to be a generalization of an exact ODE. A physical system whose total energy when in state $x(t)$ is the same as when in state $x(t_0)$ is conservative; the total energy is the integral. No real physical system has this property, but for many the change in energy as time goes on is negligible and the system can be treated as if it were conservative. Undamped spring systems, undamped pendulums, resistance-free electrical circuits, the Lotka-Volterra predator-prey system, and the rotating tennis racket of the SPOTLIGHT ON ROTATING BODIES, are all examples of conservative systems. We note in this section that conservative systems do *not* have attracting (or repelling) equilibrium points. An integral of a conservative system has some of the properties of the Lyapunov functions to be introduced in the SPOTLIGHT ON LYAPUNOV FUNCTIONS. An integral remains constant on each orbit; any orbit touching a level set of an integral remains on the level set, and there need not be any equilibrium points at all. The reason we put this section in this chapter is that (as noted before) the construction of an integral (if there is one) gives information about orbits without having to find solution formulas.

PROBLEMS

Finding Integrals. Find an integral of each system. What are the stability properties of each equilibrium point?

1. $x' = -y, \quad y' = 25x$

2. $x' = y^3, \quad y' = -x^3$

3. $x' = 3x, \quad y' = -y$

4. $x' = -2x, y' = 3y$

Answer 1: The system is $x' = -y, \; y' = 25x$. We have $dy/dx = -25x/y$ or $ydy = -25xdx$; solving that equation, we obtain the level sets $K(x, y) = 25x^2 + y^2 = C$, where $K(x, y)$ is the desired integral [any nonzero constant multiple of K would do as well]. The system matrix of the original linear system has eigenvalues $\pm 5i$ and is neutrally stable at the single equilibrium point (0,0).

Answer 3: Divide the second equation of the system $x' = 3x, \; y' = -y$ by the first to obtain $dy/dx = -y/3x$. Separating the variables and solving, we have $3\ln|y| = -\ln|x| + C_1$ or $y^3x = C$, where $C = \pm e^{C_1}$. The equation $y^3x = 0$ also defines the orbits $x = 0$ and $y = 0$; we have the integral $K(x, y) = y^3x$. The original system is linear with a single equilibrium point at $(0, 0)$. Since the eigenvalues of the system matrix are 3 and -1, the origin is an unstable saddle point.

Plot Integral Curves. For each system of ODEs plot several distinct integral curves. Insert arrowheads to show the direction of motion.

5. Problem 1   **6.** Problem 2   7. Problem 3   8. Problem 4

Answer 5: See the figure.

Answer 7: See the figure.

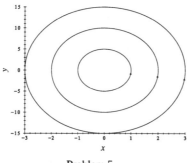

Problem 5.

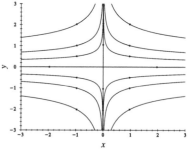

Problem 7.

Conservative? Nonconservative?. Decide whether each system is conservative. If it is conservative, find an integral. If the system is not conservative, explain why not.

9. $x' = y, y' = -10x$ **10.** $x' = y, y' = -10x - y$ **11.** $x' = y, y' = -x$

12. $x' = y, y' = -x - y$ **13.** $x' = 3x - 2y, y' = 5x - 3y$ **14.** $x' = 3x - 2y, y' = 5x - 2y$

Answer 9: Conservative with integral $K = y^2 + 10x^2$ obtained by separating variables in the ODE and integrating.

Answer 11: Conservative with integral $K = x^2 + y^2$ obtained by separating variables in the ODE and integrating.

Answer 13: The eigenvalues of the system matrix are $\pm i$, so we have neutral stability and the orbits form a family of tilted ellipses (see Example 6.5.5). Here is one way to find these ellipses and, so, an integral $K(x, y)$. Pick a point with simple coordinates, e.g., $x = 1/2$, $y = 0$, as the initial point and solve the ODE $x' = 3x - 2y, y' = 5x - 3y$. Plot the orbit on a grid (see the figure). The equation of a tilted ellipse is $x^2 + axy + by^2 = C$ for some constants a, b, and C. From the figure we see that $(1/2, 0)$, $(3/2, 2)$, and $(5/2, 3/2)$ are on the ellipse. Insert the coordinates of these points into the equation of the ellipse and solve for a, b, and C: $a = -6/5$, $b = 2/5$, and $C = 1/4$. To simplify, multiply the equation of the ellipse by 20 and we get the integral $K(x, y) = 20x^2 - 24xy + 8y^2$. The system is conservative.

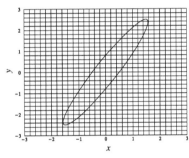

Problem 13.

15. *A Conservative Attractor?* Show that the system $x' = -x$, $y' = -2y$, is asymptotically stable at the origin and that $K = yx^{-2}$ is constant on each orbit not touching the y-axis. Why doesn't this contradict Theorem 8.3.2?

Answer 15: The system matrix of $x' = -x$, $y' = -2y$ has eigenvalues -1 and -2, and so the system is asymptotically stable at the origin. Solving $dy/dx = 2y/x$, we obtain $dy/y = 2dx/x$, $\ln|y| = 2\ln|x| + C_1$, $yx^{-2} = e^{C_1}$. $K(x, y) = yx^{-2}$ is an integral in any region not intersecting the y axis. This does not contradict the theorem that a conservative system in a region R cannot have an attractor in R since the region R in this case cannot intersect the y-axis and so does not contain the attractor at the origin.

16. *A Conservative Repeller?* Show that $x' = 2x, y' = 3y$ has a repelling equilibrium point at the origin and that $K = x^3 y^{-2}$ is constant on each orbit not touching the x-axis. Why doesn't this contradict Theorem 8.3.2?

17. *Locally Conservative Systems* The systems given in Problems 15 and 16 are not conservative. Create a definition of local conservation and show that the system of Problem 15 is locally conservative in the region $x > 0$ and in the region $x < 0$. Where is the system of Problem 16 locally conservative?

Answer 17: A system is locally conservative in a region R if there is a function $K(x)$ that meets the conditions for an integral given on page 503 in the region R (rather than in all of $\mathbb{R}^n$). So $K = yx^{-2}$ is an integral of the system of Problem 15 in the region R_1 defined by $x < 0$ and in the region R_2 defined by $x > 0$. Similarly, $K = x^3 y^{-2}$ is an integral for the system of Problem 16 in the region R_1 defined by $y > 0$ and in the region R_2 defined by $y < 0$.

18. Show that the system $x' = g(y)$, $y' = f(x)$, $z' = 0$ is conservative.

Modeling Problems.

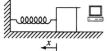

🖥 **Behavior of Springs.** Integrals help us understand the behavior of undamped springs.

19. *Hard Spring* Find an integral and plot orbits of $x' = y$, $y' = -10x - x^3$. What are the stability properties of the system at the origin? Is the system conservative?

🖎 Hard and soft springs first came up in Section 3.8.

20. *Soft Spring* Repeat Problem 19 for $x' = y$, $y' = -10x + x^3$. Discuss the stability properties at each of the three equilibrium points and plot orbits near each of the three points. Is the system conservative?.

Answer 19: Dividing the two equations of the system $x' = y$, $y' = -10x - x^3$, we have that $dy/dx = (-10x - x^3)/y$, so $y\,dy + (10x + x^3)dx = 0$, or $y^2/2 + 5x^2 + x^4/4 = C$. We see that the function $K = y^2/2 + 5x^2 + x^4/4$ is an integral. The level sets are oval-like curves enclosing the origin, and the system has a neutrally stable equilibrium point at the origin. This may be shown by setting $K = C$, where C is a positive constant, and solving for y: $y = \pm(2C - 10x^2 - x^4/2)^{1/2}$. The graph Γ of $K = C$ is a curve that is symmetric about both axes, cuts the y-axis at $y = \pm\sqrt{2C}$, and the x-axis at the two real roots of $2C - 10x^2 - x^4/2$, which are $\pm(-10 + (100 + 4C)^{1/2})^{1/2}$. Moreover, the upper half of the curve Γ rises steadily from 0 to $\sqrt{2C}$ as x increases from the negative root to zero and then falls back to 0 as x increases through zero to the positive root. By calculating dy/dx, we see that Γ has a vertical slope at the two extreme values of x. The origin is enclosed by these oval cycles. See the figure.

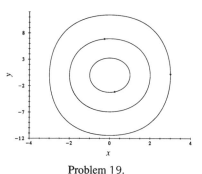

Problem 19.

Skew Symmetry. Suppose that the real $n \times n$ matrix A is *skew-symmetric*, that is, $A^T = -A$.

21. Show that the system $x' = Ax$ is conservative and neutrally stable at the origin if A is skew-symmetric. [*Hint*: look at the function $K = x^T x = ||x||^2$.]

22. Suppose that each eigenvalue of a skew-symmetric matrix A is zero or pure imaginary. Show that the surface $||x|| = $ constant is an integral surface that consists of equilibrium points and nonconstant orbits defined by solutions of $x' = Ax$ whose components are linear combinations of periodic functions.

Answer 21: Let $K(x) = x^T x$. Then, $K' = (x^T)'x + x^T x' = (Ax)^T x + x^T Ax = x^T A^T x + x^T Ax = 0$ since $A^T = -A$. K remains constant along each orbit of $x' = Ax$. Since $K(x)$ is the square of the distance from the point x to the origin, K is not constant on any ball $||x||^2 \leq C$, C a positive constant; so $K(x)$ is an integral, and the system is conservative. The level sets $K(x) = C$, C any positive constant, are the "spheres" in $\mathbb{R}^n$ of radius $\sqrt{C}$ centered at the origin: $K(x) = x^T x = x_1^2 + \cdots + x_n^2 = C$. Each orbit lies entirely on one of these spheres, and the origin is not an attractor. The system is neutrally stable at the origin with $\delta = \varepsilon$ in the definition of stability.

Hamiltonian Systems. A *Hamiltonian system*[1] has the form

$$x_i' = \frac{\partial H}{\partial y_i}, \qquad y_i' = -\frac{\partial H}{\partial x_i}, \qquad i = 1, \ldots, k$$

where the *Hamiltonian* $H(x_1, \ldots, x_k, y_1, \ldots, y_k)$ is a differentiable and real-valued scalar function on $\mathbb{R}^{2k}$ and H is nonconstant on every open ball in $\mathbb{R}^{2k}$.

23. Show that H is an integral of the Hamiltonian system.

24. Explain why a Hamiltonian system can't be asymptotically stable at an equilibrium point.

25. Show that the system $x' = y$, $y' = -f(x)$ is Hamiltonian and has no asymptotically stable equilibrium points. [*Hint*: look at the function $y^2/2 + \int_0^x f(s)\,ds$.]

26. What does the result in Problem 25 imply about the possibility of finding a function $f(x)$ such that all orbits of the system $x' = y$, $y' = -f(x)$ tend toward an equilibrium point as $t \to +\infty$?

Answer 23:   By the Chain Rule, we have that $dH(x_1, \ldots, x_k; y_1, \ldots, y_k)/dt = \sum_{i=1}^{k}[(\partial H/\partial x_i)x_i' + (\partial H/\partial y_i)y_i'] = 0$ since $x_i' = \partial H/\partial y_i$ and $y_i' = -\partial H/\partial x_i$. Since H is assumed to be nonconstant on every region, H is an integral of the Hamiltonian system.

Answer 25:   The system is $x' = y$, $y' = -f(x)$. Let $H = y^2/2 + \int_0^x f(s)ds$. Then $\partial H/\partial y = y$ and $\partial H/\partial x = f(x)$. So the system has the form $x' = y = \partial H/\partial y$, $y' = -f(x) = -\partial H/\partial x$; the system is Hamiltonian, and so cannot have an asymptotically stable equilibrium point.

 27. *Exact Planar Systems*   The planar system $x' = N(x, y)$, $y' = -M(x, y)$ is *exact* in a rectangle R of the xy-plane if the functions N and M belong to $\mathbf{C}^1(R)$ and $\partial N/\partial x = \partial M/\partial y$ on R. Show that if the system is exact, then it is a Hamiltonian system. Conversely, show that every planar Hamiltonian system is exact. Then show that $x' = y^2 + e^x \cos y + 2\cos x$, $y' = -e^x \sin y + 2y \sin x$ is exact and find a Hamiltonian. Plot orbits in the region $-20 \le x \le 10$, $-4 \le y \le 6$. Why is it hard to use a numerical solver for orbits in the region $x \ge 5$? Now show that the *teddy bear* system of Example 2 in the WEB SPOTLIGHT ON EXACT ODEs is exact, find a Hamiltonian, and plot some of the bears. Finally, make up your own Hamiltonian system with the weirdest orbits possible. [*Hint*: start with any twice continuously differentiable function $H(x, y)$.]

Answer 27:   Group project.

Spotlight on Lyapunov Functions

Comments:

At the end of the 19th century A. M. Lyapunov introduced a series of tests that could be used to determine the stability properties of a system of ODEs without using solution formulas. These tests involve the use of a scalar function of the state variables and its derivative following the motion. These Lyapunov functions are often based on the total energy of the system, if that can be defined, or else a quadratic form in the state variables.

Lyapunov functions provide an alternative to the ε-δ arguments of Section 8.2 to determine the stability properties of a system of ODEs at an equilibrium point. These real-valued functions of the state variables tell us a lot about what is going on in the system, but without having to find solution formulas. Given a system of ODEs with an equilibrium point, it is always possible (but *not* always easy) to find a Lyapunov function whose behavior reveals whether the system is stable, asymptotically stable, or unstable at the equilibrium point. Lyapunov functions are never unique, and this is both an advantage and a difficulty, the former because it allows a lot of flexibility, the latter

---

[1] These systems are named in honor of the Irish mathematician and astronomer William Rowan Hamilton (1805–1865).

because there may be no clue about how to construct a Lyapunov function. Although total energy is commonly used as a Lyapunov function for models of physical systems, quadratic forms in the state variables are usually easier to use. In this section we emphasize quadratic forms. There are many other ways to construct Lyapunov functions. Building suitable Lyapunov functions has become a cottage industry for scientists and engineers involved in designing complex systems whose stability may be a matter of life or death, systems ranging from nuclear reactors to the control systems for jets and spacecraft.

PROBLEMS

Testing for Stability.   Use a Lyapunov function of the form $V = ax^2 + cy^2$ and determine whether the system is asymptotically stable or unstable at the origin.

1. $x' = -x - x^3, \ y' = -y$ **2.** $x' = x + x^2, \ y' = -y$

  Answer 1:   The system is $x' = -x - x^3, \ y' = -y$. If $V = x^2 + y^2$, then $V' = 2x(-x - x^3) + 2y(-y) = -2x^2 - 2y^2 - 2x^4$. Since V is positive definite and V' is negative definite, the system is asymptotically stable at the origin.

Use Lyapunov Functions to Test Stability.   Test stability at the origin by using $V = ax^2 + cy^2$ or $ax^2 + 2bxy + cy^2$ for suitably chosen values for a, b, c. Distinguish between neutral and asymptotic stability.

3. $x' = y, \ y' = -9x$ **4.** $x' = -x + 3y, \ y' = -3x - y$

5. $x' = y, \ y' = -2x - 3y$ **6.** $x' = 2x + 2y, \ y' = 5x + y$

7. $x' = -4y - x^3, \ y' = 3x - y^3$ **8.** $x' = -x^3 + x^3y - x^5, \ y' = y + y^3 + x^4$

9. $x' = -2x - xe^{xy}, \ y' = -y - ye^{xy}$ **10.** $x' = (-x + y)(x^2 + y^2), \ y' = -(x + y)(x^2 + y^2)$

  Answer 3:   The system is $x' = y, \ y' = -9x$. If $V = ax^2 + cy^2$, $V' = 2axx' + 2cyy' = 2axy - 18cxy = 2(a - 9c)xy$. If we set $a = 9c > 0$ (e.g., $a = 9, \ c = 1$), then $V = 9x^2 + y^2$ is positive definite and $V' = 0$, which is (trivially) negative semidefinite. The system is stable at the origin. It is not asymptotically stable, however, since $V = 9x^2 + y^2$ is constant along each orbit (since $V' = 0$), and so $9x^2 + y^2 = C$ gives elliptical orbits, none of which tends to the origin, unless $C = 0$. Note that $9x^2 + y^2$ is an integral of the separable ODE, $9x\,dx + y\,dy = 0$, which we get from the system by dividing y' by x' and rearranging terms.

  Answer 5:   The system is $x' = y, \ y' = -2x - 3y$. If $V = ax^2 + cy^2$, then $V' = 2(a - 2c)xy - 6cy^2$, which is indefinite (or semidefinite) for small values of $|x|$ and $|y|$. So let's try a Lyapunov function of the form $V = ax^2 + 2bxy + cy^2$ and see if we can do better. Then $V' = -4bx^2 + 2(a - 3b - 2c)xy + 2(b - 3c)y^2$. If we select $b = c = 1, \ a = 5$, then $V = 5x^2 + 2xy + y^2$ is positive definite (Theorem 2) and $V' = -4x^2 - 4y^2$ is negative definite. The system is asymptotically stable at the origin.

  Answer 7:   The system is $x' = -4y - x^3, \ y' = 3x - y^3$. If $V = ax^2 + cy^2$, then $V' = 2ax(-4y - x^3) + 2cy(3x - y^3) = (-8a + 6c)xy - 2ax^4 - 2cy^4$. If a and c are chosen as positive constants for which $-8a + 6c = 0$ (e.g., $a = 3, \ c = 4$), then $V = 3x^2 + 4y^2$ is positive definite, and $V' = -2ax^4 - 2cy^4$ is negative definite. The system is asymptotically stable at the origin.

  Answer 9:   The system is $x' = -2x - xe^{xy}, \ y' = -y - ye^{xy}$. If $V = ax^2 + cy^2$, $V' = 2axx' + 2cyy' = -4ax^2 - 2cy^2 - e^{xy}(2ax^2 + 2cy^2) = -4x^2 - 2y^2 - 2(x^2 + y^2)e^{xy}$ [if we set $a = c = 1$], which is negative definite, while V is positive definite. The system is asymptotically stable at the origin.

⌨ **Level Sets of Strong Lyapunov Functions.**   Plot several level sets of the Lyapunov functions near the origin. Then plot several orbits of the indicated system; note how the orbits cut across the level sets and tend to the origin as t increases if the origin is asymptotically stable.

11. $V = x^2 + y^2$;   Problem 4.   **12.** $V = 5x^2 + 2xy + y^2$;   Problem 5.

13. $V = 3x^2 + 4y^2$;   Problem 7.   **14.** $V = x^2 + y^2$;   Problem 8.

15. $V = x^2 + y^2$;   Problem 10.

Answer 11: $x' = -x + 3y$, $y' = -3x - y$; $V = x^2 + y^2 = 0.5, 2, 6$. See the figure (the dashed curves are the given level sets of V).

Answer 13: $x' = -4y - x^3$, $y' = 3x - y^3$; $V = 3x^2 + 4y^2 = 2, 8, 16$. See the figure (the dashed curves are the given level sets of V).

Answer 15: $x' = (-x + y)(x^2 + y^2)$, $y' = -(x + y)(x^2 + y^2)$; $V = x^2 + y^2 = 0.5, 2, 6$. See the figure (the dashed curves are the given level sets of V).

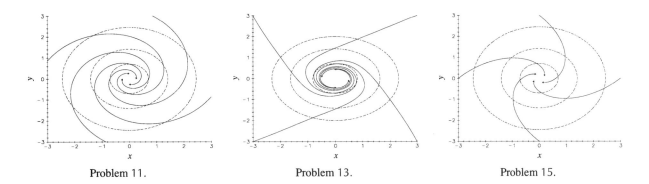

Problem 11. Problem 13. Problem 15.

16. $x' = \varepsilon x - y + x^2$, $y' = x + \varepsilon y + x^2$, where ε is a positive constant.

More Lyapunov Functions. Use $V = ax^{2m} + by^{2n}$, m and n positive integers, to determine stability properties at the origin. Plot some level sets of V and some orbits of the systems. [*Hint*: calculate V', and then pick strategic values for a, b, m, n.]

17. $x' = -x^3 + y^3$, $y' = -x^3 - y^3$ **18.** $x' = -2y^3$, $y' = 2x - y^3$

Answer 17: The system is $x' = -x^3 + y^3$, $y' = -x^3 - y^3$. If $V = ax^{2m} + by^{2n}$, then $V' = 2amx^{2m-1}(-x^3 + y^3) - 2bny^{2n-1}(x^3 + y^3)$. Choose $a = b = 1$ and $m = n = 2$. Then $V = x^4 + y^4$ and $V' = -4(x^6 + y^6)$. The system is asymptotically stable at the origin since V is positive definite and V' is negative definite. See the figure where the dashed curves are the level sets $V = 1/16, 1, 20$.

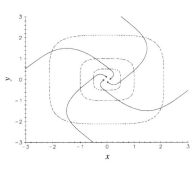

Problem 17.

19. *Neutral Stability* Show that $x' = (x - y)^2(-x + y)$, $y' = (x - y)^2(-x - y)$ is neutrally stable at the origin. [*Hint*: use $V = x^2 + y^2$. Note the line of equilibrium points.]

Answer 19: The system is $x' = (x - y)^2(-x + y)$, $y' = (x - y)^2(-x - y)$. Every point on the line $y = x$ is an equilibrium point; so the origin cannot be asymptotically stable since it is not an attractor. If $V = x^2 + y^2$, then $V' = 2xx' + 2yy' = -2(x - y)^2(x^2 + y^2) \le 0$; i.e., V' is negative semidefinite, while

V is positive definite. The system is neutrally stable at the origin since it is stable, but not asymptotically stable.

Definiteness.

20. Show the validity of Theorem 2 for the planar quadratic form $V = ax^2 + 2bxy + cy^2$.

21. Consider the system $x' = y - xf(x, y)$, $y' = -x - yf(x, y)$.

  (a) Show that the system is stable at the origin if $f(x, y)$ is positive semidefinite.

  (b) Show that the system is asymptotically stable at the origin if $f(x, y)$ is positive definite.

  (c) Show that the system is unstable at the origin if $f(x, y)$ is negative definite.

  (d) Illustrate **(b)** and **(c)** above by using the respective functions $f = |x| + |y|$ and $f = -\cos(x^2 + y^2)$. Plot orbits of each system in a region containing the origin.

  Answer 21: The system is $x' = y - xf(x, y)$, $y' = -x - yf(x, y)$. Take $V(x, y) = x^2 + y^2$, which is positive definite. The derivative of V following the motion of the system is $V' = -2(x^2 + y^2)f(x, y)$. The conclusion in each part follows from the fact that $-2(x^2 + y^2)$ is negative definite, so the definiteness properties of V' are opposite to those of f.

  (a) Since f is positive semidefinite, V' is negative semidefinite and the system is stable at the origin.

  (b) Since f is positive definite, V' is negative definite, and the system is asymptotically stable at the origin.

  (c) Since f is negative definite, V' is positive definite, and the system is unstable at the origin.

  (d) See Graph 1 where $f(x, y) = |x| + |y|$ and Graph 2 where $f(x, y) = -\cos(x^2 + y^2)$. The level sets in Graph 1 are $V = 0.25, 1, 4$ and in Graph 2 are $V = 1/16, 9/16$. Note that $r = \sqrt{\pi/2}$ is an orbit in Graph 2 since $r' = r\cos(r^2)$, $\theta' = -1$ is the polar-coordinate system equivalent to the given xy-system if $f = -\cos(r^2)$.

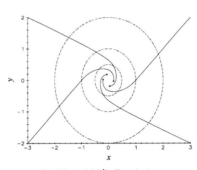
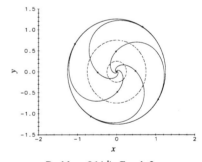

Problem 21(d), Graph 1.   Problem 21(d), Graph 2.

Modeling Problems.

22. *V-Functions and the Stability of a Damped Simple Pendulum*   Consider the damped pendulum system $x' = y$, $y' = -(g/L)\sin x - cy/m$, where g, L, c, and m are positive constants.

  (a) Show that this system is asymptotically stable at every point $(2k\pi, 0)$.[*Hint:* put $V(x, y) = (4g/L)[1 - \cos(x - 2k\pi)] + 2y^2 + (2c/m)(x - 2k\pi)y + (c^2/m^2)(x - 2k\pi)^2$.]

  (b) Show that this system is unstable at every point $(x_0, 0)$, where $x_0 = (2k + 1)\pi$. [*Hint:* use $V = -y\sin(x - x_0) + (c/m)(1 - \cos(x - x_0))$.]

23. *Lotka–Volterra Model*   The system $x' = (a - by)x$, $y' = (-c + kx)y$, where a, b, c, and k are positive, models the interactions of a predator-prey community. Show that the system is neutrally stable at the equilibrium point $(c/k, a/b)$ by using the function $Q(x, y) = (y^a e^{-by})(x^c e^{-kx})$. [*Hint:* let $V = Q(c/k, a/b) - Q(x, y)$. See Problem 11(a), Section 2.6.]

Answer 23: The system is $x' = (a - by)x$, $y' = (-c + kx)y$, where a, b, c, k are positive constants. The function Q is given by $(y^a e^{-by})(x^c e^{-kx})$. It was shown in Problem 11(a) in Section 2.6 that Q attains a strict local maximum at the equilibrium point $P(c/k, a/b)$. So $V = Q(c/k, a/b) - Q(x, y)$ is positive definite in a region containing P. Moreover, $V' = 0$ for all (x, y) in the first quadrant since $Q(x, y)$ is an integral of the equivalent first-order ODE, $(-c + kx)y\,dx - (a - by)x\,dy = 0$. Therefore, the system is stable. Since the orbits are shown in Problem 11, Section 2.6 to be cycles, the system is neutrally stable, and *not* asymptotically stable.

BackGround Material: Verification of Part 1 of Theorem 8.2.1

We want to show that the system $x' = Ax + P(x)$ is asymptotically stable at the equilibrium point 0 if the eigenvalues of the system matrix A have negative real parts and the "perturbation" $P(x)$ has order at least 2 at $x = 0$. We use the following fact (known as Lyapunov's Lemma) which we don't prove:

If the eigenvalues of A have negative real parts then the matrix equation

$$A^T B + BA = -I \qquad \text{(i)}$$

has a matrix solution B such that the scalar function $V(x) = x^T Bx$ is positive definite.

Now we will show that $V(x)$ is a strong Lyapunov function for $x' = Ax + P(x)$ in a neighborhood of $x = 0$. By Lyapunov's Lemma we know that V is positive definite. We will show that V' is negative definite, so by Lyapunov's First Theorem that means that the system $x' = Ax + P(x)$ is asymptotically stable at the origin. First note that because $P(x)$ has order at least 2 at the origin, then (for some positive constant c) $||P(x)|| \le c||x||^2$ for all x in a neighborhoood of the origin.

Now let's calculate the derivative of $V(x) = x^T Bx$ following the motion of $x' = Ax + P$:

$$V' = (x')^T Bx + x^t Bx' = (x^T A^T + P^T)Bx + x^T B(Ax + P)$$
$$= x^T(A^T B + BA)x + P^T Bx + x^T BP = x^T(-I)x + g(x) \qquad \text{(ii)}$$
$$= -||x||^2 + g(x)$$

where $g(x)$ is the scalar function $P^T Bx + x^T BP$ and where we have used (i). We can use the Triangle Inequality, the Cauchy-Schwarz Inequality, properties of the matrix norms, and the estimate on $||P(x)||$ to estimate $|g(x)|$:

$$|g(x)| = |P^T Bx + x^T BP| \le |P^T Bx| + |x^T BP| = 2|P^T Bx|$$
$$\le 2||P|| \cdot ||Bx|| \le 2c||x||^2||Bx|| \qquad \text{(iii)}$$
$$\le 2c||x||^2||B|| \cdot ||x|| \le C||x||^3$$

where C is a positive constant. An upper bound for V' is found from (ii) and (iii):

$$V'(x) \le -||x||^2[1 - C||x||] \qquad \text{(iv)}$$

If x is restricted to the open ball centered at $x = 0$ defined by $|x| < 1/C$, then (iv) implies that V' is negative definite on the open ball. $V(x)$ is a (local) strong Lyapunov function at the origin for the system $x' = Ax + P$, so that system is (locally) asymptotically stable. Note that $V(x)$ also is a (global) strong Lyapunov function for the linear system $x' = Ax$.

Set $u = x - p$ if the equilibrium point is at $x = p$ and $x' = A(x - p) + P(x)$, where $P(p) = 0$ and $P(x)$ has order at least 2 at $x = p$. Then replace the system by $u' = Au + Q(u)$ where $Q(u) = P(u + p)$, and apply the above argument.

Spotlight on Rotating Bodies

Comments:

Toss a book (not this one) into the air and try to get it to rotate around a body axis. You will soon find that you can get it to rotate about the shortest axis and also the longest axis, but not about the axis of medium length. This leads to a neat analysis of the rate equations for the components of the angular velocity vector, the construction of an integral, and the detection of stable and unstable modes of rotation. The chapter cover picture is connected with this problem.

PROBLEMS

1. *Tennis Anyone?*   Zoom in on orbits near each equilibrium point of system (2) on the inertial ellipsoid $KE = 12$ and explain the behavior of the orbits. What are the corresponding motions of the tennis racket?

Answer 1:   Group problem.

2. *Tennis Racket*   Carry out a general treatment of the tennis racket model

$$\omega_1' = \frac{I_2 - I_3}{I_1}\omega_2\omega_3, \qquad \omega_2' = \frac{I_3 - I_1}{I_2}\omega_1\omega_3, \qquad \omega_3' = \frac{I_1 - I_2}{I_3}\omega_1\omega_2$$

where I_1, I_2, and I_3 are positive constants. Address the following points:

- Assume that $I_2 < I_1 < I_3$ and find integrals whose level sets are cylinders, hyperboloids of one sheet, hyperboloids of two sheets, and cones.

- Assume that $I_1 = I_2 < I_3$ and give a full analysis of the system in this case.

- Why are nonconstant orbits periodic (except for the separatrices that connect unstable equilibrium points)?

- Use your solver to plot surfaces, orbits, and solution curves.

- Toss a racket into the air and match what you see with the model predictions.

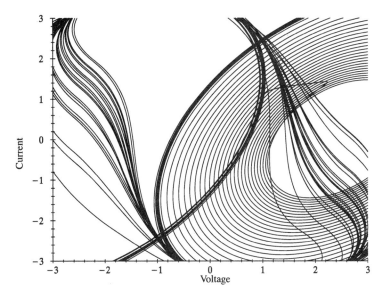

Chaotic current and voltage in a nonlinear circuit? Take a look at the scroll circuit in the WEB SPOTLIGHT ON CHAOS IN A NONLINEAR CIRCUIT.

Nonlinear Systems: Cycles and Chaos

Cycles, limit cycles, cycle-graphs, the long-term behavior of a bounded orbit of a planar autonomous system, bifurcations (including the Hopf bifurcation), and the Lorenz system and chaos are the subject of this chapter. We prove a few things, state without proof several basic results, and rely on pictures to help readers understand the ideas. Everything is carried out with just two state variables (usually) and autonomous systems (always) in the first three sections, and this is for two reasons: (1) solution and orbital behavior is easier to see, and (2) Theorem 9.2.1, which gives a precise description of the asymptotic behavior of planar orbits, fails miserably if there are more than two state variables or if the system is nonautonomous. The Lorenz system (Section 9.4) is autonomous, has three state variables, and displays apparently chaotic behavior. Some of the most interesting nonlinear systems appear in the group problems of Sections 9.3–9.4 and the SPOTLIGHT ON CHAOTIC SYSTEMS. Any one of these would make an excellent group project. The book and chapter cover figures show the apparently chaotic (but organized) wanderings of voltage and current in the nonlinear scroll circuit (see the WEB SPOTLIGHT ON CHAOS IN A NONLINEAR CIRCUIT). Take a look at the WEB SPOTLIGHT ON CHAOS IN NUMERICS for chaotic behavior in as simple a dynamical system as Euler's approximation applied to the logistic differential equations.

The dynamics of the scroll circuit and the Lorenz system raise deep (and as yet unanswered) questions about just what chaos is. This is a wide-open area of intellectual and practical research. We expect that in a few years our approach will seem quaint and a bit old-fashioned, but that often happens when ideas are still in the development stage.

9.1 Cycles

Comments:

Cyclical behavior is all around us, and it is not hard to pick up on the idea of a periodic solution as a nonconstant equilibrium state. Limit cycles are a little harder to grasp, which is why we give several examples early on in the section. Since circular cycles are most easily described in polar coordinates, there are a lot of r's and θ's here. The van der Pol system is derived as an extension from a simple linear RLC circuit system to a circuit system with a nonlinear resistor (e.g., a diode). We introduce these circuit models as systems (rather than as second-order ODEs) because that is the most natural way to do the modeling. It is worth emphasizing that the van der Pol cycles are isolated, attracting, and with periods that change with the parameter. If you haven't done much with circuits, you can do the van der Pol system entirely graphically for various values of the parameter and various forms for the function f. Numerical orbits and component curves are pretty convincing here! The other feature is that the van der Pol cycle is a "relaxation oscillation" for large values of the parameter.

In the WEB SPOTLIGHT ON QUADRATIC RATES AND LIMIT CYCLES we touch on the ODE part of problem # 16 of a famous list of 23 problems posed in 1900 by David Hilbert. Most of Hilbert's 23 problems were solved during the 20th century, but this part of the 16th remains open. You might take a look at *Honors Class: Hilbert's Problems and Their Solvers*, Benjamin Yandell (Peters, AK, Limited, 2001).

PROBLEMS

Cycles and Limit Cycles. Find the equilibrium points and the cycles of the following systems written in polar coordinates. As in Examples 9.1.1 and 9.1.2, draw labeled state lines for r. Sketch the cycles and other orbits in the xy-plane, $x = r\cos\theta$, $y = r\sin\theta$. Use arrowheads to show the direction of increasing time. Is the equilibrium point at the origin asymptotically stable, neutrally stable, or unstable? Determine whether each cycle is a limit cycle, and if it is, whether it attracts or repels.

1. $r' = 4r(4-r)(5-r)$, $\theta' = 1$

2. $r' = r(r-1)(2-r^2)(3-r^2)$, $\theta' = -3$

3. $r' = r(1-r^2)(4-r^2)$, $\theta' = 1-r^2$ [*Hint:* r' and θ' are 0 if $r = 1$.]

4. $r' = r(1-r^2)(9-r^2)$, $\theta' = 4-r^2$ [*Hint:* $\theta' = 0$ but $r' \neq 0$ if $r = 2$.]

5. $r' = r\cos\pi r$, $\theta' = 1$

Answer 1: The system is $r' = 4r(4-r)(5-r)$, $\theta' = 1$. The equilibrium point at the origin is unstable since r' is positive for $0 < r < 4$. This system has the attracting limit cycle $r = 4$ (r' is positive for $0 < r < 4$ and negative for $4 < r < 5$) and the repelling limit cycle $r = 5$ (r' is positive for $r > 5$). See the stateline and the figure.

Answer 3: The system is $r' = r(1-r^2)(4-r^2)$, $\theta' = 1-r^2$. Note that the orbits turn clockwise if $r > 1$ (since $\theta' < 0$), and counterclockwise if $0 < r < 1$. The equilibrium points are the origin (unstable since r' is positive for $0 < r < 1$) and all points on the circle $r = 1$ (since $r' = 0$, and $\theta' = 0$ there). Since $r' > 0$ for $0 < r < 1$ and $r' < 0$ for $1 < r < 2$, the circle $r = 1$ of equilibrium points attracts. In fact, each equilibrium point on the circle is stable. Here's why. Note that $dr/d\theta = r(4-r^2)$. Separating variables and using partial fractions, we have (after a calculation which we omit) $\theta = (1/8)\ln|r^2/(4-r^2)| + C$.

So as $r \to 1$, $\theta \to (1/8)\ln(1/3) + C$; that is, by choosing C appropriately, $\theta \to$ any given number θ_0. So the circle $r = 1$ consists of equilibrium points, each of which is approached by two orbits, one with $r \to 1^+$, the other with $r \to 1^-$. These equilibrium points are neutrally stable. This system has the repelling limit cycle $r = 2$ ($r' > 0$ for $r > 2$ and $r' < 0$ for $1 < r < 2$). See the state line and the figure (where the circle of neutrally stable equilibrium points is dashed).

Answer 5: The system is $r' = r\cos\pi r$, $\theta' = 1$. The equilibrium point at the origin is unstable since r' is positive for $0 < r < 1/2$. For $n = 0, 1, 2, \ldots$, this system has the attracting limit cycles $r = 2n + 1/2$ because $r' > 0$ for r near but less than $2n + 1/2$, and $r' < 0$ for r near but larger than $2n + 1/2$; similarly, the limit cycles $r = 2n + 3/2$ are repelling. See the state line below and the figure.

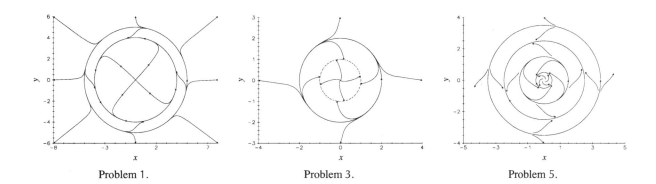

Problem 1.   Problem 3.   Problem 5.

Limit Cycles. Find all limit cycles and identify each as an attractor or a repeller. Use polar coordinates as in Example 9.1.1 and draw labeled state lines for r.

6. $x' = y - x(x^2 + y^2)$, $y' = -x - y(x^2 + y^2)$

7. $x' = x + y - x(x^2 + y^2)$, $y' = -x + y - y(x^2 + y^2)$

8. $x' = 2x - y - x(3 - x^2 - y^2)$, $y' = x + 2y - y(3 - x^2 - y^2)$

9. $r' = r(1 - r^2)(4 - r^2)(9 - r^2)/1000$, $\theta' = 1$. Plot orbits in the rectangle $|x| \le 6$, $|y| \le 4$, where $x = r\cos\theta$, $y = r\sin\theta$.

Answer 7: The system is $x' = x + y - x(x^2 + y^2)$, $y' = -x + y - y(x^2 + y^2)$. Similar to **(a)**, the original system becomes $r' = r - r^3$, $\theta' = -1$. Because $r' > 0$ if $0 < r < 1$, and $r' < 0$ if $r > 1$, this system has an unstable equilibrium point at the origin, and a unique attracting limit cycle of period 2π defined by $x^2 + y^2 = 1$.

Answer 9: See the state line. Can you see the two attracting and one repelling limit cycles in the figure that shows orbits?

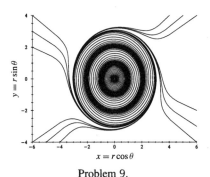

Problem 9.

Center-Spiral Equilibrium Points.

10. *Center-Spiral* Consider the system in polar coordinates $r' = r^3 \sin(1/r)$, $\theta' = 1$, where r' is defined to be zero at $r = 0$. Show that the corresponding nonlinear xy-system has infinitely many circular limit cycles around the neutrally stable equilibrium point at the origin, the sequence of shrinking cycles converges onto the origin, the cycles alternately attract and repel, and all other orbits spiral away from one cycle and toward another as t increases. The origin of the xy-system is a *center-spiral*. Explain the name. [*Hint*: plot xy-orbits in an $r \cos \theta$, $r \sin \theta$ plane.]

11. Explain why the origin of the xy-plane is a center-spiral equilibrium point for the system $r' = r \sin(\pi/r)$, $\theta' = -2$, where r' is defined to be zero if $r = 0$.

Answer 11: The origin is a neutrally stable, center-spiral point because it is enclosed by the family of alternately attracting and repelling limit cycles $r = 1/n$, $n = 1, 2, 3, \cdots$, where r' changes sign as r decreases through the values $1/n$. For r between $1/(2n)$ and $1/(2n+1)$, $r' > 0$, while $r' < 0$ for r between $1/(2n-1)$ and $1/(2n)$. Since $\theta' = -2$ the orbits in each ring bounded by adjacent cycles is a clockwise-turning spiral.

Unusual Cycles.

12. *Nonisolated, Nonlinear Cycles and a Center* Show that the nonconstant orbits of the system $x' = y^3$, $y' = -x^3$ are cycles that enclose the equilibrium point at the origin and fill the xy-plane. Plot the orbits and component curves corresponding to initial points $(0, 0)$, $(0.5, 0)$, $(1, 0)$, $(2, 0)$. Are the periods of distinct cycles the same?

13. *Semistable Cycles* Some cycles repel on one side and attract on the other; they are one kind of *semistable* cycle. Find all cycles of $r' = r(1 - r^2)^2(4 - r^2)(9 - r^2)$, $\theta' = 1$ and identify each as attracting, repelling, or semistable. Sketch orbits and draw a labeled state line for r.

Answer 13: The system $r' = r(1 - r^2)^2(4 - r^2)(9 - r^2)$, $\theta' = 1$ has a repelling limit cycle, $r = 3$, and an attracting limit cycle, $r = 2$; $r = 1$ is a semistable cycle (labeled AR on the state line), attracting interior orbits and repelling exterior orbits. This follows because $r' > 0$ if $0 < r < 1$, $1 < r < 2$, or $r > 3$, while $r' < 0$ if $2 < r < 3$. See the figure and the state line.

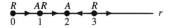

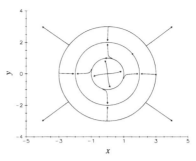

Problem 13.

14. *A Strange Cycle* Explain why the system $r' = r(r-1)^2 \sin[\pi/(r-1)]$, $\theta' = 1$, where r' is defined to be zero if $r = 1$, has a cycle $r = 1$, which is not a limit cycle, and every neighborhood of which contains other cycles as well as spirals between successive cycles. [*Hint*: as $r \to 1$, $(r-1)^2 \sin[\pi/(r-1)] \to 0$; $r' = 0$ at $r = 1$ and at $r = 1 \pm 1/n$.]

15. *Cycles in Space* Describe the orbits near the cycle $r = 1$, $z = 0$ of the system (in cylindrical coordinates) $r' = r(1 - r^2)$, $\theta' = 25$, $z' = \alpha z$, where α is a constant. Consider separately the three cases $\alpha < 0$, $\alpha = 0$, $\alpha > 0$. Plot orbits in xyz-space for $\alpha = -0.5, 0, 1$.

Answer 15: The original system in r, θ, z coordinates is $r' = r(1 - r^2)$, $\theta' = 25$, $z' = \alpha z$. The system in xyz-coordinates is $x' = x - 25y - x(x^2 + y^2)$, $y' = 25x + y - y(x^2 + y^2)$, $z' = \alpha z$. Viewed solely from the xy-plane, the unit circle is an attracting limit cycle. It is an attracting limit cycle in $\mathbb{R}^3$ as well if $\alpha < 0$. See Graph 1 (where $\alpha = -0.5$) for two nonconstant orbits, one with $x_0^2 + y_0^2 < 1$ and the other with $x_0^2 + y_0^2 > 1$. If $\alpha = 0$, then z is constant and orbits lie in planes $z = z_0$. The unit circle in each plane is a limit cycle that attracts each nonconstant orbit in the plane. However, the cycle $r = 1$, $z = 0$ is *not* an attractor because it does not attract orbits from other planes. See Graph 2 for orbits in the planes $z = 10, 40$. If $\alpha > 0$ and $x = x(t)$, $y = y(t)$, $z = z(t)$ defines a nonconstant solution, then $x^2(t) + y^2(t) \to 1$ as $t \to +\infty$, while $z(t) \to +\infty$ if $z(0) > 0$. Graph 3 shows the straight line orbit of $x = y = 0$, $z(t) = 0.01e^t$ and two other orbits with $z_0 > 0$, one with $x_0^2 + y_0^2 > 1$ and the other with $x_0^2 + y_0^2 < 1$. In this case the unit circle attracts orbits in the xy-pane, but repels orbits outside the plane.

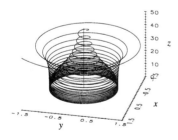

Problem 15, Graph 1.

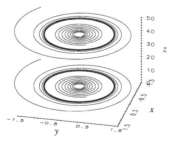

Problem 15, Graph 2.

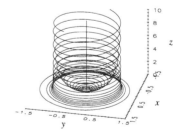

Problem 15, Graph 3.

Van der Pol Systems. Verify that each system in Problems 16–22 satisfies the conditions of Theorem 9.1.1. For each value of μ plot the limit cycle and some orbits that are attracted to it. Estimate the period and the x- and the y-amplitude of each cycle; verify that the larger the value of μ, the longer the period and the larger the y-amplitude of the cycle. The x-amplitude?

16. $x' = y - \mu(x^3 - 10x)$, $y' = -x$; $\mu = 0.1, 2$

17. $x' = y - \mu x(|x| - 1)$, $y' = -x$; $\mu = 0.5, 5, 50$

18. $x' = y - \mu x(x^4 + x^2 - 1)/10$, $y' = -x$; $\mu = 0.1, 1$

19. $x' = y - \mu x(2x^2 - \sin^2 \pi x - 2)$, $y' = -x$; $\mu = 0.5, 5$

20. $x' = y - \mu(x - |x+1| + |x-1|)$, $y' = -x$; $\mu = 0.5, 5, 50$

21. $x' = y - \mu(x^3 - x)$, $y' = -x$; $\mu = 0.1, 3, 5, 7, 10$

22. $x' = y - \mu(|x|x^3 - x)$, $y' = -x$; $\mu = 0.5, 1, 3, 5, 7, 10.$

Answer 17:   The system is $x' = y - \mu x(|x| - 1)$, $y' = -x$. Then $f(x) = x(|x| - 1)$ and $a = 1$. The values of μ, approximate respective periods, y-amplitudes, and x-amplitudes are

μ	period	y-amp	x-amp
0.5	7	1.2	1.4
5	10	2	1.5
50	55	10.3	1.4

See Graphs 1, 2, 3, where $\mu = 0.5, 5$, and 50 respectively.

Answer 19:   The system is $x' = y - \mu x[2x^2 - \sin^2(\pi x) - 2]$, $y' = -x$. Then $f(x) = x[2x^2 - \sin^2(\pi x) - 2]$ and $a = 1$. The values of μ, approximate respective periods, y-amplitudes, and x-amplitudes are

μ	period	y-amp	x-amp
0.5	8	1.4	1.4
5	10	7	1.4

See Graphs 1, 2, where $\mu = 0.5$ and 5, respectively.

Answer 21:   The function $f(x)$ is $x^3 - x$ and $a = 1$. The values of μ, approximate respective periods, y-amplitudes, and x-amplitudes are

μ	period	y-amp	x-amp
0.1	7	1.2	1.2
5	13	2.5	1.2
10	20	4.4	1.2

See Graphs 1, 2, 3, where $\mu = 0.1, 5$, and 10, respectively.

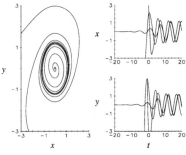

Problem 17, Graph 1.

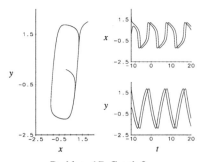

Problem 17, Graph 2.

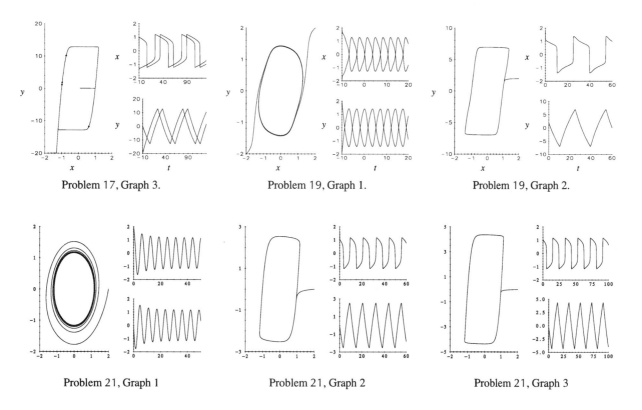

Problem 17, Graph 3.   Problem 19, Graph 1.   Problem 19, Graph 2.

Problem 21, Graph 1   Problem 21, Graph 2   Problem 21, Graph 3

23. *Bifurcation in a van der Pol System*   Use a numerical solver to show that as μ decreases from +1 to -1, the equilibrium point of system (7) changes its stability at $\mu = 0$. Describe what happens to the limit cycle as μ sweeps from +1 to -1.

Answer 23:   The system is $x' = y - \mu f(x)$, $y' = -x$ with $\mu = 1$, 0.5, 0, -0.5, -1 and $f(x) = x^3/3 - x$. See Graphs 1–3 for $\mu = 1$, 0, -1, respectively. Graphics show a repelling spiral point at $(0,0)$ for each $\mu > 0$ with orbits that spiral outward toward a limit cycle. At $\mu = 0$, the origin is a linear center and all nonconstant orbits are cycles. For each $\mu < 0$ some orbits spiral inward from a repelling limit cycle and toward the origin. In general, we can say that as μ changes from positive values to negative values, the corresponding attracting limit cycles for $\mu > 0$ pass through a neutrally stable cycle (at $\mu = 0$) and to repelling limit cycles for $\mu < 0$. The orbital picture for negative μ is the mirror image of that for positive μ but with orbits spiraling away from a repelling limit cycle as t increases.

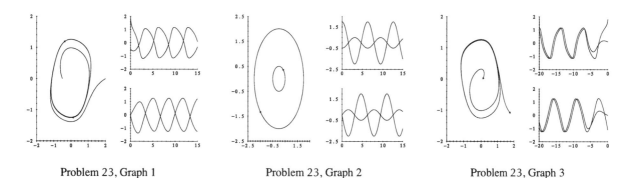

Problem 23, Graph 1   Problem 23, Graph 2   Problem 23, Graph 3

Lénard Equation. The nonlinear, second-order ODE $x'' + f(x)x' + g(x) = 0$, where $f(x)$ and $g(x)$ are

☞ Named for the French mathematician and applied physicist, Alfred Liénard (1869–1958)

continuous and piecewise smooth, is the *Liénard equation*.

24. Show that if $y = x' + F(x)$, where $F(x) = \int_0^x f(s)\,ds$, then the Liénard equation can be written in *Liénard system* form as $x' = y - F(x)$, $y' = -g(x)$. Verify that the van der Pol system is a Liénard system. [*Hint*: the xy-plane here is the *Liénard plane*.]

25. Show that $V = y^2/2 + G(x)$, where $G(x) = \int_0^x g(s)\,ds$, is a weak Lypanov function for the Liénard system of Problem 24 if $g(x)F(x)$ is positive for $x \neq 0$. Explain why the origin of the xy-plane is a stable equilibrium point of the Liénard system. [*Hint*: see the SPOTLIGHT ON LYAPUNOV FUNCTIONS.]

26. Plot the orbit of the periodic solution of $x'' + (x^2 - 1)x' + x = 0$ both in the xx'-plane and in the Liénard xy-plane, where y is defined as in Problem 24.

27. The *Rayleigh equation* is $z'' + \mu[(z')^2 - 1]z' + z = 0$. Differentiate the Rayleigh equation with respect to t, then set $x = \sqrt{3}z'$, and show that $x'' + \mu(x^2 - 1)x' + x = 0$. Explain why the ODE in x reduces to a Liénard system if we set $y = x' + \mu(x^3/3 - x)$. Show that the Rayleigh equation has a unique attracting limit cycle for each $\mu > 0$. Plot the cycle and orbits through the points $(1.5, 1.5)$ and $(0.5, 0.5)$ in the zz'-plane for $\mu = 0.1, 1, 5, 10$.

Answer 25: Group problem.

Answer 27: The Rayleigh equation is $z'' + \mu[(z')^2 - 1]z' + z = 0$. Let $x = \sqrt{3}z'$, $x' = \sqrt{3}z''$ and $x'' = \sqrt{3}z'''$. We have

$$x'' + \mu(x^2 - 1)x' + x = \sqrt{3}z''' + \mu(3(z')^2 - 1)\sqrt{3}z'' + \sqrt{3}z' = \sqrt{3}[z'' + \mu((z')^2 - 1)z' + z]' = 0$$

Therefore, if z satisfies the Rayleigh equation, $x = \sqrt{3}z'$ satisfies the van der Pol equation. Let $y = x' + \mu(x^3/3 - x)$; then we have $y' = x'' + \mu(x^2 - 1)x'$. Substituting y' into the ODE, $x'' + \mu(x^2 - 1)x' + x = 0$, we see that $y' = -x$. So, we have

$$\frac{dx}{dt} = y - \mu(x^3/3 - x) \qquad \frac{dy}{dt} = -x$$

which has the same form as the van der Pol system in Liénard coordinates. So the Rayleigh equation has a unique attracting limit cycle. For orbits in the Liénard xy-plane, see Graphs 1–4, for $\mu = 1/10$, 1, 5, and 10, respectively, and for representative initial points. You will see similar graphs if you use the inital points $(1.5, 1.5)$ and $(0.5, 0.5)$. Observe that the orbits in Graph 4 would look much better if the number of points plotted per unit time had been increased.

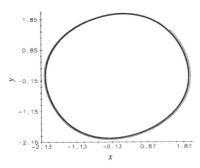

Problem 27, Graph 1.

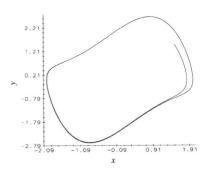

Problem 27, Graph 2.

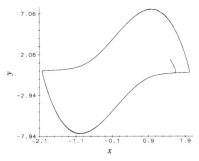

Problem 27, Graph 3.

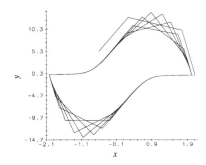

Problem 27, Graph 4.

Background Material: Proof of the van der Pol Cycle Theorem (Theorem 9.1.1)

The form of the rate equation of the van der Pol system

$$x' = y - \mu f(x), \quad y' = -x \tag{i}$$

and the given conditions on μ and f imply that:

- If $x = x(t)$, $y = y(t)$ is a solution of (i), so also is $x = -x(t)$, $y = -y(t)$. Consequently, orbits of (i) in the left half-plane are simply the reflections through the origin of orbits in the right half-plane. For this reason, we only need to carry out our analysis in the right half-plane.

- If there is a cycle, it must enclose the equilibrium point at the origin.

- As noted in Section 7.2, Sign Analysis shows that each orbit Γ of (i) in the right half-plane and above the graph K of $y = \mu f(x)$ moves to the right from initial point A and falls as time increases. Such an orbit crosses K in the vertical downward direction and as τ increases moves to the left and intersects the negative y-axis at a point G. Let D be the point on K above the point $(\beta, 0)$ where $\beta > a$.

  Sketch a figure in the xy-plane and label the curve K and the points A, G, and D. The van der Pol graph shown uses $\mu = 1/4$, $f(x) = x^3/3 - x$. Use this graph to locate the points A, B, C, D, E, F, G as well as $x = a = \sqrt{3}$ and $x = \beta$ mentioned in the construction.

- According to the reflection property, Γ is a cycle if and only if the distance from O (the origin) to A equals the distance from O to G.

- If Γ is a cycle when $\beta = \beta_0$ then that cycle is unique and attracting if and only if for $\beta > \beta_0$ the distance $|OA|$ from O to A exceeds $|OG|$, while for $0 < \beta < a$ we have $|OA| < |OG|$. This follows (but not very easily) from the reflection properties of orbits and the fact that distinct orbits cannot intersect.

  These properties are used below.

Set $E = x^2/2 + y^2/2$; we see that $dE/dt = x\,dx/dt + y\,dy/dt = x(y - \mu f) + y(-x) = -\mu x f$, Since $dy/dt = -x$, so $-x\,dt = dy$ and we have along Γ

$$dE = \mu f(x)\,dy$$

If we set $\delta E = E_G - E_A$, the difference between the values of E at G and at A, then δE is a well-defined, continuous function of β, the x-coordinate of the point D on Γ. We have that

$$\delta E(\beta) = \frac{1}{2}|OG|^2 - \frac{1}{2}|OA|^2 = \oint_{AG} dE = \oint_{AG} \mu f(x)\,dy$$

The theorem is proven if we show that:

- $\delta E(\beta)$ is positive and bounded for $0 < \beta < a$,

- $\delta E(\beta)$ is strictly decreasing for $\beta \geq a$,

- $\delta E(\beta) \rightarrow -\infty$ as $\beta \rightarrow +\infty$.

These three properties and the continuity of $\delta E(\beta)$ imply that there is exactly one point β_0 for which $\delta E = 0$. From the formula for $\delta E(\beta)$ and the reflection property for orbits, we see that the orbit Γ corresponding to β_0 must be a cycle since $|OG| = |OA|$. Moreover, the cycle is attracting because the sign of the values of $\delta E(\beta)$ changes from $+$ to $-$ as β increases through β_0. So, we need only prove the three properties of δE listed above.

First, suppose that $0 < \beta < a$. Along the arc of Γ from A to B we have that both $f(x)$ and dy are negative because in this case the arc lies entirely in the vertical strip $0 < x < a$. Since $\mu > 0$ we have that $\delta E(\beta) > 0$. Because $\delta E(\beta)$ is continuous for all $\beta \geq 0$, the Maximum Value Theorem (Theorem B.2.1) implies that $\delta E(\beta) > 0$ has a positive upper bound E_{max} on the interval $0 \leq \beta \leq a$.

Second, suppose that $\beta > a$, and divide the line integral in the formula for $\delta E(\beta)$ into three parts:

$$\delta E(\beta) = \oint_A^G \mu f \, dy = \oint_A^B + \oint_F^G + \oint_B^F = L_1(\beta) + L_2(\beta)$$

where L_1 is the sum of the line integrals from A to B (B is the point on the orbit above $x = a$, and F is the point on the orbit below $x = a$) and from F to G, and $L_2(\beta)$ is the line integral from B to F. We shall show that $L_1(\beta)$ is a positive and decreasing function of $\beta \geq a$ and that $L_2(\beta)$ is negative and decreasing on the same interval. $L_1(\beta)$ is positive because μf and dy are negative (except possibly at A, B, F, and G) on the arcs of Γ involved in calculating the value of $L_1(\beta)$. $L_1(\beta)$ decreases as β increases because

$$\mu f \, dy = \mu f \frac{dy}{dx} dx = -\frac{x \, dx}{y - \mu x f(x)}$$

and, as β increases, the arcs of Γ from A to B and from F to G move outward. This means that as β increases the denominator $y - \mu f(x)$ increases for each fixed x, $0 < x < a$, and the quotient decreases, as must $L_1(\beta)$.

Turning to the function $L_2(\beta)$, $\beta > a$, we see that L_2 is negative because dy is negative on the arc of Γ from B to F while $\mu f(x)$ is positive (except at B and F). Finally, $L_2(\beta)$ is a decreasing function of β, $\beta \geq a$, because dy is negative (as before) and $\mu f(x)$ is positive and increasing for each x, $x \geq a$. So $\delta E(\beta)$ decreases from a positive value E_0 at $\beta = a$ as β increases, $\beta \geq a$.

$$x' = y - 0.25(x^3/3 - x), \qquad y' = -x$$

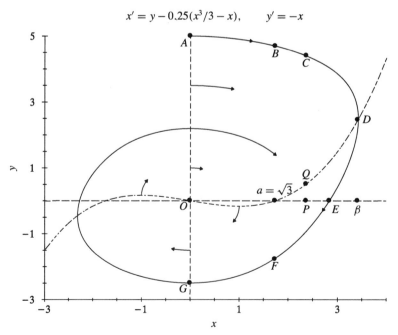

Graphs of $y = \mu f(x)$ and of van der Pol orbits.

Third, to show that $\delta E(\beta) \le E_0$, we only need show that $L_2(\beta) \to -\infty$ as $\beta \to +\infty$. Fix the point P on the x-axis as shown in the figure. We have that

$$L_2(\beta) = \oint_B^F < \oint_C^E \mu f(x)\, dy \le -|PQ| \cdot |PC|$$

because $\mu F\, dy$ is everywhere negative along the arc from B to F (except at B and F) and the magnitude of a line integral is at least as large as the product of the minimal magnitude of the integrand μf ($|PQ|$ in this case) along the arc from C to E and the arc length (which is at least $|PC|$ here). But $|PC| \to \infty$ as $\beta \to +\infty$. So, the above inequality for $L_2(\beta)$ implies that $L_2(\beta) \to -\infty$, and we see that the van der Pol system must have a cycle.

9.2 Solution Behavior in Planar Autonomous Systems

Comments:

The nature of the long-term behavior of a bounded solution of an autonomous planar system is the core of this section, but that only makes sense after the idea of a rather mysterious object called a cycle-graph has been explored. This is why we have several examples of cycle-graphs in the text and problem set. The reason for the emphasis on limit sets is that for applications in engineering and science one is often more interested in the long-term behavior of solutions than in the short-term behavior, and the limit sets show the asymptotic nature of orbits. The main result is Theorem 9.2.1 that says that a bounded orbit must tend to an equilibrium point, or to a cycle, or to a cycle-graph, and there are no other possibilities. In other words, chaotic motion *cannot* happen in this setting of a planar autonomous system. This is in marked contrast to what we will see in Section 9.4. Bendixson's Negative Criterion (Theorem 9.2.4) is a little gem; it is useful and has an elegant proof that uses Green's theorem. In doing Problem 27 people have created marvelous ODE drawings ranging from the classical kitty shown in the margin to something that might have been done by Picasso in one of his more avant-garde phases.

PROBLEMS

Limit Sets. Use analytical or graphical techniques to find the positive and the negative limit sets of the orbits through the listed initial points. Sketch some orbits for Problems 3, 4 and 6.

1. $x' = y$, $y' = -x$; $(0,0)$, $(1,1)$ **2.** $x' = y$, $y' = x$; $(1,1)$, $(1,-1)$, $(1,0)$

3. *Van der Pol* $x' = y - (x^3/3 - x)$, $y' = -x$; $(0.1, 0)$, $(3, 0)$ [*Hint*: see Section 9.1.]

4. *Undamped Simple Pendulum* $x' = y$, $y' = -\sin x$; $(0, 1)$, $(0, \sqrt{2})$, $(0, 2)$

5. *Polar Form* $r' = r(1 - r^2)(4 - r^2)$, $\theta' = 5$; $r_0 = 1/2, 3/2, 5/2$; $\theta_0 = 0$

6. $x' = [x(1 - x^2 - y^2) - y][(x^2 - 1)^2 + y^2]$, $y' = [y(1 - x^2 - y^2) + x][(x^2 - 1)^2 + y^2]$; $(1/2, 0)$, $(0, \pm 1)$, $(3/2, 0)$ [*Hint*: the unit circle is a cycle-graph consisting of two equilibrium points joined by two orbital arcs.]

Answer 1: The system is $x' = y$, $y' = -x$. Divide y' by x' to obtain $dy/dx = -x/y$. So, $x^2 + y^2 = C^2$ defines a circular orbit (i.e., a cycle) about the origin (which is then neutrally stable) for each value of C. Furthermore, the general solution of the original system is $x = C\cos(t + \phi)$ and $y = -C\sin(t + \phi)$, where C and ϕ are arbitrary constants. Consequently, the solutions through the points (0,0) and (1,1) are, respectively, $x = 0$, $y = 0$, and $x = \cos(t - \pi/4)$, $y = -\sin(t - \pi/4)$. The negative and the positive limit sets of the orbit $\Gamma_1 = \{(0, 0)\}$ through $x = 0$, $y = 0$ are Γ_1 itself, which is the only equilibrium point; so $\alpha(\Gamma_1) = \omega(\Gamma_1) = \Gamma_1$. The negative and the positive limit sets of the second orbit Γ_2 are Γ_2 itself, which is a cycle; so $\alpha(\Gamma_2) = \omega(\Gamma_2) = \Gamma_2 =$ unit circle. There are no cycle-graphs.

Answer 3: The van der Pol system is $x' = y - (x^3/3 - x)$, $y' = -x$. The negative and the positive limit sets of the orbit Γ_1 through $(0.1, 0)$ are, respectively, the origin and the cycle shown in the figure so $\alpha(\Gamma_1)$ is the origin and $\omega(\Gamma_1)$ is the cycle. For the orbit Γ_2 though $(3, 0)$, the negative and positive limit sets are, respectively, the empty set (because the orbit becomes unbounded as t decreases) and the van der Pol cycle shown in the figure. The origin is the only equilibrium point, and the cycle is the only cycle. There are no cycle-graphs.

Answer 5: The (polar) system is $r' = r(1 - r^2)(4 - r^2)$, $\theta' = 5$. The negative and the positive limit sets of the orbit through $r = 1/2$, $\theta = 0$ are, respectively, the origin and the cycle $r = 1$, because r' is positive for $0 < r < 1$ and $r' = 0$, $\theta' \neq 0$ on $r = 1$. The negative and the positive limit sets of the orbit through $r = 3/2$, $\theta = 0$ are, respectively, the cycles $r = 2$ and $r = 1$, because $r' < 0$ for $1 < r < 2$. The negative and the positive limit sets of the orbit through $r = 5/2$, $\theta = 0$ are, respectively, the cycle $r = 2$ and the empty set (because $r(t) \rightarrow +\infty$ as t increases) since $r' > 0$ for $r > 2$. These results follow from determining the sign of r' in the regions $0 < r < 1$, $1 < r < 2$, $r > 2$. There are no cycle-graphs.

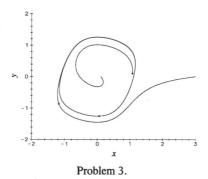

Problem 3.

Bendixson Negative Criterion. Show that each ODE has no cycles and no cycle-graphs. [*Hint*: convert each ODE to a system.]

7. *Damped Pendulum* $mLx'' + cLx' + mg\sin x = 0$; m, L, c, and g are positive constants.

8. *Damped Nonlinear Spring* $mx'' + ax' + bx + cx^3 = 0$, where m, a, and b are positive constants and c is any constant.

9. $x'' + (2 + \sin x)x' + g(x) = 0$, where $g(x)$ is any continuously differentiable function.

10. $x'' + f(x)x' + g(x) = 0$; f and g are continuously differentiable and $f(x)$ has a fixed sign.

Answer 7: The corresponding system is $x' = y$, $y' = -(g\sin x)/L - cy/m$. Then $\partial[y]/\partial x + \partial[-(g\sin x)/L - cy/m]/\partial y = -c/m < 0$. According to Bendixson's Negative Criterion, there is no cycle or cycle-graph anywhere in the xy-plane.

Answer 9: The corresponding system is $x' = y$, $y' = -g(x) - (2 + \sin x)y$. Then $\partial[y]/\partial x + \partial[-g(x) - (2 + \sin x)y]/\partial y = -(2 + \sin x) < 0$. According to Bendixson's Negative Criterion, there is no cycle or cycle-graph anywhere in the xy-plane.

Bendixson Negative Criterion. Show that each system has no cycle or cycle-graph in the region indicated.

11. $x' = 2x - y + 36x^3 - 15y^2$, $y' = x + 2y + x^2y + y^5$; xy-plane

12. $x' = 12x + 10y + x^2y + y\sin y - x^3$, $y' = x + 14y - xy^2 - y^3$; the disk $x^2 + y^2 \leq 8$

13. $x' = x - xy^2 + y\sin y$, $y' = 3y - x^2y + e^x \sin x$; the interior of the disk $x^2 + y^2 \leq 4$

Answer 11: The system is $x' = 2x - y + 36x^3 - 15y^2$, $y' = x + 2y + x^2y + y^5$. Here

$$\partial(2x - y + 36x^3 - 15y^2)/\partial x + \partial(x + 2y + x^2y + y^5)/\partial y = 4 + 109x^2 + 5y^4 > 0$$

According to Bendixson's Negative Criterion, there is no cycle or cycle-graph anywhere in the xy-plane.

Answer 13: The system is $x' = x - xy^2 + y\sin y$, $y' = 3y - x^2y + e^x \sin x$, and

$$\partial(x - xy^2 + y\sin y)/\partial x + \partial(3y - x^2y + e^x \sin x)/\partial y = 4 - (x^2 + y^2) > 0$$

if $x^2 + y^2 < 4$. According to Bendixson's Negative Criterion, there is no cycle or cycle-graph anywhere in the disk $x^2 + y^2 < 4$.

No Limit Sets. Explain why all orbits of each system are unbounded. [*Hint*: use Theorem 9.2.2.]

14. $x' = x - y + 10$, $y' = x^2 + y^2 - 1$ **15.** $x' = y$, $y' = -\sin x - y + 2$

16. $x' = x^2 + 2y^2 - 4$, $y' = 2x^2 + y^2 - 16$

Answer 15: The system is $x' = y$, $y' = -\sin x - y + 2$. Since there is no simultaneous real solution for the equations, $y = 0$, $\sin x + y - 2 = 0$, there is no equilibrium point for this autonomous system. By the Unbounded Orbits Theorem (Theorem 9.2.2), every orbit becomes unbounded as t increases and as t decreases. See the figure.

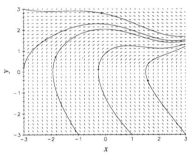

Problem 15.

Any Cycles? Determine whether the following systems have cycles. Find all cycles (graphically or analytically), if any exist. If there are no cycles, explain why not.

17. $x' = e^x + y^2$, $y' = xy$ [*Hint*: does x ever decrease?]

18. $x' = 2x^3y^4 + 5$, $y' = 2ye^x + x^3$ [*Hint*: use Bendixson's Negative Criterion.]

19. *A System in Polar Coordinates* $r' = r\sin(r^2)$, $\theta' = 1$

Answer 17: The system is $x' = e^x + y^2$, $y' = xy$. Since x' is always positive, no orbit can return to its starting point as t increases, and there are no cycles.

Answer 19: The (polar) system is $r' = r\sin(r^2)$, $\theta' = 1$. Note that when $r = \sqrt{n\pi}$, $n = 0, 1, \ldots$, then $r' = 0$. Also, $r' > 0$ for $2k\pi < r^2 < (2k+1)\pi$ and $r' < 0$ for $(2k+1)\pi < r < (2k+2)\pi$, $k = 0, 1, \ldots$. So $r^2 = 2k\pi$, $k = 1, 2, \cdots$, defines repelling limit cycles and $r^2 = (2k+1)\pi$ defines attracting limit cycles for values of $k = 0, 1, \ldots$.

Poincaré–Bendixson Investigation.

20. *Green's Theorem and Averages*   Prove that the average value of the function $F(x, y) = \partial f/\partial x + \partial g/\partial y$ on the region R inside a cycle Γ of the system $x' = f(x, y)$, $y' = g(x, y)$ must be zero. [*Hint*: the *average value* of F on R is $\int\int_R F(x, y)\,dx\,dy/A$, where A is the area of R. Use Green's Theorem (Theorem B.2.13 in Appendix B.2).]

21. *Contradicting Bendixson?*   The quantity $\partial f/\partial x + \partial g/\partial y$ for the system $x' = f = x - 10y - x(x^2 + y^2)$, $y' = g = 10x + y - y(x^2 + y^2)$ is $2 - 4x^2 - 4y^2$, which is negative in the region R, $x^2 + y^2 > 0.5$. Show that the unit circle is a cycle that lies entirely in R. Why does this *not* contradict Bendixson's Negative Criterion?

Answer 21: The system is $x' = x - 10y - x(x^2 + y^2) = f(x, y)$, $y' = 10x + y - y(x^2 + y^2) = g(x, y)$, and the region R is described by $x^2 + y^2 > 0.5$, that is the region outside the circle $x^2 + y^2 = 0.5$. The region R is not simply connected, so Bendixson's Negative Criterion does not apply. In polar coordinates the system is $r' = r(1 - r^2)$, $\theta' = 10$, so $r = 1$ is a cycle in R.

Cycle-Graphs. The following problems explore the nature of cycle-graphs.

22. *Lazy-Eight Cycle-Graph*   Show that the system $x' = y + x(1 - x^2)(y^2 - x^2 + x^4/2)$, $y' = x - x^3 - y(y^2 - x^2 + x^4/2)$ has equilibrium points at $(0, 0)$ and $(\pm 1, 0)$. Use a numerical solver and plot the orbits through the point $(-3, 2)$ forward in time. Repeat with the orbits through $(\pm 0.5, 2)$, but carry these orbits forward and backward in time. Explain what you see. [*Hint*: look at the graph in the margin on page 533 and at "The Lazy-Eight Cycle Graph" file in the "Golden ODEs" folder in the ODE Architect Library.]

23. *Triangle Cycle-Graph*   System (1) has a triangular cycle-graph that is the positive limit set of orbits inside the triangle. Let's explore the regions *outside* the triangle.

    (a) Plot a comprehensive portrait of the orbits of the system in the region $|x| \le 2$, $|y| \le 2$. Mark the six equilibrium points in that region. On the basis of what the nearby orbital arcs look like, label each point as stable or unstable, and name each as a node (stable, unstable?), a saddle, or a spiral point (stable, unstable?).

    (b) Use the Jacobian matrix technique of Section 8.2 to verify that the equilibrium points $(0, 0)$, $(1/4, 1/4)$, $(4/3, 4/3)$ have the stability characteristics claimed in **(a)**.

    (c) *Cycle-Graph Sensitivity: Bifurcation*   Replace the coefficient 3.75 in the first equation of system (1) by the parameter c, but leave the second equation as is. For $c = 4$ and 3.5, plot portraits of the orbits. On the basis of your graphs, what do you think happens to the orbits as c increases through the critical value $c = 3.75$?

Answer 23:

(a) The equilibrium points $(0, 0)$, $(1, 0)$, and $(0, 1)$ appear to be saddle points, while $(1/4, 1/4)$ seems to be an unstable spiral point, $(4/3, 4/3)$ an unstable spiral point, and $(7/4, -3/4)$ a stable node. See the figure.

(b) The Jacobian matrix is

$$
\begin{bmatrix} \partial F_1/\partial x & \partial F_1/\partial y \\ \partial F_2/\partial x & \partial F_2/\partial y \end{bmatrix} = \begin{bmatrix} 1 - 2x - 15y/4 + 4xy + y^2 & -15x/4 + 2x^2 + 2xy \\ 15y/4 - 4xy - y^2 & -1 + 2y + 15x/4 - 2x^2 - 2xy \end{bmatrix}
$$

The respective Jacobian matrices and characteristic polynomials at $(0, 0)$, $(1/4, 1/4)$, and $(4/3, 4/3)$ are

$$
\begin{bmatrix} 1 & 0 \\ 0 & -1 \end{bmatrix}, \quad \begin{bmatrix} -2/16 & -11/16 \\ 10/16 & 3/16 \end{bmatrix}, \quad \begin{bmatrix} 20/9 & 19/9 \\ -35/9 & -4/9 \end{bmatrix}
$$

and $\lambda^2 - 1$, $\lambda^2 - \lambda/16 + 13/32$, and $\lambda^2 - 16\lambda/9 + 585/81$. The eigenvalues of the first matrix are ± 1 [and so a nonlinear saddle at $(0, 0)$]; for the second and third matrices the eigenvalues are complex conjugates with positive real parts [nonlinear unstable spiral points at $(1/4, 1/4)$, $(4/3, 4/3)$].

(c) See Graphs 1 and 2 where $c = 4$ (Graph 1) and $c = 3.5$ (Graph 2). A close observation of the orbits as they are being generated suggests that for $c > 3.75$ one of the spirals from the unstable spiral point at the upper right tends to the equilibrium point $(0, 1)$ on the y-axis as $t \to +\infty$. This means that the "saddle connection" on $x + y = 1$ (when $c = 3.75$) is broken. Similarly, when $c < 3.75$, an outward spiral from the unstable spiral point at the lower left approaches $(1,0)$, again breaking the original saddle connection.

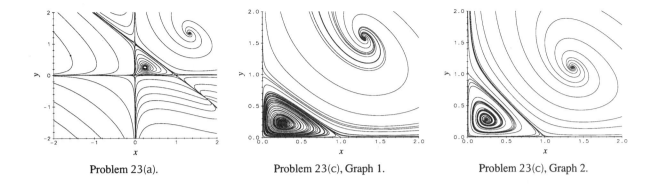

Problem 23(a).   Problem 23(c), Graph 1.   Problem 23(c), Graph 2.

24. *Dulac's Criterion for No Cycles*   Suppose that R is a simply connected region in the plane. Suppose that f, g, and K are continuously differentiable in R. Suppose that $\partial(Kf)/\partial x + \partial(Kg)/\partial y$ has a fixed sign in R. Show that the system $x' = f$, $y' = g$ can't have a cycle or a cycle-graph in R. [*Hint*: read the proof of Bendixson's Negative Criterion. Then adapt the proof to $\partial(Kf)/\partial x + \partial(Kg)/\partial y$.]

Modeling Problems.

25. *Competing Species*   Explain the meaning of each term in the rate functions using the vocabulary of a model for competing species: $x' = x(1 - 0.1x - 0.1y)$, $y' = y(2 - 0.05x - 0.025y)$. Plot orbits in the population quadrant, and then make a complete analysis of the ultimate fate of each species in terms of the initial values of your orbits.

Answer 25:   The system $x' = x(1 - 0.1x - 0.1y)$, $y' = y(2 - 0.05x - 0.025y)$ is a model for competitive interaction with overcrowding. First note that the only equilibrium points in the population quadrant are $(0, 0)$, $(10, 0)$ and $(0, 80)$. The fourth equilibrium point is $(70, -60)$. The mass-action terms $-(x^2 + xy)/10$ and $-(xy + y^2/2)/20$ model the negative effects of large populations on the species' growth rates. See the figure. It appears from the figure that the y species "wins" the competition because all orbits inside the first quadrant tend to the point $x = 0$, $y = 80$, corresponding to the extinction of the x-species. So the positive limit set of every orbit inside the population quadrant appears to be the equilibrium point $(0, 80)$.

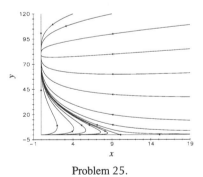

Problem 25.

26. *Ragozin's Negative Criterion for Interacting Species*   Suppose that a two-species interaction is modeled by the system $x' = xF(x, y)$, $y' = yG(x, y)$. The two species are *self-regulating* if $\partial F/\partial x$ and $\partial G/\partial y$ are each negative throughout the population quadrant. If the x-species, say, is self-regulating, then the negativity of $\partial F/\partial x$ implies that the per unit growth rate F diminishes as x increases. Use Dulac's Criterion (Problem 24) to show that the system of ODEs of a pair of self-regulating species has no cycles in the population quadrant. [*Hint*: let $K = (xy)^{-1}$ for $x > 0$, $y > 0$.]

27.   Use ODEs to draw the face of a cat.

> **Answer 27:**   Group project. In carrying out this project, you may want to search the text and the Resource Manual for pictures having bits and pieces that would help to build the face, e.g., a cycle for an eye, a line segment for a whisker, and so on. Then build a set of x and y rate functions using if... then... else statements to place these facial features where you want them. Alternatively, use ODEA and an ODE that yields the circular orbit around the face, then edit out that ODE (but don't erase the orbit!) and enter a new ODE that yields, say, the mouth.

Scaling Time.

28.   Suppose that $x(t)$, $y(t)$ is a solution of the system $x' = f(x, y)$, $y' = g(x, y)$. Rescale the time variable from t to s by setting $dt/ds = h(x(t), y(t))$ for a given function $h(x, y)$ that has a fixed sign. The scaled system has the form $dx/ds = fh$, $dy/ds = gh$.

(a) If $h(x, y)$ is positive everywhere, explain why the two systems have identical orbits with the same orientation induced by time's advance. What if $h(x, y)$ is everywhere negative?

(b) Suppose that $h(x, y)$ is positive everywhere except that h is zero at a point p that lies on a nonconstant orbit Γ of the unscaled system. What happens to the corresponding orbit of the scaled system?

(c) Suppose that $f = x - 10y - x(x^2 + y^2)$ and $g = 10x + y - y(x^2 + y^2)$, while the factor $h = 1 - \exp[-10(x - 1)^2 - 10y^2]$. Explain why the unscaled system has a limit cycle (Graphs 1 and 2) and the scaled system has a cycle-graph (with one vertex and one edge) on the unit circle. Plot the orbit through the point $(0.5, 0)$ and the corresponding component curves for each system (Graphs 1 and 3). Do the cycle and the cycle-graph attract? Explain what you see in these graphs.

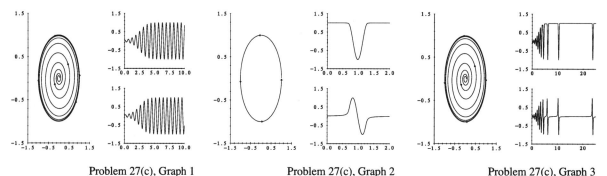

Problem 27(c), Graph 1   Problem 27(c), Graph 2   Problem 27(c), Graph 3

9.3 Bifurcations

Comments:

In the last two or three decades, mathematicians, scientists, and engineers have discovered bifurcations in all sorts of dynamical systems. And it is true that bifurcation analysis often is the easiest way to handle a system where a change in a parameter leads to a sudden change in orbital or solution behavior. Nevertheless, it seems to us that it may be a case of "everything looks like a nail to a person with a hammer." With that caveat, we present in the text and in the problems some of the bifurcations most commonly observed in nonlinear systems (saddle-node, transcritical, pitchfork, Hopf). Of these, the Hopf bifurcation that spawns a limit cycle from an equilibrium point is the most interesting and useful. We present the satiable predator model, but be warned that it is not a very realistic model of population dynamics. Incidentally, you might also think of bifurcations in the context of elementary algebra; for example, the real roots of the quadratic $x^2 + 2ax + 1$ merge and then disappear into the complex plane as the coefficient a decreases from 2 to 1, and then to 0.

PROBLEMS

Saddle-node, Transcritical, Pitchfork Bifurcations. Describe the bifurcations in each system. Find the values of c at which a bifurcation occurs, and identify the bifurcation as of saddle-node, transcritical, or pitchfork type. Saddle-node bifurcations are discussed in the text. As a parameter is changed in a *transcritical bifurcation*, an asymptotically stable node and a saddle point move toward each other, merge, and then emerge with the node now a saddle, and the saddle a node. As a parameter is changed in a *pitchfork bifurcation*, an asymptotically stable equilibrium point suddenly splits into three equilibrium points, two of which are asymptotically stable and the third unstable. Plot graphs of orbits for values of c before, at, and after the bifurcation. Sketch bifurcation diagrams. [*Hint*: use values of c near zero. See Section 2.9.]

1. $x' = c + 10x^2$, $y' = x - 5y$ **2.** $x' = cx - x^2$, $y' = -2y$

3. $x' = cx + 10x^2$, $y' = x - 2y$ **4.** $x' = cx - 10x^3$, $y' = -5y$

5. $x' = cx + x^5$, $y' = -y$

Answer 1: Saddle-node bifurcation. The system is $x' = c + 10x^2$, $y' = x - 5y$. When c is positive, there are no equilibrium points (Graph 1). If $c = 0$, there is a nonelementary saddle-node equilibrium point at the origin because 0 and -2 are the eigenvalues of the Jacobian matrix of this system at the origin (Graph 2). As in Example 9.3.1, the nature of each equilibrium point before and after the bifurcation is determined by the eigenvalues of the Jacobian matrix at that point. According to the Stability table on page 499, the system and its linearization have the same type of equilibrium point as long as the real parts of the eigenvalues of the Jacobian matrix are nonzero. From Graph 3, we see that as c changes from negative values through zero and into the positive range, an asymptotically stable node at $(-\sqrt{-c/10}, -\sqrt{-c/10}/5)$ and an unstable saddle at $(\sqrt{-c/10}, \sqrt{-c/10}/5)$. appear. Thus, we see that as c increases through 0, an asymptotically stable node and an unstable saddle move toward each other, collide (when $c = 0$), and vanish when $c > 0$. When $c = 0$, a bifurcation occurs at the saddle-node at the origin. See Graphs 1–3 where $c = 1, 0, -1$, respectively, and Graph 4 for the associated bifurcation diagram (solid curve corresponds to asymptotically stable equilibrium points, dashed to unstable equilibrium points).

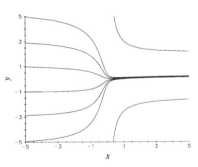

Problem 1, Graph 1.

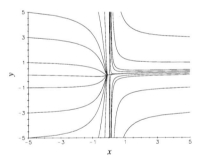

Problem 1, Graph 2.

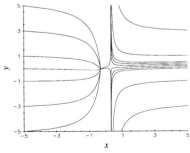

Problem 1, Graph 3.

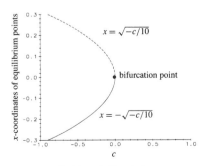

Problem 1, Graph 4.

Answer 3:   Transcritical bifurcation. The system is $x' = cx + 10x^2$, $y' = x - 2y$. The system has an asymptotically stable node at the point $(-c/10, -c/20)$ and a saddle at the origin if c is positive. From Graphs 1–3, we see that if c decreases to 0, then the two equilibrium points merge into a saddle-node at the origin. As c moves away from 0 and becomes negative, the equilibrium point $(-c/10, 0)$ and the origin exchange orbital characters, the origin becoming the asymptotically stable node and $(-c/10, -c/20)$ the saddle. A transcritical bifurcation occurs at $c = 0$. See Graphs 1–3 where $c = -20, 0,$ and 20, respectively, and Graph 4 for the associated transcritical bifurcation diagram, solid for asymptotically stable node, dashed for unstable saddle.

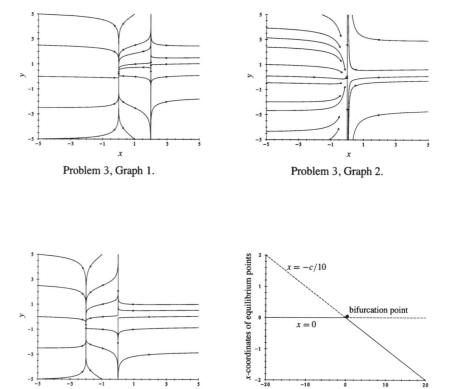

Problem 3, Graph 1.   Problem 3, Graph 2.

Problem 3, Graph 3.   Problem 3, Graph 4.

Answer 5:   Pitchfork bifurcation. The system is $x' = cx + x^5$, $y' = -y$. The system has unstable saddles at $(\pm(-c)^{1/4}, 0)$ and an asymptotically stable node at the origin if c is negative. From Graphs 1–3, we see that if c increases to 0, then the three equilibrium points merge into a nonlinear saddle at the origin. Also, as c moves away from 0 and becomes positive, the equilibrium points $(\pm(-c)^{1/4}, 0)$ vanish and the origin becomes a nonlinear saddle. A bifurcation occurs at $c = 0$. See Graphs 1–3 where $c = -1, 0,$ and 1, respectively, and Graph 4 for the associated pitchfork bifurcation diagram, where the dashed line is the x-coordinate of a *stable* equilibrium point and the solid curves the x-coordinates of *unstable* equilibrium points.

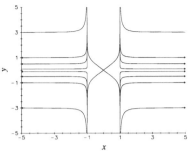

Problem 5, Graph 1.

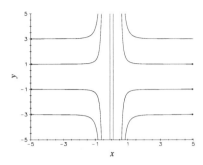

Problem 5, Graph 2.

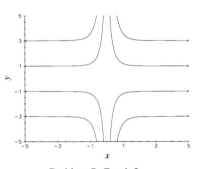

Problem 5, Graph 3.

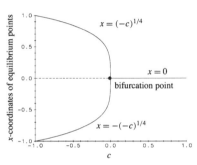

Problem 5, Graph 4.

Hopf Bifurcation. Show that each system in Problems 6–9 experiences a Hopf bifurcation at $c = 0$. Plot orbits before, at, and after the bifurcation. Draw bifurcation diagrams for Problems 6 and 7 showing the y-amplitude of the limit cycle and the y-coordinate of each equilibrium point as functions of c. [*Hint*: write the ODEs in polar coordinates.]

6. $x' = cx + 2y - x(x^2 + y^2)$, $y' = -2x + cy - y(x^2 + y^2)$

7. $x' = cx - 3y - x(x^2 + y^2)^3$, $y' = 3x + cy - y(x^2 + y^2)^3$

8. $x' = y - x^3$, $y' = -x + cy - y^3$

9. *Rayleigh's Equation*   The Rayleigh ODE $z'' + c[(z')^2 - 1]z' + z = 0$, which is equivalent to the system $x' = y$, $y' = -x + c(1 - x^2)y$ if we set $z = x$, $z' = y$.

Answer 7:   The system is $x' = cx - 3y - x(x^2 + y^2)^3$, $y' = 3x + cy - y(x^2 + y^2)^3$. In polar coordinates, the system becomes $r' = r(c - r^6)$, $\theta' = 3$. For $c > 0$ the limit cycle is defined by $r = c^{1/6}$, $\theta = 3t$. In the xy-system we have $\alpha(c) = c$, $\beta(c) = 3$, $\alpha(0) = 0$, $\alpha'(0) = 1$; so the conditions of Theorem 9.3.1 are satisfied. See Graphs 1–3, where $c = -1$, 0, and 1, respectively, and Graph 4 for the associated bifurcation diagram. The solid lines indicate the r-coordinate (or r-amplitude) of an attracting equilibrium point (or cycle), and the dashed line the r-coordinate of an unstable equilibrium point.

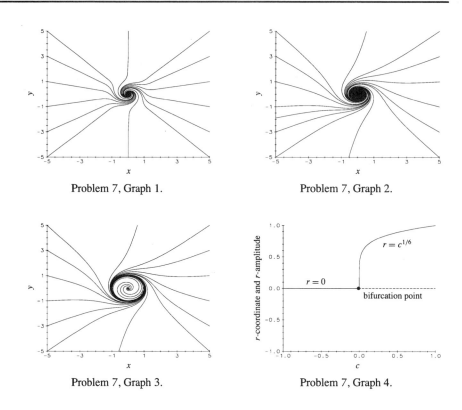

Problem 7, Graph 1.

Problem 7, Graph 2.

Problem 7, Graph 3.

Problem 7, Graph 4.

Answer 9: The Rayleigh system is $x' = y$, $y' = -x + c(1 - x^2)y$. This is *not* a case of Hopf bifurcation (contrary to the statement in the text of the first printing) because at $c = 0$ the system is a harmonic oscillator with a neutrally stable equilibrium point at the origin; all nonconstant orbits are cycles. Nevertheless, for all positive values of c there is a unique attracting limit cycle, and for all negative values of c a unique repelling limit cycle. So a bifurcation occurs at $c = 0$, but it is not a standard Hopf bifurcation. Graphs 1–3 show orbits and component graphs for $c = 1$, 0, and -1, respectively.

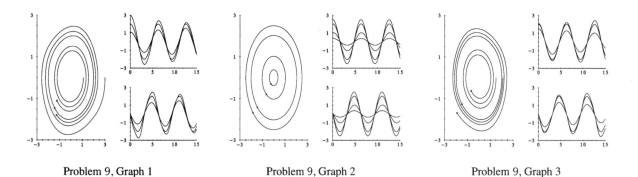

Problem 9, Graph 1   Problem 9, Graph 2   Problem 9, Graph 3

Hopf Bifurcation Investigation.

10. *Subcritical Bifurcation* Show that the origin for $x' = -cx + y + x(x^2 + y^2)$, $y' = -x - cy + y(x^2 + y^2)$ bifurcates from an unstable spiral point to a stable spiral point surrounded by a repelling limit cycle as c increases through zero. Plot orbits and a bifurcation diagram.

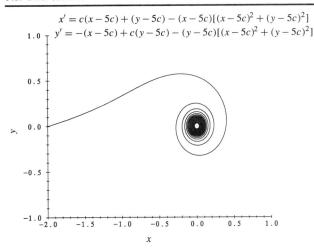

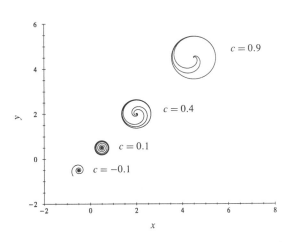

FIGURE 9.3.13: Orbit at the Hopf bifurcation: $c = 0$ [Problem 11(**b**)].

FIGURE 9.3.14: Orbits before and after the Hopf bifurcation [Problem 11(**c**)].

11. *Hopf Bifurcation: Moving Equilibrium Point* The system

$$x' = c(x - 5c) + (y - 5c) - (x - 5c)[(x - 5c)^2 + (y - 5c)^2]$$

$$y' = -(x - 5c) + c(y - 5c) - (y - 5c)[(x - 5c)^2 + (y - 5c)^2]$$

has the unique equilibrium point $P(5c, 5c)$ for each value of c.

(**a**) Rewrite the system in terms of u, v coordinates centered at P: $u = x - 5c$, $v = y - 5c$. Then write the new system in polar coordinates based at $u = 0$, $v = 0$.

(**b**) Use (**a**) and show that the system has a supercritical Hopf bifurcation at $c = 0$. Show that for $c > 0$, the circle of radius $\sqrt{c}$ centered at $P(5c, 5c)$ [in x-, y-coordinates] is an orbit. See Figures 9.3.13 and 9.3.14.

(**c**) Plot orbits for $c = -0.1, 0.1, 0.4, 0.9$ and describe what is happening. [*Hint*: see Figure 9.3.14 where we plot orbits for each of these values of c on a single graph.]

Answer 11: The system is $x' = c(x - 5c) + (y - 5c) - (x - 5c)[(x - 5c)^2 + (y - 5c)^2]$, $y' = -(x - 5c) + c(y - 5c) - (y - 5c)[(x - 5c)^2 + (y - 5c)^2]$. The equilibrium point P has coordinates $x = y = 5c$.

(**a**) Let $u = x - 5c$, $v = y - 5c$. We obtain $u' = cu + v - u(u^2 + v^2)$ and $v' = -u + cv - v(u^2 + v^2)$, In polar form, $r' = cr - r^3$, $\theta' = -1$.

(**b**) Just as in Example 9.3.2, the original system has a supercritical Hopf bifurcation at $c = 0$. In addition, $r = \sqrt{c}$ is an orbit of the above system for $c > 0$, as we see from the rate equation for r. The original system has as an orbit the circle of radius $r = \sqrt{c}$ centered at $P(5c, 5c)$.

(**c**) See Graphs 1–5, for $c = -0.5, 0, 0.5, 1$, and 1.5, respectively. See also Figures 9.3.13 and 9.3.14 in the text. The asymptotically stable equilibrium point at P [for $c < 0$] moves upward and to the right as c increases. At $c = 0$, the point is still asymptotically stable (Graph 2 and Figure 9.3.13), but there is a Hopf bifurcation as c increases beyond 0, a bifurcation to an attracting limit cycle around the destabilized P for $c > 0$. The amplitude of the limit cycle expands as c increases, while the cycle moves upward and to the right. Figure 9.3.14 shows orbits both before and after the Hopf bifurcation. The pictures in Figure 9.3.14 resemble an expanding two-dimensional smoke ring that expands and drifts upwards and to the right as c increases.

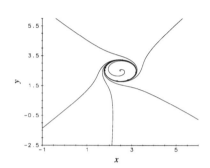

Problem 11(c), Graph 1.

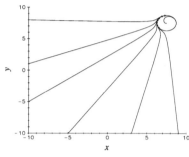

Problem 11(c), Graph 2.

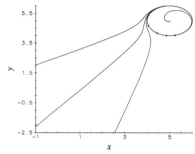

Problem 11(c), Graph 3.   Problem 11(c), Graph 4.   Problem 11(c), Graph 5.

Modeling Problems.

12. *Satiable Predation and Bifurcation*   This problem continues the exploration of Example 9.3.4. Let $a = 0.5$, $d = e = f = 1$, $c = 0.3$, $k = 0.9$, and let b be the bifurcation parameter. Explore what happens as b increases from 0.75 to 3. Any Hopf bifurcations?

13. *The Autocatalator and Bifurcation*   The autocatalator models a chemical reaction in which the concentration of a precursor W decays exponentially, generating a new species X in the process. Species X decays to Y and at the same time reacts autocatalytically with Y, the latter reaction creating more Y than is consumed. Y decays in turn to Z. The rate equations for the concentrations w, x, y, z are

$$w' = -aw, \quad x' = aw - bx - \alpha xy^2, \quad y' = bx + \alpha xy^2 - cy, \quad z' = cy$$

where the rate coefficients a, b, c, and α are positive parameters and the independent variable is time. In what appears to be a kind of Hopf bifurcation, the concentrations of the chemical intermediates X and Y undergo violent oscillations for certain ranges of the parameters and certain levels of species W. The goal of this project is to understand the behavior of the reaction, given various sets of data and parameters. Address the following points:

☞ See Section 7.1
for more on the
autocatalator.

- Treat the system as a nonautonomous planar system by setting $w = w(0)e^{-at}$ and ignoring the rate equation for z. This reduction may help with your solver graphics.

- Why is it reasonable to set $x(0) = y(0) = z(0) = 0$?

- After some initial oscillations in x and y, all four concentrations behave as expected as time increases if $w(0) = 500$, $a = 0.002$, $b = 0.14$, $\alpha = c = 1.0$, $0 \le t \le 1500$. What levels do the four concentrations approach as t becomes large?

- Repeat with $b = 0.08$ instead of 0.14, and $w(0) = 5000$. What is happening here?

- Repeat with $b = 0.02$ and $a = 0.001, 0.003, 0.008$, $\alpha = c = 1.0$; set $w(0) = 5000$. Now what is happening?

- In all of the above, plot component curves. Plot projected orbits in xyz-space or in any other space of three variables selected from t, w, x, y, z. Plot orbits in xy-space. Explain each graph in terms of the behavior of the concentrations.

- Set $a = 0.002$, $b = 0.08$, $\alpha = c = 1.0$, $w(0) = 500$. For various values of b above and below 0.08, plot $x(t)$ against t and $x(t)$ against $y(t)$. Explain what happens in terms of a Hopf bifurcation for the system $x' = C - bx - \alpha xy^2$, $y' = bx + \alpha xy^2 - cy$, where C is a fixed positive constant.

Answer 13: Group project.

9.4 Chaos

Comments:

Finally, the system that started it all! Lorenz's name is firmly attached to the theory of chaos, and deservedly so. It was his remarkable (but characteristically low-key) paper in 1963 that began the current multitude of studies on the implications and the very definition of chaotic dynamics. It should be noted, though, that Mary Cartwright and J. E. Littlewood did pioneering work on the subject in England in 1945. Our focus here (and in the several group projects in the problem set) is on chaotic behavior in ODEs; we have previously noted chaotic dynamics in a discrete map in the WEB SPOTLIGHT ON CHAOS IN NUMERICS. We prove what we can about the behavior of the Lorenz system, but ultimately must use pictures to illustrate the chaotic motion. People often have a hard time seeing where the chaos is, because the pictures seem to show fairly regular oscillations first near one equilibrium point and then near another. However, it is almost impossible to predict just when a given orbit will be near one equilibrium point and when it will move close to the other; moreover, the number of oscillatory whirls about an equilibrium point before moving away also seems to be unpredictable. Chaos is, as noted in the section, a feature of a set of orbits, a so-called strange attractor. These strange attractors are a generalization to higher dimensions of the attracting equilibrium point, cycle, or cycle-graph limit sets of any bounded orbit of a planar autonomous system. We only give a brief mention of what strange attractors and chaotic dynamics on a strange attractor actually mean in a mathematical (rather than an informal "hand-waving") sense. The reason is that the definitions are still in a state of flux—and the students should be so informed. **Warning**: Time-state curves and orbits of the Lorenz system, as well as the Rössler system and the Duffing equation in the SPOTLIGHT ON CHAOTIC SYSTEMS are sensitive to changes in initial data. Since numerically computed solutions are never exact, this means that these solutions may be reasonably accurate only over a short time span. However, the orbits and solutions computed in this section are thought to be near a strange attractor that attracts all nearby solutions. So the computed solutions, orbits, and Poincaré time sections shown here are each most likely composites of several orbits or time-state curves that (at least for the end segment of the solution interval) are essentially in the strange attractor. Although we may not be looking at the actual orbit or time-state curve through the initial point, we do get a pretty good view of the strange attractor.

PROBLEMS _____

Lorenz System. In the following problems you are asked to verify assertions made in this section about the equilibrium points of the Lorenz system (1).

1. Show that the origin and the points P_1, P_2 given in (3) are the only equilibrium points (the latter two only for $r > 1$).

2. Construct the Jacobian matrix of the Lorenz system at the origin and show that the eigenvalues are as given in (2). Prove that all eigenvalues of this Jacobian matrix are negative if $0 < r < 1$ but that one eigenvalue is positive if $r > 1$.

3. Construct the Jacobian matrix of the Lorenz system at P_1 and at P_2 for $r > 1$. Show that its eigenvalues are roots of the cubic given in (4). Explain why one root is negative for all $r > 1$, while the other two roots are negative or have negative real parts if and only if r satisfies (5).

4. Show that a cubic $\lambda^3 + A\lambda^2 + B\lambda + C$, $B > 0$, has a pair of pure imaginary roots if $C = AB$. Apply this result to the Lorenz system with $\sigma = 10$, $b = 8/3$ to find the value of r for which the Jacobian matrix at P_1 and P_2 has pure imaginary eigenvalues.

Answer 1: The equilibrium points P_1 and P_2 are found by equating the right sides of the equation of the Lorenz system (1) to zero and solving for x, y, and z.

Answer 3: At P_1 (and P_2), the eigenvalues of the Jacobian matrix are the roots of $\lambda^3 + (1 + b + \sigma)\lambda^2 + (\sigma + r)b\lambda + 2\sigma b(r - 1)$. According to Problem 17 in Section 6.6 all roots of this cubic have negative real parts if and only if the coefficients are positive (which they are because b and σ are positive and $r > 1$) *and* if $(1 + b + \sigma)(\sigma + r)b > 2\sigma b(r - 1)$. This equality reduces to $\sigma(\sigma + b + 3) > r(\sigma - b - 1)$, i.e., $r < r_c = \sigma(\sigma + b + 3)/(\sigma - b - 1)$ if $\sigma > b + 1$. P_1 and P_2 are locally asymptotically stable if $1 < r < r_c$ since all the eigenvalues have negative real parts for these values of r, while if $r > r_c$ both P_1 and P_2 are unstable.

 5. The aim of this group project is to generate a gallery of portraits of orbits, projections, and their component curves for several values of r (set $\sigma = 10$, $b = 8/3$). Explain what you see. Look particularly for periodic orbits, period-doubling sequences, and strange attractors. Be sure to check sensitivity to changes in initial data. Find a copy of Colin Sparrow's book (see footnote 6) and try to duplicate some of his pictures. If your ODE solver is not adequate, replace the Lorenz system by the discrete approximation that uses Euler's Method: $x_{n+1} = x_n + h(-\sigma x_n + \sigma y_n)$, $y_{n+1} = y_n + h(rx_n - y_n - x_n z_n)$, $z_{n+1} = z_n + h(-bz_n + x_n y_n)$, where h is the step size. The Euler approximation is surprisingly revealing for $h = 0.01$. Try to duplicate some of the figures in the text using your own solver. We guarantee that, although your graphs start out looking like ours, after a while they will be very different. This is a reflection of the sensitivity of the solutions of the Lorenz system to the slightest change, whether in initial data, parameters, or in the numerical method itself. It is not that your solver is more accurate than ours (or the reverse); it is only that it is different.

Answer 5: Group project.

Background Material: Verification of the Lorenze Squeeze

To show the squeeze property we argue informally along the following lines. Suppose S is a bounded region of state space with positive volume $V(S)$ and bounding surface ∂S, and suppose that ∂S is smooth, connected, and closed. For example, S could be a solid ball with spherical surface ∂S. Let ∂S_t denote the bounding surface of S_t, and let $V(S_t)$ denote the volume of the region S_t. As t increases from 0, S_t and ∂S_t will look less and less like $S = S_0$ and $\partial S = \partial S_0$. Interpreting the Lorenz equations as the equations of motion of the particles of a gas or a fluid, we have that

$$\frac{dV(S_t)}{dt} = \left\{ \begin{array}{c} \text{net volume of fluid} \\ \text{crossing } S_t \text{ in the} \\ \text{unit outward normal} \\ \text{direction per unit of} \\ \text{time} \end{array} \right\} = \iint_{\partial S_t} \mathbf{F} \cdot \mathbf{n} \, dA \qquad \text{(i)}$$

where $\mathbf{F}$ is the vector field defined by the right-hand side of the Lorenz system [text system (1)], $\mathbf{n}$ is a unit outward normal vector field to the bounding surface ∂S_t, and the integral is a surface integral. By the Divergence Theorem of vector calculus (Theorem B.2.14), the surface integral becomes the volume integral of the divergence over S_t:

$$\iint_{\partial S_t} \mathbf{F} \cdot \mathbf{n} \, dA = \iiint_{S_t} \nabla \cdot \mathbf{F} \, dV \tag{ii}$$

The divergence of $\mathbf{F}$ is given by

$$\nabla \cdot \mathbf{F} = \frac{\partial F_1}{\partial x} + \frac{\partial F_2}{\partial y} + \frac{\partial F_3}{\partial x}$$

where F_1, F_2, F_3 are given by the right-hand sides of the equations of text system (1). We have that $\nabla \cdot \mathbf{F} = -\sigma - 1 - b$. From (i) and (ii) we have that

$$\frac{dV(S_t)}{dt} = \iiint_{S_t} -(\sigma + 1 + b) \, dV = -(\sigma + 1 + b)V(S_t) \tag{iii}$$

Solving the scalar ODE $V' = -(\sigma + 1 + b)V$, with initial condition $V(0) = V(S)$ when $t = 0$, we have that

$$V(S_t) = V(S)e^{-(\sigma+1+b)t} \tag{iv}$$

We see that as $t \to \infty$, $V(S_t) \to 0$.

Spotlight on Chaotic Systems

PROBLEMS

 1. *Rössler System*   The Rössler system[1] is

$$x' = -y - z, \qquad y' = x + ay, \qquad z' = b - cz + xz$$

where a, b, and c are positive parameters. O. E. Rössler wanted to build the simplest possible system that would model the phenomena displayed by the slightly more complicated Lorenz system. The system (now named after the inventor) is the result. In the parameter region of interest, this model of a model has two equilibrium points and one nonlinear term. There appears to be a folded strange attractor, spiraling, chaotic wandering, and period doubling. There are sequences of values of a that (for $b = 2$ and $c = 4$) correspond to period doubling and apparently terminate with the creation of a strange attractor. Study the orbits and component curves of the Rössler system as one of the three parameters is varied. For most of your study, fix the parameters b and c at the respective values 2 and 4, and vary a. Look for period-doubling sequences, perhaps interspersed with values of a for which there are orbits of other periods. Check for chaotic spiraling, and especially for the folded strange attractor. Plot orbits in xyz-space from different points of view. Plot orbital projections on various state-variable planes and use component plots to detect periodic orbits. If possible, plot only over an end segment of the time interval for which the system is actually solved (e.g., if the system is solved for $0 \le t \le 1000$, plot for $900 \le t \le 1000$). This way any transient behavior is likely to disappear and the orbit will be nearly in the strange attractor. Other suggestions:

---

[1]We adapted the description of the Rössler system, the scroll circuit (see the WEB SPOTLIGHT ON CHAOS IN A NONLINEAR CIRCUIT), and Duffing's equation with permission from R. L. Borrelli, C. Coleman, and W. E. Boyce, *Differential Equations Laboratory Workbook* (New York: John Wiley & Sons, 1992).

- Plot orbits for $x(0) = z(0) = 0$, $y(0) = 2$, $b = 2$, $c = 4$, $a = 0.410, 0.400, 0.395$. Explain what you see. The IVP is solved for $0 \leq t \leq 1000$, but the orbits shown are plotted only for $800 \leq t \leq 1000$, so initial transients have died out.

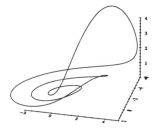

$a = 0.410$

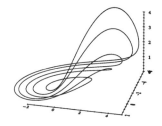

$a = 0.400$

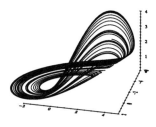

$a = 0.395$

- Plot orbits and component curves for the sequence of values of $a = 0.3, 0.35, 0.375, 0.386, 0.3909, 0.398, 0.4, 0.411$. Use $x_0 = 0$, $y_0 = 2$, and $z_0 = 0$ and identify periodic orbits and their periods, period doubling, chaotic spiraling, and any other significant phenomena. What happens if you use different initial data?

- Find the equilibrium points of the Rössler system if a, b, and c are positive constants. For $b = 2$, $c = 4$, and the values of a given above, study orbital behavior near each equilibrium point.

- Explain why, if $|z|$ is very small, the subsystem $x' = -y$, $y' = x + ay$ has an unstable spiral point at the origin if $0 < a < 2$. Explain why this leads to a spreading apart of nearby orbits (a central part of chaotic wandering). This spreading is not unbounded, however. Explain from the third equation of the Rössler system how if $x < c$, the coefficient of z is negative and the z-system stabilizes near $b/(c - x)$. If $x > c$, and $b > 0$, the z-system diverges. Now look at the x-system and explain why orbits alternately lie in a plane parallel to the xy-plane, then are thrown upward, and in turn are folded back and reinserted into the plane, but closer to the origin.

- Vary b and c as well as a and analyze the effect on the orbits. Any period doubling, folding, chaotic wandering? Explain.

Answer 1: Group project.

  2. *Duffing's Equation*   Chaos in autonomous differential systems needs at least three dimensions, and the autonomous Lorenz, Rössler, and scroll systems all have three state variables. However, a somewhat different kind of chaotic wandering may appear with just two state variables as long as the system is nonautonomous; time itself is the third dimension. Duffing's ODE and system and the ODE and system of the driven pendulum (Problem 3) provide examples of apparently chaotic behavior in two state dimensions with a sinusoidal driving term that is periodic in time.

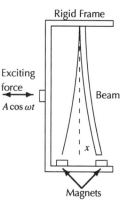

Model the motion of a long, thin and elastic steel beam with its top end embedded in a pulsating frame and the bottom end hanging just above two magnets (see the margin figure) by the *driven damped Duffing's equation*, $x'' + cx' - ax + bx^3 = A \cos \omega t$, where a, b, c, A, and ω are nonnegative constants. The equivalent *driven damped Duffing's system* is $x' = y$, $y' = ax - bx^3 - cy + A \cos \omega t$. Orbits of the system display a remarkable variety of behaviors, ranging from oscillatory, to periodic, to chaotic. For some sets of values of the parameters and some sets of initial data, orbits approach a strange attractor. **Warning!** Numerical solutions of chaotic systems over long time spans are almost certain to be inaccurate. Nevertheless, the shadowing property mentioned in footnote 7 gives us some confidence that what the numerical computations reveal actually do approximate reality. The sources cited below give additional background.[2]

---

[2] J. Guckenheimer and P. Holmes, *Nonlinear Oscillations, Dynamical Systems, and Bifurcations of Vector Fields* (New York: Springer-Verlag, 1990); P. Holmes and F. C. Moon, *J. Appl. Mech.* **108** (1983), pp. 1021–1032; J. M. T. Thompson and H. B. Stewart, *Nonlinear Dynamics and Chaos* 2nd ed., (New York: John Wiley & Sons, 2001).

- Analyze and graph the orbits of an undriven, undamped Duffing system: $x' = y$, $y' = x - x^3$. Find the three equilibrium points. Find an orbital figure-eight cycle-graph that consists of the origin and two other orbits. Why are all but five orbits periodic? See Figure (**a**). Explain why the period of the orbit inside a lobe tends to $\sqrt{2}\pi$ as the amplitude of the orbit tends to zero while the period tends to $+\infty$ as the orbit tends to the boundary of the lobe. Analyze the periods of the orbits outside the lobes.

- Add damping and obtain the system $x' = y$, $y' = x - x^3 - 0.15y$. Orbits are graphed for $x_0 = 1$, $0.8 \le y_0 \le 1.0$ in Figure (**b**). Some orbits tend to $(-1, 0)$, others to $(1, 0)$ as $t \to +\infty$. Fix x_0 at 1 and plot orbits for several values of y_0 from 0 to 5.0. For each y_0 solve over the t-interval $0 \le t \le 200$. Do you see any pattern?

- Now add a periodic driving force: $x' = y$, $y' = x - x^3 - 0.15y + 0.3\cos t$. Solve the system with initial data $x(0) = -1$, $y(0) = 1$ over the interval $0 \le t \le 1000$, but plot over $500 \le t \le 1000$. Describe the chaotic tangle [see Figure (**c**)]. Are there any periodic orbits? [*Hint*: try an initial point $x(0) \approx -0.2$, $y(0) \approx 1.4$.]

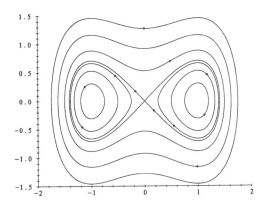

(**a**)

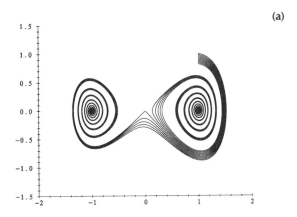

(**b**)

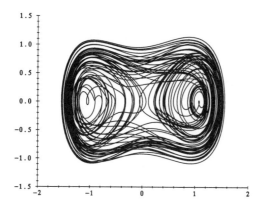

(c)

 3. *Driven Pendulum*   The system, $x' = y$, $y' = -ay - b\sin x + c\cos(\omega t) + d$, models a sinusoidally driven pendulum with a constant torque. Explain the model. Search for chaotic wandering, and plot and discuss your results.

Answer 3:   Group project.

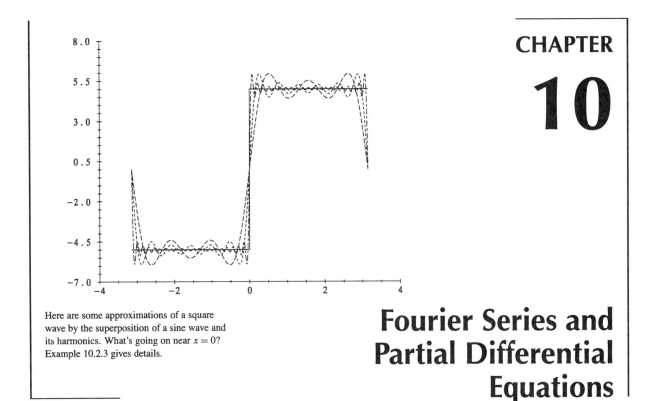

CHAPTER

10

Here are some approximations of a square wave by the superposition of a sine wave and its harmonics. What's going on near $x = 0$? Example 10.2.3 gives details.

Fourier Series and Partial Differential Equations

This chapter begins with the derivation of a boundary/initial value problem for the PDE which models a vibrating guitar string and the development of a method for constructing a solution formula for that problem. In between we lay the groundwork for deriving solution formulas for a class of linear boundary/initial value problems. The intervening sections contain material on Fourier series and Sturm-Liouville problems that is needed to solve the boundary/initial value problems of this chapter.

Section 10.1 was written to be as self-contained as possible in order to provide a brief introduction and example of the Method of Separation of Variables. One could quit at this point and still come away with a good idea of what this method is all about. The SPOTLIGHT ON APPROXIMATION OF FUNCTIONS sets the stage for Fourier Series in order to give a clean introduction to Fourier Trigonometrical Series in Sections 10.2 and 10.3. Sturm-Liouville Problems are treated in a unified way in Section 10.5. It is a short section and should be read before reading Section 10.6 on eigenfunction expansions and the Method of Separation of Variables. In Section 10.4 we discuss a model PDE for heat conduction, and apply the eigenfunction techniques for the heat equation to discuss the changing temperatures in a rod and in an underground storage cellar. In Section WEB SPOTLIGHT ON LAPLACE'S EQUATION we consider Laplace's PDE as a model for steady-state temperatures in a body, given the temperatures at all the boundary points. Fourier series, eigenfunction expansions, and Separation of Variables underlie all we do in this chapter.

10.1 Vibrations of a Guitar String

Comments:

We use the PDE of a vibrating string as the introduction to PDEs, separation of variables, and Fourier series because the twanging of a guitar string is familiar to everyone. Our approach is to start with a derivation of the wave equation for the transverse motion of the string, introduce the three different kinds of linear boundary conditions and the corresponding boundary operators, and then focus on separating the PDE to get the standing waves modulated by sinusoidal functions of time. We work out the details of a boundary problem that involves fixed endpoints (e.g., the guitar string fastened at each end). We show how the coefficients are calculated by the Fourier method, but dodge all questions of convergence.

Everything is straightforward, but the problems tend to be somewhat lengthy. If you don't have access to a CAS, use integral tables or other sources to calculate the integrals. You need to develop early on a standard format for doing a separation of variables construction. Otherwise you will lose your way in the lengthy process. Be sure to look at the steps leading up to (7) where the variables are separated, the separation constant λ is introduced, and separate ODEs are found (one with boundary conditions).

PROBLEMS

Classify the PDE.   Identify each PDE as a Laplace, wave, or heat equation.

   1. $10u_{xx} - u_{tt} = 0$ **2.** $3u_t - 5(u_{xx} + u_{yy})$ **3.** $u_{xx} = -u_{yy} - u_{zz}$

   Answer 1: $10u_{xx} - u_{tt} = 0$, or $u_{tt} = 10u_{xx}$. This is a wave equation.

   Answer 3: $u_{xy} = -u_{yy} - u_{zz}$, or $u_{xx} + u_{yy} + u_{zz} = 0$, or $\nabla^2 u = 0$. This is a three space dimensional Laplace equation.

Solutions.   Verify directly that the listed functions are solutions of the given PDEs.

   4. $u_1 = \sin(x - 5t)$, $u_2 = 8\sin 3x \cos 15t$, $u_3 = \cos(10t - 2x)$; $u_{tt} = 25u_{xx}$

   5. $u_1 = e^{2y} \cos 2x$, $u_2 = e^{5y} \sin 5x$, $u_3 = 3u_1 + 20u_2$; $\nabla^2 u = 0$

   6. $u_1 = e^{-4t} \sin 2x$, $u_2 = e^{-25t} \cos 5x$, $u_3 = 7u_1 - 2u_2$; $u_t = u_{xx}$

   7. $u_1 = F(x + 8t) + G(x - 8t)$, where F and G are any twice continuously differentiable functions of one variable; $u_{tt} - 64u_{xx} = 0$

   Answer 5: We show by direct substitution that these functions solve the PDE $u_{xx} + u_{yy} = 0$. For $u_1 = e^{2y} \cos 2x$:

$$u_{xx} = -4e^{2y} \cos 2x, \quad u_{yy} = 4e^{2y} \cos 2x$$

For $u_2 = e^{5y} \sin 5x$:

$$u_{xx} = -25e^{5y} \sin 5x, \quad u_{yy} = 25e^{5y} \sin 5x$$

For $u_3 = 3u_1 + 20u_2$:

$$u_{xx} = -12e^{2y} \cos 2x - 500e^{5y} \sin 5x \quad u_{yy} = 12e^{2y} \cos 2x + 500e^{5y} \sin 5x$$

   Answer 7: We show by direct substitution that this function solves the PDE. Let $r = x + 8t$ and

$s = x - 8t$. Then

$$u_t = \frac{dF}{dr} \cdot \frac{dr}{dt} + \frac{dG}{ds} \cdot \frac{ds}{dt} = F'(r) \cdot 8 + G'(s) \cdot (-8)$$

$$u_{tt} = 8F''(r)\frac{dr}{dt} - 8G''(s)\frac{ds}{dt} = 64F''(r) + 64G''(s)$$

$$u_x = \frac{dF}{dr} \cdot \frac{dr}{dx} + \frac{dG}{ds} \cdot \frac{ds}{dx} = F'(r) + G'(s)$$

$$u_{xx} = F''(r)\frac{dr}{dx} + G''(s)\frac{ds}{dx} = F''(r) + G''(s)$$

Thus we have

$$u_{tt} - 64u_{xx} = 64F''(r) + 64G''(s) - 64(F''(r) + G''(s)) = 0$$

Separation of Variables: Guitar String. Suppose that the guitar string modeled by IBVP (9) has length $L = \pi$, $c = 1$, $u(x, 0) = f(x)$, and $u_t(x, 0) = 0$. Find the formal solution in the form of the series (17) with specific values for the coefficients A_n given in (18).

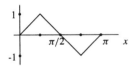

8. $f(x) = \sin x$

9. $f(x) = 0.1 \sin 2x$

10. $f(x) = \sum_{n=1}^{10} (\sin nx)/n$

11. $f(x) = \begin{cases} x/10, & 0 \le x \le \pi/2 \\ \pi/10 - x/10, & \pi/2 \le x \le \pi \end{cases}$

12. $f(x) = x(\pi - x)$

13. The graph of $f(x)$ is sketched in the margin.

Answer 9: Let $f(x) = 0.1 \sin 2x$. Then by text formula (20) $A_2 = (2/\pi) \int_0^\pi 0.1 \sin 2x \sin 2x \, dx = (2/\pi) \cdot 0.1 \cdot (\pi/2) = 0.1$, and $A_n = 0$ for $n = 1, 3, 4, \ldots$, so $u(x, t) = 0.1 \sin 2x \cos 2t$. Notice that $B_n = 0$, for $n = 1, 2, \cdots$, because $g(x) = 0$ for $0 \le x \le L$.

Answer 11: The solution of the guitar string problem is given by the series

$$u(x, t) = \sum_{n=1}^{\infty} A_n \sin nx \cos nt$$

where (see the Table of Integrals on the inside back cover of the text)

$$A_n = \frac{2}{\pi} \int_0^\pi f(x) \sin nx \, dx$$

$$= \frac{2}{\pi} \int_0^{\pi/2} \frac{x}{10} \sin nx \, dx + \frac{2}{\pi} \int_{\pi/2}^\pi \left[\frac{\pi}{10} - \frac{x}{10} \right] \sin nx \, dx$$

$$= \frac{1}{5\pi} \left[\frac{-x \cos nx}{n} + \frac{\sin nx}{n^2} \right]\Big|_0^{\pi/2} + \frac{1}{5} \left[\frac{-\cos nx}{n} \right]\Big|_{\pi/2}^\pi - \frac{1}{5\pi} \left[\frac{-x \cos nx}{n} + \frac{\sin nx}{n^2} \right]\Big|_{\pi/2}^\pi$$

Note that $A_n = 0$, if n is even, and that if $n = 2k + 1$, $k = 0, 1, 2, \cdots$ then $\sin(2k + 1)\pi/2 = (-1)^k$, and we can also write the solution as

$$u(x, t) = \sum_{k=0}^{\infty} \frac{2(-1)^k}{5\pi(2k + 1)^2} \sin(2k + 1)x \cos(2k + 1)t$$

Answer 13:   For the graph in the margin, we have

$$f(x) = \begin{cases} \dfrac{4x}{\pi} & 0 \le x \le \pi/4 \\[2mm] -\dfrac{4x}{\pi}+2 & \pi/4 < x \le 3\pi/4 \\[2mm] \dfrac{4x}{\pi}-4, & 3\pi/4 < x \le \pi \end{cases}$$

Then the solution of the guitar problem is

$$u(x,t) = \sum_{n=1}^{\infty} A_n \sin nx \cos nt$$

where (see the Table of Integrals on the inside back cover of the textbook)

$$A_n = \frac{2}{\pi}\int_0^{\pi} f(x)\sin nx\,dx$$

$$= \frac{8}{\pi^2}\int_0^{\pi/4} x\sin nx\,dx - \frac{8}{\pi^2}\int_{\pi/4}^{3\pi/4} x\sin x\,dx + \frac{4}{\pi}\int_{\pi/4}^{3\pi/4}\sin nx\,dx$$

$$+\frac{8}{\pi^2}\int_{3\pi/4}^{\pi} x\sin nx\,dx - \frac{8}{\pi}\int_{3\pi/4}^{\pi}\sin nx\,dx$$

$$= \frac{8}{\pi^2}\left[\frac{-x\cos nx}{n}+\frac{\sin nx}{n^2}\right]\Big|_0^{\pi/4} - \frac{8}{\pi^2}\left[\frac{-x\cos nx}{n}+\frac{\sin nx}{n^2}\right]\Big|_{\pi/4}^{3\pi/4}$$

$$+\frac{4}{\pi}\left[\frac{-\cos nx}{n}\right]\Big|_{\pi/4}^{3\pi/4} + \frac{8}{\pi^2}\left[\frac{-x\cos nx}{n}+\frac{\sin nx}{n^2}\right]\Big|_{3\pi/4}^{\pi} - \frac{8}{\pi}\left[\frac{-\cos nx}{n}\right]\Big|_{3\pi/4}^{\pi}$$

$$= \frac{16}{\pi^2 n^2}\left(\sin(n\pi/4)-\sin(3n\pi/4)\right)$$

Hence the solution is

$$u(x,t) = \frac{16}{\pi^2}\sum_{n=1}^{\infty}\left[\frac{\sin\dfrac{n\pi}{4}-\sin\dfrac{3n\pi}{4}}{n^2}\right]\sin nx \cos nt$$

Partial Sum: Guitar String.   Graph the partial sum of the first four terms of the series for $u(x,t)$ at $t=0$.

14.   Problem 11   **15.**   Problem 13

Answer 15:   See the figure for the graph of the first four nonzero terms of the series at $t=0$ of the function $u(x,t)$ in Problem 13:

$$\frac{16}{\pi^2}\left(2\sin 2x - \frac{\sin 6x}{18} + \frac{\sin 10x}{50} - \frac{\sin 14x}{98}\right)$$

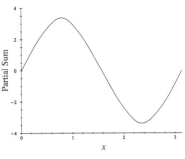

Problem 15.

 A computer algebra system (CAS) would come in handy here to do the calculation and graphing.

16. *Two Profiles: Guitar String*   For IBVP (23) graph an approximate profile of the string at each of the given times. [*Hint*: use partial sums of the series in (25).]

(a) $t_1 = 0$ (b) $t_2 = 0.2$ (c) $t_3 = 0.3$ (d) $t_4 = 0.4$

Properties of Solutions: Guitar String.

17. Show that the initial shape $f(x)$ of the plucked guitar string returns after the passage of $T = 2L/c$ units of time. [*Hint*: evaluate the series solution (21) at $t = 0$ and then at $T = 2L/c$.]

> **Answer 17:**   We saw that the solution $u(x, t)$ of the plucked guitar string of length L and initial data $u(x, 0) = f(x), u_t(x, 0) = g(x), 0 \le x \le L$, is
>
> $$u(x, t) = \sum_{n=1}^{\infty} \left(A_n \sin \frac{n\pi x}{L} \cos \frac{n\pi ct}{L} + B_n \sin \frac{n\pi x}{L} \sin \frac{n\pi ct}{L} \right)$$
>
> where
>
> $$A_n = \frac{2}{L} \int_0^L f(x) \sin \frac{n\pi x}{L} \, dx, \quad B_n = \frac{2}{n\pi c} \int_0^L g(x) \sin \frac{n\pi x}{L} \, dx$$
>
> Now observe that
>
> $$u(x, 0) = \sum_{n=1}^{\infty} A_n \sin \frac{n\pi x}{L} = f(x)$$
>
> Assume that the series converges. Here is the profile of the string at $T = 2L/c$:
>
> $$u(x, 2L/c) = \sum_{n=1}^{\infty} \left(A_n \sin \frac{n\pi x}{L} \cos \frac{n\pi c(2L/c)}{L} + B_n \sin \frac{n\pi x}{L} \sin \frac{n\pi c(2L/c)}{L} \right)$$
>
> $$= \sum_{n=1}^{\infty} A_n \sin \frac{n\pi x}{L} = f(x)$$
>
> This establishes the claim.

18. Show that we can write the solution (21) of the plucked guitar string problem as the sum of two traveling waves, one moving to the right and one to the left, both with speed c. [*Hint*: use the trigonometric identity $\sin \alpha \sin \beta = [\cos(\alpha - \beta) - \cos(\alpha + \beta)]/2$ with $\alpha = n\pi x/L$ and $\beta = n\pi ct/L$ for each term in series (21).]

19. Consider the guitar string in IBVP (9), where c is a positive constant. Explain in mathematical terms why the tauter the string, the higher the pitch. Now replace the string by a denser string and pluck it again. In mathematical terms, what do you hear now?

> **Answer 19:**   Since c^2 is the ratio of the tension τ in the string to the density ρ of the string, c increases if τ is increased (i.e., the strip is made tauter), and c decreases if the mass (and so, presumably, the density) is increased. The frequency of a standing wave is nc, so the frequency (i.e., the pitch) is higher if the tension is increased and lower if the density is increased.

General Solution of the One-Dimensional Wave Equation.

20. Show that the linear change of variables $r = x + ct$, $s = x - ct$, where c is a positive constant, takes the PDE $u_{tt} = c^2 u_{xx}$ into the equivalent PDE $u_{rs} = 0$ [*Hint:* write the change of variables formula as $r = r(x, t)$, $s = s(x, t)$. We can think of the dependent variable u as a function of x and t, denoted by $u(x, t)$, or as a function of r and s, denoted by $u(r, s)$. These two functions $u(x, t)$ and $u(r, s)$ are related by the substitution $u(x, t) = u(r(x, t), s(x, t))$. By the chain rule $u_x = u_r r_x + u_s s_x = u_r + u_s$, and $u_t = u_r r_t + u_s s_t = c u_r - c u_s$. Use the chain rule again to find u_{xx} and u_{tt}.]

21. *Traveling Waves* Show that the general solution of the PDE $u_{rs} = 0$ is $u(r, s) = F(r) + G(s)$. Show that the general solution of the PDE $u_{tt} = c^2 u_{xx}$ is $u(x, t) = F(x + ct) + G(x - ct)$, where F and G are twice continuously differentiable functions of a single variable. [*Hint:* use the result of Problem 20 and the Antidifferentiation Theorem.]

Answer 21: From Problem 20 we saw that the PDE $u_{tt} = c^2 u_{xx}$ is equivalent to the PDE $u_{rs} = 0$ if the independent variables in the two PDEs are related as follows: $r = x + ct$, and $s = x - ct$. Now let u be a solution of the PDE $u_{rs} = 0$. Then the Antiderivative Theorem and the fact that $(u_r)_s = 0$ for all s and any r implies that u_r is a constant for each fixed value of r. hence, there is a function $f(r)$ such that $u_r = f(r)$. Let $F(r)$ be an antiderivative of $f(r)$, then since $(u - F(r))_r = 0$, it follows from the Antiderivative Theorem that $u - F(r)$ is a constant for each fixed s, but this constant is a function of s which we denote by $G(s)$. Thus, the general solution of the PDE $u_{rs} = 0$ is $u = F(r) + G(s)$ where F and G are arbitrary twice continuously differentiable functions. Hence, $u(x, t) = F(x + ct) + G(x - ct)$ is the general solution of the PDE $u_{tt} = c^2 u_{xx}$.

Initial Value Problem.

☞ The WEB SPOTLIGHT ON INITIAL VALUE PROBLEMS has more on this topic.

22. *D'Alembert's Formula* Suppose that c is a positive constant and $f(x)$ and $g(x)$ are any twice continuously differentiable functions for all x. Show that the initial value problem

$$u_{tt} - c^2 u_{xx} = 0, \qquad \text{for all } x, \ t > 0$$
$$u(x, 0) = f(x), \quad u_t(x, 0) = g(x), \quad \text{for all } x$$

has the unique solution

☞ This is *D'Alembert's Formula.*

$$u(x, t) = \frac{f(x + ct) + f(x - ct)}{2} + \frac{1}{2c} \int_{x-ct}^{x+ct} g(s)\, ds$$

[*Hint:* from Problem 21, we know that the general solution of the wave equation is $u(x, t) = F(x - ct) + G(x + ct)$, where F and G are any twice continuously differentiable functions. From the initial conditions show that $F'(x) = (1/2)f'(x) - (1/(2c))g(x)$ and $G'(x) = (1/2)f'(x) + (1/(2c))g(x)$ for all x. Integrate both sides of each equation from zero to x to find $F(x)$ and $G(x)$. Observe that $F(0) + G(0) = f(0)$.]

23. *Infinitely Long Vibrating String* Suppose that $g(x) = 0$, x in $\mathbb{R}$, $c = 1$ and $f(x) = x$ for $0 \le x \le \pi/2$, $f(x) = \pi - x$ for $\pi/2 \le x \le \pi$, and $f(x) = 0$ for all other values of x. Plot the profiles of the solution $u(x, t)$ in Problem 22 for $t = 0, 1, 3, 5, 10$.

Answer 23: With the given data, the solution to the infinite string problem is given the solution formula in Problem 22:

$$u(x, t) = \frac{1}{2}\left(f(x + t) + f(x - t)\right)$$

The requested profiles are $u(x, 0) = \frac{1}{2}(f(x) + f(x)) = f(x)$, $u(x, 1) = \frac{1}{2}(f(x+1) + f(x-1))$, $u(x, 3) = \frac{1}{2}(f(x+3) + f(x-3))$, $u(x, 5) = \frac{1}{2}(f(x+5) + f(x-5))$, $u(x, 10) = \frac{1}{2}(f(x+10) + f(x-10))$. See the figure for the graphs of $u(x, 0) = f(x)$, $u(x, 1)$ and $u(x, 5)$ which are, respectively, the solid, long-dashed, and short-dashed curves.

[*Hint:* use graph paper and plot these profiles by hand. Here is how to plot $u(x, 1)$, for example: take the triangular pulse and divide it by 2 to obtain $g(x) = f(x)/2$. Shift $g(x)$ to the right by one unit and plot it. Add these two shifted graphs together graphically to obtain the profile $u(x, 1)$.]

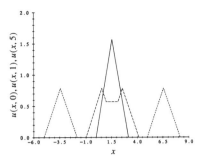

Problem 23.

Boundary Value Problem.

24. Verify directly that the eigenvalue problem (14) has no nontrivial solutions when either $\lambda = 0$, or $\lambda > 0$: [*Hint*: use formulas for all solutions of $X'' - \lambda X = 0$ in each case.]

Orthogonality Relations.

25. Verify the integration formulas (20). [*Hint*: use the following trigonometric identity: $\sin \alpha \sin \beta = [\cos(\alpha - \beta) - \cos(\alpha + \beta)]/2$, for all α, β.]

Answer 25: Using the identity $\sin \alpha \sin \beta = [\cos(\alpha - \beta) - \cos(\alpha + \beta)]/2$, we see that

$$\int_0^L \sin \frac{n\pi x}{L} \sin \frac{m\pi x}{L} = \frac{1}{2} \int_0^L \cos \frac{(n-m)\pi x}{L} dx - \frac{1}{2} \int_0^L \cos \frac{(n+m)\pi x}{L} dx$$

If $m = n = 0$, the integrand and the integral are zero. Now, if $n \neq m$, then the right-hand side becomes

$$\frac{1}{2} \frac{L}{(n-m)\pi} \sin \frac{(n-m)\pi x}{L} \Big|_0^L - \frac{1}{2} \frac{L}{(n+m)\pi} \sin \frac{(n+m)\pi x}{L} \Big|_0^L = 0$$

On the other hand, if $n = m \neq 0$, then the right-hand side of the identity becomes

$$\frac{1}{2} \int_0^L dx - \frac{L}{4n\pi} \sin \frac{2n\pi x}{L} \Big|_0^L = \frac{L}{2}$$

Uniqueness Theorem.

 26. Suppose that R is the region $0 < x < L$, $t > 0$, and that $F(x, t)$ is continuous on the closed region $0 \leq x \leq L$, $t \geq 0$, that $\varphi(t)$ and $\mu(t)$ are continuous on $t \geq 0$, and that $f(x)$ and $g(x)$ are continuous on $0 \leq x \leq L$. Show that the problem below can't have more than one solution $u(x, t)$ that is twice continuously differentiable in R with first derivatives which are continuous to the boundary of R:

(PDE) $u_{tt} = c^2 u_{xx} + F(x, t)$ in R
(BC) $u(0, t) = \varphi(t),$ $u(L, t) = \mu(t),$ $t \geq 0$
(IC) $u(x, 0) = f(x),$ $u_t(x, 0) = g(x),$ $0 \leq x \leq L$

[*Hint*: follow the outline below.]

- Suppose the given problem has two solutions u and v. Put $U = u - v$ and show that U satisfies the same problem, but with the data F, φ, μ, f, and g all set equal to zero.

- Define the function $W(t) = \frac{1}{2} \int_0^L (U_t^2 + c^2 U_x^2) \, dx$, for any $t \geq 0$. Use the fact that $(U_x U_t)_x = U_{xx} U_t + U_x U_{tx}$ to show that

$$W'(t) = c^2 [U_x(L, t) U_t(L, t) - U_x(0, t) U_t(0, t)]$$

- Show that $U_t(L, t) = 0$ and $U_t(0, t) = 0$ for $t \geq 0$, so $W'(t) = 0$ for $t \geq 0$. [*Hint*: use the definition of partial derivatives.]

- Show that $U_t(x, t) = 0$ and $U_x(x, t) = 0$ for all (x, t) in R, so $U = $ constant on R.

- Show that actually $U = 0$ on R and conclude that $u = v$ on R.

Modeling Problem.

 27. *Damped Vibrations of a Guitar String*   IBVP (9) does not take into account forces such as air resistance that resist the transverse motion of the string. Add a term to (PDE) that models resistance. [*Hint*: add a term such as ru_t, where r is a small positive constant.] Then use the Method of Separation of Variables to find a formal solution of your modified version of IVP (9). Put damping into the model of Example 10.1.2 for the plucked string released from rest and find the corresponding solution formula. Interpret the behavior of the string (as described by the formula) as time increases. Is that behavior realistic? Then specialize to the initial and boundary data of Example 10.1.2, select a value for the coefficient r and plot partial sums of the series solution for several values of t. What do the plots reveal about the motion of the damped guitar string?

Answer 27:   A separated solution of the PDE $u_{tt} + b^2 u_t - a^2 u_{xx} = 0$ has the form $u = X(x)T(t)$ where

$$X(x)T''(t) + b^2 X(x)T'(t) - a^2 X''(x)T(t) = 0$$

Dividing by $a^2 XT$ and separating, we have that

$$\frac{X''}{X} = \frac{T'' + b^2 T'}{a^2 T} = \lambda$$

where λ is a separation constant. Imposing the boundary conditions we have the Sturm-Liouville Problem

$$X'' = \lambda X, \quad X(0) = X(L) = 0$$

and the ODE

$$T'' + b^2 T' - \lambda a^2 T = 0$$

We have that $X_n = C_n \sin(n\pi x/L)$, $\lambda_n = -n^2\pi^2/L^2$, $n = 1, 2, \ldots$. Recalling that we require that $b^2 < 2\pi a/L$, then for all integers $n = 1, 2, \ldots$, the roots of the quadratic $r^2 + b^2 r - \lambda_n a^2$ are the complex conjugates $-b^2/2 \pm i\beta_n$, where $\beta_n = \sqrt{-\lambda_n a^2 - b^4/4}$. Then we have that

$$T_n(t) = e^{-b^2 t/2} [A_n \cos \beta_n t + B_n \sin \beta_n t]$$

and the solution has the form

$$u(x, t) = e^{-b^2 t/2} \sum_{1}^{\infty} \sin \frac{n\pi}{L} x [A_n \cos \beta_n t + B_n \sin \beta_n t]$$

Now set $t = 0$ to obtain

$$u(x, 0) = f(x) = \sum_{1}^{\infty} A_n \sin \frac{n\pi}{L} x$$

and

$$u_t(x, 0) = 0 = \sum_{1}^{\infty} \sin \frac{n\pi}{L} x \left[-\frac{b^2 A_n}{2} + \beta_n B_n \right]$$

We have that

$$A_n = \frac{2}{L} \int_0^L f(x) \sin(n\pi x/L)\, dx$$

$$B_n = b^2 A_n / (2\beta_n)$$

and we are done.

Here is an interesting aside to the answer of Problem 18

Background Material: The Wave Equation

Say that the formal series $\sum_{n=1}^{\infty} A_n \sin(n\pi x/L)$ converges to $f(x)$ on $0 \le x \le L$, and that the series $\sum_{n=1}^{\infty} (n\pi c/L) B_n \sin(n\pi x/L)$ converges to $g(x)$ on $0 \le x \le L$.

If $f_{\text{odd}}(x)$ and $g_{\text{odd}}(x)$ are the odd extensions of $f(x)$ and $g(x)$, respectively, onto $-L \le x \le L$, and if $\tilde{f}(x)$ and $\tilde{g}(x)$ are the periodic extensions, respectively, of f_{odd} and g_{odd} into $\mathbb{R}$, then

$$\sum_{n=1}^{\infty} A_n \sin \frac{n\pi x}{L} = \tilde{f}(x), \quad \text{all } x \text{ in } \mathbb{R}$$

$$\sum_{n=1}^{\infty} \frac{n\pi c}{L} B_n \sin \frac{n\pi x}{L} = \tilde{g}(x), \quad \text{all } x \text{ in } \mathbb{R}$$

This follows from the fact that $\sin(n\pi x/L)$ is odd on $-L \le x \le L$ and periodic on $\mathbb{R}$. Hence,

$$\frac{1}{2} \sum_{n=1}^{\infty} A_n \sin \frac{n\pi}{L}(x+ct) = \frac{1}{2}\tilde{f}(x+ct), \quad \frac{1}{2} \sum_{n=1}^{\infty} A_n \sin \frac{n\pi}{L}(x-ct) = \frac{1}{2}\tilde{f}(x-ct), \quad \text{all } x \text{ and } t$$

Also for all x and t,

$$\frac{1}{2} \sum_{n=1}^{\infty} B_n \left(\cos \frac{n\pi}{L}(x-ct) - \cos \frac{n\pi}{L}(x+ct) \right)$$

$$= \frac{1}{2} \sum_{n=1}^{\infty} \frac{n\pi}{L} B_n \int_{x-ct}^{x+ct} \sin s \, ds$$

$$= \frac{1}{2c} \int_{x-ct}^{x+ct} \sum_{n=1}^{\infty} \frac{n\pi c}{L} B_n \sin s \, ds = \frac{1}{2c} \int_{x-ct}^{x+ct} \tilde{g}(s) \, ds$$

Hence, we can write the solution $u(x, t)$ of the guitar string problem IBVP (9) as

$$u(x, t) = \frac{1}{2} \left(\tilde{f}(x+ct) + \tilde{f}(x-ct) \right) + \frac{1}{2c} \int_{x-ct}^{x+ct} \tilde{g}(s) \, ds$$

10.2 Fourier Trigonometric Series

Comments:

The basic theorems of "classical" Fourier Trigonometric Series are introduced in this section and in its problem set for the orthogonal set $\Phi = \{1, \cos \frac{\pi x}{T}, \sin \frac{\pi x}{T}, \ldots, \cos \frac{n\pi x}{T}, \sin \frac{n\pi x}{T}, \ldots\}$ in the Euclidean space PC$[-T, T]$. Pointwise convergence, uniform convergence, decay estimates—all of this is here along with the verification. This section and the previous two form a unit that can be used for self-study of the basics of the mathematical approach to Fourier Series.

PROBLEMS

Find Fourier Series Without Integration. Use trigonometric identities, common sense, or previous results to find the Fourier series on $[-\pi, \pi]$ of each function below *without* using the Fourier–Euler formulas (8).

1. $5 - 4\cos 6x - 7\sin 3x - \sin x$

2. $\cos^2 7x + (\sin \frac{1}{2}x)(\cos \frac{1}{2}x) - 2\sin x \cos 2x$

3. $\sin^3 x$

Answer 1:   The rearrangement $5 - \sin x - 7\sin 3x - 4\cos 6x$ already has the form of a Fourier Series (7), with $T = \pi$. So $A_0 = 5$, $B_1 = -1$, $B_3 = -7$, $A_6 = -4$, and all other coefficients are zero.

Answer 3:   Using trigonometric identities in the table on the inside back cover, we see that $\sin^3 x = \sin x(1 - \cos 2x)/2 = (\sin x - \sin x \cos 2x)/2 = (\sin x)/2 - (\sin 3x - \sin x)/4 = (3\sin x)/4 - (\sin 3x)/4$ which has the form of text formula (7) with $T = \pi$. So $B_1 = 3/4$, $B_3 = -1/4$, and all other Fourier coefficients are zero.

Fourier Series.   Calculate FS[f] for each function $f(x)$, $-\pi \leq x \leq \pi$. See the margin figures for Problems 10 and 11.

<u>4.</u> $2x - 2$ **5.** x^2 **6.** $a + bx + cx^2$, a, b, c constants

7. $\sin \pi x$ **8.** $|x| + e^x$ **9.** $|x(x^2 - 1)|$

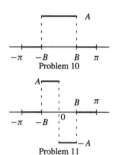

Problem 10

Problem 11

10. *Rectangular Pulse* For $0 < B \leq \pi$, $f(x) = \begin{cases} A, & |x| < B \\ 0, & B \leq |x| \leq \pi \end{cases}$

11. *Alternating Pulse* For $0 < B \leq \pi$, $f(x) = \begin{cases} 0, & B \leq |x| \leq \pi \\ A, & -\quad B < x < 0 \\ -A, & 0 < x < B \end{cases}$

Answer 5:   For $f(x) = x^2$, note that $A_0 = \int_{-\pi}^{\pi} x^2 dx/(2\pi) = \pi^2/3$, and for $k \geq 1$,

$$A_k = \frac{1}{\pi}\int_{-\pi}^{\pi} x^2 \cos kx\, dx = \frac{2}{\pi}\int_0^{\pi} x^2 \cos kx\, dx = \frac{4(-1)^k}{k^2}$$

where we have used the Table of Integrals on the inside back cover of the textbook. All the coefficients $B_k = 0$, $k \geq 1$ since $f(x) = x^2$ is an even function on $[-\pi, \pi]$. We have

$$FS[x^2] = \frac{\pi^2}{3} + 4\sum_{k=1}^{\infty} \frac{(-1)^k}{k^2}\cos kx$$

Answer 7:   Here $f(x) = \sin \pi x$. All the coefficients A_k, $k \geq 0$, vanish since $f(x) = \sin \pi x$ is an odd function on $[-\pi, \pi]$. Using an identity from the table on the inside back cover for $\sin \alpha \sin \beta$, we have

$$B_k = \frac{1}{\pi}\int_{-\pi}^{\pi} \sin \pi x \sin kx\, dx = \frac{1}{\pi}\int_0^{\pi} [\cos(\pi - k)x - \cos(\pi + k)x]dx$$

$$= \frac{1}{\pi}\left[\frac{\sin(\pi - k)x}{\pi - k} - \frac{\sin(\pi + k)x}{\pi + k}\right]\Big|_{x=0}^{x=\pi} = \frac{(-1)^k 2k \sin \pi^2}{\pi(\pi^2 - k^2)}$$

We have

$$FS[\sin \pi x] = \frac{2\sin \pi^2}{\pi}\sum_{k=1}^{\infty}\frac{(-1)^k k}{\pi^2 - k^2}\sin kx$$

Answer 9:   Since $f(x) = |x^3 - x|$, $-\pi \leq x \leq \pi$, is an even function about $x = 0$, it follows that FS[f] contains only cosine terms. Note that

$$A_0 = \frac{1}{2\pi}\int_{-\pi}^{\pi} |x^3 - x|dx = \frac{1}{2\pi}\int_0^1 (x - x^3)\, dx - \frac{1}{2\pi}\int_1^{\pi}(x^3 - x)\, dx = \frac{2 - 2\pi^2 + \pi^4}{4\pi}$$

Similarly, splitting the integral at $x = 1$ we have that

$$A_k = \frac{2}{\pi}\int_0^{\pi} |x^3 - x|\cos kx\, dx$$

$$= \frac{2}{\pi k^4}\left[(12 - 4k^2)\cos k + 12k\sin k - 2k^2 - 6 + (-1)^k k^2(3\pi^2 - 1) - 6(-1)^k\right]$$

With these values for A_0, A_k, we have

$$\text{FS}[|x^3 - x|] = A_0 + \sum_1^\infty A_k \cos kx$$

Answer 11: Note that the given alternating pulse f is an odd function, so FS[f] contains only sine terms. Observe that for $k \geq 1$

$$B_k = \frac{2}{\pi} \int_0^B (-A) \sin kx \, dx = \frac{2A}{\pi} \frac{\cos kx}{k} \Big|_0^B = \frac{2A}{k\pi}(-1 + \cos kB)$$

so

$$\text{FS}[f] = \frac{2A}{\pi} \sum_{k=1}^\infty \frac{-1 + \cos kB}{k} \sin kx$$

Graphs of Partial Sums of Fourier Series. Graph $f(x)$ and the partial sum of the first four nonzero terms of FS[f](x). In Problem 14 take $a = b = c = 1$. In Problems 18 and 19 take $A = 1$, $B = \pi/2$). Is the partial sum a good approximation to $f(x)$?

12. Problem 4	**13.** Problem 5	**14.** Problem 6	**15.** Problem 7
16. Problem 8	**17.** Problem 9	**18.** Problem 10	**19.** Problem 11

Answer 13: See the figure.

Answer 15: See the figure.

Answer 17: See the figure.

Answer 19: See the figure.

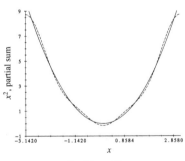

Problem 13.

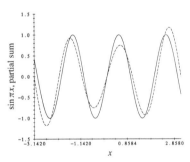

Problem 15.

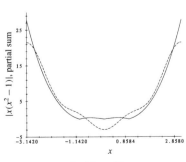

Problem 17.

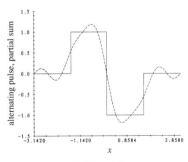

Problem 19.

Fourier Series.   Find the Fourier series of each function on the indicated interval.

20. $f(x) = \sin x, \quad -\pi \le x \le \pi$

21. $f(x) = \sin x, \quad -\pi/2 \le x \le \pi/2$ [*Hint*: use (7), (8) with $L = \pi/2$.]

22. $f(x) = \sin x, \quad -3\pi/2 \le x \le 3\pi/2$ [*Hint*: use (7), (8) with $L = 3\pi/2$.]

Answer 21:   According to text formulas (7), (8) with $T = \pi/2$, the Fourier Series of the function $f(x) = \sin x, -\pi/2 \le x \le \pi/2$, has the form

$$\text{FS}[f] = A_0 + \sum_{k=1}^{\infty} [A_k \cos 2kx + B_k \sin 2kx]$$

where

$$A_0 = \frac{1}{\pi} \int_{-\pi/2}^{\pi/2} \sin x \, dx, \quad A_k = \frac{2}{\pi} \int_{-\pi/2}^{\pi/2} \sin x \cos 2kx \, dx \quad k \ge 1,$$

$$B_k = \frac{2}{\pi} \int_{-\pi/2}^{\pi/2} \sin x \sin 2kx \, dx, \quad k \ge 1$$

Since f is odd about $x = 0$, $A_k = 0$ for all $k \ge 0$. Using an identity from the table on the inside back cover for $\sin \alpha \sin \beta$, we have

$$B_k = \frac{2}{\pi} \int_0^{\pi/2} (\cos(2k-1)x - \cos(2k+1)x) \, dx$$

$$= \frac{2}{\pi} \left\{ \frac{\sin(2k-1)x}{2k-1} - \frac{\sin(2k+1)x}{2k+1} \right\} \Big|_0^{\pi/2}$$

$$= \frac{2}{\pi} \left\{ \frac{-(-1)^k}{2k-1} - \frac{(-1)^k}{2k+1} \right\} = \frac{8(-1)^k k}{\pi(1 - 4k^2)}$$

The corresponding Fourier Series is

$$\text{FS}[\sin x] = \frac{8}{\pi} \sum_{k=1}^{\infty} \frac{(-1)^k k}{1 - 4k^2} \sin 2kx$$

which is very different from the Fourier Series found in Problem 20. The change of interval from $[-\pi, \pi]$ to $[-\pi/2, \pi/2]$ caused the orthogonal set to change [from $\sin kx$, $\cos kx$, and so on, to $\sin 2kx$, $\cos 2kx$, and so on], and this makes all the difference.

Properties of Fourier Series.

23.   Let $f(x)$ be the function $|x|$ on $[-\pi, \pi]$.

     (a) Find FS[f]. Then use Theorem 10.2.2 with $x_0 = 0$ to show that

$$\frac{\pi^2}{8} = \sum_{k=0}^{\infty} \frac{1}{(2k+1)^2}$$

     (b) Discuss the convergence properties of FS[f] on all of $\mathbb{R}$.

Answer 23:

(a) The function $f(x) = x$, $|x| \le \pi$ has the Fourier series (see Example 10.2.2 with $L = \pi$)

$$\text{FS}[|x|] = \frac{\pi}{2} - \frac{4}{\pi} \sum_{m=0}^{\infty} \frac{\cos(2m+1)x}{(2m+1)^2}$$

The Pointwise Convergence Theorem 10.2.2 implies that $\text{FS}[|x|]_{x=0}$ converges to $\frac{1}{2}[|0^+| + |0^-|] = 0$, and the asserted relation follows.

(b) Since the periodic extension $\tilde{f}$ of $f(x) = |x|$, $-\pi \leq x \leq \pi$, into $\mathbb{R}$ is in $\mathbf{C}^0(\mathbb{R})$ and is also piecewise smooth, the Pointwise Convergence Theorem (Theorem 10.2.2) states that FS[$|x|$] converges uniformly on $\mathbb{R}$ to $\tilde{f}$.

Method of Separation of Variables: Guitar String. Solve the IBVP

$$
\begin{aligned}
u_{tt} &= c^2 c_{xx}, & 0 < x < L, t > 0 \\
u(0,t) &= u(L,t) = 0, & t \geq 0 \\
u(x,0) &= f(x), \quad u_t(x,0) = g(x), & 0 \leq x \leq L
\end{aligned}
$$

for the given initial conditions f, g. Interpret it in terms of a vibrating guitar string.

24. $f(x) = 0$, $\quad g(x) = 3\sin(\pi x/L)$, $\quad 0 \leq x \leq L$

25. $f(x) = g(x) = \begin{cases} x, & 0 \leq x \leq L/2 \\ L - x, & L/2 \leq x \leq L \end{cases}$

26. $f(x) = x(L - x) = -g(x)$, $\quad 0 \leq x \leq L$

Answer 25: Here $f(x) = g(x) = \begin{cases} x, & 0 \leq x \leq L/2 \\ L - x, & L/2 \leq x \leq L \end{cases}$. Using (17) and (22) in Section 10.1, we have that for $n \geq 1$,

$$
A_n = \frac{2}{L}\int_0^{L/2} x\sin\frac{n\pi x}{L}dx + \frac{2}{L}\int_{L/2}^L (L-x)\sin\frac{n\pi x}{L}dx = \frac{4L}{n^2\pi^2}\sin\frac{n\pi}{2}
$$

and

$$
B_n = \frac{2}{n\pi c}\int_0^{L/2} x\sin\frac{n\pi x}{L}dx + \frac{2}{n\pi c}\int_{L/2}^L (L-x)\sin\frac{n\pi x}{L}dx = \frac{4L^2}{n^3\pi^3 c}\sin\frac{n\pi}{2}
$$

The solution to the problem is

$$
u(x,t) = \sum_{n=1}^{\infty}\left(\frac{4L}{n^2\pi^2}\cos\frac{n\pi ct}{L} + \frac{4L^2}{n^3\pi^3 c}\sin\frac{n\pi ct}{L}\right)\sin\frac{n\pi}{2}\sin\frac{n\pi x}{L}
$$

Separated Solutions and Separation Constants.

27. Construct separated solutions $X(x)T(t)$ associated with the IBVP

$$
\begin{aligned}
u_{tt} - c^2 u_{xx} &= 0, & 0 < x < L, \ t > 0; \\
u(0,t) = 0, \ u_x(L,t) &= -hu(L,t), & t \geq 0, \ h \text{ a positive constant}; \\
u(x,0) = f(x), \ u_t(x,0) &= g(x), & 0 \leq x \leq L.
\end{aligned}
$$

and show that the separation constants $\lambda_n = -s_n^2/L^2$, $n = 1, 2, \ldots$, where s_n is the nth consecutive positive zero of the function $s + hL\tan s$ (do not evaluate s_n). Find a corresponding basis of eigenfunctions for PC[0, L].

Answer 27: The problem is $u_{tt} - c^2 u_{xx} = 0$, $0 < x < L$, $t > 0$; $u(0,t) = 0$, $u_x(L,t) = -hu(L,t)$, $t \geq 0$; $u(x,0) = f(x)$, $u_t(x,0) = g(x)$, $0 < x < L$.

(a) The SL problem is as follows: find all constants λ with the property that $X'' = \lambda X$ has a nontrivial solution X in $\mathbf{C}^2[0, L]$ such that $X(0) = 0$ and $X'(L) + hX(L) = 0$.

(b) First, try $\lambda = k^2$, $k > 0$. Then $X'' = k^2 X$ has the general solution $X = A\cosh kx + B\sinh kx$. The boundary conditions imply that $A = 0$ and $B(k\cosh kL + h\sinh kL) = 0$. Since B cannot vanish, it must be that $k > 0$ satisfies the equation $k\cosh kL + h\sinh kL = 0$. But this is impossible since h, $\cosh kL$, and $\sinh kL$ are all positive. There are no positive eigenvalues. Next try $\lambda = 0$. The general solution of $X'' = 0$ is $X = Ax + B$. The boundary conditions imply that $A = B = 0$, so $\lambda = 0$ is also not an eigenvalue. Now put $\lambda = -k^2$, $k > 0$. The general solution of $X'' + k^2 X = 0$ is $X = A\sin kx + B\cos kx$. The condition $X(0) = 0$ implies that $B = 0$, and the condition at $x = L$ implies that $k\cos kL + h\sin kL = 0$. Putting $s = kL$, the defining relation for k becomes the relation $s + hL\tan s = 0$ for s. Writing the relation as $\tan s = -s/(hL)$ and graphing each side against s for $s > 0$, we obtain the plot shown in Fig. 2(b) [$hL = 1$ in the plot]. From the graph we see that the relation $s + hL\tan s = 0$ has an infinite sequence of positive zeros $s_1, s_2, \ldots$. So $\lambda_n = -n^2 = -s_n^2/L^2$, $n = 1, 2, \ldots$, are the eigenvalues of our Sturm-Liouville Problem. Note that $(n - 1/2)\pi < s_n < n\pi$ for all n, and $s_n - (n - 1/2)\pi \to 0$ as $n \to \infty$.

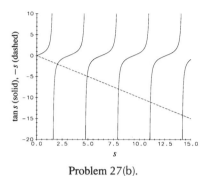

Problem 27(b).

28. *Damped Wave Equation*   The PDE $u_{tt} + b^2 u_t - c^2 u_{xx} = 0$, $0 < x < L$, $t > 0$, models a damped vibrating string. Find the formal solution $u(x, t)$ of the IBVP

$$
\begin{aligned}
&u_{tt} + b^2 u_t - c^2 u_{xx} = 0, && 0 < x < L, \quad t > 0 \\
&u(0, t) = u(L, t) = 0, && t \geq 0 \\
&u(x, 0) = f(x), \quad u_t(x, 0) = 0, && 0 \leq x \leq L
\end{aligned}
$$

where $b^2 < 2\pi c/L$, c and L are positive constants and f and g belong to PC[0, L].

10.3 Half-Range and Exponential Fourier Series

Comments:

The Fourier techniques of half-range and complex exponential expansions are "bread-and-butter" methods in doing applications. It is worthwhile pointing out the advantages of using the complex basis functions e^{ikx}: the Fourier-Euler coefficients formulas are a lot easier to remember, and one doesn't have to compute Fourier sine and cosine coefficients separately. The flip side of the coin, however, is that without the often tedious simplification at the end of the process, the Fourier Series of a real function may appear to be complex-valued.

PROBLEMS

Fourier Sine and Cosine Series.   Find the Fourier Sine Series FSS[f] on the interval [0, π] for each function $f(x)$. Then find the Fourier cosine series FCS[f] on [0, π].

1. $f(x) = 1$ **2.** $f(x) = \sin x$ **3.** $f(x) = \pi - x$ **4.** $f(x) = \sin 2x$

Answer 1:   The function $f(x) = 1$.

$$
\text{FSS}[1] = \sum_{k=0}^{\infty} b \sin kx, \quad b_k = \frac{2}{\pi} \int_0^{\pi} 1 \cdot \sin kx \, dx
$$

So $b_k = -2(\cos kx)/k\pi \Big|_{x=0}^{x=\pi} = 2[1 - \cos k\pi]/k\pi$, and we have $b_k = 0$ if k is even, and $b_k = 4/k\pi$ if k is odd. Then

$$\text{FSS}[1] = \frac{4}{\pi} \sum_{\text{odd } k} \frac{1}{k} \sin kx$$

We have FCS[1]=FS[1]=1 since $\Phi_c = \{1, \cos x, \ldots, \cos nx, \ldots\}$.

Answer 3: For $f(x) = \pi - x, 0 \leq x \leq \pi$, we have that

$$\text{FSS}[f] = \sum_{n=1}^{\infty} b_n \sin nx$$

where for $n \geq 1$,

$$b_n = \frac{2}{\pi} \int_0^{\pi} (\pi - x) \sin nx \, dx$$

$$= \frac{2}{\pi} \left[\frac{-\pi \cos nx}{n} \right]_0^{\pi} - \frac{2}{\pi} \left[\frac{-x \cos nx}{n} + \frac{\sin nx}{n^2} \right]_0^{\pi}$$

$$= \frac{-2(-1)^n}{n} + \frac{2}{n} + \frac{2(-1)^n}{n} = \frac{2}{n}$$

$$\text{FSS}[f] = \sum_{n=1}^{\infty} \frac{2}{n} \sin nx$$

Now

$$\text{FCS}[f] = a_0 + \sum_{n=1}^{\infty} a_n \cos nx$$

where

$$a_0 = \frac{1}{\pi} \int_0^{\pi} (\pi - x) \, dx = \frac{1}{\pi} \left(\pi x - \frac{x^2}{2} \right) \Big|_0^{\pi} = \frac{\pi}{2}$$

and for $n \geq 1$,

$$a_n = \frac{2}{\pi} \int_0^{\pi} (\pi - x) \cos nx \, dx$$

$$= 2 \left[\frac{\sin nx}{n} \right]_0^{\pi} - \frac{2}{\pi} \left[\frac{x \sin nx}{n} + \frac{\cos nx}{n^2} \right]_0^{\pi}$$

$$= \frac{-2(-1)^n}{\pi n^2} + \frac{2}{\pi n^2} = \frac{2}{\pi n^2}(-(-1)^n + 1) = \begin{cases} 0, & \text{if } n = \text{ even} \\ \dfrac{4}{\pi n^2}, & \text{if } n = \text{ odd} \end{cases}$$

$$\text{FCS}[f] = \frac{\pi}{2} + \sum_{k=1}^{\infty} \frac{4\cos(2k-1)x}{\pi(2k-1)^2}$$

Fourier Exponential Series. Find the Fourier exponential series of $f(x)$, $|x| \leq \pi$.

5. $f(x) = \cos x$ **6.** $f(x) = x^2$ **7.** $f(x) = e^{ix}$ **8.** $f(x) = e^{\pi i x}$

9. $f(x) = |x| + ix$ [*Hint*: show that FS[$|x|$] = $(\pi/2) - \pi^{-1} \sum_{k=1}^{\infty} (e^{i2kx} + e^{-i2kx})/(2k^2)$.]

Answer 5: We want to find the Fourier Exponential Series of $f(x) = \cos x$. FS[f] = $\sum_{k=-\infty}^{\infty} c_k e^{ikx}$, where

$$c_k = \frac{1}{2\pi} \int_{-\pi}^{\pi} \cos x e^{-ikx} \, dx = \frac{1}{2\pi} \int_{-\pi}^{\pi} \cos x (\cos kx - i \sin kx) \, dx$$

which is zero, unless $k = 1$ or -1. We have that $c_1 = \pi/(2\pi) = 1/2 = c_{-1}$. So $FS[f] = (e^{ix} + e^{-ix})/2 = \cos x$, which we should have expected in the first place.

Answer 7: For $f(x) = e^{ix}$, $FS[f] = \sum_{n=-\infty}^{\infty} c_n e^{inx}$ where

$$c_n = \frac{1}{2\pi} \int_{-\pi}^{\pi} e^{ix} e^{-inx}\, dx = \frac{1}{2\pi} \int_{-\pi}^{\pi} e^{ix(1-n)}\, dx$$

If $n = 1$, then $c_1 = 1$, and if $n \neq 1$, then

$$c_n = \frac{1}{2\pi} \left(\frac{e^{ix(1-n)}}{i(1-n)} \right)\Big|_{-\pi}^{\pi} = \frac{1}{2\pi}\left[\frac{e^{i\pi(1-n)} - e^{-i\pi(1-n)}}{i(1-n)} \right] = \frac{\sin(1-n)\pi}{\pi(1-n)} = 0$$

Hence, $FS[f] = e^{ix}$.

Answer 9: Here $f(x) = |x| + ix$. We shall find $FS[x]$ and $FS[|x|]$ and combine the results to find $FS[|x| + ix]$. First, observe that $(2\pi)^{-1} \int_{-\pi}^{\pi} x\, dx = 0$, and that for $k \neq 0$,

$$\frac{1}{2\pi} \int_{-\pi}^{\pi} x e^{-ikx}\, dx = -\frac{1}{2\pi i k} \left\{ x e^{-ikx}\Big|_{-\pi}^{\pi} - \int_{-\pi}^{\pi} e^{-ikx}\, dx \right\} = \frac{i}{k}(-1)^k$$

So $FS[x] = i\sum_{k=1}^{\infty}((-1)^k/k)(e^{ikx} - e^{-ikx})$. Next, $(2\pi)^{-1}\int_{-\pi}^{\pi}|x|\,dx = (\pi)^{-1}\int_0^{\pi}x\,dx = \pi/2$, and that for $k \neq 0$,

$$\frac{1}{2\pi}\int_{-\pi}^{\pi}|x|e^{-ikx}\,dx = -\frac{1}{2\pi}\int_{-\pi}^{\pi}\{|x|\cos kx - i|x|\sin kx\}\,dx$$

$$= \frac{1}{\pi}\int_0^{\pi} x\cos kx\,dx = \frac{(-1)^k - 1}{\pi k^2}.$$

So, $FS[|x|] = \pi/2 + \pi^{-1}\sum_{k=1}^{\infty}[(-1)^k - 1]k^{-2}(e^{ikx} + e^{-ikx})$. Hence,

$$FS[|x| + ix] = \frac{\pi}{2} + \sum_{k=1}^{\infty}\left[\frac{(-1)^k - 1}{\pi k^2}(e^{ikx} + e^{-ikx}) - \frac{(-1)^k}{k}(e^{ikx} - e^{-ikx})\right]$$

Convergence of Points of Discontinuity. Find the value of the Fourier series of the given function f at the given points.

10. $FCS[f](x)$, f is given in Example 10.3.2, $x = n, n = \pm 1, \pm 2, \cdots$

11. $FSS[f](x)$, f is given in Example 10.3.1, $x = 2k+1, k = 0, \pm 1, \pm 2, \cdots$.

12. $FS[g](x)$, g in Example 10.3.3, $x = n\pi, n = 0, \pm 1, \pm 2, \cdots$.

13. $FSS[f](x)$, f in Example 10.3.3, $x = n\pi, n = 0, \pm 1, \pm 2, \cdots$.

Answer 11: Let $FSS[f]$ be the Fourier Sine series in Example 10.3.1 for the function $f(x) = 0$, $0 \leq x \leq 1$, and $f(x) = 1, 1 < x \leq 2$. Extend f into $[-2, 2]$ as an odd function (which we denote by f_{odd}). Then $FS[f_{odd}] = FSS[f]$ and hence we can use Theorem 10.2.2 to examine the convergence of $FSS[f]$. Looking at the graph of $\tilde{f}_{odd}$ we see that

$$FSS[f](n) = \begin{cases} -1/2, & n = -1+4k, & k = 0, \pm 1, \cdots \\ 0, & n = 4k, & k = 0, \pm 1, \cdots \\ 1/2, & n = 1+4k, & k = 0, \pm 1, \cdots \\ 0, & n = 2+4k, & k = 0, \pm 1, \cdots \end{cases}$$

Answer 13: From Example 10.3.3, $f(x) = e^x, 0 \leq x \leq \pi$. Extend f into $[-\pi, \pi]$ as an odd function (which we denote by f_{odd}). Then $FS[f_{odd}] = FSS[f]$ and hence we can use Theorem 10.2.2 to examine the pointwise convergence of $FSS[f]$. Looking at the graph of $\tilde{f}_{odd}$ we have

$$FSS[f](n\pi) = 0, \quad n = 0, \pm 1, \cdots$$

More Fourier Series.

14. *Another Kind of Fourier Sine Series*   Suppose that f is in PC$[0, c]$ and $f(x) = f(c - x)$ for all x, $c/2 < x < c$ (i.e., f is *even* about $x = c/2$).

    (a) Verify that FSS$[f]$ contains only terms of the form $\sin((2k + 1)\pi x/c)$.

    (b) Reflect the function $\sin x$, $0 \le x \le \pi/4$, about $x = \pi/4$ to obtain a function f defined on $[0, \pi/2]$ that is even about $x = \pi/4$. Use (a) to find FSS$[f]$.

15. *Another Kind of Fourier Cosine Series*   Suppose that f is in PC$[0, c]$, $f(x) = -f(c - x)$, for $c/2 < x < c$ (i.e., f is *odd* about $x = c/2$).

    (a) Verify that FCS$[f]$ contains only terms of the form $\cos((2k + 1)\pi x/c)$.

    (b) Use (a) to find FCS$[f]$ for the function f defined by reflecting x^2, $0 \le x < 1$, as an odd function about $x = 1$.

    Answer 15:   Part (a) is proven by direct calculation. The calculation of FCS$[f]$ in (b) is carried out in a straightforward manner after noting the simplification resulting from (a). In particular, FCS$[f] = \sum_{k=0}^{\infty} A_k \cos[(2k + 1)\pi x/2]$, where $A_k = 2\int_0^1 x^2 \cos[(2k + 1)\pi x/2]\, dx = (-1)^k (2a^2 - 4)a^{-3}$, $a = (k + 1/2)\pi$.

Modeling Problems.   [*Hint*: see Example 10.3.6.]

16. *Electrical Circuit*   Find the steady-state charge in the capacitor in an RLC loop if $R = 10\ \Omega$, $L = 0.5$ H, $C = 10^{-4}$ F, and $E(t) = 10\,\mathrm{SqWave}(t, 2\pi, \pi)$ volts.

17. *Driven Hooke's Law Spring*   In a spring-block system, let the spring constant be 1.01 newtons/m and the mass of the block be 1 kilogram. Suppose that the block is acted on by a viscous damping force with coefficient equal to 0.2 newtons per meter per sec. Suppose that the block is driven by the periodic force $f(t) = \mathrm{SqWave}(t, 2\pi, \pi)$ newtons. Find a Fourier series for the steady-state motion of the system.

    Answer 17:   The spring motion is modeled by $my'' + cy' + ky = y'' + 0.2y' + 1.01y = f(t)$. The steady-state motion of the given spring-mass system can be calculated by using precisely the same method outlined in Problem 5 above. Using the fact that $f = \mathrm{SqWave}(t, 50, 2\pi)$ and

    $$FS[f] = \frac{1}{2} + \frac{1}{2\pi i} \sum_{k=1}^{\infty} \frac{1 - (-1)^k}{k}(e^{ikx} - e^{-ikx})$$

    we can replace $y(t)$ by $\sum_{k=-\infty}^{\infty} c_k e^{ikt}$ in the ODE to obtain

    $$\sum_{k=-\infty}^{\infty} [(ik)^2 + 0.2ik + 1.01]c_k e^{ikt} = FS[f]$$

    where FS$[f]$ is given above. Matching coefficients, we have

    $$c_0 = \frac{1}{2.02}, \quad c_k = \begin{cases} 0, & k \ne 0,\ k \text{ even} \\ [\pi ik(-k^2 + 0.2ik + 1.01)]^{-1}, & k \text{ odd} \end{cases}$$

    and the solution is $y = \sum_{-\infty}^{\infty} c_k e^{ikt}$.

10.4 Temperature in a Thin Rod

Comments:

Wave motion has been the physical phenomenon supporting our discussion of PDEs until now. We show in this section that conduction of heat is modeled by a linear PDE as well, but it is not the wave equation. The heat (or diffusion) PDE is derived from physical principles concerning the flow of thermal energy from warmer to cooler regions in a body. Natural boundary conditions are introduced that model heat sources or sinks or insulating material placed around the boundary of the body. Separation of variables is applied to construct a series solution for the flow of heat (measured by changing temperatures) in a circular rod whose lateral walls are wrapped in thermal insulation.

We show how to determine the optimal depth of a storage cellar to protect its contents from the temperature variations at the earth's surface. Finally, we compare and contrast properties of solutions of the heat equation with those of solutions of the wave equation. In particular, the heat operator smooths out any "corners" or discontinuities in the initial heat curve, while the wave equation simply propagates the initial kinks and jumps. Moreover, thermal energy moves "infinitely" fast, while a signal propagates with a finite speed. Some of the properties, incidentally, show that the wave and heat equations cannot be exact models of the corresponding real phenomena.

PROBLEMS

Temperature in a Rod; Zero Boundary Conditions. Use the Method of Separation of Variables to construct a series solution of the heat equation $u_t - Ku_{xx} = 0$, for $0 < x < L$, $t > 0$, and the following initial and boundary conditions. Describe how $u(x,t)$ behaves as $t \to +\infty$.

1. $u(0,t) = u(L,t) = 0$, $\quad u(x,0) = \sin(2\pi x/L)$

2. $u(0,t) = u(L,t) = 0$, $\quad u(x,0) = x$

3. $u(0,t) = u(L,t) = 0$, $\quad u(x,0) = u_0 > 0$, $\quad u_0$ a nonzero constant

4. $u(0,t) = u(L,t) = 0$, $\quad u(x,0) = \begin{cases} u_0, u_0 \neq 0 & 0 \le x \le L/2 \\ 0, & L/2 < x \le L \end{cases}$

5. $u(0,t) = 0$, $\quad u_x(L,t) = 0$, $\quad u(x,0) = \sin(\pi x/(2L))$ [*Hint:* see Example 10.4.2].

6. $u(0,t) = 0$, $\quad u_x(L,t) = 0$, $\quad u(x,0) = x$

7. $u_x(0,t) = u_x(L,t) = 0$, $\quad u(x,0) = x$

Answer 1: Let us consider the separated solution $= X(x)T(t)$. Inserting $X(x)T(t)$ into the PDE and applying the boundary conditions, we see that $X(x)$ and $T(t)$ must be solutions of

$$\frac{d^2 X}{d^2 x} - \lambda X = 0, \quad X(0) = 0, \quad X(L) = 0$$

$$\frac{dT}{dt} - \lambda KT = 0$$

As in (12) and (13) in the text, we have that $\lambda_n = -(n\pi/L)^2$, $X_n(x) = A_n \sin(n\pi x/L)$, $T_n(t) = B_n \exp[-K(n\pi/L)^2 t]$, $n = 1, 2, \ldots$, where A_n and B_n are any constants. The solutions of the PDE, $u_t - Ku_{xx} = 0$ and the boundary conditions $u(0,t) = u(L,t) = 0$ are given by

$$u_n = X_n(x)T_n(t) = C_n \sin \frac{n\pi x}{L} \exp\left[-K\left(\frac{n\pi}{L}\right)^2 t\right], \quad n = 1, 2, \ldots$$

where $C_n = A_n B_n$ is an arbitrary constant. Since the PDE and the boundary conditions are homogeneous in this problem,

$$u(x,t) = \sum_{n=1}^{\infty} C_n \sin \frac{n\pi x}{L} \exp\left[-K\left(\frac{n\pi}{L}\right)^2 t\right]$$

is also a solution of the PDE and the boundary conditions. Now we only need to determine the constants C_n so that $u(x,t)$ satisfies the initial condition. That is, we must choose C_n so that

$$u(x,0) = \sum_{n=1}^{\infty} C_n \sin \frac{n\pi x}{L} = \sin \frac{2\pi x}{L}, \quad 0 \le x \le L$$

So $C_1 = C_3 = \cdots = 0$ and $C_2 = 1$. The solution is

$$u(x, t) = \sin \frac{2\pi x}{L} \exp\left[-K\left(\frac{2\pi}{L}\right)^2 t\right]$$

This represents the temperature as one hump of a sine curve whose amplitude decays exponentially in time to zero.

Answer 3: Since $f(x) = u_0$,

$$C_n = \frac{2}{L} \int_0^L u_0 \sin \frac{n\pi x}{L} dx = \frac{2u_0}{n\pi}[1 - (-1)^n]$$

The series solution is

$$u(x, t) = \frac{2u_0}{\pi} \sum_{n=1}^{\infty} \frac{1 - (-1)^n}{n} \sin \frac{n\pi x}{L} \exp\left[-K\left(\frac{n\pi}{L}\right)^2 t\right]$$

$$= \frac{4u_0}{\pi} \sum_{\text{odd } n} \frac{1}{n} \sin \frac{n\pi x}{L} \exp\left[-K\left(\frac{n\pi}{L}\right)^2 t\right]$$

Note that for all $t \geq 0$ the series solution for u is zero at $x = 0, L$. As time goes on, the temperatures everywhere decline to the zero values maintained at the endpoints.

Answer 5:   Here the boundary conditions are $u(0, t) = 0$ and $u_x(L, t) = 0$ [perfect insulation at $x = L$]. Let's consider the separated solution $= X(x)T(t)$. Inserting $X(x)T(t)$ into the PDE and boundary conditions, we see that $X(x)$ and $T(t)$ must be solutions of

$$X''(x) - \lambda X(x) = 0, \quad X(0) = 0, \quad X'(L) = 0$$

$$T'(t) - \lambda K T(t) = 0$$

We have $\lambda_n = -[(2n - 1)\pi/(2L)]^2$ and

$$X_n(x) = A_n \sin \frac{(2n - 1)\pi x}{2L}, \quad T_n(t) = B_n \exp\left[-K\left(\frac{2n - 1}{2L}\pi\right)^2 t\right], \quad n = 1, 2, \ldots$$

where A_n and B_n are any constants. The separated solutions of the PDE, $u_t - K u_{xx} = 0$, and the boundary conditions $u(0, t) = u_x(L, t) = 0$ are given by

$$u_n = X_n(x)T_n(t) = C_n \sin \frac{(2n - 1)\pi x}{2L} \exp\left[-K\left(\frac{2n - 1}{2L}\pi\right)^2 t\right], \quad n = 1, 2, \ldots$$

where $C_n = A_n B_n$ is an arbitrary constant. Since the PDE and the boundary conditions are homogeneous in this problem,

$$u(x, t) = \sum_{n=1}^{\infty} C_n \sin \frac{(2n - 1)\pi x}{2L} \exp\left[-K\left(\frac{2n - 1}{2L}\pi\right)^2 t\right]$$

is also a solution of the PDE and the boundary conditions. Now we only need to determine constants C_n so that $u(x, t)$ satisfies the initial condition. That is, we must choose C_n so that

$$u(x, 0) = \sum_{n=1}^{\infty} C_n \sin \frac{(2n - 1)\pi x}{2L} = \sin \frac{\pi x}{2L}, \quad 0 \leq x \leq L$$

We have $C_1 = 1, C_2 = C_3 = \cdots = 0$. The solution is

$$u(x, t) = \sin \frac{\pi x}{2L} \exp\left[-K\left(\frac{\pi}{2L}\right)^2 t\right]$$

and we have the initial temperature profile decaying exponentially to 0. Note that the condition $u_x(L, t) = 0$ means that a perfect insulator is clamped to the far end of the rod.

Answer 7: Here we have perfect insulation at both ends: $u_x(0, t) = u_x(L, t) = 0$. Let us consider the separated solution $= X(x)T(t)$. Inserting $X(x)T(t)$ into the PDE and the boundary conditions, we see that $X(x)$ and $T(t)$ must be solutions of

$$\frac{d^2 X}{d^2 x} - \lambda X = 0, \quad X'(0) = 0, \quad X'(L) = 0$$

$$\frac{dT}{dt} - \lambda K T = 0$$

We have $\lambda_n = -(n\pi/L)^2$ and

$$X_n(x) = A_n \cos \frac{n\pi x}{L}, \qquad T_n(t) = B_n \exp[-K(\frac{n\pi}{L})^2 t], \quad n = 0, 1, 2, \dots$$

where A_n and B_n are any constants. The separated solutions of the PDE: $u_t - K u_{xx} = 0$ with boundary conditions $u_x(0, t) = u_x(L, t) = 0$ are given by

$$u_n = X_n(x)T_n(t) = C_n \cos\left(\frac{n\pi x}{L}\right) \exp\left[-K\left(\frac{n\pi}{L}\right)^2 t\right], \qquad n = 0, 1, 2, \dots$$

where $C_n = A_n B_n$ is an arbitrary constant. Since the PDE and the boundary conditions are homogeneous in this problem,

$$u(x, t) = C_0 + \sum_{n=1}^{\infty} C_n \cos\left(\frac{n\pi x}{L}\right) \exp\left[-K\left(\frac{n\pi}{L}\right)^2 t\right]$$

is also a solution of PDE. Now we only need to determine the constants C_n so that $u(x, t)$ satisfies the initial condition. That is, we must choose C_n so that

$$u(x, 0) = \sum_{n=0}^{\infty} C_n \cos \frac{n\pi x}{L} = x, \quad 0 \le x \le L$$

Therefore, the coefficients C_n are given by

$$C_n = \frac{\langle f, \cos \frac{n\pi x}{L}\rangle}{\|\cos \frac{n\pi x}{L}\|^2} = \begin{cases} \dfrac{1}{L}\displaystyle\int_0^L x\, dx = \dfrac{L}{2}, & n = 0 \\[3mm] \dfrac{2}{L}\displaystyle\int_0^L x \cos \dfrac{n\pi x}{L}\, dx = \dfrac{2L}{n^2\pi^2}[(-1)^n - 1], & n > 0 \end{cases}$$

The series solution is

$$u(x, t) = \frac{L}{2} - \frac{4L}{\pi^2} \sum_{\text{odd } n} \frac{1}{n^2} \cos\left(\frac{n\pi x}{L}\right) \exp\left[-K\left(\frac{n\pi}{L}\right)^2 t\right]$$

Note that since both ends of the rod and the cylindrical walls are perfectly insulated, no thermal energy can escape. This means that as time goes on the energy tends towards a level equilibrium. In fact, as $t \to +\infty$ the temperature tends to the uniform value of $L/2$.

Temperature in a Rod; Nonzero Boundary Conditions.

8. *Constant End Temperatures*   Follow the outline below to find the solution of the problem

$$\begin{aligned} u_t - K u_{xx} &= 0, & 0 < x < 1, & \quad t > 0 \\ u(0, t) &= 10, & u(1, t) = 20, & \quad t \ge 0 \\ u(x, 0) &= 0, & 0 < x < 1 & \end{aligned}$$

(a) First find a function $A(x)$ such that $A''(x) = 0$, $A(0) = 10$, $A(1) = 20$. Then find a solution $v(x, t)$ of the IBVP

$$\begin{aligned} v_t - K v_{xx} &= 0, & 0 < x < 1, & \quad t > 0 \\ v(0, t) &= 0, & v(1, t) = 0, & \quad t \ge 0 \\ v(x, 0) &= -A(x), & 0 \le x \le 1 & \end{aligned}$$

(b) Show that $u(x, t) = A(x) + v(x, t)$ solves the given IBVP. Show that $u = A(x)$ is the *steady-state solution* because $v(x, t) \to 0$ as $t \to +\infty$.

Modeling Problem.

 9. *Time's Arrow*

(a) Prove that the series (16) diverges for every $x, 0 < x < 2$, if $t = t_0 < 0$.

(b) Explain this in terms of time's arrow.

(c) Find an initial data function $f(x) \neq 0$ so that (13) and (14) give solutions defined for all time, even for $t < 0$. [*Hint*: consider $f(x) = \sin(\pi x/L)$.]

Answer 9:

(a) If $t = t_0 < 0$, then $-n^2 \pi^2 t_0 > 0$. So the term $e^{-n^2 \pi^2 t_0/4}/n^2 \to +\infty$ as $n \to \infty$ (use L'Hôpital's Rule). The series (1) diverges for every $x, 0 < x < 2$, if $t = t_0 < 0$.

(b) The heat equation is valid only for t increasing from the initial time $t = 0$. If the initial data are not smooth, then only for $t > 0$ is the equation meaningful. The direction of time from angular to smooth temperature profiles determines time's advance.

(c) Consider $f(x) = \sin(\pi x/L)$. By text equation (15),

$$C_n = \frac{2}{L} \int_0^L f(x) \sin \frac{n\pi x}{L} dx = \frac{2}{L} \int_0^L \sin \frac{\pi x}{L} \sin \frac{n\pi x}{L} dx = \begin{cases} 1 & n = 1 \\ 0 & n > 1 \end{cases}$$

So

$$u(x, t) = \sin \frac{\pi x}{L} \exp\left[-K\left(\frac{\pi}{L}\right)^2 t\right]$$

which is a solution of the boundary/initial value problem (8) valid for all values of t.

10.5 Sturm–Liouville Problems

Comments:

The separation step in the technique of separation of variables for a boundary-initial value problem for a PDE always generates a Sturm-Liouville problem. This typically involves solving an ordinary differential operator equation $Ly = \lambda y$, subject to some conditions on the solution $y(x)$ at $x = a$ and $x = b$, say. More precisely, the problem is to find the values of λ so that the problem $Ly = \lambda y$ with boundary conditions has a nontrivial solution. We must find the eigenvalues of the operator L whose domain is restricted by the boundary conditions. The Sturm-Liouville Theory provides the mathematical foundations for all of this. Everything needed in Sections 10.6–WEB SPOTLIGHT ON LAPLACE'S EQUATION and the ideas that underlie what is in Section WEB SPOTLIGHT ON STEADY TEMPERATURES for the separation step in solving PDE problems can be found right here.

PROBLEMS

Sturm–Liouville Problems. In each of the regular Sturm–Liouville systems below, identify the associated operator M. Find the eigenvalues and eigenspaces of M, and state the orthogonality and basis properties of the eigenfunctions.

1. $\quad y'' = \lambda y;$ $y(0) = 0,$ $y(\pi/2) = 0$

2. $\quad y'' = \lambda y;$ $y(0) = 0,$ $y'(L) = 0$

3. $\quad y'' = \lambda y;$ $y'(0) = 0,$ $y'(L) = 0$

4. $\quad y'' = \lambda y;$ $y(-L) = y(L),$ $y'(-L) = y'(L), \quad L > 0$

5. $\quad y'' = \lambda y;$ $y(0) = 0,$ $y(\pi) + y'(\pi) = 0$

6. $\quad y'' - 4y' + 4y = \lambda y;$ $y(0) = 0,$ $y(\pi) = 0$

7. $\quad y'' = \lambda y;$ $y(0) = 0,$ $y(1) - y'(1) = 0$

8. $\quad y'' = \lambda y;$ $2y(0) + y'(0) = 0,$ $y(1) = 0$

Answer 1: $L[y] = y''$, $\mathrm{Dom}(L) = \{y \text{ in } \mathbf{C}^2[0, \pi/2]: y(0) = y(\pi/2) = 0\}$. From the Nonpositivity of Eigenvalues Theorem (Theorem 10.5.2) all the eigenvalues of L are nonpositive. First try $\lambda = 0$. The eigenfunctions (if any) must have the form $y = Ax + B$. The endpoint conditions imply that $A = B = 0$, so $\lambda = 0$ cannot be an eigenvalue. Now try $\lambda = -k^2$, $k \geq 0$. The eigenfunctions (if any) must have the form $y = A \sin kx + B \cos kx$. The condition $y(0) = 0$ implies that $B = 0$. The condition $y(\pi/2) = 0$ implies that $\sin(k\pi/2) = 0$. This can only happen if $k\pi/2 = n\pi$, $n = 1, 2, \ldots$. So $\lambda_n = -4n^2$ is an eigenvalue for each $n = 1, 2, \ldots$, and the corresponding eigenspace V_{λ_n} has $y_n = \sin 2nx$ as a basis. The Orthogonality of Eigenspaces Theorem 10.5.2 implies that the eigenspaces $V_{\lambda_1}, V_{\lambda_2}, \ldots$ are orthogonal with respect to the standard scalar product in $PC[0, \pi/2]$ and that $\Phi = \{\sin 2x, \cdots, \sin 2nx, \cdots\}$ is a basis for $PC[0, \pi/2]$.

Answer 3: $L[y] = y''$, $\mathrm{Dom}(L) = \{y \text{ in } \mathbf{C}^2[0, T]: y'(0) = y'(T) = 0\}$. First, $\lambda = 0$ is an eigenvalue and $y = 1$ a corresponding eigenvector. As in Problem 2, the other eigenvalues must be negative, $\lambda = -k^2$, and $y_k = A_k \cos kx + B_k \sin kx$. Imposing the boundary conditions, we see that $B_k = 0$, and the eigenvalues are $\lambda_n = -(n\pi/T)^2$, $n = 0, 1, 2, \ldots$ with each corresponding eigenspace V_{λ_n} spanned by the eigenfunction $y_n = \cos(n\pi x/T)$. The eigenspaces V_{λ_n} are orthogonal with respect to the standard scalar product in $PC[0, T]$, and the orthogonal set $\Phi = \{\cos(n\pi x/T) : n = 0, 1, 2, \ldots\}$ is a basis for $PC[0, T]$.

Answer 5: $L[y] = y''$, $\mathrm{Dom}(L) = \{y \text{ in } \mathbf{C}^2[0, \pi]: y(0) = y(\pi) + y'(\pi) = 0\}$. First, find eigenvalues of L of the form $\lambda = k^2$, for some $k > 0$. Such eigenvalues would have eigenfunctions (if any) of the form $y = A \sinh kx + B \cosh kx$. The condition $y(0) = 0$ implies that $B = 0$. The condition $y(\pi) + y'(\pi) = 0$ implies that $\sinh k\pi + k \cosh k\pi = 0$ or $\tanh r = -\pi/r$, where $k\pi = r > 0$. The sketch in Fig. 1(e), Graph 1 indicates that there is no value $r > 0$ such that $\tanh r = -\pi/r$, so no positive eigenvalues. (Note that the Nonpositivity of Eigenvalues Theorem 10.5.2 would have given the result immediately). Next try $\lambda = 0$. The eigenfunctions (if any) must have the form $y = Ax + B$. The endpoint conditions imply that $A = B = 0$, so $\lambda = 0$ cannot be an eigenvalue. Finally, let's try $\lambda = -k^2$, $k \geq 0$. In this case any eigenfunction would have the form $y = A \cos kx + B \sin kx$. The condition $y(0) = 0$ implies that $A = 0$, and the other boundary condition implies that $\sin k\pi + k \cos k\pi = 0$ or $\tan r = -\pi/r$, for $r > 0$, where $k\pi = r$. The sketch in Graph 2 indicates that there is an infinite sequence $r_1, r_2, \ldots$, for which $\tan r_n = -\pi r_n, r_n > 0$. Note that $(n - 1/2)\pi < r_n < n\pi$, for all n, and in fact $r_n \rightarrow (n + 1/2)\pi$ as $n \rightarrow \infty$. So $\lambda_n = -k_n^2 = -(r_n/\pi)^2$ are the eigenvalues of L, and the corresponding eigenspace V_{λ_n} is spanned by $y_n = \sin k_n x = \sin(r_n x/\pi)$. Because L is a symmetric operator, the eigenspaces are orthogonal with the standard scalar product in $PC[0, \pi]$. Finally, the Sturm-Liouville Theorem 10.5.3 implies that $\Phi = \{\sin(r_n x/\pi) : n = 1, 2, \ldots\}$ is a basis for $PC[0, \pi]$.

Answer 7:   We solve the SLP $y'' = \lambda y$, $y(0) = 0$, $y(1) - y'(1) = 0$, by a case argument:

Case 1 $\lambda < 0$. Put $\lambda = -k^2$, with $k > 0$. Then the general solution of $y'' + k^2 y = 0$ is $y = A \cos kx + B \sin kx$, for arbitrary constants A and B. Imposing the boundary conditions we see that A and B must satisfy the linear system

$$A = 0$$
$$A \cos k + B \sin k - k(-A \sin k + B \cos k) = 0$$

Does this system have a nontrivial solution for A and B? It does if and only if $\sin k - k \cos k = 0$

has a solution with $k > 0$. By overlaying the graphs of $z = \tan k$ and $z = k$ in the zk-plane we see that these graphs intersect infinitely often on the positive k axis. Label the k-values at those intersection points by size: $k_1, k_2, \cdots$ with $\pi/2 < k_1 < 3\pi/2$. Thus $\lambda_n = -k_n^2$, $n = 1, 2, \cdots$ and corresponding eigenfunctions are $y_n = \sin k_n x$.

Case 2 $\lambda = 0$. The general solution of $y'' = 0$ is $y = Ax + B$, where A and B are arbitrary constants. The boundary conditions imply that

$$B = 0$$
$$A + B - A = 0$$

This has the nontrivial solution $B = 0$, and A is arbitrary. Therefore $\lambda_0 = 0$ is an eigenvalue and a corresponding eigenfunction is $y_0(x) = x$.

Case 3 $\lambda > 0$. Put $\lambda = k^2$, $k > 0$. The general solution of $y'' - k^2 y = 0$ is $y = A \cosh kx + B \sinh kx$, where A and B are arbitrary constants. Imposing the boundary conditions we have the linear system

$$A = 0$$
$$A \cosh k + B \sinh k - k(A \sinh k + B \cosh k) = 0$$

This sytem has a nontrivial solution if and only if $\sinh k - k \cosh k = 0$ has a solution $k > 0$. Equivalently, this "eigenvalue" equation becomes $\tanh k = k$. Overlaying the graphs of $z = \tanh k$ and $z = k$ in the zk-plane we see that the only intersection point is at $k = 0$; i.e., there are no positive values of k such that $\tanh k = k$. Therefore, this SLP has no positive eigenvalues.

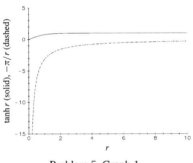

Problem 5, Graph 1.

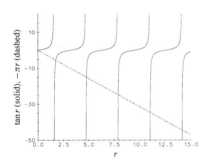

Problem 5, Graph 2.

9. *Sturm–Liouville Problems on Complex Euclidean Spaces* An operator M defined on a subspace S of a complex Euclidean space E that takes values in E is *symmetric* if $\langle u, Mv \rangle = \langle Mu, v \rangle$, for all u, v in S. Show that all the eigenvalues of M are real numbers. [*Hint*: recall that $\langle r, \lambda s \rangle = \overline{\lambda} \langle r, s \rangle$, for all r, s in E and all scalars λ.]

Answer 9: Let λ be an eigenvalue of the symmetric operator M, and suppose that v is a corresponding eigenfunction. Then $\langle v, Mv \rangle = \langle Mv, v \rangle$. Now since $Mv = \lambda v$, we have that $\langle v, \lambda v \rangle = \langle \lambda v, v \rangle$, or, after factoring out λ,

$$\overline{\lambda} \langle v, v \rangle = \lambda \langle v, v \rangle$$

or

$$(\overline{\lambda} - \lambda) \|v\|^2 = 0$$

Since $\|v\| \neq 0$, it follows that $\overline{\lambda} - \lambda = 0$. Hence λ is real.

10. Show that for any real number h, the Sturm–Liouville problem

$$y'' = \lambda y, \quad y(0) = 0, \quad y(1) - h y'(1) = 0$$

cannot have more than one positive eigenvalue.

Existence and Uniqueness for a Boundary Value Problem.

☞ BVP is an abbreviation for *boundary value problem*.

11. Consider the BVP $y'' + y = f(x)$, $y(0) + y'(0) = 0$, $y(\pi) + y'(\pi) = 0$.

(a) Show that the Sturm–Liouville problem $y'' + y = \lambda y$, $y(0) + y'(0) = 0$, $y(\pi) + y'(\pi) = 0$ has the eigenvalues $\lambda_n = 1 - n^2$, $n = 1, 2, \ldots$, and the corresponding eigenfunctions $\sin nt - n \cos nt$, $n = 1, 2, \ldots$. [*Hint:* see Example 10.3.1.]

(b) Show that BVP has a solution if and only if

$$\int_0^\pi f(s)(\sin s - \cos s)\, ds = 0$$

(c) Show that BVP either has no solution or infinitely many solutions of the form

$$y = A(\sin x - \cos x) + \sum_{k=2}^{\infty} \frac{a_k}{1 - k^2}(\sin kx - k \cos kt)$$

for any value of A and suitably chosen values of the constants a_k.

Answer 11:

(a) First we search for eigenvalues λ for which $1 - \lambda > 0$. So let's put $1 - \lambda = k^2$ where $k > 0$. Then the Sturm-Liouville Problem becomes

$$y'' + k^2 y = 0, \quad y(0) + y'(0) = 0, \quad y(\pi) + y'(\pi) = 0$$

The general solution of the ODE is $y = A \cos kx + B \sin kx$, where A and B are arbitrary reals. The boundary conditions imply that $k > 0$ must be chosen such that the linear equations

$$A + kB = 0$$
$$A \cos k\pi + B \sin k\pi - kA \sin k\pi + kB \cos k\pi = 0$$

have a nontrivial solution for A and B. This will happen if and only if k makes the system determinanent vanish: $(1 + k^2) \sin k\pi = 0$. This implies that $k = 1, 2, \cdots, n, \cdots$. Thus the eigenvalues in this case are $\lambda_n = 1 - n^2$, $n = 1, 2, \cdots$, and the corresponding eigenfunctions are $y_n(x) = \sin nx - n \cos nx$.

Next, we ask if $\lambda = 1$ is an eigenvalue. In that case $y'' = 0$ which has the general solution $y = Ax + B$, where A and B are arbitrary reals. The boundary conditions imply that $2A + B = 0$, and $(1 + \pi)A + B = 0$. The linear equations have only the trivial solution $A = B = 0$, and so $\lambda = 1$ is not an eigenvalue.

Finally let's examine the case where $1 - \lambda = -k^2$, $k > 0$. The ODE $y'' - k^2 y = 0$ has the general solution $y = Ae^{kx} + Be^{-kx}$, where A and B are arbitrary reals. The boundary conditions imply that

$$A(1 + k) + B(1 - k) = 0$$
$$Ae^{k\pi}(1 + k) + Be^{-k\pi}(1 - k) = 0$$

The determinant of this linear system is $(1 - k^2)(e^{-k\pi} - e^{k\pi})$ which vanishes for a value of $k > 0$ if and only if $k = 1$. Hence, $\lambda_0 = 2$ is an eigenvalue and a corresponding eigenfunction is $y_0(x) = e^{-x}$.

(b) To find a solution of the given BVP we use the fact the collection of eigenfunctions $\{y_0, y_1, y_1, \cdots\}$ is a basis for PC$[0, \pi]$. We use a method of undetermined coefficients to find a solution, and let L be the operator with action $L[y] = y'' + y$ and domain $\{y$ in C$^2[0, \pi] : y(0) + y'(0) = 0$, $y(\pi) + y'(\pi) = 0\}$. Suppose that y is a solution of $L[y] = f$, where $f(x)$ is in C$^0[0, \pi]$. Write y as the convergent orthogonal series $y = A_0 y_0 + A_1 y_1 + \cdots$, where the coefficients A_n are to be determined. Next, write the givenfunction $f(x)$ as the orthogonal series $f = a_0 y_0 + a_1 y_1 + \cdots$, where the coefficients are given by the Fourier-Euler formulas

$$a_n = \frac{\langle f, y_n \rangle}{\langle y_n, y_n \rangle}$$

using the standard scalar product $\langle \cdot, \cdot \rangle$ on $C^0[0, \pi]$. Hence, we require that

$$L\left[\sum_{n=0}^{\infty} A_n y_n\right] = \sum_{n=0}^{\infty} a_n y_n$$

Now assume that the operations of L and Σ can be interchanged. Then, since $L[y_n] = \lambda_n y_n$, we have

$$\sum_{n=0}^{\infty} A_n \lambda_n y_n = \sum_{n=0} a_n y_n$$

Since $\{y_0, y_1, \cdots\}$ is a basis for $C^0[0, \pi]$ we can equate the coefficients of each y_n on both sides of the equation to obtain

$$A_n \lambda_n = a_n, \quad n = 0, 1, \cdots \tag{i}$$

But $\lambda_1 = 0$ and $y_1 = \sin x - \cos x$, hence a necessary condition that $L[y] = f$ is that $a_1 = 0$ which, in turn implies that

$$\int_0^{\pi} f(x)(\sin x - \cos x)\, dx$$

as asserted. If $a_1 = 0$, then A_1 is arbitrary and can be taken to be zero and hence $L[y] = f$ has the solution $y = A_0 y_0 + A_2 y_2 + \cdots$, where coefficients are given by the relations (i).

(c) Recall that $\lambda_0 = 2$, and $\lambda_n = 1 - n^2$, for $n = 1, 2, \cdots$ and so in particular $\lambda_1 = 0$. Thus $L[y_1] = 0$ and from **(b)** we see that

$$y = \frac{a_0}{2} y_0 + A_1 y_1 + \sum_{n=2}^{\infty} \frac{a_1}{1 - n^2} y_n$$

is a solution of $L[y] = f$ for any value of A_1. Therefore, either $L[y] = f$ has no solution or infinitely many solutions.

 12. *Uniqueness Theorem for IBVPs*   Suppose that R is the region $0 < x < L, t > 0$, and $F(x, t)$ is continuous to the boundary of R. Suppose that $\phi(t)$, $\mu(t)$ are continuous on $t \geq 0$; and that $f(x)$ and $g(x)$ are continuous on $0 \leq x \leq L$. Consider the IBVP

(PDE)	$u_{tt} = c^2 u_{xx} + F(x, t),$	in R
(BC)	$\alpha u_x(0, t) + \beta u(0, t) = \phi(t),$	$t \geq 0$
	$\gamma u_x(L, t) + \delta u(L, t) = \mu(t),$	$t \geq 0$
(IC)	$u(x, 0) = f(x), \quad u_x(x, 0) = g(x),$	$0 \leq x \leq L$

where $\alpha, \beta, \gamma, \delta$ are constants, and $\alpha\beta \leq 0$, $\gamma\delta \geq 0$. Show that this IBVP cannot have more than one solution $u(x, t)$ that is twice continuously differentiable in R and (together with its first derivatives) is continuous to the boundary of R. [*Hint*: look at the outline in Problem 26 of Section 10.1.]

10.6   The Method of Eigenfunction Expansions

Comments:

Now we can approach a boundary/initial value problem for a PDE in a systematic way, using the techniques introduced earlier. A systematic approach is very helpful because there are so many separate steps in constructing the series solution of the problem that a "do-it-by-the-numbers" approach is preferred. This does not mean doing a problem without thought, because every new problem brings new geometry, data sets, or coefficients in the PDE. However, the template we have arranged in the form of six steps (numbered as 1–6 in Section 10.1) helps in avoiding needless errors and pointless searching for what to do next. The whole process has at its core the Method of Eigenfunction Expansions, and we make note of that important fact. Finally, we show how to solve boundary/initial value problems for PDEs when the PDE, the boundary conditions, or both, are not homogeneous. Here is where eigenfunction expansions play a central role in constructing the solution. We distinguish between a "formal solution" of a boundary/initial value problem where we find the solution as a series but don't worry about the convergence, from a "classical solution" where we add enough extra smoothness conditions on the data that all the series converge to the desired sum. On a first "read-through" you might ignore the distinction and focus on the formal solution. You might also skip the verification of the theorems.

PROBLEMS

Shifting Data.

1. The initial/boundary value problem below models the vertical displacement $u(x, t)$ of a taut flexible string tied at both ends with vanishing initial data and acted on by gravity.

$$u_{tt} - c^2 u_{xx} = g, \qquad\qquad 0 < x < L, \quad t > 0$$
$$u(0, t) = 0, \quad u(L, t) = 0, \qquad t \geq 0$$
$$u(x, 0) = 0, \quad u_t(x, 0) = 0, \qquad 0 \leq x \leq L$$

Follow the outline below to shift the driving term in the PDE into an initial condition.

(a) Find a function $v(x)$ such that $-c^2 v_{xx} = g$, $0 < x < L$, $t > 0$; $v(0) = 0$, $v(L) = 0$, $t > 0$. [*Hint*: try $v(x) = Ax + Bx^2$, where A, B are constants. The function $v(x)$ is the steady-state sag of the string under the force of gravity.]

(b) Let $w = u - v$ and show that w satisfies the same equations as u, but with g replaced by zero and the condition $u(x, 0) = 0$ replaced by $w(x, 0) = -gx(L - x)/2c^2$.

(c) Find $u(x, t)$. [*Hint*: $u = w + v$.]

Answer 1:

(a) Since the driving force $F(x, t) = g$, we will try to "shift" F off of the PDE onto the initial data by using a function $v(x)$ that satisfies $v_{tt} - c^2 v_{xx} = g$, that is, $-c^2 v_{xx} = g$, and the boundary conditions $v(0) = 0$, $v(a) = 0$. Solving the ODEs for $v(x)$ we have the general solution

$$v = -\frac{g}{2c^2} x^2 + Ax + B$$

If v is to satisfy the boundary conditions, then $B = 0$ and $A = gL/(2c^2)$. The function $v = gx(L - x)/(2c^2)$ satisfies all the required conditions.

(b) Let $u(x, t)$ be the solution of the given problem. Put $w(x, t) = u(x, t) - v(x)$, where v is as in **(a)**. Since $u_{tt} - c^2 u_{xx} = g$ and $v_{tt} - c^2 v_{xx} = g$, it follows that

$$w_{tt} - c^2 w_{xx} = (u_{tt} - c^2 u_{xx}) - (v_{tt} - c^2 v_{xx}) = g - g = 0$$

We have that

$$w(0, t) = u(0, t) - v(0) = 0 - 0 = 0$$
$$w(L, t) = u(L, t) - v(L) = 0 - 0 = 0$$
$$w(x, 0) = u(x, 0) - v(x) = -gx(L - x)/2c^2$$
$$w_t(x, 0) = u_t(x, 0) - 0 = 0$$

(c) Since $w = u - v$ solves the problem

$$w_{tt} - c^2 w_{xx} = 0, \qquad\qquad 0 < x < L, \quad t > 0$$
$$w(0, t) = w(L, t) = 0, \qquad t \geq 0$$
$$w(x, 0) = -gx(L - x)/2c^2, \qquad 0 < x < L$$
$$w_t(x, 0) = 0, \qquad\qquad 0 < x < L$$

☞ *g* is the constant gravitational acceleration at the earth's surface.

we may use (21) and (22) in Section 10.1 to find w, with $f(x) = -gx(L-x)/2c^2$ and $g(x) = 0$. We have $B_n = 0$ for all n, and

$$A_n = \frac{2}{L}\int_0^L -\frac{g}{2c^2}x(L-x)\sin\frac{n\pi x}{L}dx = \frac{2gL^2}{n^3\pi^3c^2}((-1)^n - 1)$$

so

$$w(x, t) = \sum_{n=1}^\infty A_n \cos\frac{n\pi ct}{L}\sin\frac{n\pi x}{L}$$

From the way w was constructed in (a) and (b), we see that $u(x, t) = w(x, t) + v(x)$ is the solution to the original problem.

2. *Shifting Boundary Data onto Initial Data*   We clamp a string of unit length with $c^2 = 1$ at one end, driven by $\sin\pi t/2$ at the other end. The model for the motion is

$$u_{tt} - u_{xx} = 0, \quad 0 < x < 1, \quad t > 0;$$

$$u(0, t) = \sin\pi t/2, \quad u(1, t) = 0, \quad t \ge 0;$$

$$u(x, 0) = f(x), \quad u_t(x, 0) = g(x), \quad 0 \le x \le 1$$

The following steps show how to shift the boundary data $\sin\pi t/2$ into an initial condition.

(a) Suppose that v is any solution of $v_{tt} - v_{xx} = 0$, $v(0, t) = \sin\pi t/2$, $v(1, t) = 0$, $t \ge 0$. Set $w = u - v$. Show that w is a solution of the same problem as u except we replace the boundary condition by $w(0, t) = 0$ and the data $f(x)$ and $g(x)$ by $f(x) - v(x, 0)$ and $g(x) - v_t(x, 0)$, respectively.

(b) Find $v(x, t)$ in the form $X(x)T(t)$. [*Hint*: let $v = X(x)\sin(\pi t/2)$.]

(c) Solve the problem if $g(x) = 0$ for all x.

3. Find any solution v of the problem $v_{tt} - v_{xx} = 0$, $v(0, t) = \sin(3\pi t/2L)$, $v(L, t) = 0$. Then, use v to shift the boundary data into the initial conditions and solve the problem:

$$u_{tt} - u_{xx} = 0, \qquad\qquad\qquad 0 < x < L, \quad t > 0$$

$$u(0, t) = \sin(3\pi t/2L), \quad u(L, t) = 0, \qquad t \ge 0$$

$$u(x, 0) = u_t(x, 0) = 0, \qquad\qquad 0 \le x \le L$$

Answer 3:   The problem is $u_{tt} - u_{xx} = 0$, $0 < x < L$, $t > 0$; $u(0, t) = \sin(3\pi t/2L)$, $u(L, t) = 0$, $t \ge 0$; $u(x, 0) = u_t(x, 0) = 0$, $0 \le x \le L$. First, consider the intermediate problem

$$v_{tt} - v_{xx} = 0, \qquad 0 < x < L, \quad t > 0$$
$$v(0, t) = \sin(3\pi t/2L), \qquad v(L, t) = 0, \quad t > 0$$

To solve this problem, recall that $X(x)T(t)$ satisfies $v_{tt} - v_{xx} = 0$ if $X'' = \lambda X$ and $T'' = \lambda T$ for some constant λ. So one approach is to take $v(x, t) = X(x)\sin 3\pi t/(2L)$ where $X'' = -(3\pi/2L)^2 X$ and $X(0) = 1$, $X(L) = 0$. The function $X(x) = \cos(3\pi x/2L)$ has these properties, so the intermediate problem has the solution

$$v(x, t) = \cos\frac{3\pi x}{2L}\sin\frac{3\pi t}{2L}$$

Now put $w = u - v$, where u is the solution to the original problem and v is the above solution to the intermediate problem. Then we can verify that w must be a solution of the problem

$$w_{tt} - w_{xx} = 0, \qquad\qquad\qquad 0 < x < L, \quad t \ge 0$$
$$w(0, t) = w(L, t) = 0, \qquad\qquad t > 0$$
$$w(x, 0) = 0, \qquad\qquad\qquad 0 < x < L$$
$$w_t(x, 0) = -(3\pi/2L)\cos(3\pi x/2L), \qquad 0 < x < L$$

Use text formula **??** to find the series for w. Note that $f(x) = 0$, $g(x) = -(3\pi/2L)\cos(3\pi x/2L)$, so

$A_n = 0$ for all n. To find the coefficients B_n, we must evaluate the integrals

$$B_n = \frac{2}{n\pi} \int_0^L \left(-\frac{3\pi}{2L} \cos \frac{3\pi x}{2L} \right) \sin \frac{n\pi x}{L} dx$$

$$= -\frac{3}{nL} \int_0^L \left\{ \sin \frac{(2n+3)\pi x}{2L} + \sin \frac{(2n-3)\pi x}{2L} \right\} dx = -\frac{24}{\pi(4n^2 - 9)}$$

The solution to the problem for w is

$$w(x,t) = -\frac{24}{\pi} \sum_{n=1}^{\infty} \frac{1}{4n^2 - 9} \sin \frac{n\pi x}{L} \sin \frac{n\pi t}{L}$$

so $u = w + v$ is the sought-for (formal) solution to the original problem.

Method of Eigenfunction Expansions.

4. Use the Method of Eigenfunction Expansions to solve

$$
\begin{aligned}
u_{tt} - u_{xx} &= 6x, & 0 < x < 1, \quad t > 0 \\
u(0,t) = u(1,t) &= 0, & t \geq 0 \\
u(x,0) = u_t(x,0) &= 0, & 0 \leq x \leq 1
\end{aligned}
$$

Modeling Problems.

 5. *Breaking a String by Shaking It*   Consider the problem of a string of length L with one end fastened and the other driven:

$$
\begin{aligned}
u_{tt} - c^2 u_{xx} &= 0, & 0 < x < L, \quad t > 0 \\
u(0,t) = 0, & \quad u(L,t) = \mu(t), & t \geq 0 \\
u(x,0) = 0, & \quad u_t(x,0) = 0, & 0 \leq x \leq L
\end{aligned}
$$

Show that there is a periodic function $\mu(t)$ such that the string will eventually break. [*Hint*: what if $\mu(t) = A \cos \omega t$ for ω near a natural frequency $n\pi c/L$?]

Answer 5:   We want to find a periodic function $\mu(t)$ to break the string modeled by $u_{tt} - c^2 u_{xx} = 0$, $0 < x < L$, $t > 0$; $u(0,t) = 0$, $u(L,t) = \mu(t)$, $t \geq 0$; $u(x,0) = 0$, $u_t(x,0) = 0$, $0 \leq x \leq L$. Consider the "shaking" function $\mu(t) = \cos \omega t$ for some constant ω. To solve the boundary/initial value problem for u, we first solve the "intermediate" problem

$$
\begin{aligned}
v_{tt} - c^2 v_{xx} &= 0, & 0 < x < L, \quad t > 0 \\
v(0,t) &= 0, & t > 0 \\
v(L,t) &= \cos \omega t, & t > 0
\end{aligned}
$$

Now the separated function $X(x)T(t)$ solves the wave equation if $X'' = \lambda X$ and $T'' = c^2 \lambda T$ for some constant λ. Say we search for v in the form $v(x,t) = X(x) \cos \omega t$. Then clearly $X(x)$ must be a solution of $X'' = -(\omega/c)^2 X$ and satisfy the conditions $X(0) = 0$, $X(L) = 1$. Observe that if $\omega L/c \neq n\pi$, then

$$X(x) = \left(\sin \frac{\omega L}{c} \right)^{-1} \sin \frac{\omega x}{c}$$

has this property. The intermediate problem has the solution

$$v(x,t) = \left(\sin \frac{\omega L}{c} \right)^{-1} \sin(\frac{\omega x}{c}) \cos \omega t$$

Now we put $w = u - v$, where u solves the original problem and v the intermediate problem, and observe that w solves the problem

$$
\begin{array}{ll}
w_{tt} - w_{xx} = 0, & 0 < x < L, \quad t \geq 0 \\
w(0, t) = w(L, t) = 0, & t > 0 \\
w(x, 0) = -(\sin(\omega L/c))^{-1} \sin(\omega x/c), & 0 < x < L \\
w_t(x, 0) = 0, & 0 < x < L
\end{array}
$$

The solution of this problem for w is given by **??**. Note that $B_n = 0$ for all n since $g(x) = 0$. To find the coefficients A_n, we must evaluate the integrals

$$
\frac{2}{L} \int_0^L \sin \frac{\omega x}{c} \sin \frac{n\pi x}{L} dx = \frac{1}{L} \int_0^L \left\{ \cos\left(\frac{\omega}{c} - \frac{n\pi}{L}\right)x - \cos\left(\frac{\omega}{c} + \frac{n\pi}{L}\right)x \right\} dx
$$

$$
= \frac{(-1)^n 2n\pi c^2}{(\omega L)^2 - (n\pi c)^2} \sin \frac{\omega L}{c} = \alpha_n
$$

Then the coefficient A_n is

$$
A_n = -\left(\sin \frac{\omega L}{c}\right)^{-1} \alpha_n = \frac{(-1)^{n+1} 2n\pi c^2}{(\omega L)^2 - (n\pi c)^2}
$$

The solution to the problem for w is

$$
w(x, t) = \sum_{n=1}^{\infty} \frac{(-1)^{n+1} 2n\pi c^2}{(\omega L)^2 - (n\pi c)^2} \sin \frac{n\pi x}{L} \cos \frac{n\pi c t}{L}
$$

so $u = w + v$ is the solution of the original problem. If we allow ω to approach the fundamental frequency $\pi c / L$, then

$$
\left(\frac{\omega L}{c}\right)^2 - \pi^2 \to 0
$$

and both $v(x, t)$ and the first term of $w(x, t)$ become unbounded (i.e., the string "breaks").

Temperatures in a Rod.

6. *Variable Boundary Temperatures* Suppose that u is a solution to the problem

$$
\begin{array}{ll}
u_t - K u_{xx} = 0, & 0 < x < 1, \quad t > 0 \\
u(0, t) = g_1(t), & u(1, t) = g_2(t), \quad t \geq 0 \\
u(x, 0) = f(x), & 0 \leq x \leq 1
\end{array}
$$

(a) Let $V = g_1(t) + [g_2(t) - g_1(t)]x$. If $U(x, t)$ satisfies the equations

$$
\begin{array}{ll}
U_t - K U_{xx} = -K V_t, & 0 < x < 1, \quad t > 0 \\
U(0, t) = U(1, t) = 0, & t \geq 0 \\
U(x, 0) = f(x) - V(x, 0), & 0 \leq x \leq 1
\end{array}
$$

then show that $u = V(x, t) + U(x, t)$ is a solution of the given IBVP.

(b) Find $U(x, t)$ if $g_1(t) = \sin t$ and $g_2(t) = 0$.

(c) Find $u(x, t)$. [*Hint:* use your answers to **(a)** and **(b)**.]

7. *Internal Sources/Sinks as Driving Terms* Use the Eigenfunction Expansion Method to solve the problem with internal heat sources/sinks modeled by the driving term $3e^{-2t} + x$ (K is a positive constant):

$$
\begin{array}{ll}
u_t - K u_{xx} = 3e^{-2t} + x, & 0 < x < 1, \quad t > 0 \\
u(0, t) = u(1, t) = 0, & t \geq 0 \\
u(x, 0) = 0, & 0 \leq x \leq 1
\end{array}
$$

Answer 7:   The problem is $u_t - Ku_{xx} = 3e^{-2t} + x$, $0 < x < 1$, $t > 0$; $u(0, t) = u(1, t)$, $t \geq 0$; $u(x, 0) = 0$, $0 \leq x \leq 1$. Let $u(x, t) = \sum_1^\infty U_n(t) \sin n\pi x$, $3e^{-2t} + x = \sum_1^\infty C_n(t) \sin n\pi x$, where

$$C_n(t) = 2 \int_0^1 (3e^{-2t} + x) \sin n\pi x\, dx = \frac{6e^{-2t} - 2(1 + 3e^{-2t})(-1)^n}{n\pi}$$

After inserting the series for u into the PDE, we see that $U_n(t)$ satisfies the first order linear IVP

$$U_n'(t) + K(n\pi)^2 + \frac{6e^{-2t} - 2(1 + 3e^{-2t})(-1)^n}{n\pi} = 0, \qquad U_n(0) = 0$$

(because $u(x, 0) = 0 = \sum_{n=1}^\infty U_n(0) \sin n\pi x$). The solution of the IVP is

$$U_n(t) = 2e^{-K(n\pi)^2 t} \int_0^t \frac{3e^{-2s} - (1 + 3e^{-2s})(-1)^n}{n\pi} e^{K(n\pi)^2 s}\, ds$$

$$= \frac{2(-1)^n}{K(n\pi)^3}[e^{-K(n\pi)^2 t} - 1] + \frac{6(-1)^n - 3}{n\pi(Kn^2\pi^2 - 2)}[e^{-K(n\pi)^2 t} - e^{-2t}]$$

So

$$U(x, t) = 2 \sum_{n=1}^\infty \left\{ \frac{(-1)^n}{K(n\pi)^3}[e^{-K(n\pi)^2 t} - 1] + \frac{3(-1)^n - 3}{n\pi(Kn^2\pi^2 - 2)}[e^{-K(n\pi)^2 t} - e^{-2t}] \right\} \sin n\pi x$$

8.   *Variable Boundary Temperatures*   Suppose that the temperature $u(x, t)$ in a rod of length 1 and diffusivity 1 satisfies the problem

$$u_t - u_{xx} = 0, \qquad\qquad 0 < x < 1, \quad t > 0$$
$$u(0, t) = te^{-t}, \qquad\qquad u(1, t) = 0, \quad t \geq 0$$
$$u(x, 0) = 0.01x(1 - x), \qquad 0 \leq x \leq 1$$

Show that $|u(x, t)| \leq 1/e$ for $0 \leq x \leq 1$, $t \geq 0$.

Spotlight on Decay Estimates

PROBLEMS

Decay Rate.   Find the best possible decay rate for $|A_n| + |B_n|$, where A_n, B_n are the Fourier coefficients of the given function $f(x)$ on $|x| \leq \pi$.

1. $f(x) = 1$ for $0 < x < \pi$, and $f(x) = 0$ for $-\pi < x < 0$.

Answer 1: $f(x)$ is in PS$[-\pi, \pi]$, and $\tilde{f}(x)$ has jump discontinuities at all multiples of π. So Theorem 1 applies and $|A_n| + |B_n| = O(1/n)$.

2. $f(x) = (x - \pi)^k(\pi - x)^k$, $k = 1, 2, \ldots$.

Spotlight on the Optimal Depth for a Wine Cellar

PROBLEMS

Wine Cellars.   Suppose that the surface temperature wave is $T_0 + A_0 \cos \omega t$, where A_0, T_0, and ω are positive constants.

1.   Find the optimal depth of the wine cellar.

2. What is that depth if ω corresponds to 1 day instead of 1 yr?

3. Find the temperature function $u(x_0, t)$ at the optimal depth x_0.

4. Solve the wine cellar problem for the surface wave $T_0 + A_1 \cos \omega_1 t + A_2 \cos \omega_2 t$, where ω_1 corresponds to 1 yr and $\omega_2 = 365\omega_1$ corresponds to 1 day.

Answer 1: Let $u = U + T_0$, where $u(x, t)$ satisfies system (1) in the text. Then, we have

$$U_t - KU_{xx} = 0, \qquad 0 < x < \infty, \qquad -\infty < t < \infty$$
$$U(0, t) = A_0 \cos \omega t, \quad -\infty < t < \infty,$$
$$|U(x, t)| < C, \qquad 0 \leq x < \infty, \qquad -\infty < t < \infty$$

since the surface temperature is given as $u(0, t) = T_0 + A_0 \cos \omega t$. As in the text, we have

$$U = \text{Re}\{A_0 e^{-\alpha x} e^{i(\omega t - \alpha x)}\} = A_0 e^{-\alpha x} \cos(\omega t - \alpha x), \qquad \alpha = \left(\frac{\omega}{2K}\right)^{1/2}$$

The temperature at depth x at time t is

$$u = T_0 + A_0 e^{-\alpha x} \cos(\omega t - \alpha x), \quad \omega = \frac{2\pi}{3.15 \times 10^7}, \quad \alpha = \left(\frac{\omega}{2K}\right)^{1/2}$$

The optimal depth x_0 reverses the "seasons" in the wine cellar in comparison to the natural seasons at the surface (see the text). We should have $\alpha x = \pi$:

$$x_0 = \frac{\pi}{\alpha} = \pi \left(\frac{2K}{\omega}\right)^{1/2} \approx 4.45 \text{ m}$$

Answer 3:   From part **(a)**, we know that the temperature function at the optimal depth x_0 is $u = T_0 + A_0 e^{-\pi} \cos(\omega t - \pi)$ at time t.

Spotlight on Approximation of Functions

Comments:

This spotlight contains the basic ideas about the Fourier expansion of a function with respect to an orthogonal set. We start with a Euclidean space and an orthogonal set in it. We show how to construct the Fourier series of a function in the Euclidean space with respect to the orthogonal set, explain mean convergence and apply that idea to the convergence of a Fourier series of a function to the function itself. The idea of a basis comes in along the way. We explain what is meant when it is said that "a partial sum of the Fourier series of a function is the best approximation in the mean to the function."

PROBLEMS

Subspaces.   A subset X of a vector space V is a *subspace* of V if for any scalar α and any pair of vectors u, v in X, the vectors αu and $u + v$ are in X. For any scalars $\alpha_1, \alpha_2, \ldots, \alpha_n$ and any vectors $u_1, u_2, \ldots, u_n$ in X, the sum $\alpha_1 u_1 + \cdots + \alpha_n u_n$ is a *finite linear combination* over X. The *span* of X, denoted by $\text{Span}(X)$, is the collection of all finite linear combinations over X. The set of all vectors in V and the set consisting of the zero vector alone are both subspaces of V (the *trivial subspaces*).

1. Show that a set of vectors X in a vector space V is a subspace if and only if $\text{Span}(X) = X$.

2. For a closed interval $[a, b]$ in $\mathbb{R}$ and any nonnegative integers m, n, suppose that the continuity sets $C^n[a, b]$ and $C^m[a, b]$ are both real or both complex function vector spaces of functions. Show that $C^n[a, b]$ is a subspace of $C^m[a, b]$ if $n \geq m$.

3. Show that $C^n[a, b]$ is a subspace of $PC[a, b]$ for $n = 0, 1, 2, \ldots$.

4. Suppose that $C^\infty[a, b]$ is the set of all real-valued functions on an interval $[a, b]$ having derivatives of all orders on $[a, b]$. Show that $C^\infty[a, b]$ is a real vector space.

Answer 1: We want to show that A is a subspace of V if and only if $\text{Span}(A) = A$. Let the set A in the linear space V be a subspace. Then A is closed under the operations of addition and multiplication by a scalar, and it follows that $\text{Span}(A) = A$. Conversely, if a subset A of a linear space V is such that $\text{Span}(A) = A$, then A is a linear space since it is closed under finite linear combinations.

Answer 3: For any integer $n = 0, 1, \cdots$, any n-times continuously differentiable function on $[a, b]$ is also piecewise continuous on $[a, b]$. Hence $C^n[a, b]$ is a subset of $PC[a, b]$. From the differentiation rules of calculus, if α, β are any constants and y, z are any two functions in $C^n[a, b]$, then $\alpha y + \beta z$ is also in $C^n[a, b]$. Hence $C^n[a, b]$ is a subspace of $PC[a, b]$.

A Weighted Scalar Product. For two functions f and g in the real space $C^0[0, 1]$, define $\langle f, g \rangle = \int_0^1 e^x f(x) g(x) \, dx$. The function e^x is a *weight* or *density* function.

5. Show that $\langle \cdot, \cdot \rangle$ defines a scalar product.

<u>**6.**</u> Calculate the scalar product of each function pair: $f = 1 - 2x$, $g = e^{-x}$; $f = x^2$, $g = e^x$.

7. Calculate the scalar product of each function pair: $f = x$, $g = 1 - x$; $f = e^{-x/2} \sin(\pi x/2)$, $g = e^{-x/2} \sin(3\pi x/2)$; $f = \cos(\pi x/2)$, $g = 1$.

Answer 5: We want to show that $\langle f, g \rangle$ is a scalar product. Since $f(x)g(x) = g(x)f(x)$ for all scalar functions f and g, and since the functions in $C^0[0, 1]$ are real, we have that

$$\int_0^1 f(x)g(x)e^x \, dx = \int_0^1 g(x)f(x)e^x \, dx$$

and $\langle f, g \rangle = \langle g, f \rangle$, so the first condition of a scalar product is met. The property of bilinearity, $\langle \alpha f + \beta g, h \rangle = \alpha \langle f, h \rangle + \beta \langle g, h \rangle$ follows immediately from the corresponding property for the integral of a product. To show positive definiteness, observe that $\langle f, f \rangle = \int_0^1 f^2(x)e^x \, dx$. If $f(x) = 0$ for all x in $[0, 1]$, then $\langle f, f \rangle = 0$. Integrals of nonnegative continuous functions satisfy the integral mean value theorem (see Theorem B.2.4), so for some x_0 in the interval $[0, 1]$, $\langle f, f \rangle = f^2(x_0)e^{x_0}$, which is certainly nonnegative. Suppose that $\langle f, f \rangle = \int_0^1 f^2(x)e^x \, dx = 0$, but that $f(x_0) \neq 0$ for some x_0, say $f(x_0) > 0$. Then there is an interval $I = [c, d]$ (maybe quite small) around x_0 and $f(x) > 0$ for all x in $[c, d]$. But then $\int_c^d f^2(x)e^x \, dx = f^2(x_1)e^{x_1}(d - c) > 0$ for some x_1 in $[c, d]$. So $\int_0^1 = \int_0^c + \int_c^d + \int_d^1 > 0$; the supposition that $f(x_0) > 0$ for some x_0 is wrong. So $\langle f, g \rangle$ meets all three conditions for an inner product.

Answer 7: Tables of integrals are used to compute the integrals, but a CAS could be used as well.

$$\langle x, 1 - x \rangle = \int_0^1 x(1 - x)e^x \, dx = 3 - e$$

$$\langle e^{-x/2} \sin(\pi x/2), e^{-x/2} \sin(3\pi x/2) \rangle = \int_0^1 \sin(\pi x/2) \sin(3\pi x/2) \, dx = 0$$

$$\langle \cos(\pi x/2), 1 \rangle = \int_0^1 \cos(\pi x/2)e^x \, dx = (2\pi e - 4)/(\pi^2 + 4)$$

Properties of a Scalar Product. Suppose that $\langle \cdot, \cdot \rangle$ is a scalar product on a real vector space V, and that $\| \cdot \|$ is the norm induced by the scalar product.

8. Show that $\|u + v\|^2 + \|u - v\|^2 = 2\|u\|^2 + 2\|v\|^2$ for all u, v in V. [*Hint*: use properties (2a)–(2c) in this spotlight.]

9. *Cauchy–Schwarz Inequality* Show that $|\langle u, v \rangle| \leq \|u\| \cdot \|v\|$ for all u, v in V. [*Hint*: use the fact that $\|\alpha u - \beta v\|^2 \geq 0$ for *all* scalars α and β.]

10. Show that $\langle u, v \rangle = (\|u + v\|^2 - \|u - v\|^2)/4$, for all u, v in V.

Answer 9: We want to show that $|\langle u, v\rangle| \le \|u\| \cdot \|v\|$. If $u = 0$ or $v = 0$, we are done. Let $u \ne 0$, $v \ne 0$. Then, for any real constants α and β, we have

$$0 \le \|\alpha u - \beta v\|^2 = \alpha^2 \|u\|^2 + \beta^2 \|v\|^2 - \alpha\beta\langle u, v\rangle - \alpha\beta\langle v, u\rangle$$

Let $|\alpha| = 1/\|u\|$, $|\beta| = 1/\|v\|$. Then we see that $2|\alpha| \cdot |\beta| \cdot |\langle u, v\rangle| \le |\alpha|^2\|u\|^2 + |\beta|^2\|v\|^2 = 2$ since $0 \le \langle |\alpha|u - |\beta|v, |\alpha|u - |\beta|v\rangle = |\alpha|^2\|u\|^2 - 2|\alpha| \cdot |\beta| \cdot \langle u, v\rangle + |\beta|^2\|v\|^2$. So $|\langle u, v\rangle| \le 1/(|\alpha| \cdot |\beta|) = \|u\| \cdot \|v\|$.

Fourier Series. Consider the orthogonal set $\Phi = \{\sin kx : k = 1, 2, \ldots\}$ in $PC[0, \pi]$. Find the Fourier series of the following functions in $PC[0, \pi]$ with respect to Φ:

11. $f(x) = x$ **12.** $f(x) = 1$ **13.** $f(x) = 1 + x/\pi$

Answer 11: The function $f(x)$ is x. $FS_\Phi[f]$ has the form $\sum_{k=1}^\infty B_k \sin kx$, where

$$B_k = \left[\int_0^\pi f(x)\sin kx\, dx\right] \Big/ \left[\int_0^\pi \sin^2 kx\, dx\right], \qquad k = 1, 2, \ldots.$$

Note that $\int_0^\pi \sin^2 kx\, dx = \pi/2$, for all $k = 1, 2, \ldots$. We have for any $k = 1, 2, \ldots$,

$$B_k = \frac{2}{\pi}\int_0^\pi x\sin kx\, dx = -\frac{2}{k\pi}\int_0^\pi x(\cos kx)'dx$$

$$= -\frac{2}{k\pi}\left\{x\cos kx\Big|_0^\pi - \int_0^\pi \cos kx\, dx\right\}$$

$$= 2(-1)^{k+1}/k$$

So we have $FS_\Phi[f] = 2\sum_{k=1}^\infty (-1)^{k+1}(\sin kx)/k$.

Answer 13: Here the function is $f(x) = 1 + x/\pi$, and $FS_\Phi[f] = \sum_{k=1}^\infty B_k \sin kx$, where

$$B_k = \frac{2}{\pi}\int_0^\pi \sin kx\, dx = -\frac{2}{k\pi}\cos kx\Big|_0^\pi = \frac{2(1 - (-1)^k)}{k\pi} = \begin{cases} 0, & \text{even } k \\ 4/k\pi, & \text{odd } k \end{cases}$$

So

$$FS_\Phi[f] = \frac{4}{\pi}\sum_{m=0}^\infty \frac{\sin(2m+1)x}{2m+1}$$

14
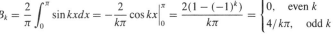

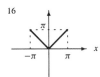

15

Calculating Fourier Series. $\Phi = \{1, \cos x, \sin x, \ldots, \cos nx, \sin nx, \ldots\}$ is an orthogonal subset, with the standard scalar product, of the Euclidean space $PC[-\pi, \pi]$. Find the Fourier series of the following functions in $PC[-\pi, \pi]$ with respect to Φ. See margin figures for graphs.

16

14. *Sawtooth* $f(x) = \begin{cases} -A(1 + x/\pi), & -\pi \le x < 0 \\ A(1 - x/\pi), & 0 < x < \pi \end{cases}$

15. *Triangular* $f(x) = \begin{cases} A(1 + x/\pi), & -\pi \le x \le 0 \\ A(1 - x/\pi), & 0 < x < \pi \end{cases}$

16. *Inverted Triangular* $f(x) = |x|, \quad |x| \le \pi$

Answer 15: The triangular function $f(x)$ is even, so FS[f] does not contain any sine terms. Also, note that $f(x)\cos kx$ is even for each $k = 0, 1, 2, \ldots$. We have

$$A_0 = \frac{2}{2\pi} \int_0^\pi A\left(1 - \frac{x}{\pi}\right) dx = \frac{A}{2},$$

$$A_k = \frac{2}{\pi} \int_0^\pi A\left(1 - \frac{x}{\pi}\right) \cos kx\, dx = \frac{2A}{k\pi} \int_0^\pi \left(1 - \frac{x}{\pi}\right)(\sin kx)'dx$$

$$= \frac{2A}{k\pi}\left\{\left(1 - \frac{x}{\pi}\right)\sin kx\Big|_0^\pi - \int_0^\pi \left(-\frac{1}{\pi}\right)\sin kx\, dx\right\}$$

$$= \frac{2A}{k^2\pi^2}\left(-\cos kx\Big|_0^\pi\right) = \frac{2A}{k^2\pi^2}(1 - (-1)^k) = \begin{cases} 0, & \text{even } k \\ \frac{4A}{k^2\pi^2}, & \text{odd } k \end{cases}$$

We have

$$\mathrm{FS}_\Phi[f] = \frac{A}{2} + \frac{4A}{\pi^2}\sum_{m=0}^\infty \frac{\cos(2m+1)x}{(2m+1)^2}$$

Euclidean Spaces.

17. *Orthogonal Projection*   Suppose that E is a Euclidean space and that S is a subspace of E spanned by the finite orthogonal set $\{u^1, u^2, \ldots, u^n\}$. For u in E the vector

$$\mathrm{proj}_S(u) = \sum_{i=1}^n [\langle u, u^i\rangle / \|u^i\|^2] u^i$$

is the *orthogonal projection* of u onto S. [*Note*: $\mathrm{proj}_S(u)$ does not depend on the orthogonal spanning set of S chosen.]

(a) Show that for each u in E we can uniquely write $u = v + w$, with v in S and w orthogonal to S.

(b) Show that for any u in E, $\|u - \mathrm{proj}_S(u)\| \leq \|u - v\|$ for all v in S, and that equality holds if and only if $v = \mathrm{proj}_S(u)$.

(c) In $\mathbb{R}^3$, find the distance from the point $(1, 1, 1)$ to the plane: $x + 2y - z = 0$.

Answer 17:

(a) For u in E, consider the vectors $v = \mathrm{proj}_S(u)$ and $w = u - \mathrm{proj}_S(u)$. Then $v + w = u$ and v is in S, where S is the span of the orthogonal set $\{u^1, \ldots, u^n\}$. Since

$$\langle u^k, w\rangle = \langle u^k, u - \sum_1^n [\langle u, u^i\rangle/\|u^i\|^2]u^i\rangle = \langle u^k, u\rangle - [\langle u, u^k\rangle/\|u^k\|^2]\langle u^k, u^k\rangle = 0$$

for each $k = 1, 2, \ldots, n$, it follows that w is orthogonal to S, and the assertion follows. The proof of uniqueness is omitted.

(b) We want to show that $\|u - \mathrm{proj}_S(u)\| \leq \|u - v\|$ for all v in S, equality holding if and only if $v = \mathrm{proj}_S(u)$. First notice that for any pair of orthogonal vectors $\{x, y\}$ in E we have that $\|x + y\|^2 = \|x\|^2 + \|y\|^2$ because $\langle x, y\rangle = \langle y, x\rangle = 0$. Now let u in E and v in S be given. Then the vectors $u - \mathrm{proj}_S(u)$, $\mathrm{proj}_S(u) - v$ are orthogonal since $\mathrm{proj}_S(u) - v$ is in S and $u - \mathrm{proj}_S(u)$ is orthogonal to S by part **(a)**. We have that

$$\|u - v\|^2 = \|(u - \mathrm{proj}_S(u)) + (\mathrm{proj}_S(u) - v)\|^2 = \|u - \mathrm{proj}_S(u)\|^2 + \|\mathrm{proj}_S(u) - v\|^2$$

$$\geq \|u - \mathrm{proj}_S(u)\|^2$$

The inequality becomes an equality if and only if $v = \mathrm{proj}_S(u)$.

(c) We want to find the distance from $(1, 1, 1)$ to the plane $x + 2y - z = 0$. The plane $x + 2y - z = 0$ contains the two vectors $u = (1, 0, 1)$ and $v = (0, 1, 2)$. Now if v is regarded as a basis for the line S through the origin parallel to v then $w = u - \mathrm{proj}_S(u)$ also lies in the plane and the pair $\{v, w\}$ is a basis for the plane P. We have that $w = (1, 0, 1) - 2(0, 1, 2)/5 = (1, -2/5, 1/5)$. To find the distance d between the point $q = (1, 1, 1)$ and the plane P we see from the statement of part **(b)** that this distance is given by $d = \|q - \text{proj}_S(q)\|$. Since

$$\text{proj}_S(q) = \frac{\langle q, v \rangle}{\|v\|^2} v + \frac{\langle q, w \rangle}{\|w\|^2} w = (1/3)(2, 1, 4)$$

and so $q - \text{proj}_S(q) = \frac{1}{3}(1, 2, -1)$, giving the result that $d = \|\frac{1}{3}(1, 2, -1)\| = \sqrt{6}/3$.

18. *Little Ell Two: l^2*   Suppose that $\mathbb{R}^\infty$ is the set of all sequences of real numbers, $(x_n)_{n=1}^\infty$, abbreviated simply as (x_n). Define addition and multiplication by real numbers in $\mathbb{R}^\infty$ termwise; that is, $(x_n) + (y_n) = (x_n + y_n)$, $\alpha \cdot (x_n) = (\alpha x_n)$. Suppose that l^2 is the collection of all sequences (x_n) in l^2 for which $\sum_{n=1}^\infty |x_n|^2$ converges; l^2 is pronounced "little ell two."

    (a) Show that for any x, y in l^2, $\sum_{i=1}^\infty x_i y_i$ converges.

    (b) Show that $\langle x, y \rangle = \sum_{i=1}^\infty x_i y_i$ is a scalar product on l^2.

19. *Best Approximation*   In the Euclidean space $PC[-\pi, \pi]$, consider the subspace S_N spanned by the set $\Phi_N = \{1, \cos x, \sin x, \ldots, \cos Nx, \sin Nx\}$. Find the element in S_N that is closest to the element $f(x) = x$ in $PC[-\pi, \pi]$.

    Answer 19:   This problem is solved by using Theorem **??** with $\Phi_N = \{1, \cos x, \sin x, \ldots, \cos Nx, \sin Nx\}$. The element which is closest to $f(x) = x$, $-\pi < x < \pi$, in the subspace S_N of $PC[-\pi, \pi]$ is given by

$$\text{proj}_{S_N}(f) = a_0 + \sum_{k=1}^N (a_k \cos kx + b_k \sin kx)$$

$$a_k = \frac{\langle f, \cos kx \rangle}{\|\cos kx\|^2} = \begin{cases} (2\pi)^{-1} \int_{-\pi}^{\pi} x \, dx, & \text{for } k = 0 \\ (\pi)^{-1} \int_{-\pi}^{\pi} x \cos kx \, dx, & \text{for } k = 1, 2, \ldots, N \end{cases}$$

$$b_k = \frac{\langle f, \sin kx \rangle}{\|\sin kx\|^2} = \frac{1}{\pi} \int_{-\pi}^{\pi} x \sin kx \, dx, \quad \text{for } k = 1, 2, \ldots, N$$

Notice that f is odd about the origin, so $f(x) \cos kx$ is also odd about the origin for all $k = 0, 1, 2, \ldots, N$. We have $a_k = 0$ for $k = 0, 1, 2, \ldots, N$. Note that $f(x) \sin kx$ is even about the origin, so

$$b_k = \frac{2}{\pi} \int_0^\pi x \sin kx \, dx = -\frac{2}{k\pi} \int_0^\pi x (\cos kx)' \, dx$$

$$= -\frac{2}{k\pi} \left[x \cos kx \Big|_0^\pi - \int_0^\pi \cos kx \, dx \right] = \frac{(-1)^{k+1} 2}{k}, \quad \text{for } k = 1, 2, \ldots, N,$$

So $\text{proj}_{S_n}(f) = 2 \sum_{k=1}^N (-1)^{k+1} k^{-1} \sin kx$ is the element in S_N closest to $f(x) = x$.

20. *Not All Trigonometric Series Are Fourier Series*   Show that the trigonometric series

$$\sum_1^\infty \frac{1}{n^{1/4}} \cos nx, \qquad \sum_1^\infty (\sin n) \sin nx, \qquad \sum_2^\infty \frac{1}{\ln(n)} \sin nx$$

are *not* the Fourier series of any functions in $PC[-\pi, \pi]$. [*Hint*: use the Parseval Relation (Theorem 7).]

21. Evaluate $\lim_{n \to \infty} \int_{-\pi}^{\pi} e^{\sin x} (x^5 - 7x + 1)^{52} \cos nx \, dx$. [*Hint*: see Decay of Coefficients Theorem 7 and relate the integral to a Fourier coefficient of a portion of the integrand.]

    Answer 21:   The integral is the coefficient $A_n = \int_{-\pi}^{\pi} e^{\sin x} (x^5 - 7x + 1)^{52} \cos nx \, dx$ of $\cos nx$ in the Fourier Series of the function $f(x) = \pi e^{\sin x} (x^5 - 7x + 1)^{52}$, $|x| < \pi$, and so by the Decay Theorem 10.2.2 A_n must decay to zero as $n \to \infty$.

22. Show that $\{1, \cos x, \ldots, \cos nx, \ldots\}$ is *not* a basis of $PC[-\pi, \pi]$.

23. Show that $\{1, \cos x, \ldots, \cos nx, \ldots\}$ *is* a basis of the subspace of all even functions in $PC[-\pi, \pi]$.

Answer 23: Let E be the subspace of even functions in PC$[-\pi, \pi]$, Since

$$\{1, \cdots, \cos nx, \cdots, \sin x, \cdots; \sin nx, \cdots\}$$

is a basis for PC$[-\pi, \pi]$ it follows that for any $f(x)$ in E,

$$f = \sum_{n=0}^{\infty} \frac{\langle f, \cos nx \rangle}{\| \cos nx \|^2} \cos nx + \sum_{n=1}^{\infty} \frac{\langle f, \sin nx \rangle}{\| \sin nx \|^2} \sin nx$$

Since f is an even function it follows that $\langle f, \sin nx \rangle = 0$, for $n = 1, 2, \cdots$. Hence every even function on $[-\pi, \pi]$ can be represented as an orthogonal series over $\Phi = \{1, \cdots, \cos nx, \cdots\}$. Thus Φ is a basis for E.

24. *Scalar Product Continuity*   Show the continuity of the scalar product (Theorem 4). [*Hint*: write $\langle f_n, g_m \rangle - \langle f, g \rangle = \langle f_n, g_m - g \rangle + \langle f_n - f, g \rangle$ and use the Cauchy-Schwarz inequality.]

25. *Uniqueness*   Suppose that Φ is a basis for a Euclidean space E. Show that if two vectors f and g in E have the same Fourier series over Φ, then $f = g$.

Answer 25:   We want to show that two functions f and g with the same Fourier series with respect to an orthogonal basis Φ of a Euclidean space are "equal" in that space, where by equal we mean that $\|f - g\| = 0$. Let $w = f - g$. Since $\Phi = \{\phi_1, \phi_2, \ldots\}$ is a basis for E, the Fourier series of w over Φ must converge to w in the mean; that is,

$$w_N = \sum_{k=1}^{N} \frac{\langle w, \phi_k \rangle}{\|\phi_k\|^2} \phi_k \to w \text{ as } N \to \infty$$

By the continuity of the scalar product, $\langle w_N, w \rangle \to \|w\|^2$, as $N \to \infty$. But since $\langle w, \phi_k \rangle = 0$ for all $k = 1, 2, \ldots$, we see that $w_N = 0$ for all N, so $\langle w_N, w \rangle \to 0$ as $N \to \infty$. But $\langle w_N, w \rangle \to \langle w, w \rangle$ by Theorem 4, so $\langle w, w \rangle = \|w\|^2$, so $\|w\|^2 = 0$, which in turn implies that $f = g$.

26. *A Euclidean Space of Functions*   The collection of real-valued functions on $\mathbb{R}$ defined by $E = \{f$ in $C^0(\mathbb{R}) : \int_{-\infty}^{\infty} |f|^2 \, dx < \infty\}$ is a linear space.

    (a) Show that $\langle f, g \rangle = \int_{-\infty}^{\infty} fg \, dx$ is a scalar product on E.

    (b) Consider the sequence $f_n(x) = n^{-1/2} e^{-x^2/n^2}$, $n = 1, 2, \ldots$. Show that $\{f_n\}$ converges uniformly to the zero function $f(x) = 0$, all x.

    (c) Show that the sequence $\{f_n\}$, where f_n is given in **(b)**, does *not* converge in the mean in E. [*Hint*: if $\{f_n\}$ were mean-convergent, the limit must be the zero function (why?), but Theorem 4 contradicts this.]

Some Useful Properties of Functions.

27. *Lagrange's Identity*   Show that

$$\frac{1}{2} + \sum_{k=1}^{n} \cos ks = \frac{\sin(n + \frac{1}{2})s}{2 \sin(s/2)}$$

[*Hint*: for $s \neq 2m\pi$, and m an integer, use the trigonometric identity

$$2 \sin(s/2) \cos ks = \sin[k + (1/2)]s - \sin[k - (1/2)s]$$

]

Answer 27:   Using a trigonometric formula from the inside back cover of the textbook, we verify that

$$2 \sin \frac{s}{2} \cos ks = \sin(k + \frac{1}{2})s - \sin(k - \frac{1}{2})s$$

for all s and all k. Sum this identity over the set $k = 1, 2, \ldots, n$; observing that the right-hand side is a telescoping series, we obtain that

$$2 \sin \frac{s}{2} \sum_{k=1}^{n} \cos ks = \sin(n + \frac{1}{2})s - \sin \frac{s}{2}$$

or

$$2 \sin \frac{s}{2} \left(\frac{1}{2} + \sum_{k=1}^{n} \cos ks \right) = \sin(n + \frac{1}{2})s$$

from which Lagrange's identity

$$\frac{1}{2} + \sum_{k=1}^{n} \cos ks = \sin(n + 1/2)s / [2 \sin(s/2)]$$

follows for $s \neq 2k\pi$. It is also valid if $s = 2k\pi$, as may be verified by L'Hôpital's Rule applied to $\lim_{s \to 2k\pi} \sin((n+1/2)s) / \sin(s/2)$.

28. *Integrating a Periodic Function*   Show that $\int_{-T}^{T} f(x)\,dx = \int_{a}^{a+2T} f(s)\,ds$ for all real numbers a if f is periodic on $\mathbb{R}$ with period $2T$ and piecewise continuous on $[-T, T]$. [*Hint*: break up the integral on the right and change the integration variable.]

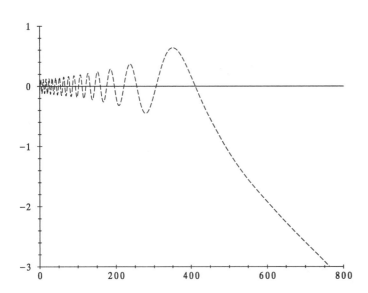

This graph shows the changing position of a weight acted on by an aging spring. See Examples 11.1.1 and 11.2.7 as well as the WEB SPOTLIGHT ON THE EXTENDED METHOD OF FROBENIUS.

Series Solutions

Chapters 3, 4, and 5 give the general theory, solution techniques, and properties of solutions for a linear second-order ODE, but with little said about how to find solution formulas if the coefficients are nonconstant. The solutions of nonconstant-coefficient, second-order linear ODEs cannot usually be found in closed form, but they can be written as power series or "Frobenius" series, and that is the topic of this chapter. It is this kind of second-order ODE that often arises when separating variables in certain second-order PDEs written in polar, cylindrical, or spherical coordinates, (e.g., see the WEB SPOTLIGHT ON STEADY TEMPERATURES). That is the reason we have placed this chapter here. The chapter could be covered immediately following Chapter 4, and that is when we often cover it in our courses. In Section 11.1 (and Appendix B.1) we review properties of power series; of particular importance are the techniques of reindexing and the Identity Theorem. The notion of an ordinary point for a second-order linear ODE and the techniques needed for constructing convergent power series solutions in the neighborhood of an ordinary point are introduced in Sections 11.1 and 11.2 via examples. The two best known (and widely applied) ODEs with nonconstant coefficients are the Legendre and the Bessel ODEs; these are treated in the SPOTLIGHT ON LEGENDRE POLYNOMIALS, the SPOTLIGHT ON BESSEL FUNCTIONS, and the WEB SPOTLIGHT ON THE EXTENDED METHOD OF FROBENIUS. Regular singular points are defined in Section 11.3 and discussed mostly in the context of the Euler ODEs. In Sections 11.4, the SPOTLIGHT ON BESSEL FUNCTIONS, and the WEB SPOTLIGHT ON THE EXTENDED METHOD OF FROBENIUS we construct series solutions near regular singular points. These methods are straightforward, but involve much algebraic manipulation; a CAS may be appropriate here. We model the behavior of an aging spring in Section 11.1 by a second-order linear ODE with nonconstant coefficients and then return to that model in Sections 11.2 and the WEB SPOTLIGHT ON THE EXTENDED METHOD OF FROBENIUS.

The material of this section is now "classical" in the sense that it was largely completed by the end of the 19th century. It still is widely used in the applications of ODEs, especially when working with those ODEs that arise when separating variables in a PDE.

11.1 The Method of Power Series

Comments:

Series expansions are central in analysis and, for at least 150 years, in ODEs as well. In this section we take the first steps in developing series techniques for solving a second-order linear ODE whose coefficients are polynomials or convergent power series in the independent variable, expressing the solution as a series in powers of the independent variable (usually x in this chapter, but occasionally t as in the aging spring model). The reason for this choice of variables is mainly historical: those were the variables used when the PDEs for heat flow in a rod and the displacement of a vibrating string were originally reduced to ODEs by the technique of separating variables (Chapter 10).

Series techniques are used throughout the chapter, so it is a good idea to brush up on these techniques; power series, Taylor series, the Ratio Test, geometric series, reindexing, and the Identity Theorem are given in Appendix B.1. A good CAS will handle much of the necessary bookkeeping when working with series and may even express the solution of an ODE directly in series form. However, it is good to do some of the problems "by hand" to solidify understanding of how series can be manipulated. We introduce the model of the aging spring. We return to the aging spring problem in Section 11.2 and the WEB SPOTLIGHT ON THE EXTENDED METHOD OF FROBENIUS.

PROBLEMS

Interval of Convergence. Use the Ratio Test to determine the radius of convergence R for each series. [*Hint:* the Ratio Test is in Appendix B.1.]

1. $\sum_{n=0}^{\infty} \dfrac{x^n}{n!}$ **2.** $\sum_{n=0}^{\infty} \dfrac{nx^n}{2^n}$ 3. $\sum_{n=1}^{\infty} \dfrac{x^{2n+1}}{(2n+1)!}$ 4. $\sum_{n=1}^{\infty} n^2 x^n$

Answer 1: The series is $\sum x^n/n!$ and the ratio of the magnitude of successive terms is

$$\left| \frac{x^{n+1}/(n+1)!}{x^n/n!} \right| = \frac{1}{n+1}|x| \to 0 \quad \text{as} \quad n \to \infty$$

So the interval of convergence is $-\infty < x < \infty$.

Answer 3: In this case the series is $\sum x^{2n+1}/(2n+1)!$ and the ratio of the magnitude of successive terms is

$$\left| \frac{x^{2n+3}/(2n+3)!}{x^{2n+1}/(2n+1)!} \right| = \frac{1}{(2n+3)(2n+2)}|x^2| \to 0 \quad \text{as} \quad n \to \infty$$

So the convergence interval is $-\infty < x < \infty$.

Taylor Series. Expand each function in a Taylor series about x_0, and find the radius of convergence R. Plot the graphs of the function and its polynomial approximations using the first two, the first three, and the first four nonzero terms of the series.

☞ See Appendix B.1 for the Taylor series.

5. e^x, $x_0 = 0$, $|x| \le 2$ 6. $\sin x$, $x_0 = 0$, $|x| \le \pi$

7. $(1+x)^{-1}$, $x_0 = 0$, $|x| < 1$ 8. $\sqrt{x}$, $x_0 = 1$, $|x - 1| < 1$

Answer 5: The Taylor series about $x_0 = 0$ for e^x is $\sum_0^{\infty} x^n/n!$ since $d^n(e^x)/dx^n = e^x$ and $e^0 = 1$. The Ratio Test shows that the series converges for all x. See the figure; the solid curve is the graph of the function, the dashed curves are the graphs of the first two, three and four of the nonzero terms of the series.

Answer 7:   We can find the Taylor series for $1/(1 + x)$ about $x_0 = 0$ either from the Taylor formula with $x_0 = 0$, or more easily from the geometric series expansion: $1/(1 + x) = \sum_{n=0}^{\infty}(-1)^n x^n$. The Ratio Test implies that the series converges for $|x| < 1$. See the figure; the solid curve is the graph of the function, the dashed curves are the graphs of the first two, three and four of the nonzero terms of the series.

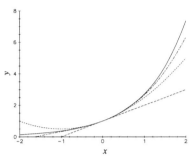

Problem 5.   Problem 7.

Reindexing.   Reindex each of the following as a series of form $\sum_{n=0}^{\infty}[\cdots]x^n$:

Note: the index range for n in your answer need not start at 0.

9. $\displaystyle\sum_{n=0}^{\infty}\frac{2(n+1)}{n!}x^{n+1}$

10. $\displaystyle\sum_{n=2}^{\infty}n(n-1)a_n x^{n-2}$

11. $\displaystyle\sum_{n=1}^{\infty}(-1)^{n-1}\frac{x^{n+1}}{n(n+1)}$

12. $\displaystyle\sum_{n=-1}^{\infty}(-1)^n n x^{n+2}$

Answer 9:   The power of x must be lowered one step, so the lower limit of the range is raised one step: $\sum_{n=0}^{\infty}2(n+1)x^{n+1}/n! = \sum_{n=1}^{\infty}2nx^n/(n-1)!$.

Answer 11:   [**Note:** the index range for n in your answer need not start at 0.] The power of x must be lowered one step and the lower range on the sum is raised one step: $\sum_{n=1}^{\infty}(-1)^{n-1}x^{n+1}/(n(n+1)) = \sum_{n=2}^{\infty}(-1)^n x^n/((n-1)n)$. Note that $(-1)^{n-2} = (-1)^n$, and we have selected the simpler form.

Taylor Series.   Use the Taylor series of e^x, $\sin x$, $\cos x$, and $1/(1 + x)$ about the base point $x_0 = 0$ to obtain the Taylor series of each function. [*Hint*: see Appendix B.1.]

13. $e^x + e^{-x}$ **14.** $\sin x - \cos x$ **15.** $\sin x \cos x$ **16.** $e^x/(1 + x)$

Answer 13: $e^x + e^{-x} = \sum_{n=0}^{\infty}x^n/n! + \sum_{n=0}^{\infty}(-x)^n/n! = \sum_{k=0}^{\infty}2x^{2k}/(2k)!$ for all x, since the odd-indexed terms cancel.

Answer 15:   By a formula on the inside back cover of the textbook, $\sin x \cos x = (\sin 2x)/2$, so we have the Taylor series about $x_0 = 0$:
$(1/2)\sum_{k=0}^{\infty}(-1)^k(2x)^{2k+1}/(2k+1)!$, all x.

Recursion Formulas.   Use the Identity Theorem (Appendix B.1) to find a recursion formula for a_n.

17. $\displaystyle\sum_{n=1}^{\infty}(na_n - n + 2)x^n = 0$

18. $\displaystyle\sum_{n=1}^{\infty}[(n+1)a_n - a_{n-1}]x^n = 0$

19. $\displaystyle\sum_{n=0}^{\infty}[(n+2)(n+1)a_{n+2} - a_n]x^n = 0$

20. $\displaystyle\sum_{n=0}^{\infty}\left[(n+2)(n+1)a_{n+2} - na_n\right] = 0$

Answer 17:   By the Identity Theorem (see item 5 in Appendix B.1), $na_n - n + 2 = 0$ for $n = 1, 2,$ $\ldots$. Hence, $a_n = 1 - 2/n$.

Answer 19:   By the Identity Theorem we have $(n+2)(n+1)a_{n+2} = a_n$, or $a_{n+2} = a_n/(n+2)(n+1)$,

$n = 0, 1, 2, \ldots$. So

$$a_{n+2} = \frac{1}{(n+2)(n+1)} a_n = \frac{1}{(n+2)(n+1)} \left(\frac{1}{n(n-1)} a_{n-2} \right) = \cdots$$

$$= \frac{1}{n+2} \cdot \frac{1}{n+1} \cdot \frac{1}{n} \cdot \frac{1}{n-1} \cdots \frac{1}{2} a_0 = \frac{1}{(n+2)!} a_0$$

if n is even. If n is odd, we get $a_{n+2} = a_1/(n+2)!$. In any case a_n is a_0 or a_1 divided by $n!$ if $n \geq 2$. Note that we have continued to solve the recursion in order to express the general coefficient as a multiple of a_0 or a_1.

Series Solutions.

21. Show that $y = \sum_0^\infty x^n/n!$ is the solution of $y'' - y = 0$, $y(0) = 1$, $y'(0) = 1$ by direct substitution. How is this solution expressed in terms of elementary functions?

22. Solve the IVP $y'' - 4y = 0$, $y(0) = 2$, $y'(0) = 2$ in the form of a series. Express the solution in terms of exponentials, and describe the connection between this form of the solution and the series. Plot the solution for $|x| \leq 1$. Overlay the graphs of the polynomial approximations that use the first two, three, and four nonzero terms of the series.

Answer 21: We want to show that $y = \sum_0^\infty x^n/n!$ solves $y'' - y = 0$, $y(0) = 1$, $y'(0) = 1$. If $y = \sum_{n=0}^\infty x^n/n!$, then $y'' = \sum_{n=0}^\infty n(n-1)x^{n-2}/n! = \sum_{n=2}^\infty x^{n-2}/(n-2)!$ since the series converges for all x (by the Ratio Test), so can be differentiated term by term. (Note the first two terms in the first series for y'' have been dropped since they are zero.) Then, $y'' - y = \sum_{n=2}^\infty x^{n-2}/(n-2)! - \sum_{n=0}^\infty x^n/n!$. Reindexing the first series, we have that $y'' - y = \sum_{n=0}^\infty x^n/n! - \sum_{n=0}^\infty x^n/n! = 0$ for all x. Since the exact solution is e^x, we see that $e^x = \sum_{n=0}^\infty x^n/n!$, which is hardly surprising since that is the Taylor series for e^x about $x_0 = 0$.

Using Recursion Formulas.

23. Find the coefficient a_{100} in the series $\sum_{n=0}^\infty a_n x^n$ if we know that $a_0 = a_1 = 1$ and that $\sum_{n=0}^\infty [(n+1)^2 a_{n+2} - n^2 a_{n+1} + (n-1)a_n]x^n = 0$.

Answer 23: We want to find a_{100} if it is known that $\sum_0^\infty [(n+1)^2 a_{n+2} - n^2 a_{n+1} + (n-1)a_n]x^n = 0$. The recursion identity here is $(n+1)^2 a_{n+2} - n^2 a_{n+1} + (n-1)a_n = 0$, that is, $a_{n+2} = [n^2 a_{n+1} - (n-1)a_n]/(n+1)^2$, $n = 0, 1, 2, \ldots$, where $a_0 = a_1 = 1$. A computer may be programmed to calculate the coefficients using the identity and the initial data $a_0 = a_1 = 1$. In fact, $a_{100} \approx -1.0629 \times 10^{-6}$.

Power Series Solutions: IVPs. Find the general solution in power series form. Then find the solution that meets the initial conditions $y(0) = 1$ and $y'(0) = 1$.

24. $y'' + xy' + y = 0$ **25.** $y'' - 2xy' + y = 0$

26. $5y'' - 2xy' + 10y = 0$ **27.** $y'' - xy' + 2y = 0$

28. $y'' - xy' + 4y = 0$ **29.** $(1 + x^2)y'' + 6xy' + 4y = 0$

Answer 25: Let $y = \sum_{n=0}^\infty a_n x^n$. Then, after some reindexing,

$$y'' - 2xy' + y = \sum_{n=0}^\infty \left((n+2)(n+1)a_{n+2}x^{n+2} - 2na_n + a_n \right) x^n = 0$$

The recursion formula is

$$a_{n+2} = \frac{(2n-1)a_n}{(n+2)(n+1)}, \quad n = 0, 1, 2, \cdots$$

So

$$a_{2k+2} = \frac{(4k-1)a_{2k}}{(2k+2)(2k+1)} = \cdots = \frac{(4k-1)\cdots 3 \cdot (-1) \cdot a_0}{(2k+2)!}$$

$$a_{2k+3} = \frac{(4k+1)a_{2k+1}}{(2k+3)(2k+2)} = \cdots = \frac{(4k+1)\cdots 5 \cdot 1 \cdot a_1}{(2k+3)!}$$

The general solution is

$$y = \sum_{k=-1}^{\infty} \left(a_{2k+2}x^{2k+2} + a_{2k+3}x^{2k+3} \right)$$

where a_0 and a_1 are arbitrary constants. Let $y(0) = 1 = a_0$, $y'(0) = 1 = a_1$; then

$$y = 1 + x - \frac{x^2}{2} + \frac{x^3}{6} - \frac{x^4}{8} + \cdots$$

Answer 27:   Let

$$y = \sum_{n=0}^{\infty} a_n x^n$$

After some reindexing, we get

$$y'' - xy' + 2y = \sum_{n=0}^{\infty} \left[(n+2)(n+1)a_{n+2} - na_n + 2a_n \right] x^n = 0$$

The recursion formula is

$$a_{n+2} = \frac{(n-2)a_n}{(n+2)(n+1)}, \quad n = 0, 1, 2, \cdots$$

Note that $a_4 = 0 \cdot a_2$, and so $a_6 = 2a_4/30 = 0$, and so on; so $a_{2k} = 0$, $k = 2, 3, \cdots$. The first few terms of the general solution are

$$y = a_0 + a_1 x - a_0 x^2 - a_1 x^3/6 + 0x^4 - a_1 x^5/120 + 0x^6 - \cdots$$
$$= a_0(1 - x^2) + a_1(x - x^3/6 - x^5/120 - \cdots)$$

where a_0 and a_1 are arbitrary constants. If we set $y(0) = a_0 = 1$ and $y'(0) = a_1 = 1$, then

$$y = \left(1 - x^2\right) + \left(x - x^3/6 - x^5/120 - \cdots\right)$$

Answer 29:   Let

$$y = \sum_{n=0}^{\infty} a_n x^n$$

After some reindexing, we have

$$(1 + x^2)y'' + 6xy' + 4y = \sum_{n=0}^{\infty} \left[(n+2)(n+1)a_{n+2} + (n(n-1) + 6n + 4)a_n \right] x^n = 0$$

The recursion relation is

$$a_{n+2} = -\frac{n^2 + 5n + 4}{(n+2)(n+1)}a_n = -\frac{n+4}{n+2}a_n$$

So the general solution is

$$y = a_0 + a_1 x - 2a_0 x^2 - 5a_1 x^3/3 + 3a_0 x^4 + 7a_1 x^5/3 - \cdots$$
$$= a_0(1 - 2x^2 + 3x^4 - 4x^6 + \cdots + (-1)^k(k+1)x^{2k} + \cdots)$$
$$+ a_1(x - 5x^3/3 + 7x^5/3 + \cdots + (-1)^k(2k+3)x^{2k+1}/3 + \cdots)$$

where a_0 and a_1 are arbitrary constants. If we set $y(0) = a_0 = 1$ and $y'(0) = a_1 = 1$, then

$$y = 1 + x - 2x^2 - 5x^3/3 + 3x^4 + 7x^5/3 - 4x^6 + \cdots + (-1)^k \left((k+1)x^{2k} + (2k+3)x^{2k+1}/3 \right) + \cdots$$

Modeling Problem.

 30. *Aging Spring*   Create models of an aging spring for which the value of the spring coefficient is zero after a finite time. Use graphs of solution curves in your analysis.

- What would you expect to happen to the spring in the long run? Compare your result with what happens in a model ODE with zero spring constant.

- What is the effect of including damping in the aging spring model?

11.2   Series Solutions near an Ordinary Point

Comments:

Ordinary points and singular points are defined for the ODE $y'' + P(x)y' + Q(x)y = 0$, and power series solutions are constructed in powers of $x - x_0$, where x_0 is an ordinary point. In most of the examples and in the problems, x_0 is taken to be zero for simplicity; if $x_0 \neq 0$ then the variable change $z = x - x_0$ could be applied and you look for a series solution of the transformed ODE based at $z_0 = 0$. The Procedure for using series to solve ODEs in this case is outlined in the text.

Two of the famous "named" ODEs of classical analysis are introduced in the text and the problem set: the Legendre and Airy ODEs. A series solution of the ODE for the aging spring is obtained and its usefulness examined. The Convergence Theorem for the series solutions about an ordinary point is presented, but not proved. In the text and in the problems when dealing with a specific power series we (mostly) rely on the Ratio Test, rather than the Convergence Theorem, to determine intervals of convergence.

PROBLEMS

Ordinary Points, Singular Points.   Determine whether x_0 is an ordinary or a singular point.

1. $y'' + k^2 y = 0$, x_0 arbitrary (k is a constant)   **2.** $y'' + \dfrac{1}{1+x} y = 0$, $x_0 = -1$

3. $y'' + \dfrac{1}{x} y' - y = 0$, $x_0 = 0$ **4.** $(\sin x)y'' - (\cos x)y = 0,\ x_0 = 0$

Answer 1:   The ODE is $y'' + k^2 y = 0$. Since $P(x) = 0$ and $Q(x) = k^2$ are real analytic at every point, every point x_0 is ordinary.

Answer 3:   The ODE is $y'' + y'/x - y = 0$. Since $P(x) = 1/x$ is undefined at $x_0 = 0$, $P(x)$ is not real analytic at $x_0 = 0$, so 0 is a singular point.

Singular Points.   Find all singular points (if any) of the following ODEs. [*Hint*: for 8–10 use the fact that the quotient of two polynomials with no common factors is real analytic everywhere except at the roots of the denominator.]

5. $y'' + (\sin x)y' + (\cos x)y = 0$ **6.** $y'' - (\ln|x|)y = 0$

7. $y'' + |x|y' + e^{-x}y = 0$ **8.** $(1-x)y'' + y' + (1-x^2)y = 0$

9. $(1 - x^2)y'' + xy' + y = 0$ **10.** $x^2 y'' + xy' - y = 0$

Answer 5:   The ODE is $y'' + (\sin x)y' + (\cos x)y = 0$. Since $\sin x$ and $\cos x$ are real analytic for all x, every point is ordinary and there are no singular points.

Answer 7:   The ODE is $y'' + |x|y' + e^{-x}y = 0$. The coefficient function $|x|$ is not real analytic at $x = 0$ since it is not differentiable there. It is real analytic at all other values of x. The function e^{-x} is real analytic everywhere so $x_0 = 0$ is the only singular point.

Answer 9: The ODE is $(1 - x^2)y'' + xy' + y = 0$. After normalizing the ODE, we see that $P(x) = (1 - x^2)^{-1}x$ and $Q(x) = (1 - x^2)^{-1}$. Both $P(x)$ and $Q(x)$ are real analytic at every x_0, except $x_0 = \pm 1$ which are the two singularities.

Series Solutions. Find the recursion formula and use it to find the first five nonzero terms in the power series $\sum a_n x^n$ of the general solution. Determine the radius R of convergence of the series.

11. $y'' + xy' + 3y = 0$ **12.** $(x^2 + 1)y'' - 6y = 0$

13. $(x^2 - 1)y'' + 4xy' + 2y = 0$ **14.** $(x^2 + 2)y'' + 3xy' - y = 0$

15. $(x^2 - 1)y'' + xy' - y = 0$ **16.** $y'' + (\sin x)y = 0$

Answer 11: The ODE is $y'' + xy' + 3y = 0$. Let the general solution be $y = \sum_{n=0}^{\infty} a_n x^n$, so $y' = \sum_{n=1}^{\infty} n a_n x^{n-1}$, $y'' = \sum_{n=2}^{\infty} n(n-1)a_n x^{n-2}$. Then $y'' + xy' + 3y = 0$ becomes

$$\sum_{n=2}^{\infty} n(n-1)a_n x^{n-2} + \sum_{n=1}^{\infty} n a_n x^n + \sum_{n=0}^{\infty} 3 a_n x^n = 0$$

In order to include all three sums in a single summation, we must reindex the first sum and lower the range on the second sum from 1 to 0 (easy to do since $n = 0$ only gives the term 0 in the sum):

$$\sum_{n=0}^{\infty} [(n+2)(n+1)a_{n+2} + n a_n + 3 a_n] x^n = 0$$

By the Identity Theorem,

$$a_{n+2} = -(n+3)a_n/[(n+2)(n+1)]$$

So, $y = a_0 + a_1 x - 3a_0 x^2/2 - 2a_1 x^3/3 + 5a_0 x^4/8 + \cdots$ where a_0 and a_1 are arbitrary constants.

If n is even [odd], we can push the recursion formula for the coefficients back to a_0 [a_1]. This is most easily done by considering the two cases separately, first n even and then n odd:

$$a_{2k} = -\frac{2k+1}{(2k)(2k-1)}a_{2k-2} = \left(-\frac{2k+1}{(2k)(2k-1)}\right)\left(-\frac{2k-1}{(2k-2)(2k-3)}a_{2k-4}\right)$$

$$= \cdots = \frac{(-1)^k(2k+1)(2k-1)\cdots 3}{(2k)!}a_0$$

Since

$$(2k+1)(2k-1)\cdots 3 = \frac{(2k+1)(2k)(2k-1)(2k-2)\cdots 3 \cdot 2}{(2k)(2k-2)\cdots 2} = \frac{(2k+1)!}{2^k k!}$$

we have that

$$a_{2k} = (-1)^k \frac{(2k+1)!}{(2k)!2^k k!} = (-1)^k \frac{2k+1}{2^k k!}a_0$$

In exactly the same way, it may be shown that

$$a_{2k+1} = -\frac{2k+2}{(2k+1)(2k)}a_{2k-1} = \cdots = (-1)^k \frac{(2k+2)(2k)\cdots 4}{(2k+1)!}a_1 = (-1)^k \frac{2^k(k+1)!}{(2k+1)!}a_1$$

We may write the general solution as

$$y = a_0 \sum_0^{\infty} a_{2k} x^{2k} + a_1 \sum_0^{\infty} a_{2k+1} x^{2k+1} = a_0 \sum_{k=0}^{\infty} (-1)^k \frac{(2k+1)}{2^k k!}x^{2k} + a_1 \sum_{k=0}^{\infty} (-1)^k \frac{2^k(k+1)!}{(2k+1)!}x^{2k+1}$$

where a_0 and a_1 are arbitrary constants.

We may test the convergence of the series by the Ratio Test. For the first series we have that the magnitude of the ratio of successive terms is $|(2k+1)x^2/[2k(2k-1)]|$, which $\to 0$ as $k \to \infty$. The series converges for all x. It may be shown in much the same way that the second series also converges for all x.

Answer 13: Let $y = \sum_{n=0}^{\infty} a_n x^n$. After some reindexing, we see that

$$(x^2 - 1)y'' + 4xy' + 2y = \sum_{n=0}^{\infty}\left[-(n+2)(n+1)a_{n+2} + (n(n-1)+4n+2)a_n\right]x^n = 0$$

The recursion relation is

$$a_{n+2} = \frac{n(n-1)+4n+2}{(n+2)(n+1)}a_n = a_n, \quad n = 0,1,2,\cdots$$

So the general solution is (by geometric series)

$$y = a_0 + a_1 x + a_0 x^2 + a_1 x^3 + \cdots$$
$$= a_0(1 + x^2 + x^4 + \cdots + x^{2k} + \cdots) + a_1 x(1 + x^2 + \cdots + x^{2k} + \cdots)$$
$$= \frac{a_0}{1-x^2} + \frac{a_1 x}{1-x^2},$$

where a_0 and a_1 are arbitrary constants. The radius of convergence is $R = 1$, that is $|x| < 1$.

Answer 15: Let $y = \sum_{n=0}^{\infty} a_n x^n$. After some reindexing, we have

$$(x^2 - 1)y'' + xy' - y = \sum_{n=0}^{\infty}\left[-(n+2)(n+1)a_{n+2} + (n(n-1)+n-1)a_n\right]x^n = 0$$

The recursion relation is

$$a_{n+2} = -\frac{n(n-1)+n-1}{(n+2)(n+1)}a_n = -\frac{n-1}{n+2}a_n, \quad n = 0,1,2,\cdots$$

First, note that $a_3 = 0$, and so $a_{2k+1} = 0, k = 1,2,\cdots$. So, the general solution is

$$y = a_1 x + a_0(1 + x^2/2 - x^4/8 + x^6/16 - 5x^8/128 + \cdots)$$

where a_0 and a_1 are arbitrary constants. The magnitude of the ratio of successive terms in the series is

$$\left|\frac{n-1}{n+2}\right|x^2, \quad \text{which} \to |x|^2 \text{ as } n \to \infty$$

So $R = 1$ and the series converges for $|x| < 1$.

17. *Airy's Equation* The ODE $y'' - xy = 0$ is *Airy's equation*. We call certain solutions Airy functions and use them to model the diffraction of light.

(a) Find all the solutions of the equation $y'' - xy = 0$ as power series in x.

(b) Plot the solution satisfying $y(0) = 1$, $y'(0) = 0$ over the interval $-20 \le x \le 2$.

(c) Plot the sums of the first two and the first four nonvanishing terms of the series expansion of the solution of the IVP, $y'' - xy = 0$, $y(0) = 1$, $y'(0) = 0$. Compare the results with the curve in (b).

(d) Why would you expect the solution in (b) to oscillate for $x < 0$ but not for $x > 0$? [*Hint:* look at solutions of the IVPs $y'' + \alpha y = 0$ and $y'' - \alpha y = 0$, $y(0) = 1$, $y'(0) = 0$, where α is a positive constant.]

Answer 17:

(a) Let $y = \sum_{n=0}^{\infty} a_n x^n$, $y'' = \sum_{n=2}^{\infty} n(n-1)a_n x^{n-2}$, $-xy = -\sum_{n=0}^{\infty} a_n x^{n+1}$. So

$$y'' - xy = 2a_2 + \sum_{n=1}^{\infty}[(n+2)(n+1)a_{n+2} - a_{n-1}]x^n = 0$$

where the term $2a_2$ has been split off from the summation since the new index range is from 1 to ∞ rather than from 0 to ∞. By the Identity Theorem we see that $a_2 = 0$ and that

$$a_{n+2} = \frac{a_{n-1}}{(n+2)(n+1)}, \quad n \ge 1$$

Since $a_2 = 0$, we have $a_5 = a_8 = \cdots = a_{3k-1} = \cdots = 0$. From the recursion formula we can express coefficients of the form a_{3k} [a_{3k+1}] as multiples of a_0 [a_1]. The general solution is

$$y = a_0 \left[1 + \sum_{k=1}^{\infty} \frac{x^{3k}}{(3k)(3k-1)(3k-3)(3k-4)\cdots 3 \cdot 2} \right]$$

$$+ a_1 \left[x + \sum_{k=1}^{\infty} \frac{x^{3k+1}}{(3k+1)(3k)(3k-2)(3k-3)\cdots 4 \cdot 3} \right]$$

where a_0 and a_1 are arbitrary.

(b) See the figure, where a numerical solver has been used to solve the IVP, $y'' - xy = 0$, $y(0) = 1$, $y'(0) = 0$.

(c) Here $a_0 = 1$ and $a_1 = 0$ since $y(0) = 1$ and $y'(0) = 0$. We have plotted the solution curve, and the sum $p_3(x)$ of the first two nonzero terms $p_3 = 1 + x^3/6$ and the sum $p_9(x)$ of the first four nonzero terms $p_9 = p_3 + x^6/180 + x^9/12960$ of the series expansion given in **(a)** with $a_0 = 1$, $a_1 = 0$ in the figure. Note how poor the approximations p_3 and p_9 are for $x < -4$, although p_9 is a good approximation for $-2.5 \le x \le 0$. See the figure.

(d) From the figure for **(b)**, we see that the solution of $y'' - xy = 0$, $y(0) = 1$, $y'(0) = 0$ is strictly increasing for $x > 0$ and oscillatory for $x < 0$. Is there a plausibility argument that this behavior should be expected? One way is to pretend that for all negative (or positive) x, the coefficient of x in Airey's ODE is a constant with the same sign as x. The IVPs, $y'' + \alpha y = 0$ and $y'' - \alpha y = 0$, $y(0) = 1$, $y'(0) = 0$, α a positive constant, have the unique solutions $y = \cos \sqrt{\alpha} x$ and $y = (e^{\sqrt{\alpha}x} + e^{-\sqrt{\alpha}x})/2$, respectively. The function $y = \cos \sqrt{\alpha} x$, $x \le 0$, is oscillatory, and $y = (e^{\sqrt{\alpha}x} + e^{-\sqrt{\alpha}x})/2$, $x \ge 0$, is strictly increasing. So we may expect that the solution of $y'' - xy = 0$, $y(0) = 1$, $y'(0) = 0$ is oscillatory for $x < 0$ and strictly increasing for $x > 0$.

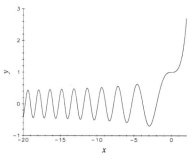

Problem 17(b).

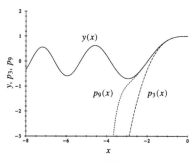

Problem 17(c).

Series Solutions of Nonlinear First-Order IVPs. Assume a solution of the form $y = \sum_0^{\infty} a_n x^n$ and find a_0, $a_1, \ldots, a_4$ for the following IVPs that involve nonlinear ODEs:

18. $y' = 1 + xy^2$, $y(0) = 0$ **19.** $y' = x^2 + y^2$, $y(0) = 0$

Answer 19:   The IVP is $y' = x^2 + y^2$, $y(0) = 0$. Since $y(0) = 0$, we have that $y = a_1 x + a_2 x^2 + \cdots$, $y' = a_1 + 2a_2 x + 3a_3 x^2 + \cdots$, and $y^2 = x^2(a_1 + a_2 x + \cdots)^2 = x^2(a_1^2 + 2a_1 a_2 x + \cdots)$. We have that

$$0 = y' - x^2 - y^2 = a_1 + 2a_2 x + (3a_3 - 1 - a_1^2)x^2 + (4a_4 - 2a_1 a_2)x^3 + \cdots.$$

We see that $a_1 = a_2 = 0$, $a_3 = 1/3, a_4 = 0$. So $y = \frac{1}{3}x^3 + \cdots$.

Estimating Series Solutions.

20. *Calculate the Leading Coefficients*   Calculate the first four nonzero terms in the power series expansion of the general solution of the equation $y'' + (1 - e^{-x^2})y = 0$.

21. *Estimate the Value of a Solution* Estimate $y(1)$ to three decimal places if $y(x)$ is the solution of the IVP $y'' + x^2y' + 2xy = 0$, $y(0) = 1$, $y'(0) = 0$.

Answer 21: The IVP is $y'' + x^2y' + 2xy = 0$, $y(0) = 1$, $y'(0) = 0$. Let $y = \sum_{n=0}^{\infty} a_n x^n$. Then $y'' = \sum_{n=2}^{\infty} n(n-1)a_n x^{n-2} = \sum_{n=0}^{\infty} (n+2)(n+1)a_{n+2}x^n$, $x^2 y' = \sum_{n=1}^{\infty} n a_n x^{n+1} = \sum_{n=2}^{\infty} (n-1)a_{n-1}x^n$, and $2xy = \sum_{n=0}^{\infty} 2a_n x^{n+1} = \sum_{n=1}^{\infty} 2a_{n-1}x^n$. Pulling the first two terms in the sum for y'' outside the summation sign as well as the first term in the summation for $2xy$, we obtain

$$y'' + x^2y' + 2xy = 2a_2 + (6a_3 + 2a_0)x + \sum_{n=2}^{\infty} [(n+2)(n+1)a_{n+2} + (n+1)a_{n-1}]x^n = 0$$

The initial conditions imply that $a_0 = 1$, $a_1 = 0$. The above expansion and the Identity Theorem imply that

$$a_2 = 0, \ a_3 = -\frac{1}{3}, \ a_{n+2} = -\frac{1}{n+2}a_{n-1}, \ n \geq 2$$

So $a_2 = 0$, $a_3 = -1/3$, $a_4 = a_5 = 0$, $a_6 = 1/18$, and so on. In general, $a_{3k+1} = a_{3k+2} = 0$, while $a_{3k} = (-1)^k[3k(3k-3)\cdots 3]^{-1} = (-1)^k \cdot 1/(3^k k!)$. The series for $y(x)$ is then $\sum_{k=0}^{\infty} (-1)^k x^{3k}/(3^k k!)$. If $x = 1$ this is an alternating series whose terms decrease in magnitude as k increases. By the Alternating Series Test (Appendix B.1, item 10), $|y(1) - \sum_{k=0}^{4} (-1)^k/(3^k k!)| \leq 1/(3^5 5!) = 1/29160 < 10^{-4}$.

$$y(1) \approx 1 - \frac{1}{3} + \frac{1}{18} - \frac{1}{162} + \frac{1}{1944} \approx 0.7166$$

with an error of magnitude less than 10^{-4}.

11.3 Regular Singular Points: The Euler ODE

Comments:

Now that we know how to construct power series solutions of a linear ODE in powers of $x - x_0$, where x_0 is an ordinary point, it is time to address the challenging problem of constructing series solutions if x_0 is singular. That can be done if x_0 is "regular singular" (isn't that an oxymoron?), and we start out by defining just what a regular singular point is. The simplest class of ODEs with regular singular points are the Euler ODEs, and most of the section is devoted to the construction of their solutions. The nice thing about an Euler ODE is that the intricate series techniques of the preceding three sections are not needed. Students (and the instructor) will have a welcome respite from the algebraic rigors of dealing with series. The indicial polynomial appears here, and should be emphasized because it plays a critical role in the analysis of the next section. There are some interesting pictures of solution behavior of an Euler ODE near its regular singularity at the origin.

Each ODE in Problems 9–14 may be written as an Euler equation $x^2 y'' + p_0 xy' + q_0 y = 0$, and the solutions are determined by the roots r_1 and r_2 of the indicial polynomial $r^2 + (p_0 - 1)r + q_0$. The three types of solutions are given by formulas (12) in the text. In each case we list the indicial polynomial, its roots, and then the general solution; c_1 and c_2 denote arbitrary constants. It is assumed throughout that $x > 0$, and absolute value signs around x are not needed. See the appropriate figures for plots of some solutions of each equation. If you want to include negative values of x, replace x by $|x|$ in all formulas.

It should be noted that the basic Existence and Uniqueness Theorem (Theorem 3.1.1) doesn't apply on an interval that includes a singularity, and in this case an IVP may have no solution at all or else infinitely many (see Example 11.3.6).

In Problems 1–8 we use the fact that the product of $f \cdot g$ of two polynomials is real analytic at every x_0, and the quotient of two polynomials f/g with no common roots is real analytic for every

x that is *not* a root of $g(x)$. We use the notation $a_2, a_1, a_0, P, Q, p, q$ introduced near the beginning of this section.

PROBLEMS

Regular and Irregular Singularities. Determine if the given point is a regular or an irregular singular point for the corresponding ODE.

1. $x^2 y'' + xy' + y = 0$, $x_0 = 0$

2. $xy'' + (1-x)y' + xy = 0$, $x_0 = 0$

3. $x(1-x)y'' + (1-2x)y' - 4y = 0$, $x_0 = 1$

4. $x^2 y'' + 2y'/x + 4y = 0$, $x_0 = 0$

Answer 1: The singularity is $x_0 = 0$; $a_2 = x^2$, $a_1 = x$, $a_0 = 1$; $P = 1/x$, $Q = 1/x^2$; $p = 1$, $q = 1$. Since p and q are real analytic, 0 is a regular singular point. The standard form of the ODE at $x_0 = 0$ is $x^2 y'' + xy' + y = 0$.

Answer 3: The singularity under study here is $x_0 = 1$ ($x_5 = 0$ is another singularity, but you aren't asked to classify it); $a_2 = x(1-x)$, $a_1 = 1-2x$, $a_0 = -4$; $P = (1-2x)/(x(1-x))$, $Q = -4/(x(1-x))$; $p = (x-1)P = (2x-1)/x$, $q = (x-1)^2 Q = 4(x-1)/x$. Since p and q are quotients of polynomials and $x_0 = 1$ is not a root of the denominators, p and q are real analytic at 1, and 1 is a regular singularity. The standard form of the ODE at $x_0 = 1$ is $(x-1)^2 y'' + ((x-1)(2x-1)/x) y' + (4(x-1)/x) y = 0$.

Classify Singular Points. Find and classify all singular points of the following equations:

5. $(1-x^2)y'' - xy' + 2y = 0$

6. $x(1-x^2)^3 y'' + (1-x^2)^2 y' + 2(1+x)y = 0$

7. $y'' + \left(\dfrac{x}{1-x}\right)^2 y' + (1+x)^2 y = 0$

8. $x^3(1-x^2)y'' - x(x+1)y' + (1-x)y = 0$

Answer 5: The singularities are $x_0 = -1$ and $x_1 = 1$; $a_2 = 1-x^2$, $a_1 = -x$, $a_0 = 2$; $P = -x/(1-x^2)$, $Q = 2/(1-x^2)$. First, we look at the singularity $x_0 = -1$ and we see that $p = (x+1)P = x/(x-1)$, $q = (x+1)^2 Q = 2(x+1)/(1-x)$. Since p and q are quotients of polynomials and $x_0 = -1$ is not a root of the denominators, p and q are real analytic at $x_0 = -1$; -1 is a regular singularity of the ODE. The standard form of the ODE at $x_0 = -1$ is $(x+1)^2 y'' + ((x+1)x/(x-1)) y' + (2(x+1)/(1-x)) y = 0$.

A similar analysis at $x_1 = 1$ shows that 1 is also a regular singularity. The standard form of the ODE at $x_1 = 1$ is $(x-1)^2 y'' - (x(x-1)/(x+1)) y' - (2(x-1)/(x+1)) y = 0$.

Answer 7: Here $x_0 = 1$ is the singularity; $a_2 = 1$, $a_1 = x^2/(1-x)^2$, $a_0 = (1+x)^2$; $P = a_1$, $Q = a_0$; $p = (x-1)P = x^2/(x-1)$, $q = (x-1)^2 Q = (x-1)^2(1+x)^2$, But $P(1)$ is not defined, so $x_0 = 1$ is an irregular singularity of the ODE.

Euler ODEs. Find the general, real-valued solution (for $x > 0$) of each ODE. Plot some solutions.

9. $x^2 y'' - 6y = 0$

10. $x^2 y'' + xy' - 4y = 0$

11. $x^2 y'' + xy' + 9y = 0$

12. $x^2 y'' + xy'/2 - y/2 = 0$

13. $xy'' - y' + (5/x)y = 0$

14. $x^2 y'' + 7xy' + 9y = 0$

Answer 9: $r^2 - r - 6$; $r_1 = -2, r_2 = 3$; $y = c_1 x^{-2} + c_2 x^3$. See the figure. The solid curve corresponds to $c_1 = 1$, $c_2 = 0$; the long-dashed curve the other way around. Neither c_1 nor c_2 is zero for the other curves.

Answer 11: $r^2 + 9$; $r_1 = 3i, r_2 = -3i$; $y = c_1 \cos(3\ln x) + c_2 \sin(3\ln x)$. See the figure. The solid curve corresponds to $c_1 = 1$, $c_2 = 0$; the long-dashed curve the other way around. Neither c_1 nor c_2 is zero for the other curves.

Answer 13: $r^2 - 2r + 5$; $r_1 = 1 + 2i = \bar{r}_2$; $y = x[c_1 \cos(2\ln x) + c_2 \sin(2\ln x)]$. See the figure. The solid curve corresponds to $c_1 = 1$, $c_2 = 0$; the long-dashed curve the other way around. Neither c_1 nor c_2 is zero for the other curves.

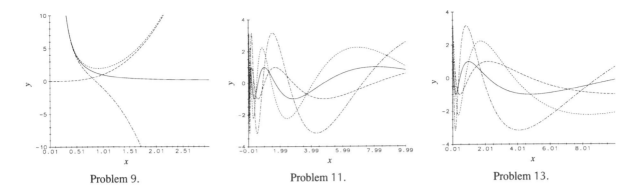

Problem 9.   Problem 11.   Problem 13.

Nonhomogeneous Euler ODEs. Find the general solution for $x > 0$ of each ODE. [*Hint*: use the Method of Undetermined Coefficients to find a particular polynomial solution.]

15. $x^2 y'' - 4xy' + 6y = 2x + 5$ **16.** $x^2 y'' - 6y = 2x^2$

Answer 15: The indicial polynomial of the Euler operator is $r^2 - 5r + 6$ with roots $r_1 = 2, r_3 = 3$. A particular solution will have the form $y_d = a + bx$. So $-4xb + 6a + 6bx = 2x + 5$ and $a = 5/6, b = 1$. The general solution is $y = c_1 x^2 + c_2 x^3 + 5/6 + x$, where c_1 and c_2 are any constants.

Decaying Solutions of Euler ODEs.

17. Show that every solution $y(x)$ of the ODE $x^2 y'' + p_0 xy' + q_0 y = 0$, where p_0, q_0 are real, and has the property that $|y(x)| \to 0$ as $x \to \infty$ if and only if $p_0 > 1$ and $q_0 > 0$.

Answer 17: The solutions of $x^2 y'' + p_0 xy' + q_0 y = 0$ are given by the formulas in (12), and we see that every solution tends to 0 as $x \to +\infty$ if and only if r_1 and r_2 are negative or else have negative real parts. This is clear if r_1 and r_2 are real and distinct. If $r_1 = r_2 < 0$, then the term $x^{r_1} \ln x = \ln(x)/x^{-r_1} \to 0$ as $x \to +\infty$ by L'Hôpital's Rule. If $r_1 = a + bi = \bar{r}_2$ and $a < 0$, $b \neq 0$, then $|x^a \cos(b \ln x)| \leq |x|^a \to 0$ as $x \to +\infty$, and similarly for $x^a \sin(b \ln x)$. The roots r_1 and r_2 of the indicial polynomial $r^2 + (p_0 - 1)r + q_0$ have negative real parts if and only if $p_0 > 1$ and $q_0 > 0$, as can be seen by using the quadratic formula for the roots.

The nth-order Euler ODE. The *nth-order Euler ODE* is

$$x^n y^{(n)} + a_{n-1} x^{n-1} y^{(n-1)} + \cdots + a_1 xy' + a_0 y = 0, \quad x > 0 \tag{i}$$

where $a_0, a_1, \ldots, a_{n-1}$ are real constants. The *indicial polynomial* is

$$p(r) = r(r-1) \cdots (r-n+1) + r(r-1) \cdots (r-n+2)a_{n-1} + \cdots + ra_1 + a_0$$

18. Show that $y = x^{r_1}$, $x > 0$, is a solution of ODE (i) if and only if r_1 is a root of $p(r)$.

19. Use the results of Problem 18 to find three independent solutions of the third-order Euler ODE $x^3 y''' + 4x^2 y'' - 2y = 0$, $x > 0$.

Answer 19: The indicial polynomial for $x^3 y''' + 4x^2 y'' - 2y = 0$ is $r(r-1)(r-2) + 4r(r-1) - 2 = r^3 + r^2 - 2r - 2 = (r+1)(r^2 - 2)$, and the roots are -1 and $\pm\sqrt{2}$. The functions $x^{-1}, x^{-\sqrt{2}}, x^{\sqrt{2}}$ are three independent solutions for $x > 0$.

11.4 Series Solutions near Regular Singular Points

Comments:

Now we take up the techniques for constructing a pair of independent solutions of a homogeneous second-order linear ODE with nonconstant coefficients near a regular singularity. We address the simplest case when the roots of the indicial polynomial are real and do *not* differ by an integer. We also show that even if the roots do differ by an integer we can always construct a solution corresponding to the larger root. The process by which Frobenius series expansions of solutions are constructed is similar to the methods used in Section 11.2 to find power series expansions of solutions near an ordinary point, but with the twist that the powers in the series need not be integers. As before, the successful construction of the series requires patience, algebraic "smarts," and careful bookkeeping. Once again, a CAS may be helpful. We have noticed over the years that readers have a lot of trouble with this material. But those who finally "see" how to do it are quite proud of their accomplishment.

PROBLEMS

Frobenius Series. Find the recursion formula for the coefficients for the solutions $y_1 = x^{r_1} \sum_0^\infty a_n x^n$ and $y_2 = x^{r_2} \sum_0^\infty b_n x^n$, where r_1 and r_2 are the indicial roots, $r_2 < r_1$, $r_1 - r_2$ is not an integer. Solve the recursion formula for a_n in terms of a_0 in Problems 1 and 3. In each problem, $a_0 = 1$.

1. $9x^2 y'' + 3x(x+3)y' - (4x+1)y = 0$ **2.** $4x^2 y'' + x(2x+9)y' + y = 0$

3. $x^2 y'' + x(1-x)y' - 2y = 0$ **4.** $x^2(1-x^2)y'' + x(x-1)y' + 8y/9 = 0$

Answer 1:   The ODE is $9x^2 y'' + 3x(x+3)y' - (4x+1)y = 0$. The equation in standard form is $x^2 y'' + x(1+x/3)y' + (-1/9 - 4x/9)y = 0$. So $p_0 = 1$ and $q_0 = -1/9$. The indicial polynomial is $r^2 + (p_0 - 1)r + q_0 = r^2 - 1/9$, and its roots are $r_1 = 1/3$, $r_2 = -1/3$ and $r_1 - r_2$ is not an integer. Let $y = \sum_{n=0}^\infty a_n x^{n+r}$ be a solution. Then

$$y' = \sum_{n=0}^\infty (n+r)a_n x^{n+r-1}, \quad y'' = \sum_{n=0}^\infty (n+r)(n+r-1)a_n x^{n+r-2}$$

Inserting these series into the ODE in its original form, reindexing, and rearranging, we have that

$$\sum_{n=0}^\infty [9(n+r)(n+r-1) + 9(n+r) - 1]a_n x^{n+r} + \sum_{n=1}^\infty [3(n+r-1) - 4]a_{n-1} x^{n+r} = 0$$

Note that the term corresponding to $n = 0$ in the first summation above is zero because $r^2 = 1/9$. The index in that sum actually runs from $n = 1$ to ∞, and the two summations can be combined because they have the same index range and involve the same power of x. Applying the Identity Theorem to the coefficient of x^{n+r}, we obtain the recursion relation:

$$[9(n+r)(n+r-1) + 9(n+r) - 1]a_n = -[3(n+r-1) - 4]a_{n-1}$$

and, solving this relation for a_n,

$$a_n = -\frac{3(n+r-1) - 4}{9(n+r)^2 - 1}a_{n-1}$$

So for $r = 1/3$,

$$a_n = -(n-2)a_{n-1}/[n(3n+2)], \ n \geq 1$$

Setting $a_0 = 1$, we have that $a_1 = 1/5$, while $a_n = 0$, $n \geq 2$. We have a solution corresponding to $r = 1/3$:

$$y_1 = x^{1/3} + \frac{1}{5}x^{4/3}$$

For $r = -1/3$,

$$a_n = -(3n - 8)a_{n-1}/[3n(3n - 2)]$$

$$= a_0(-1)^n(3n - 8)(3n - 11)\cdots(-2)(-5)/[3^n n!(3n - 2)(3n - 5)(3n - 8)\cdots 1]$$

$$= (-1)^n \frac{10}{3^n n!(3n - 2)(3n - 5)}a_0, \quad n \geq 1$$

So a second independent solution (corresponding to $r = -1/3$) is

$$y_2 = x^{-1/3} \sum_{n=0}^{\infty} a_n x^n$$

where $a_0 = 1$ and a_n is given above. The series themselves converge for all x but the general solution $y = c_1 y_1 + c_2 y_2$ is not defined at $x = 0$, because $x^{1/3}$ is not differentiable at 0 and $x^{-1/3}$ is not continuous at 0.

Answer 3: The ODE is $x^2 y'' + x(1 - x)y' - 2y = 0$. The indicial polynomial is $r^2 - 2$ and its roots are $r_1 = \sqrt{2}$, $r_2 = -\sqrt{2}$. The recursion formula is

$$a_n(r) = (n + r - 1)a_{n-1}/[(n + r)^2 - 2]$$

So

$$a_n(\pm\sqrt{2}) = \frac{(n \pm \sqrt{2} - 1)(n \pm \sqrt{2} - 2)\cdots(\pm\sqrt{2})a_0}{n!(n \pm 2\sqrt{2})(n - 1 \pm 2\sqrt{2})\cdots(1 \pm 2\sqrt{2})}$$

A pair of independent solutions is

$$y_1 = x^{\sqrt{2}} \sum_{n=0}^{\infty} a_n(\sqrt{2})x^n \quad \text{and} \quad y_2 = x^{-\sqrt{2}} \sum_{n=0}^{\infty} a_n(-\sqrt{2})x^n$$

where $a_0 = 1$ and $a_n(\pm\sqrt{2})$ is given above.

5. Solve the ODE $3x^2 y'' + 5xy' - e^x y = 0$ by expanding e^x in a Taylor series about $x_0 = 0$. Use the formula for the product of two series (Appendix B.1). Find the first three terms in the Frobenius series corresponding to indicial root $r_1 = 1/3$. Repeat for $r_2 = -1$.

Answer 5: The ODE is $3x^2 y'' + 5xy' - e^x y = 0$. The Taylor series for e^x about $x_0 = 0$ is $1 + x + x^2/2! + x^3/3! + \cdots$. So $p_0 = 5/3$, $q_0 = -1/3$ for the ODE in standard form

$$x^2 y'' + 5xy'/3 - e^x y/3 = x^2 y'' + 5xy'/3 + (-1/3 - x/3 - \cdots)y = 0$$

The indicial polynomial is $r^2 + 2r/3 - 1/3$ and $r_1 = 1/3, r_2 = -1$. There are independent solutions of the form

$$y_1 = x^{1/3} + a_1 x^{4/3} + a_2 x^{7/3} + \cdots, \quad y_2 = x^{-1} + b_1 + b_2 x + \cdots$$

where we have set $a_0 = b_0 = 1$ in each series. Although we could use Frobenius Theorem I to determine the coefficients, it is probably simpler to make a direct substitution of y_1 and then y_2 into the ODE and use the Identity Theorem to determine the coefficients. For y_1 we have that

$$3x^2 y_1'' + 5xy_1' - e^x y_1 = (7a_1 - 1)x^{4/3} + (20a_2 - a_1 - 1/2)x^{7/3} + \cdots = 0$$

So $a_1 = 1/7$, $a_2 = 9/280$ and $y_1 = x^{1/3} + x^{4/3}/7 + 9x^{7/3}/280 + \cdots$. In the same way we see that $y_2 = x^{-1} - 1 - x/8 + \cdots$

Frobenius Series and Nonuniqueness.

6. Find a solution $y(x)$ of $xy'' - y' - 4x^5 y = 0$ for which $y(0) = y'(0) = 0$, but $y(x)$ is not identically zero.

7. Show that the ODE $x^3 y'' + y = 0$ has no nontrivial solutions of the form $y = x^r \sum_0^{\infty} a_n x^n$. Why does this not contradict the results of this section?

Answer 7:   Suppose $x^3 y'' + y = 0$ has a solution $y = \sum_{n=0}^{\infty} a_n x^{n+r}$ for some constant r and constants a_n, where the series $\sum_{n=0}^{\infty} a_n x^n$ converges in a nontrivial interval I containing 0. Then, after a certain amount of reindexing, we obtain a power series

$$x^3 y'' + y = a_0 x^r + \sum_{n=1}^{\infty} [(n+r-1)(n+r-2)a_{n-1} + a_n] x^{n+r} = 0$$

So $a_0 = 0$, and

$$a_n = -(n+r-1)(n+r-2)a_{n-1}$$

Since $a_0 = 0$, this implies that $a_1 = a_2 = \cdots = a_n = \cdots = 0$, and the only solution in power series form is the trivial solution $y = 0$. This does not contradict the results of this section because the ODE $x^3 y'' + y = 0$ cannot be written in the form $x^2 y'' + xp(x)y' + q(x)y = 0$ where $p(x)$ and $q(x)$ are real analytic.

Driven ODEs.

8. Find the first four nonzero terms of a power series solution of the ODE

    $$x^2 y'' + x(1-x)y' - 2y = \ln(1+x)$$

    Then find the general solution. [*Hint*: first find the general solution of the undriven ODE $x^2 y'' + x(1-x)y' - 2y = 0$ (see Problem 3). Next expand $\ln(1+x)$ in a Taylor series about $x_0 = 0$ (see Appendix B.1). Then assume a particular solution of the form $y_d = \sum_0^{\infty} a_n x^n$ and use the Method of Power Series to find $a_0, a_1, \cdots, a_4$.]

Laguerre's Equation. $xy'' + (1-x)y' + py = 0$ is *Laguerre's equation of order p, $p = 0, 1, \ldots$*.

9. Show that zero is a regular singular point of Laguerre's equation.

10. Find the recursion formula for the coefficients of a Frobenius series solution. Solve to find the coefficient a_n of the series as a multiple of a_0. Set $a_0 = 1$ and write out the corresponding Frobenius series solution of Laguerre's equation of order p.

11. Show that if p is a nonnegative integer, then there are polynomial solutions of degree p.

12. The *Laguerre polynomials* are those polynomial solutions $y = L_p(x)$ of Laguerre's equation of order p, $p = 0, 1, 2, \ldots$, for which $L_p(0) = 1$. Find the first five polynomials.

13. Show that the family of Laguerre polynomials $\{L_p : p = 0, 1, 2, \ldots\}$ is orthogonal on the interval $[0, \infty)$ with a weight function $\rho(x) = e^{-x}$; that is, show that $\int_0^{\infty} L_p(x)L_q(x)e^{-x}dx = 0$, where p and q, $p \neq q$, are integers.

Answer 9:   Laguerre's ODE of order p is $xy'' + (1-x)y' + py = 0$ and 0 is a singularity. The normalized form of Laguerre's equation is $y'' + x^{-1}(1-x)y' + px^{-1}y = 0$. Since $P = x^{-1}(1-x)$, $Q = px^{-1}$, $xP = 1-x$ and $x^2 Q = p$, xP and $x^2 Q$ are real analytic at 0, so 0 is a regular singularity.

Answer 11:   Suppose p is a nonnegative integer, $p = n$ say. Then the factor $n - p = 0$ appears as a factor of x^k for all $k \geq n+1$. So y_1 is a polynomial of degree p.

Answer 13:   Let L_p and L_q be the Laguerre polynomials of degree p and q, respectively, $p \neq q$. We have

$$xL_p'' + (1-x)L_p' + pL_p = 0, \quad xL_q'' + (1-x)L_q' + qL_q = 0$$

If the first equation is multiplied by L_q and the second by L_p, subtraction yields

$$x(L_p'' L_q - L_q'' L_p) + (1-x)(L_q L_p' - L_p L_q') + (p-q)L_p L_q = 0$$

and so

$$e^{-x}x(L_p'' L_q - L_q'' L_p) + e^{-x}(1-x)(L_q L_p' - L_q' L_p) + (p-q)e^{-x}L_p L_q = 0$$

$$[e^{-x}x(L_q L_p' - L_q' L_p)]' + (p-q)e^{-x}L_p L_q = 0$$

Integrating both sides from 0 to $+\infty$, we have that

$$[e^{-x}x(L_q L_p' - L_q' L_p)]\Big|_0^{+\infty} = (q - p)\int_0^{+\infty} e^{-x}L_p L_q\, dx$$

Since the left side of this equality is 0 and since $p \neq q$, we have that $\int_0^{+\infty} e^{-x}L_p L_q\, dx = 0$. This means $\{L_p : p = 0, 1, 2, \ldots\}$ is orthogonal on the interval $(0, \infty)$ with weight $\rho(x) = e^{-x}$. The left-hand side of this equality tends to 0 as x tends to $+\infty$ since $x^k e^{-x} \to 0$ as $x \to +\infty$ for any real number k.

Frobenius Series and Legendre's Equation. The Legendre equation of order r is

$$(1 - x^2)y'' - 2xy' + r(r + 1)y = 0$$

14. Find the indicial roots of the Legendre equation at the regular singularity $x = 1$.

15. Find a series solution for the Legendre equation in powers of $x - 1$.

Answer 15: Frobenius series techniques may be used to find a solution $z = \sum_{n=0}^{\infty} a_n s^n$ of

$$(s + 2)s\frac{d^2 z}{ds^2} + 2(s + 1)\frac{dz}{ds} - r(r + 1)z = 0$$

which is the ODE obtained from the Legendre equation if x is replaced by $s + 1$ and the result multiplied by -1. The recursion formula turns out to be

$$a_{n+1} = \frac{r(r + 1) - n(n + 1)}{2(n + 1)^2}a_n$$

Setting $a_0 = 1$ and replacing s by $x - 1$ we have that

$$y_1(x) = 1 + \frac{r(r + 1)}{2}(x - 1) + \frac{r(r + 1)[r(r + 1) - 2]}{16}(x - 1)^2 + \cdots$$

$$+ \frac{1}{2^n (n!)^2}r(r + 1)[r(r + 1) - 2]\cdots[r(r + 1) - n(n + 1)](x - 1)^n + \cdots$$

which is a series solution in powers of $(x - 1)$, and converges for $|x - 1| < 2$.

Spotlight on Legendre Polynomials

Comments:

The Legendre polynomials $\{P_n : n = 0, 1, 2, \ldots\}$ are the best known of the sets of orthogonal polynomials, and so we have chosen to give them and their ODEs a spotlight of their own. This section, its problem set, and the full solutions of the problems should be regarded as an abbreviated compendium of properties of Legendre polynomials, other solutions of the Legendre ODEs, and other sets of orthogonal functions.

The Hermite ODEs and polynomials H_n and the Chebyshev ODEs and polynomials T_n make their appearance in the problem set. The reason the Chebyshev polynomials are labeled T_n and not C_n illustrates the difficulty in using the Latin alphabet to transliterate Cyrillic letters (Chebyshev was a well-known 19th century Russian mathematician). Sometimes one sees Tschebysheff, or Tchebyshev, or Chebyshov. Philip J. Davis has written an amusing and mathematically informative book based on his difficulties with these various transliterations [*The Thread: A Mathematical Gem*; Boston, Birkhaüser, 1983].

PROBLEMS

Legendre Polynomials and Their Properties.

1. *Recursion Relation*   Use the Legendre recursion relation to derive formulas for P_4, P_5 and P_6, given the formulas for P_2 and P_3.

Answer 1:   Since $P_n = (2 - 1/n)xP_{n-1} - (1 - 1/n)P_{n-2}$, $n = 2, \cdots$, and $P_0 = 1$, $P_1 = x$, we see that $P_2 = (3x^2 - 1)/2$ and $P_3 = (5x^3 - 3x)/2$. So $P_4 = (7/4)xP_3 - 3P_2/4$, $P_5 = (9/5)xP_4 - 4P_3/5$, $P_6 = (11/6)xP_5 - 5P_4/6$. We omit any further calculations.

2. *Values of Legendre Polynomials*   Show that

$$P_{2n+1}(0) = 0, \qquad P_{2n}(0) = (-1)^n \frac{(2n)!}{2^{2n}(n!)^2}, \qquad P_n(-1) = (-1)^n$$

[*Hint*: use induction on n and an identity in Table 1 to prove the third formula.]

3. *Rodrigues's Formula*   Use this formula to calculate P_3 and P_4.

Answer 3:   By Rodrigues's formula in Table 1 we see that:

$$P_3 = \frac{1}{2^3 \cdot 3!}\frac{d^3}{dx^3}\left[(x^2 - 1)^3\right] = \frac{1}{48}[x^6 - 3x^4 + 3x^2 + 1]''' = (5x^3 - 3x)/2$$

$$P_4 = \frac{1}{2^4 4!}\left[(x^2 - 1)^4\right]^{(4)} = \frac{1}{384}[x^8 - 4x^6 + 6x^4 - 4x^2 + 1]^{(4)} = (35x^4 - 30x^2 + 3)/8$$

4. *Orthogonality*   Carry out the integration and verify that

$$\int_{-1}^1 P_3^2(x)dx = 2/7 \quad \text{and} \quad \int_{-1}^1 P_3(x)P_4(x)dx = 0$$

5. *An Orthogonality Property*   Show that $\int_{-1}^1 P_n(x)Q(x)\,dx = 0$ if $n > 0$ and $Q(x)$ is *any* polynomial of degree $k < n$. [*Hint*: use induction on k to show that there are constants $c_0, \ldots, c_k$ such that $Q(x) = c_0 P_0 + c_1 P_1(x) + \cdots + c_k P_k(x)$. Then use Theorem 2.]

Answer 5:   Suppose that $Q(x)$ is any polynomial of degree $k < n$. We want to show that the integral $\int_{-1}^1 P_n(x)Q(x)\,dx = 0$. We prove by induction that there are constants $c_0, \ldots, c_k$ such that $Q(x) = \sum_{j=0}^k c_k P_k(x)$. This is certainly the case for $k = 0$ since $Q(x) = Q(0) \cdot 1 = Q(0) \cdot P_0(x)$. Suppose the result is true for all polynomials Q of degree $\leq k$ (the induction hypothesis). Let Q be a polynomial of degree $k + 1$. Let $Q = a_0 + \cdots + a_{k+1}x^{k+1}$ and $P_{k+1} = b_0 + \cdots + b_{k+1}x^{k+1}$. Then $Q(x) = a_{k+1}P_{k+1}(x)b_{k+1}^{-1} + R(x)$, where $R(x)$ is a polynomial of degree k or less. $R(x) = \sum_{j=0}^k \alpha_j P_j(x)$ for some constants α_j by the induction hypothesis, so $Q(x)$ is a linear combination of $P_0(x), \ldots, P_{k+1}(x)$ and we are done. Note that the induction proof did not use the specific form of the Legendre polynomials, only that $P_n(x)$ is a polynomial of degree n.

Now to the problem itself: suppose $Q(x)$ is any polynomial of degree of $k < n$. Then $Q = c_0 P_0 + \cdots + c_k P_k$ and

$$\int_{-1}^1 P_n(x)Q(x)\,dx = \sum_{j=0}^k c_j \int_{-1}^1 P_n(x)P_j(x)\,dx = 0$$

by Theorem 2.

6. *Legendre's ODE and the Wronskian Method*   Legendre's equations of orders $n = 0, 1$ have the solutions $P_0 = 1$ and $P_1 = x$, respectively. Use the Wronskian Reduction Method (Problem 14 in Section 3.7) to find second independent solutions for $n = 0, 1$ in terms of elementary functions. Plot the second solutions for $|x| < 1$.

Orthogonal Polynomials.   Suppose that $\{R_n(x) : n = 0, 1, 2, \ldots\}$ is a family of polynomials indexed by degree (i.e., $\deg R_n = n$). Suppose that $\rho(x)$ is positive and continuous on an open interval $I = (a, b)$, and that $\lim_{x \to a^+} \rho(x)$ and $\lim_{x \to b^-} \rho(x)$ exist (although the limiting values may be infinite). The family of polynomials is *orthogonal on I with respect to the weight ρ* if

$$\int_a^b R_n(x) R_m(x) \rho(x)\, dx = 0, \quad n \neq m$$

7. What are the values of a, b, and ρ for the orthogonal family of Legendre polynomials?

8. Show that if $\{R_n : n = 0, 1, 2, \ldots\}$ is any family of polynomials indexed by degree and orthogonal on an interval I with respect to a weight function ρ, then there are constants a_n, b_n, c_n such that the *three-term recursion relation*

$$R_n = (a_n x + b_n) R_{n-1} + c_n R_{n-2}, \quad n = 2, 3, \ldots$$

holds. [*Hint*: $R_n = \alpha_n x R_{n-1} + \alpha_{n-1} R_{n-1} + \alpha_{n-2} R_{n-2} + \cdots + \alpha_0 R_0$ for some constants α_i. Then calculate $\int_a^b R_n R_j \rho\, dx, n \geq 2, 0 \leq j \leq n-2$.]

9. Use the hint in Problem 8 to verify the recursion formula (11) for Legendre polynomials.

Answer 7: Theorem 2 states that

$$< P_n, P_m > = \int_{-1}^1 P_n(x) P_m(x)\, dx = \begin{cases} 0, & n \neq m \\ 2/(2n+1), & n = m \end{cases}$$

So, $a = -1$ and $b = 1$, $\rho(x) = 1$ for this orthogonal family.

Answer 9: The recursion formula for Legendre polynomials is

$$P_n = (2 - 1/n) x P_{n-1} - (1 - 1/n) P_{n-2}$$

which coincides with the formula

$$R_n = (a_n x + b_n) R_{n-1} + c_n R_{n-2}$$

of Problem 8 if $a = 2 - 1/n$, $b_n = 0$, and $c_n = -(1 - 1/n)$.

Hermite's Equation and Polynomials. *Hermite's equation* of order n, where n is a nonnegative integer, is $y'' - 2xy' + 2ny = 0$.

☞ Take a look at Examples 11.1.4–11.1.6 for Hermite's equation of order 1 and its solutions.

10. Show that the equation has solutions that are polynomials of degree n, and these polynomials are even functions if n is even, odd functions if n is odd.

11. We determine the *Hermite polynomials* H_n by selecting (for each n) the polynomial solution with leading coefficient $2^n : H_n(x) = 2^n x^n + (\cdot) x^{n-2} + \cdots$. Show that the family of Hermite polynomials is orthogonal on the real line with respect to the weight function $\rho = e^{-x^2}$.

12. The three-term recursion formula for the Hermite polynomials is

$$H_n = 2x H_{n-1} - 2(n-1) H_{n-2}$$

Use $H_0 = 1$, $H_1 = 2x$ and this formula to construct $H_2, \ldots, H_5$. Plot $H_2, \ldots, H_5$ for $|x| \leq 3$, and make a conjecture about the number and the nature of the roots of $H_n(x)$.

Answer 11: Next we show orthogonality over $\mathbb{R}$ with weight e^{-x^2}. Since $H_n'' - 2x H_n' + 2n H_n = 0$ and $H_m'' - 2x H_m' + 2m H_m = 0$, we have $[H_n'' H_m - H_m'' H_n] - 2x [H_n' H_m - H_m' H_n] + 2(n-m) H_n H_m = 0$. So we have

$$\frac{d}{dx} \left[e^{-x^2} (H_n' H_m - H_m' H_n) \right] + 2(n-m) e^{-x^2} H_n H_m = 0$$

that is,

$$e^{-x^2} [H_n' H_m - H_m' H_n] \Big|_{x=-\infty}^{x=+\infty} + 2(n-m) \int_{-\infty}^{+\infty} e^{-x^2} H_n H_m\, dx = 0$$

Since $x^k e^{-x^2} \to 0$ as $x \to \pm\infty$ for k any integer, the first term on the left vanishes. Since $n \neq m$, we have that $\int_{-\infty}^{+\infty} e^{-x^2} H_n H_m\, dx = 0$. This means that the family of Hermite polynomials is orthogonal on the real line with respect to the density $\rho = e^{-x^2}$.

Spotlight on Bessel Functions

Comments:

The classic examples of ODEs with a regular singularity are the Bessel equations, and this Spotlight is mostly about these equations and their solutions. The construction of the Bessel functions of the first kind follows the Frobenius steps outlined in Section 11.4. Bessel functions of the first kind resemble decaying sinusoids in many ways. For use in solving PDEs in the WEB SPOTLIGHT ON STEADY TEMPERATURES, we discuss the positive zeros λ_n of Bessel functions J_p, the resemblance between $J_p(\lambda_n x)$ and $\sin(n\pi x)$, and the orthogonality properties. See also the WEB SPOTLIGHT ON THE EXTENDED METHOD OF FROBENIUS.

PROBLEMS

Properties of $J_3(x)$.   Use a few terms of the series for $J_3(x)$ to determine some of its properties.

1.   Find the first four terms in the series for $J_3(x)$.

2.   Let $f(x)$ be the ninth-degree polynomial of the first four terms of the series for $J_3(x)$. Explain why $|J_3(x) - f(x)| < 0.02$ for $0 \le x \le 4$.

3.   Explain why the first positive zero of $J_3(x)$ is larger than 4.

4.   Explain why the graph of $J_3(x)$ is concave upward at $x = 0$.

Answer 1:   Using (6), we have that

$$J_3(x) = \left(\frac{x}{2}\right)^3 \left[\frac{1}{3!} - \frac{1}{4!}\left(\frac{x}{2}\right)^2 + \frac{1}{2!5!}\left(\frac{x}{2}\right)^4 - \frac{1}{3!6!}\left(\frac{x}{2}\right)^6 + \cdots\right]$$

$$= x^3/48 - x^5/768 + x^7/30720 - x^9/2211840 + \cdots$$

Answer 3:   For $0 < x \le 4$ the series in Answer 1, $\frac{1}{3!} - \frac{1}{4!}\left(\frac{x}{2}\right)^2 + \cdots$, is a convergent alternating series of positive terms of nonincreasing magnitudes. If $a_0 - a_1 + a_2 - \cdots$ is any alternating series with these properties, then the sum S of the series is bounded by:

$$(a_0 - a_1) + (a_2 - a_3) + \cdots + (a_{2k} - a_{2k+1}) \le S \le a_0 - (a_1 - a_2) - \cdots - (a_{2k-1} - a_{2k}), \quad k = 1, 2, \cdots$$

We have $f(x) \le J_3(x)$ for $|x| \le 4$, where $f(x)$ is the sum of the first four terms of $J_3(x)$. But analysis of $f(x)$ shows that $f(x)$ is positive for $0 < x \le 4$. So, the first positive zero of $J_3(x)$ must be larger than 4.

Properties of Bessel Functions of the First Kind of Integer Order.

5.   *Convergence*   Use the Ratio Test to show that the series for $J_n(x)$ converges for all x.

Answer 5:   The magnitude of the ratio of the $(k+1)^{st}$ and k^{th} terms in the series (6) for $J_n(x)$ is

$$\frac{x^2}{4(k+1)(k+n+1)}$$

Since the above ratio $\to 0$ as $k \to \infty$ for each x, the series for $J_p(x)$ converges for all x.

6.   *Graphs*   Plot J_0, J_1, J_2, and J_3. $0 \le x \le 20$.

7.   *Zeros of Bessel Functions*   Locate the zeros of J_0, J_1, J_2 in Figure 1, and make a conjecture about the relative positions of the positive zeros of J_n and of J_{n+1}.

Answer 7:   The graphs in Figure 1 lead us to guess that the positive zeros of J_n and J_{n+1} interlace. That is, between two successive zeros of J_n there is a zero of J_{n+1}, and between two successive zeros of J_{n+1} there is a zero of J_n.

8. *Recursion Formula* Use formula (11) to express J_2, J_3, and J_4 in terms of J_0 and J_1.

9. Plot $J_1(x)$ and find (graphically) the first pair of its successive zeros λ_n, λ_{n+1} for which $|(\lambda_{n+1} - \lambda_n) - \pi| < 0.0001$.

   Answer 9: See the figure for the graph of $J_1(x)$, $0 \le x \le 30$. From the graph the 8th and 9th positive zeros λ_8, λ_9 appear to be about 25.9 and 29, respectively. The difference is slightly less than π. If your computer approximates the values of Bessel functions and their zeros, check to see if $|\lambda_9 - \lambda_8 - \pi| \le 0.0001$.

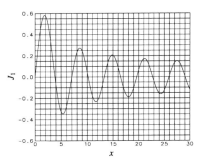

Problem 9.

Integrating $J_n(x)$. We can use the recursion formulas (11)–(14) and integration by parts to derive integration formulas for Bessel functions of the first kind. The symbol C in the formulas below denotes an arbitrary constant of integration.

10. Show that $\int^x J_1(t)\, dt = -J_0(x) + C$ and that $\int^x t J_0(t)\, dt = x J_1(x) + C$.

11. Show that $\int^x J_0(t) \sin t\, dt = x J_0(x) \sin x - x J_1(x) \cos x + C$ and

$$\int^x J_0(t) \cos t\, dt = x J_0(x) \cos x + x J_1(x) + x J_1(x) \sin x + C$$

[*Hint*: show that the derivative of the right side is the given integrand.]

12. Show that

$$\int^x t^5 J_2(t)\, dt = x^5 J_3(x) - 2x^4 J_4(x) + C$$

[*Hint*: $\int^x t^5 J_2\, dt = \int^x t^2 (t^3 J_2)\, dt = \int^x t^2 (t^3 J_3)'\, dt$.]

Answer 11: The identity $\int^x J_0(t) \sin t\, dt = x J_0(x) \sin x - x J_1(x) \cos x + C$ holds if and only if

$$J_0(x) \sin x = (x J_0(x))' \sin x + x J_0(x) \cos x - (x J_1(x))' \cos x + x J_1(x) \sin x$$

which we obtain from the identity by differentiation. The differentiated identity holds if and only if

$$J_0(x) \sin x = [(x J_0(x))' + x J_1(x)] \sin x$$

[since Problem 10 implies $x J_0(x) = (x J_1(x))'$] which in turn holds if and only if

$$J_0 = x(J_0' + J_1) + J_0$$

It follows from Problem 10 that $J_0' + J_1 = 0$. So the identity is true.

General Solution of Bessel's Equation.

13. Use the Wronskian Reduction of Order Method (Problem 14, Section 3.7) to show that we can write the general solution of Bessel's equation of order n in the form

$$y = c_1 J_n(x) + c_2 J_n(x) \int^x \frac{ds}{s[J_n(s)]^2}, \qquad c_1 \text{ and } c_2 \text{ any constants}$$

Answer 13: According to Problem 14, Section 3.7, a second independent solution y_2 of Bessel's equation of order n satisfies the first-order linear equation $J_n y_2' - J_n' y_2 = x^{-1}$, where the term on the right side is $\exp[-\int^x a(s)\,ds]$ and $a(x) = x^{-1}$ (the normalized form of Bessel's equation is $y'' + x^{-1}y' + (1 - n^2 x^{-2})y = 0$). So $y_2' - J_n' y_2 / J_n = 1/(x J_n)$ and the integrating factor of the first-order linear ODE is $1/J_n$. We have $(y_2/J_n)' = 1/(x J_n^2)$ and we can take $y_2 = J_n(x) \int^x (s J_n^2(s))^{-1}\,ds$. The general solution of Bessel's equation of order n is $y = c_1 J_n(x) + c_2 J_n(x) \int^x (s J_n^2(s))^{-1}\,ds$, c_1 and c_2 any constants.

APPENDIX A

Basic Theory of Initial Value Problems

A.1 Uniqueness

PROBLEMS

1. Consider the ODE $y' = f(t, y)$, where $f = |y|$.

   (a) Show that the function f does *not* satisfy the hypotheses of the Uniqueness Principle in any region containing all or part of the t-axis in the ty-plane.

   (b) Show that the IVP $y' = |y|$, $y(t_0) = 0$, has a unique solution even so. Does a contradiction exist?

   (c) Show that we can replace the hypothesis "$\partial f/\partial y$ is continuous" in the Uniqueness Principle with "$f(t, y)$ satisfies a *Lipschitz condition* in y." Use this conclusion to prove again that $y' = |y|$, $y(t_0) = 0$, has a unique solution.

   Answer 1:
   (a) The function $f = |y|$ is not differentiable at $y = 0$ since $\lim_{h \to 0}(f(h) - f(0))/h = \lim_{h \to 0}((|h| - 0)/h) = \pm 1$, depending upon whether h goes to zero through positive or negative values. So f doesn't satisfy the hypotheses of Theorem A.1.1.

   (b) Since one of the hypotheses does not hold in any region containing a portion of the t-axis, the Uniqueness Principle cannot be used to show that the problem $y' = |y|$, $y(t_0) = 0$, has a unique solution. The problem has the trivial solution $y(t) = 0$, all t. To show that there are no other solutions, we proceed as follows. Note that if $y_0 \neq 0$, then the problem $y' = |y|$, $y(t_0) = y_0$, meets the conditions of the Uniqueness Principle in the region $y > 0$ and also in the region $y < 0$. The unique solution is $y = y_0 e^{\pm(t - t_0)}$, $-\infty < t < +\infty$, where "+" is used if $y_0 > 0$, and "−" is used if $y_0 < 0$. Any solution $y_1(t)$ of $y' = |y|$, $y(t_0) = 0$ (except $y = 0$) must intersect at least one of these exponential solutions at some t-value. But if it does so, then by uniqueness in the region above and below the t-axis, it would have to coincide with that exponential solution for all t, so could never vanish. This contradiction shows that $y(t) = 0$, all t, is the unique solution of the given problem.

   (c) The proof of the Uniqueness Principle in the textbook actually works when f satisfies a Lipschitz condition. From the Triangle Inequality (Section SPOTLIGHT ON VECTORS, MATRICES, INDEPENDENCE) we have that $||y_1| - |y_2|| \leq |y_1 - y_2|$, for all y_1, y_2. So $f(y) = |y|$ satisfies a Lipschitz condition [see (4) in the text] with $L = 1$; it follows that the IVP of part **(b)** has a unique solution.

2. Suppose that m and n are positive integers without common factors (i.e., m and n are *relatively prime*). The IVP below has a unique solution for some values of m and n and infinitely many solutions for other values:

$$y' = |y|^{m/n}, \qquad y(0) = 0$$

   (a) Show that the IVP has the unique solution $y(t) = 0$, if $m \geq n$.

   (b) Show that the IVP has infinitely many solutions, if $m < n$.

A.2 The Picard Process for Solving an Initial Value Problem

PROBLEMS

Picard Iterates.

1. This problem involves Picard iterates for the IVP $y' = -y$, $y(0) = 1$.

   (a) Find the Picard iterates y_1, y_2, y_3.

   (b) Plot the graphs of these iterates for $|t| < 2$.

   (c) Find a formula for the nth iterate y_n by induction.

   (d) Find the exact solution $y(t)$ and sketch its graph along with those of y_0, y_1, y_2, and y_3.

   (e) Show that $y_n(t)$ is a partial sum of the Taylor series for the solution $y(t)$ of the IVP.

   Answer 1: We want to find the Picard iterates for $y' = -y$, $y(0) = 1$.

   (a) Using iteration scheme (8), we have the Picard iterates $y_0(t) = 1$, all t

$$y_1(t) = 1 + \int_0^t (-y_0(s))\,ds = 1 + \int_0^t (-1)\,ds = 1 - t$$

$$y_2(t) = 1 + \int_0^t (-y_1(s))\,ds = 1 + \int_0^t -(1-s)\,ds = 1 - t + t^2/2$$

$$y_3(t) = 1 + \int_0^t (-y_2(s))\,ds = 1 + \int_0^t -(1-s+s^2/2)\,ds = 1 - t + t^2/2 - t^3/6$$

   These iterates are defined for all t.

   (b) See the figure for the graphs of y_0, y_1, y_2 and y_3. The dashed line is the solution curve for the IVP.

   (c) Following the pattern of y_0, y_1, y_2, y_3, we guess that the n-th Picard iterate is given by $y_n(t) = \sum_{j=0}^n (-1)^j t^j/j!$. This is proved by induction on n: The formula holds for $n = 1$ (see **(a)**). Suppose it holds for n (the induction hypothesis). We shall show it holds for $n + 1$. From the iteration scheme (8) we have

$$y_{n+1} = 1 + \int_0^t (-y_n(s))\,ds = 1 + \int_0^t \left(-\sum_{j=0}^n \frac{(-1)^j s^j}{j!} \right) ds = 1 - \sum_{j=0}^n \int_0^t \frac{(-1)^j s^j}{j!}\,ds$$

$$= 1 + \sum_{j=0}^n \frac{(-1)^{j+1} t^{j+1}}{(j+1)!} = \sum_{j=0}^{n+1} \frac{(-1)^j t^j}{j!}$$

   where we have used the induction hypothesis and reindexed a sum. So the formula holds for $n + 1$, if it holds for n. By induction, the formula holds for all n, and all t.

   (d) The exact solution is $y = e^{-t}$, $-\infty < t < \infty$, as we see by using an integrating factor. See the dashed line in the figure for **(b)**.

   (e) The Taylor series for e^{-t} is $\sum_{j=0}^\infty (-1)^j t^j/j!$ about $t_0 = 0$ since if $f(t) = e^{-t}$ then $f^{(j)}(0) = (-1)^j$, and the Taylor series for $f(t)$ is $\sum_{j=0}^\infty f^{(j)} t^j/j!$. So $y_n(t)$ is just the partial sum $\sum_{j=0}^n (-1)^j t^j/j!$ of the Taylor series for e^{-t}. In this problem, Picard iteration constructs the Taylor series for the solution by adding one term to the series at each iteration.

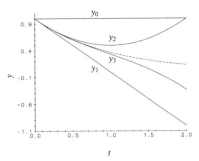

Problem 1(b).

2. Repeat Problem 1 for the IVP $y' = 2ty$, $y(0) = 2$.

3. This problem involves Picard iterates for the IVP $y' = y^2$, $y(0) = 1$.

(a) Find the Picard iterates y_1, y_2, and y_3.

(b) Find the solution $y(t)$ directly and plot y_0, y_1, y_2, y_3, and y on their maximal domains.

(c) Show by induction that the nth Picard iterate $y_n(t)$ is a polynomial of degree $2^n - 1$.

(d) Show that we can define every Picard iterate for all t, but that the exact solution is defined only for $t < 1$. Then show that the Taylor series of the exact solution converges only for $|t| < 1$. So we may define Picard iterates, the exact solution, and the Taylor series of the exact solution on different intervals.

Answer 3:   The IVP is $y' = y^2$, $y(0) = 1$.

(a) Using the iteration scheme (8) to calculate the Picard iterates, we have

$$y_0(t) = 1,$$

$$y_1(t) = 1 + \int_0^t y_0^2(s)\,ds = 1 + \int_0^t 1\,ds = 1 + t$$

$$y_2(t) = 1 + \int_0^t y_1^2(s)\,ds = 1 + \int_0^t (1+s)^2 ds = 1 + t + t^2 + t^3/3$$

$$y_3(t) = 1 + \int_0^t y_2^2(s)\,ds = 1 + \int_0^t (1 + s + s^2 + s^3/3)^2\,ds$$

$$= 1 + t + t^2 + t^3 + 2t^4/3 + t^5/3 + t^6/9 + t^7/63$$

All iterates are defined for all t.

(b) See the figure for the graphs of y_0, y_1, y_2, and y_3. The exact solution (shown by the dashed line) is $y = (1 - t)^{-1}$ where $t < 1$.

(c) Certainly, $y_1(t) = 1 + t$ is a polynomial of degree 1. Assume that $y_n(t)$ is a polynomial of degree $2^n - 1$ (the induction hypothesis). To show that $y_{n+1}(t)$ is a polynomial of degree $2^{n+1} - 1$, we note that

$$y_{n+1}(t) = 1 + \int_0^t y_n^2(s)\,ds = 1 + \int_0^t (c_0 + \cdots + c_k s^{2^n-1})^2\,ds$$

$$= 1 + \int_0^t (c_0^2 + \cdots + c_k^2 s^{2^{n+1}-2})\,ds = \cdots + c_k^2 (2^{n+1} - 1)^{-1} t^{2^{n+1}-1}$$

where we have set $y_n(t) = c_0 + \cdots + c_k t^{2^n-1}$. So $y_{n+1}(t)$ is a polynomial of the desired degree and the proof is complete.

(d) Since polynomials are defined for all values of their variables, every Picard iterate is defined for all t, even though the exact solution is defined only for $t < 1$. On the other hand, the Taylor series for the exact solution $y = (1 - t)^{-1}$ is the geometric series $\sum_{j=0}^{\infty} t^j$, which by the Ratio Test converges only for $-1 < t < 1$.

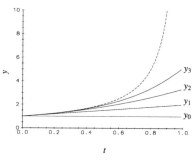

Problem 3(b).

4. The Picard iteration scheme is sturdy enough to survive a bad choice for the first approximation $y_0(t)$. Show that the Picard iterates for the IVP $y' = -y$, $y(0) = 1$, converge to the exact solution e^{-t} even if we use the wrong starting point $y_0(t) = 0$ to start the iteration scheme. [*Hint*: the Picard iteration scheme becomes in this case $y_0(t) = 0$, $y_{n+1}(t) = 1 + \int_0^t (-y_n(s)) \, ds$, for $n = 0, 1, 2, \ldots$.]

5. Repeat Problem 4 but with the absurd first approximation $y_0(t) = \sin t$. Show that the sequence of iterates still converges to e^{-t}.

    Answer 5: In this case we have $y' = -y$, $y(0) = 1$, but we take $y_0 = \sin t$ as the zeroth Picard iterate:

    $$y_0(t) = \sin t$$

    $$y_1(t) = 1 + \int_0^t (-y_0(s)) \, ds = 1 + \int_0^t (-\sin s) \, ds = \cos t,$$

    $$y_2(t) = 1 + \int_0^t (-y_1(s)) \, ds = 1 + \int_0^t (-\cos s) \, ds = 1 - \sin t$$

    If we continue this process, being careful to separate the terms as above, we see that each $y_n(t)$ is the sum of two terms, the first being the partial sum $\sum_{j=0}^{n-1} (-1)^j t^j/j!$ of the Taylor series of e^{-t}, while the magnitude of the second term is the magnitude of the difference between $\pm \sin t$ [or $\pm \cos t$] and its corresponding Taylor series through the term t^{n-1}. Since the latter Taylor series tends to $\sin t$ [or $\cos t$] as $n \to \infty$, these terms tend to zero as $n \to \infty$ and the sequence of iterates converges to e^{-t}, even with the absurd choice of $y_0(t) = \sin t$.

6. This problem involves Picard iterates for the IVP $y' = 1 + y^2$, $y(0) = 0$.

    (a) Find the Picard iterates y_1, y_2, and y_3.

    (b) Prove by induction that $y_n(t)$ is a polynomial of degree $2^n - 1$.

    (c) Find the exact solution $y(t)$ and identify the maximal interval on which it is defined.

    (d) Does $y_n(t) \to y(t)$ for all t? Explain.

7. How many Picard iterates can you calculate for $y' = \sin(t^3 + ty^5)$, $y(0) = 1$? What is the practical difficulty in finding the iterates?

    Answer 7: The IVP is $y' = \sin(t^3 + ty^5)$, $y(0) = 1$. We have $y_0(t) = 1$, $y_1(t) = 1 + \int_0^t \sin(s^3 + s) \, ds$, which cannot be expressed in terms of elementary functions. In general, the use of the Picard scheme is limited by the difficulty of carrying out the integration of any but the simplest of functions. This does not mean that the integrals fail to exist, only that they cannot be written in terms of elementary functions.

8. Let $y_0(t) = 3$, $y_1(t) = 3 - 27t$, $y_n(t) = 3 - \int_0^t y_{n-1}^3(s)\,ds$, $n = 0, 1, 2, \ldots$. Find $\lim_{n \to \infty} y_n(t)$. [*Hint*: consider a certain IVP related to the Picard iteration scheme.]

Uniform Convergence of a Sequence of Functions.

9. Show that the sequence $\{(1/n)\sin(nt)\}_{n=1}^{\infty}$ converges uniformly to the zero function on any interval I. [*Hint*: $|(1/n)\sin(nt) - 0| \leq 1/n$.]

   Answer 9: Let ε be any positive number. We must find a positive integer N, which may depend upon ε, so that for all integers $n \geq N$, $|(\sin nt)/n| < \varepsilon$, for all t in some interval I. Since $|(\sin nt)/n| \leq 1/n$, we see that we may take for N any integer larger than $1/\varepsilon$. Then if $n \geq N$, we have

   $$|\sin nt/n| \leq \frac{1}{n} \leq \frac{1}{N} < \frac{1}{1/\varepsilon} = \varepsilon$$

   So $\{(\sin nt)/n\}_1^{\infty}$ converges uniformly to the zero function on any interval I.

10. Show that the sequence $\{t^n\}_{n=1}^{\infty}$ converges uniformly to the zero function on the interval $0 \leq t \leq 0.5$, while it converges, but not uniformly, on the interval $0 \leq t \leq 1$ to the function $f(t) = 0$, $0 \leq t < 1$, $f(1) = 1$.

A.3 Extension of Solutions

PROBLEMS

1. Find a solution formula for each IVP. Determine the largest t-interval on which the solution is defined. What happens to the solution as t approaches each endpoint of the interval?

   (a) $y' = -y^3$, $\quad y(0) = 1$ **(b)** $y' = -te^{-y}$, $\quad y(0) = 2$

   Answer 1:

   (a) Separate the variables and integrate

   $$-y^{-3}\,dy = dt, \quad y^{-2}/2 = t + C, \quad y = (2t + 2C)^{-1/2}$$

   where the initial condition implies that $C = 1/2$. This solution is defined for all t for which $2t > -1$, so for $t > -1/2$. As $t \to (-1/2)^+$, $y \to +\infty$. As $t \to +\infty$, $y \to 0$.

   (b) Separate the variables and integrate

   $$e^y\,dy = -dt, \quad e^y = -t + C, \quad y = \ln|C - t|$$

   where the initial condition implies that $C = e^2$. As $t \to (e^2)^-$, then $y \to -\infty$. As $t \to -\infty$, $y \to +\infty$. This solution is defined for $t < e^2$.

2. Find an ODE $y' = f(t, y)$ whose solutions are defined only on finite t-intervals. [*Hint*: try $y' = 1 + y^2$.]